Oil-Field Chemistry

ACS SYMPOSIUM SERIES **396**

Oil-Field Chemistry

Enhanced Recovery and Production Stimulation

John K. Borchardt, EDITOR
Shell Development Co.

Teh Fu Yen, EDITOR
University of Southern California

Developed from a symposium sponsored
by the Division of Petroleum Chemistry, Inc.,
and the Division of Geochemistry
of the American Chemical Society
at the Third Chemical Congress of North America
(195th National Meeting of the American Chemical Society),
Toronto, Ontario, Canada,
June 5–11, 1988

American Chemical Society, Washington, DC 1989

CHEM
sep/ae

Library of Congress Cataloging-in-Publication Data

Oil field chemistry: enhanced recovery and production stimulation
  John K. Borchardt, editor, Teh Fu Yen, editor.

   Developed from a symposium sponsored by the Division
of Petroleum Chemistry, Inc., and the Division of
Geochemistry of the American Chemical Society at the
Third Chemical Congress of North America (195th
Meeting of the American Chemical Society), Toronto,
Ontario, Canada, June 5–11, 1988.

   p.   cm.—(ACS Symposium Series, 0097–6156; 396).
Bibliography: p.

   Includes index.

   ISBN 0–8412–1630–4
   1. Oil fields—Production methods—Congresses.

   I. Borchardt, John K., 1946–   . II. Yen, Teh Fu, 1927–
. III. American Chemical Society. Division of Petroleum
Chemistry. IV. American Chemical Society. Division of
Geochemistry. V. American Chemical Society. Meeting
(195th: 1988: Toronto, Ont.) VI. Series: ACS
symposium series; 396

TN870.C51245   1988
622'.33827—dc20
                                          89–6829
                                             CIP

ACS Symposium Series

M. Joan Comstock, *Series Editor*

1989 ACS Books Advisory Board

Foreword

The ACS SYMPOSIUM SERIES was founded in 1974 to provide a medium for publishing symposia quickly in book form. The format of the Series parallels that of the continuing ADVANCES IN CHEMISTRY SERIES except that, in order to save time, the papers are not typeset but are reproduced as they are submitted by the authors in camera-ready form. Papers are reviewed under the supervision of the Editors with the assistance of the Series Advisory Board and are selected to maintain the integrity of the symposia; however, verbatim reproductions of previously published papers are not accepted. Both reviews and reports of research are acceptable, because symposia may embrace both types of presentation.

Contents

OVERVIEWS

POLYMER PROPERTIES AND DESIGN

Preface

THE VARIOUS FIELDS OF CHEMISTRY play an important role in the discovery and exploitation of oil and gas reserves. Improved drilling and well completion fluids, cement slurries, hydraulic fracturing and acidizing fluids to improve well productivity, various chemical additives to be used in these fluids, and chemicals for enhanced oil recovery are essential to the improvement of production economics and to an increase in recoverable hydrocarbon reserves. Chemistry will become increasingly important in future hydrocarbon production with the decreased likelihood of major onshore discoveries, increased discovery and production costs associated with deep offshore wells and Arctic frontier provinces, and the decline in drilling since early 1982.

The multidisciplinary nature of oil-field chemistry includes the development of surfactants and polymers for enhanced oil recovery, drilling and completion fluids, and stimulation fluids. Study of relevant inorganic and organometallic chemistry has resulted in the development of improved polysaccharide cross-linking agents, corrosion inhibitors, improved cement slurry additives, and completion fluids. Polymer chemistry has resulted in improved cement slurry additives to control cement set time, fluid loss, compressive strength, and formation damage characteristics. Contributions of organic chemistry include development of improved antioxidants and more cost-effective monomer syntheses for the production of oil-field polymers.

The study of foam behavior in porous media; interfacial phenomena, especially the aggregation and arrangement of surfactants in oil–aqueous systems; aqueous fluid–rock mineral interactions, particularly that of strong acids with silicaceous minerals; the interaction of polymers and organic chemicals with rock; and the rheological behavior of polymer solutions and gels are all important physical chemistry and engineering research areas. Biology also plays an important role. Bacterial growth in wellbores and within formations has long been recognized as a serious problem. It causes hydrogen sulfide formation and corrosion problems that require careful operating practices and the use of bactericides. One of the most exciting recent research areas has involved the promotion and utilization of the growth of bacteria within formations to improve oil recovery.

Oil-field chemistry has undergone major changes since the publication of earlier books on this subject. Enhanced oil recovery research has shifted from processes in which surfactants and polymers are the primary promoters of increased oil production to processes in which surfactants are additives to improve the incremental oil recovery provided by steam and miscible gas injection fluids. Improved and more cost-effective cross-linked polymer systems have resulted from a better understanding of chemical cross-links in polysaccharides and of the rheological behavior of cross-linked fluids. The thrust of completion and hydraulic fracturing chemical research has shifted somewhat from systems designed for ever deeper, hotter formations to chemicals, particularly polymers, that exhibit improved cost effectiveness at more moderate reservoir conditions.

Although opinions on the timing of the next oil shortage differ, oil is a finite resource and the current oil-production surplus is definitely a temporary phenomenon. Despite some cutbacks, major oil companies, some chemical companies, and various academic groups and research institutes are all maintaining oil-field chemistry research and development programs. The long lead time required for much oil-field chemistry research, particularly enhanced oil recovery, requires that basic research, laboratory product development, and field testing all be maintained. This will enable the industry to respond when the current oil surplus vanishes and to develop chemicals and processes that will be economical, even at current oil prices.

The symposium on which this book is largely based was held with the temporary nature of the current oil-production surplus in mind. The Division of Geochemistry, Inc., provided part of the cost of printing color photographs within the text. The views and conclusions expressed herein are those of the authors.

The editors thank the authors for their contributions and the referees, the unsung heroes of science, for their diligent and timely manuscript reviews. Last, but certainly not least, we thank Cheryl Shanks of the ACS Books Department, without whose patience, help, and encouragement this volume would not be possible.

JOHN K. BORCHARDT
Shell Development Company
P.O. Box 1380
Houston, TX 77251–1380

TEH FU YEN
School of Engineering
University of Southern California
Los Angeles, CA 90089–0231

March 23, 1989

OVERVIEWS

Chapter 1

Chemicals Used in Oil-Field Operations

Westhollow Research Center, Shell Development Company, Houston, TX 77251-1380

Chemicals of various types are used in every stage of drilling, completing, and producing oil and gas wells. This review describes these chemicals, why they are used, and recent developments. These chemicals include common inorganic salts, transition metal compounds, common organic chemicals and solvents, water-soluble and oil-soluble polymers, and surfactants. As existing fields become depleted, use of chemistry to maintain production via well stimulation, more efficient secondary recovery operations, and enhanced oil recovery become ever more important.

The modern chemical industry is highly dependent on crude oil and natural gas feedstocks. Conversely, chemicals, the science of chemistry and chemical engineering join petroleum engineering to play an important roll in the production of oil and gas. The discovery rate of of major new oil fields is declining, particularly in the United States. As the petroleum industry becomes more dependent on increasing production from existing fields, the use of chemicals to more efficiently drill and operate oil and gas wells and enhance productivity from these fields will grow. Environmental considerations will probably be an increasingly important in the choice of chemicals used in well treatment fluids particularly in offshore locations. While geochemistry plays a role in the discovery of oil and gas and production chemicals are used to break produced oil-water emulsions and as friction reducers in pipelines, this review will be restricted to the chemistry and chemicals involved in drilling, completing, stimulating, and operating production and injection wells and in enhanced oil recovery.

Drilling Fluids

Drilling fluids (1-3) are often called drilling muds because of their appearance. This is due to the dispersed clays added to most

0097–6156/89/0396–0003$13.95/0
© 1989 American Chemical Society

drilling fluids. The drilling fluid is circulated down the drill
pipe, around the drill bit, and up the wellbore while drilling is
in progress. The purpose of the drilling fluid is to cool and
lubricate the drill bit, suspend formation cuttings and lift them
to the surface, and control formation pressure reducing pressure
surges up the wellbore (thereby reducing the possibility of
blowouts). By cooling the drill bit and removing the cuttings from
the bottom of the well bore, the rate of drilling can be increased.
The drilling fluid is designed to be thixotropic i.e., have high
viscosity under low shear conditions when moving up the wellbore
carrying suspended solids and have low viscosity under the high
shear conditions near the drill bit where rapid fluid movement is
necessary to cool the drill bit.

Drilling fluids usually contain water as the primary
component. However, oil-based muds may be used for high
temperature operations and for drilling highly water-sensitive
formations. Oil-based muds are of two types, oil-external
emulsions containing as much as 50% water in the internal phase or
an oil-based fluid containing little if any water.

A great many additives can be used to impart desired
properties to the drilling fluid. In general, the deeper and
hotter the well, the more chemical additives are needed to obtain
the desired fluid properties. These additives can be classified
into different types by function. These include:
Weighting materials which are used to adjust fluid density and thus
hydrostatic pressure exerted on the formation by the wellbore
fluid. The objective is to prevent sudden pressure surges or
blowouts during drilling while simultaneously avoiding excessive
fluid leak-off into formations being penetrated by the well bore.
Barium sulfate (barite) is the most commonly used weighting agent.
Other insoluble minerals used include hematite, siderite, and lead
sulfide. Salts may be dissolved in the base water to increase
fluid viscosity. The use of high density calcium chloride, sodium
and calcium bromides, and zinc bromide solutions and blends thereof
has become common in the U.S. Gulf Coast region (4-6). These
fluids are somewhat corrosive (7) and the use of corrosion
inhibitors such as thiocyanate ion has been recommended. However,
these fluids have excellent formation damage characteristics; fluid
leakoff from the wellbore into the formation has little adverse
effect on rock permeability (8,9).
Fluid loss additives such as solid particles and water-thickening
polymers may be added to the drilling mud to reduce fluid loss from
the well bore to the formation. Insoluble and partially soluble
fluid loss additives include bentonite and other clays, starch from
various sources, crushed walnut hulls, lignite treated with caustic
or amines, resins of various types, gilsonite, benzoic acid flakes,
and carefully sized particles of calcium borate, sodium borate, and
mica. Soluble fluid loss additives include carboxymethyl cellulose
(CMC), low molecular weight hydroxyethyl cellulose (HEC), carboxymethylhydroxyethyl cellulose (CMHEC), and sodium acrylate. A large
number of water-soluble vinyl copolymers and terpolymers have been
described as fluid loss additives for drilling and completion
fluids in the patent literature. However, relatively few appear to
be used in field operations.

Controlling fluid loss loss is particularly important in the case of the expensive high density brine completion fluids. While copolymers and terpolymers of vinyl monomers such as sodium poly(2-acrylamido-2-methylpropanesulfonate-co-N,N-dimethylacrylamide-co-acrylic acid) has been used (10), hydroxyethyl cellulose is the most commonly used fluid loss additive (11). It is difficult to get most polymers to hydrate in these brines (which may contain less than 50% wt. water). The treatment of HEC particle surfaces with aldehydes such as glyoxal can delay hydration until the HEC particles are well dispersed (12). Slurries in low viscosity oils (13) and alcohols have been used to disperse HEC particles prior to their addition to high density brines. This and the use of hot brines has been found to aid HEC dissolution. Wetting agents such as sulfosuccinate diesters have been found to result in increased permeability in cores invaded by high density brines (14).

Foaming agents provide another way to reduce fluid loss is in the drilling fluids. Mist or foam drilling is used in relatively shallow formations; commonly used foaming agents include C_{14-16} alpha-olefin sulfonates and alcohol ethoxysulfates. While these drilling fluids have not been used extensively in recent years, the development of improved foaming agents and systems containing water thickening polymer to stabilize the foam has been reported (15). Defoamers are used to reduce undesirable foaming which often occurs when saline drilling fluids exit the well bore. Tributylphosphate, low molecular weight aliphatic alcohols, polyglycols, fatty alcohol glycol ethers, acetylenic glycols, aluminum stearate, potassium chloride, silicone-based defoamers, and sodium alkylaromatic sulfonates have been used.

Lost circulation chemical treatments are necessary when the drill bit penetrates a "thief" zone and very large amounts of drilling fluid are lost to the formation. In this situation, the addition of water thickeners or solid particles may not be sufficient. The face of the formation can be plugged using a rapidly setting cement slurry or a process involving the in-situ gelation/precipitation of sodium silicate, treatment with Portland cement, and in less serious cases by plugging the formation face with shredded cellophane, crushed walnut and almond hulls, cedar and cane fibers, and carefully sized sodium chloride and calcium carbonate particles.

Viscosifiers are used as rheology modifiers to aid in suspending rock cuttings as they are carried to the surface. Many of the fluid loss additives described above are used in this application. Clays such as bentonite (montmorillonite) are the most commonly used rheology modifiers. The main organic polymers that are used, polysaccharides and acrylamide and acrylate polymers, often have limited temperature stability or exhibit excessive temperature thinning in deep hot wells. At concentrations below 2.8 g/L, xanthan gum is a more effective solids suspending agent than HEC, CMC, and partially hydrolyzed polyacrylamide (16).

While starches are commonly used, they are relatively poor viscosifiers. Acids and bacterial enzymes readily attack the acetal linkages resulting in facile depolymerization. Both formaldehyde and isothiazolones have been used as starch biocides (17). Development of improved high temperature water viscosifiers for drilling and other oil field applications is underway. For the

present, oil-based drilling fluids offer the best alternative for
elevated temperature applications despite their relatively high
cost.

Stabilizing agents are used to maintain drilling fluid rheological
properties at highly elevated downhole temperatures. Chromium and
chromium-free lignosulfonates, polyglycol ethers, sodium poly-
(styrene sulfonate-co-maleic anhydride), and a melanin polymer have
been used in this application. Additives such as sodium diethyldi-
thiocarbamate have been used to stabilize aqueous polysaccharides
such as xanthan gum (18).

Flocculants cause colloidal clay particles to coagulate thus
promoting separation from the drilling fluid which has been
circulated down the wellbore and returned to the surface. The
treated fluid may then be pumped back down the well bore. Sodium
chloride, hydrated lime, gypsum, sodium tetraphosphate, polyacrylamide, poly(acrylamide-co-acrylic acid), cationic polyacrylamides,
and poly(ethylene oxide) have been used commercially.

Thinners and dispersants are used to prevent excessive flocculation
of clay particles and maintain pumpability of the fluid. Tannins,
various lignosulfonate salts, sodium tetraphosphate and other
phosphates, and synthetic polymers such as sodium poly(styrene
sulfonate-co-maleic anhydride) have been used.

Friction reducers such as partially hydrolyzed polyacrylamide
may also be used in drilling fluids (19). They allow fluid to be
circulated through the well bore more easily thereby reducing
horsepower requirements for the circulating pumps and thus
decreasing well treatment costs.

Lubricants offer a means of reducing torque and increasing the
effective horsepower to the drillbit by reducing friction. Various
vegetable oils, graphite powder, soaps, asphalt blends, air-blown
asphalt colloids, diesel oil, and fatty acid esters have been used.

Pipe-freeing agents are used to reduce friction and increase
lubricity in areas of expected drill pipe sticking such as angles
in deviated wellbores. Soaps, surfactants, oils, soda lime, glass
beads, and cationic polyacrylamide have been used.

Corrosion inhibitors are used to reduce the corrosion of surface
equipment, surface casing, and the drill string by drilling and
well treatment fluids. Many different corrosion inhibitors have
been used. These include amine salts such as ammonium sulfite
-bisulfite blends, zinc carbonate, zinc chromate, hydrated lime,
fatty amine salts of alkylphosphates, cationic polar amines,
ethoxylated amines, and tertiary cyclic amines. Commercial
products are usually proprietary blends of chemicals.

Bactericides are used to control bacterial growth which can cause
corrosion, plugging of the fomation face, and alteration of drilling fluid rheological properties. Paraformaldehyde,
glutaraldehyde, sodium hydroxide, lime derivatives, dithiocarbamates, isothiazolones, and diethylamine have all been used.

pH control aids in reducing corrosion and scaling and in controlling interaction of the drilling fluid with formation minerals.
Sodium hydroxide, calcium carbonate, sodium bicarbonate, sodium
carbonate, potassium hydroxide, magnesium oxide, calcium oxide,
fumaric acid, and formic acid have all been used commercially in
this application.

Formation damage control chemicals are added to reduce the permeability damage that occurs when drilling fluid enters the formation. This also aids in preventing erosion of the formation into the wellbore. Maintaining the cylindrical geometry and uniform diameter of the welbore aids in subsequent cementing operations. Potassium chloride, ammonium chloride, sodium chloride, gypsum, sodium silicate, partially hydrolyzed polyacrylamide and poly(acrylamide-co-acrylic acid), certain polymers having quaternary ammonium groups in the repeat unit (see Chapter 10), and lignosulfonate derivatives have all been used to reduce formation damage.

Scale inhibitors are used to prevent the formation of insoluble calcium salts when the drilling fluid contacts formation minerals and saline formation waters. Commonly used scale inhibitors include sodium hydroxide, sodium carbonate, sodium bicarbonate, polyacrylates, polyphosphates, and phosphonates.

Emulsifiers have been used to prepare oil-external emulsion drilling fluids. Surfactants used as emulsifiers include fatty acid salts. fatty acid amides, petroleum sulfonates, and lignosulfonates.

Because of the relatively low cost of many of the chemicals used in drilling fluids, development of more cost effective additives is a major challenge. However, improved high temperature polymers, surfactants, and corrosion inhibitors are under development in many laboratories.

Cementing Fluids (20,21)

After completion of the drilling operation, steel casing is lowered down the well bore and into the drilling fluid. A spacer fluid is then pumped down the well bore to remove the drilling fluid and prevent contact of the drilling mud with the cement slurry. Efficient displacement of the drilling mud also promotes bonding of the cement slurry to rock surfaces.

Intermixing of the spacer and the drilling fluid should not produce solids or a high viscosity phase. Most spacers are aqueous and contain polymers to increase fluid viscosity. Spacer density is usually intermediate between that of the drilling fluid and the cement slurry (22). Salts may be added to control fluid density and pH. Surfactants are used to aid removal of drilling mud from formation surfaces. Water-wetting surfactants also aid in making the casing and exposed rock surfaces water-wet to promote good cement bonding (23). This is particularly important when using oil-based drilling fluids.

Turbulent flow at reasonable pump rates aids in removal of drilling mud from surfaces (24). Downhole devices called scratchers can be installed on the casing to scrape drilling mud residues from formation surfaces. Other devices called centralizers may be attached to the casing to center it in the wellbore.

With increased development work from offshore platforms, more non-vertical (deviated) wells are being drilled. Settling of mud solids to the low side of the well bore can result in a continuous channel of undisplaced drilling mud solids in the casing annulus

that reduce the effectiveness of cement bonding (25,26). Conversely, cement solids can settle from the slurry before it sets
resulting in a channel of water in the high side of the casing
annulus. Proper design of both the drilling fluid (particularly
through use of surfactants as dispersants) and the cement slurry
(including good control over cement set time) are necessary to
prevent this.

The cement slurry is pumped down the casing and up the annular
space between the casing and the formation. The spacer and
drilling fluid are thus displaced by the cement slurry. A
compatible fluid (one that does not substantially alter the set
time of the cement slurry) is pumped into the wellbore to displace
nearly all the cement slurry into the annular space between the
casing and the formation. The well is then shut in to allow the
cement to set. This bonds the casing to the formation and isolates
oil- and gas-bearing formations from aquifers and brine-containing
formations. Fluid communication between formations can adversely
affect production operations or lead to contamination of potable
water aquifers.

Incomplete displacement of fluid from the annular space can
result in gaps in the cement sheath through which fluids from
different formations can intermingle. In this situation, a
"squeeze cementing" treatment is required to plug these gaps.
Portland cement or rapidly setting sodium silicate slurries can be
used in this operation.

When cementing high pressure gas formations, the gas can
penetrate the cement slurry before it sets greatly weakening the
set cement (27). Various solutions to this problem have been
proposed including the use of cement slurry formulations which
expand as they harden thereby resisting gas invasion (28).
Foamed cement slurries have been used to provide a low density
cement slurry to reduce permeability damage to highly sensitive
formations through reduced fluid loss (29). Glass microspheres
have also been used to substantially reduce cement slurry density
(30, 31). Other additives which reduce cement slurry density to a
lesser extent include bentonite, fly ash, silicates, perlite,
gilsonite, diatomaceous earth, and oil emulsions (see citations in
reference 29).

Corrosion-resistant cements have been developed for use in wells
used to inject supercritical carbon dioxide for enhanced oil
recovery (32). These are based on Portland cement and high levels
(as much as 40% wt.) of additives such as fly ash. Epoxy resins
have been successfully used as cements in corrosive environments
(33).

Lignosulfonates and lignosulfonate derivatives are used extensively
as cement set time retarders (20, 21). Many of the same additives
used in drilling muds are used in cement slurries and spacer fluids
for similar purposes.

Completion Fluids and Operations (1,20,34)

After cementing the well, communication must be established with
the productive formation. This is done in an operation called
perforating. The wellbore is filled with a non-damaging fluid of

the proper density to control pressure surges while not exhibiting excessive fluid loss to the formation. A perforating tool or "gun" is lowered into the well bore and placed opposite the productive formation. The gun fires projectiles or powerful jets of gas generated in small explosions to penetrate the casing and cement sheath. Perforations are generated in a controlled pattern and spacing chosen after considering the formation properties and productive capacity. A small amount of acid may be used to wash the perforations to remove pulverized debris which reduces the fluid carrying capacity of the perforations and adjacent formation. Production tubing is then lowered into the hole and the productive portion(s) of the well isolated using sealing tools called packers (21). This is done to produce from more than one formation simultaneously and to minimize the volume of oil and gas in the wellbore during production.

Fluid loss from the wellbore to the formation may be reduced using the less permeability damaging drilling fluid loss additives described above. In saturated brines, carefully sized sodium chloride particles have been used to temporarily plug the formation face (35). The particles may be dissolved by pumping a less saline fluid down the wellbore.

Sand production from poorly consolidation formations is a significant problem in important oil producing areas such as the U.S. Gulf Coast; Kern County, California; Venezuela; Alberta; Nigeria; and Indonesia. The most commonly used technique for sand control is called gravel packing. A slurry of as much as 1.8 kg of carefully-sized sand particles per liter of aqueous fluid is pumped downhole. The sand particle size is chosen based on size analyses of the formation sand (19,36). The sand-carrying capacity of the water is enhanced by increasing its viscosity using 20-80 lb polysaccharide per 1000 gallon. The most commonly used polysaccharide is hydroxyethyl cellulose because its low content of insoluble solids minimizes permeability damage to the formation (37).

Carboxymethyl cellulose and derivitized guars are occasionally used in this application (1). Methods of stabilizing polysaccharides originally developed for hydraulic fracturing applications (see below) hold promise for increasing the range of temperatures at which polysaccharide polymers can be used in gravel packing applications.

Certain mixtures of polymers have been shown to form complexes which exhibit substantially higher than expected solution viscosity under low shear conditions. Xanthan gum blends with guar gum (38, 39), sodium poly(styrene sulfonate) (40), polyacrylamide (41), sulfonated guar gum (38), sodium poly(vinylsulfonate) (40), hydrolyzed sodium poly(styrene sulfonate-co-maleic anhydride) (38), and poly(ethylene oxide) (41) and blends of xanthan gum and locust bean gum have exhibited substantially higher than expected solution viscosity (42, 43).

An enzyme, acid, or oxidative "breaker" is added to effect a controlled depolymerization and thus a programmed loss of fluid viscosity. This depolymerization is timed to occur when the sand-laden fluid is opposite the productive formation. The sand then drops out of suspension and is packed against the formation. The sand creates a high permeability fluid pathway from the formation

into the wellbore while substantially preventing the migration of
formation particles. Downhole tools such as wire-wrapped screens
or liners are used in conjunction with gravel packing. These
devices serve to hold the sand in place.
     Since formation damage is a critical factor in successful
gravel pack treatments, continuing efforts are being made to
improve the formation damage characteristics of polysaccharide
fluids both before and after depolymerization (37). Recently,
grades of hydroxyethyl cellolose having improved formation damage
characteristics were introduced into the market place.
     Injecting epoxy, furan, or furan-formaldehyde resins into
poorly consolidated formations to consolidate them was a common
sand control practice for thin highly productive formations (44-
46). Organic solvents (46) and silane coupling agents (47) are used
to promote adhesion of the resin to the rock surface. Excess resin
is flushed deeper into the formation to minimize resin hardening in
the flow channels since this would reduce formation permeability.
     While effective, the relatively high cost of this sand control
method as compared to gravel packing has restricted its use. The
use of aqueous slurries of epoxy resins can reduce solvent costs
(44). Surfactants, particularly fluorochemicals, may hold promise
for increasing epoxy resin fluidity (49). The in situ crosslinking
of polybutadiene has been proposed as a method of reducing resin
costs (50). The gravel packing technique could be used to place
resin-coated sand against a poorly consolidated formation (51-53).
The resin is then cured resulting in a hard, but permeable mass
holding formation sand grains in place. Silica dissolution in high
temperature steam injection wells can destroy the integrity of a
gravel pack and lead to sand production when the well is placed
back on production (54). Use of a resin-coated sand could aid in
maintaining gravel pack stability and effectiveness.
     Another method of sand control is use of a silicon halide
which reacts with water at the surface of sand grains forms SiO_2
which can bond the grains together (55). Reducing the cost of sand
consolidation could be very useful since the applicability of
gravel packing methods is limited by the bottom hole circulating
temperature and the limited temperature stability of polysaccharide
polymers.

Hydraulic Fracturing (20,56)

Since hydraulic fracturing is reviewed in a subsequent chapter,
this important production stimulation technique will be only
briefly discussed. Hydraulic fracturing is a process whereby the
permeability of a formation is increased by generating high permeability cracks in the rock. Particulate suspensions (usually sand
slurries) are injected at sufficiently high rates (which require
high injection pressures) to generate fractures in the rock which
are held open by the suspended proppant in the fracturing fluid.
The majority of hydraulic fracturing treatments are performed using
water-based fluids; foams (with nitrogen or carbon dioxide as the
gas phase) have been used extensively in recent years to reduce
formation damage. Oil-external emulsions have also been used for

the same purpose. Ingredients used in hydraulic fracturing are chosen from the following:

A viscosifier, usually a polysaccharide, is used to suspend the proppant during pumping and placement in the rock fracture. Generally 2-7g of guar per liter of fracturing fluid is used. Hydrolytic depolymerization beings at 79.6°C ($\underline{34}$). The degradation rate in alkaline media, acid, and in the presence of cellulase and hemicellulase enzymes has been determined ($\underline{1}$). Most fracturing treatments employ a crosslinked polymer.

A disadvantage of guar is the relatively high level of insoluble materials normally left in the product after processing, 10-14% wt. ($\underline{57}$). Alkaline refining can reduce the insoluble materials level substantially, to ca. 3.9% wt. ($\underline{58}$). Guar derivatives such as HPG and carboxymethylhydroxypropyl guar (CMHPG) contain fewer insolubles, <2% wt. ($\underline{59,60}$). HPG also exhibits better high temperature stability and resistance to hydrolytic decomposition as compared to guar ($\underline{1}$). Locust bean gum, karaya gum, and gum traganth have been used to a limited extent as viscosifiers and fluid loss additives ($\underline{61-63}$).

Hydroxyethyl cellulose has been used in fracturing fluids not requiring the increased viscosity and longer break times provided by crosslinking. Until recently, it has been difficult to cross-link HEC ($\underline{64}$). This has limited its use in hydraulic fracturing applications despite its excellent formation damage characteristics. Hydroxyethyl celluloses containing vicinal hydroxy groups are more easily crosslinked ($\underline{65}$) and exhibit better shear stability ($\underline{66}$). Zr(IV) has been used to crosslink HEC ($\underline{64}$). Crosslinked CMHEC tends to show more shear degradation on passing through pumps and small orifices in downhole tools than crosslinked HPG fluids ($\underline{67}$).

Other viscosifiers described in the literature include acrylate copolymers such as poly(acrylamide-co-dodecylmethacrylate) ($\underline{68}$) and poly(vinyl alcohol) ($\underline{69}$). The driving force behind using synthetic polymers is presumably better high temperature stability.

The use of ionomers such as lightly sulfonated polystyrene as an oil-based fracturing fluid viscosifier has been studied ($\underline{70}$). The most commonly used oil-based viscosifiers are phosphate esters of various types ($\underline{56,71,72}$).

Fluid loss additives are used are used to reduce the rate of fluid loss from the fracture to the formation and to naturally occurring macro- and micro-fractures within the formation. Silica flour ($\underline{73,74}$), oil-soluble resins ($\underline{75}$), diesel oil emulsions (5% by volume) ($\underline{74}$) have also been used.

Proppants are solid particles used to hold open the fracture after conclusion of the well treatment. Criteria to choose the economically most effective proppant for a given set of formation conditions have been discussed ($\underline{76}$). While sand is the most commonly used proppant because of its low cost, resin-coated sand, sintered bauxite, and Al_2O_3 particles have also been used because of their greater compressive strength and resistance to dissolution at high temperature and pH ($\underline{55}$). While epoxy resins are most commonly used, the use of other resins such as phenol-formaldehyde has been described.

Sand has been treated with oil-soluble organosilicon compounds to form a hydrophobic proppant (77). A double layer resin coating has also been developed. The inner layer coating the sand particle is a cured gamma-aminopropyltriethoxysilane - hexamethylenetetramine. The outer layer is an uncured mixture of the same two chemicals which cures within the fracture to form a consolidated permeable mass holding the fracture open (78).

Crosslinking agents are used to increase solution viscosity and thus solids suspending properties of fluids at T>150°F. The most commonly used crosslinking agents are organotitanates, borates, and zirconium compounds. Organozirconates are the preferred crosslinkers for hydroxyethyl cellulose (79-81). Antimonates and aluminum compounds have also been used as polysaccharide crosslinkers. Encapsulation of crosslinkers and the use of ligands to complex with the transition metal atom have been used to delay crosslinking (82). Delayed crosslinking which occurs within the formation under lower shear conditions can provide higher and more predictable crosslinked fluid viscosity (83). Polyamines such as tetramethylenediamine can be used to accelerate crosslinking reactions (84).

A breaker: an enzyme (at T<140°F), strong oxidizing agent, or an acid, is used to depolymerize polysaccharides and break crosslinks such that viscosity declines at a controlled rate so that the proppant may be deposited in the fracture. Too rapid proppant dropout would cause a premature "sand-out" which prevents future extension of the fracture. Peroxydisulfates are the most frequently used breakers. Less reactive organic peroxides may be preferred for high temperature formations (85).

Chemical stabilizers have been used to reduce the rate of oxygen-promoted degradation of polysaccharides at T>225°F. Methanol and sodium thiosulfate are the most commonly used (86). Sodium dithiocarbamate, alkanolamines, and thiol derivatives of imidazolines, thiazolines, and other heterocyclic compounds have also been tested for this application. Calcined dolomite (87) and Cu(I) and Cu(II) salts (88) have been reported to increase the thermal stability of HEC.

Biocides are used to prevent aerobic bacterial degradation of fracturing fluids in surface mixing and storage tanks. Anaerobic bacterial growth in the wellbore and within the formation has to be controlled to prevent introduction and/or growth of these bacteria within the formation during the fracturing treatment and subsequent generation and production of hydrogen sulfide. Glutaraldehyde, chlorophenates, quaternary amines, and isothiazoline derivatives have been used (89,90). The biocide is best added to the base fluid before addition of the polysaccharide viscosifier.

pH buffers are added to the base fluid to keep the pH basic. This promotes rapid polymer particle dispersion and controls polysaccharide hydration rate to avoid formation of large, partially hydrated particles.

Other techniques to promote complete polymer hydration include vigorous mixing and slow addition of the polysaccharide. Specially designed mixing devices have been used to promote rapid particle dispersion (91). Adding already prepared dispersions of guar, HPG, and HEC in nonaqueous media is another means of promoting rapid

polymer particle dispersion and complete hydration (81,92).
Wetting polymer surfaces with ethylene glycol (92) or isopropanol
(93) has also been used as a means of promoting rapid polymer
dispersion prior to the onset of hydration.

Various particle surface treatments have been used to delay
polymer hydration until polymer particles have been thoroughly
dispersed. These include guar treatment with borax (2,94) and HEC
treatment with glyoxal (95).

Buffers also are used to maintain the proper pH for the crosslinking reaction to occur at an optimum rate. Sodium bicarbonate
and sodium carbonate are used to attain basic pH while weak acids:
acetic, fumaric, formic, and adipic, are generally used to obtain
acidic pH values.

Formation damage control additives are added to reduce permeability
damage caused by clay swelling and consequent fine particle migration (which can also occur in the absence of clay swelling).
Potassium chloride, ammonium chloride, sodium chloride, and, for
longer term treatment effectiveness, organic polymers containing
quaternary ammonium groups in the repeat unit have been used for
this application. While avoiding permeability damage to the
formation adjacent to the propped fracture is critical in determining initial hydrocarbon production rate, fluid conductivity in the
propped fracture is the primary determinant of long-term productivity (96).

Surfactants are used to stabilize water-in-oil emulsions and to
promote rapid return of injected fluids and a faster return of the
well to hydrocarbon production. Although they are expensive,
water-soluble fluorochemicals have been shown to be effective in
this application (97,98).

Foams have become widely used to limit the fluid lost to the
formation and thus reduce formation damage. Foam cell size plays a
major role in determining fluid rheology (99). Guar, HPG, and
xanthan gum stabilize the foam bubbles by increasing the viscosity
of the surrounding aqueous fluid. Both nitrogen and carbon dioxide
have been used as the internal phase of the foam (100-102) and
foams based on each exhibit similar rheological behavior in laminar
flow (102) and similar fluid loss behavior (103,104). The carbon
dioxide is pumped as a supercritical fluid which changes to a gas
downhole (105).

Acidizing Chemicals (20,106)

Acid treatments fall into three general types:

Acid washing is used to dissolve acid-soluble scales from the
well bore and to open gravel packs and perforations plugged by such
scales.

Matrix acidizing is the injection of acids into the formation
at a pressure below the formation parting pressure (the pressure at
which natural fractures are forced open by injected fluids).
Properly designed, the injected acid enters the flow channels of
the formation and flows radially outward from the wellbore dissolving mineral fine particles in the flow channels. Minerals forming
the flow channel walls also react with the acids. These processes
increase formation permeability near the wellbore. The end result

is to increase well productivity without increasing the produced
water:oil or gas:oil ratios.

So-called "wormholes" can be formed when the injected acid
primarily enters the largest diameter flow channels in carbonate
rock further widening them (107).   Acid only invades the small flow
channels a short distance greatly reducing treatment effectiveness.
High fluid loss rates, low injection rates, and reduced rates of
acid-rock reactions decrease the wormhole length.

In the third type of acidizing, fracture acidizing, acid is
injected above the parting or fracture pressure of the formation.
The acid reacts with the minerals on the exposed fracture face in a
process called etching.   With sufficient etching, the fracture does
not reseal when normal well production or injection operations are
resumed.

Acids can sometimes break emulsions within the formation
either by reducing the pH or by dissolving fine particles which can
stabilize emulsions.   Breaking the emulsion reduces fluid viscosity
thus increases the fluid carrying capacity of the flow channel.
Acids may be used as breakers to reduce the viscosity of acid-
sensitive fracturing gels.

Acids are sometimes used ahead of fracturing fluids to
dissolve mineral fine particles and allow more rapid injection of
the fracturing fluid.   When used as the initial stage of a squeeze
cementing treatment, the acid-promoted mineral and drilling mud
particle dissolution can result in increased entry of the cement
slurry into the desired portions of the formation.

Acids are selected based on the nature of the well treatment
and the mineralogy of the formation. The critical chemical factors
in properly selecting an acid are: stoichiometry (how much formation material is dissolved by a given amount of acid), the
equilibrium constant (complete reaction of the acid is desired),
and reaction rate between the acid and the formation material
(106).

Mineral acids include hydrochloric acid and blends of hydrochloric
and hydrofluoric acid (usually 12% HCl/3% HF).   Hydrochloric acid
is used to acidize carbonate formations.   Its advantages are
relatively low cost, high carbonate mineral dissolving power, and
the formation of soluble reaction products (which minimizes formation damage).   The primary disadvantage of hydrochloric acid is its
corrosive nature.

Hydrofluoric acid may be prepared by dilution of a
concentrated aqueous solution or by reaction of enough ammonium
bifluoride with aqueous 15% HCl to prepare a 12% HCl/3% HF
solution.   Hydrochloric - hydrofluoric acid blends have the major
advantage of dissolving silicaceous mineral including clays and
silica fine particles.   HCl/HF blends are quite corrosive.

Earlier corrosion inhibitors limited the maximum strength of
the acid to 15% by weight.   Improved corrosion inhibitors (see
below) have made the use of higher acid concentrations, such as 28%
HCl more common.   More dilute solutions may initially be injected
in sandstone acidizing to reduce the formation of insoluble sodium
and potassium fluorosilicates by displacing saline formation water
before injection of hydrochloric acid.

Organic acids used in carbonate rock acidizing include formic,
acetic, sulfamic, and chloroacetic acids. These have the advantage
of being less corrosive than the mineral acids. This permits use
in applications requiring a long contact time with pipe (as perforating fluids) or with aluminum- or chrome- plated pump parts. It
is also easier to retard (inhibit) reaction of organic acids with
carbonates at elevated temperatures. This permits deeper penetration of the acid treatment fluid into the formation. Organic acids
are used to a much smaller extent than mineral acids due to their
higher cost and incomplete reaction with many carbonate minerals.
Sulfamic and chloroacetic acids are seldom used except in situations such as remote well locations where their solid form (100%
activity) makes transportation costs a critical consideration. The
180°F decomposition temperature limits the use of sulfamic acid to
temperatures below ca. 160°F.

Mixed acid systems are blends of mineral acids and organic acids.
Combinations that have been used in carbonate acidizing include
acetic acid/HCl and formic acid/HCl. While these are less corrosive than hydrochloric acid alone, the organic acid may not react
completely with the rock. Blends of formic acid and hydrofluoric
acid have been used in high temperature sandstone acidizing and are
less corrosive than HCl/HF blends.

High fluid injection rates are often required. For this
reason, friction reducers are often used in acid fracturing. These
include polyacrylamide and acrylamide copolymers, guar gum,
hydroxyethyl cellulose, and karaya gum (108)

In many cases, it is desirable to retard the rate of acid -
rock reactions to permit deeper penetration of the treatment fluid
into the formation. Four techniques hve been used to accomplish
this: using retarded acids which generate HF in situ, chemically
retarding the acid by placing an organic film on rock surfaces,
using polymers to increase acid viscosity (use of so-called
"gelled" acids), and foaming or emulsifying the acid to increase
the apparent viscosity.

Retarded acids are primarily applicable to sandstone acidizing.
Fluoroboric acid slowly hydrolyzes to form the more reactive
hydrofluoric acid (109,110). The time required for this hydrolysis
process may enable deeper penetration of the HF into the formation
although one report contradicts these findings (111). Na_2TiF_6 and
similar salts also slowly generate HF in acid media (112).
Phosphorous acid addition to hydrochloric acid has been used to
reduce the HCl reaction rate with limestone (113).

Organic polymers have been used to increase the viscosity of
acids. The primary application is in fracture acidizing. Binary
and ternary acrylamide copolymers are the most commonly used
chemicals for this application. Many of these polymers degrade
rapidly in strong acids at temperatures $\geq 130^{\circ}$F; development of more
stable polymers suitable for high temperatures is desirable.
Recently developed polymers for this application include acrylamide
copolymers with:
    methacryloyltrimethylammonium chloride (114)
    2-acrylamido-2-methylpropanesulfonic acid (115)
    methacryloyloxyethyltrimethylammonium methosulfate (116).
    N-vinyl lactam (117)

Other polymers used in this application include:
    poly(vinylpyrrolidinone) (118,119)
    sodium poly(vinylsulfonate-co-vinylamide) (120)
    sodium poly(acrylamide-co-N-vinyl lactam-co-vinyl sulfonate)
    (119)
and mixtures of sodium
poly(2-acrylamido-2-methylpropanesulfonate-co-N-vinylacetamide) and
poly(acrylic acid-co-vinylformamide-co-vinylpyrrolidinone) (120).

    Despite its limited stability in acid (1), guar gum has been
used to thicken 3-15% hydrochloric acid (121). An allyl ether -
guar gum adduct has been proposed for use as an acid viscosifier
(122). Zr(IV) crosslinked CMHEC has been used to thicken
hydrochloric acid (81).

    Low viscosity oil-external retarded hydrochloric acid micro-
emulsions exhibiting quite low acid diffusion rates (ca. 1% of that
of aqueous HCl) have been developed (123,124). Foaming (125) or
emulsifying acid (106) also has the effect of limiting the contact
of the acid with formation surfaces and increasing acid viscosity
thereby reducing the rate of acid-rock chemical reactions. The
foaming agents are generally nonionic surfactants and the gas phase
is usually nitrogen. The acid is usually the internal phase of
emulsified acids and the fluid contains 10-30% of a low viscosity
hydrocarbon as the external phase. Polyacrylamide has been used to
thicken the aqueous phase of hydrochloric acid emulsions (126)
while nonionic surfactants have been used as the emulsifiers (127).
Overall, emulsified acids appear to be the most suitable for high
temperature formations.

    By adding an oil-wetting surfactant to an acid, one can
promote the temporary formation of a film on formation surfaces
thus reducing the rate of rock dissolution. Acids containing these
surfactants are known as chemically retarded acids.
Surfactants are also used to break low mobility oil emulsions.
Organic amines and quaternary ammonium salts (128), alkylphenol
ethoxylates (128), poly(ethylene oxide-co-propylene oxide-co-
propylene glycol) (129) and alkyl- or alkylaryl polyoxyalkylene
phosphate esters (130) are among the surfactants that have been
used.
Mutual solvents have been used to reduce surfactant adsorption on
formation minerals, particularly oil-wetting surfactants (131).
Ethylene glycol monobutyl ether is the most commonly used mutual
solvent.
Formation permeability damage caused by precipitation of dissolved
minerals such as colloidal silica, aluminum hydroxide, and aluminum
fluoride can reduce the benefits of acidizing (132-134). Careful
treatment design, particularly in the concentration and amount of
HF used is needed to minimize this problem. Hydrofluoric acid
initially reacts with clays and feldspars to form silicon and
aluminum fluorides. These species can react with additional clays
and feldspars depositing hydrated silica in rock flow channels
(106). This usually occurs before the spent acid can be recovered
from the formation. However, some workers have concluded that
permeability damage due to silica precipitation is much less than
previously thought (135).

Precipitation of Fe(III) compounds from acid solutions as the pH increases above 2.2 is a particular problem. Complexing agents that have been used include 5-sulfosalicylic acid and citric acid (136); dihydroxymaleic acid (137); ethylenediaminetetraacetic acid (138); lactic acid (138); blends of hydroxylamine hydrochloride, citric acid, and glucono-delta-lactone (139); nitriloacetic acid; blends of citric acid and acetic acid; lactic acid; and gluconic acid (140).

Diverting agents assist in distributing acid more uniformly through the perforated formation interval (141). These are usually oil-soluble hydrocarbon resin particles. They may be dissolved by post-acid injection of xylene or similar solvents. Oil-soluble waxes, naphthalene, and solid organic acids such as benzoic acid have also been used (142). Best results are obtained using a broad range of particle sizes.

Blends of sodium hypochlorite with 15% HCl and with 12% HCl/3% HF have been used to stimulate aqueous fluid injection wells(143). Waterflood injection wells have also been stimulated by injecting linear alcohol propoxyethoxysulfate salts in the absence of any acid (144). The oil near the well bore is mobilized thus increasing the relative permeability of the rock to water (145). Temperature effects on interfacial tension and on surfactant solubility can be a critical factor in surfactant selection for this application (146).

Corrosion inhibition is primarily associated with acidizing. Buffered hydrofluoric acid compositions have been shown to be less corrosive (147). Corrosion inhibitors are designed to reduce the rate of reaction of fluid with metal surfaces, generally by forming films on the surfaces. Acetylenic alcohols and amines are frequently components of corrosion inhibitor blends. Other compounds that have been used include nitrogen heterocyclics, substituted thioureas, thiophenols, and alpha-aminoalkyl thioethers (148).

Arsenic compounds can be very effective corrosion inhibitors but their toxicity, ineffectiveness in hydrochloric acids above 17% active and in the presence of H_2S, and their ability to poison refinery catalysts has limited their use (148). Epoxy resins have been coated onto metal surfaces and cured with a polyamine to reduce corrosion (149).

High density brine completion fluids also often require the use of corrosion inhibitors (8,9). Blends of thioglycolates and thiourea; alkyl, alkenyl, or alkynyl phosphonium salts; thiocyanate salts; mercaptoacetic acid and its salts; and the reaction products of pyridine or pyrazine derivatives with dicarboxylic acid monoanhydrides have been used as high density brine corrosion inhibitors.

Hydrogen sulfide promoted corrosion can be a serious problem (150); the best solution is prevention. Corrosion problems can be minimized by choice of the proper grades of steel or corrosion resistant alloys, usually containing chromium or nickel (150, 151) and avoiding generation of H_2S by sulfate reducing bacteria in situations where H_2S is not initially present. Cathodic protection of casing is often effective for wells less than 10,000 feet deep (150).

Scale inhibitors may also be used in acidizing. These include
alcohol ethoxysulfonic acids (152). Scale inhibitors are also used
in water and enhanced oil recovery injection wells and include low
molecular weight poly(vinylsulfonate), poly(methylmethacrylate-co-
ethylenediamine) (153), bis(phosphonomethylene)aminomethylene
carboxylic acid, and poly(acrylic acid-co-3-acrylamido-3-methylbu-
tanoic acid). Ethylenediaminetetraacetic acid and similar
complex-ing agents have been used to remove scale from formation
surfaces near wellbores.

Formation Damage Control Chemicals

The fluid flow capacity of rock, particularly the rock adjacent to
an oil or gas well is critical in determining well productivity.
The region near the wellbore acts as a choke for the entire formation; because the flow is radial more and more fluid is flowing
through a given volume of rock as the fluid approaches the well
bore. The reduction of the rock fluid carrying capacity is
referred to as formation damage.
    Formation damage may be due to invasion of rock capillaries by
solid particles in wellbore fluids (drilling and completion fluids)
and plugging of rock capillaries adjacent to fractures by fine
solid particles in fracturing fluids. These fines may be generated
when sand-laden fracturing fluid passes through small orifices such
as choke valves at high flow rates and pressures (67) or by
proppant crushing within the fracture. They may also be due to the
use of solid fluid loss additives. This type of formation damage
may be reduced by filtration of fluids before their entry into the
well bore and by proper choice and sizing of solid particles used
in drilling, gravel packing, and fracturing fluids. Acidizing the
rock immediately adjacent to the wellbore can dissolve clays,
silica particles, and precipitates plugging rock flow channels.
However, precipitation of hydrated silica, fluoroaluminates, and
iron compounds (above pH 2.2) in acidizing can cause formation
damage reducing well treatment effectiveness (see above).
    Reduced injectivity due to formation damage can be a
significant problem in injection wells. Precipitate formation due
to ions present in the injection water contacting counterions in
formation fluids, solids initially present in the injection fluid
(scaling), bacterial corrosion products, and corrosion products
from metal surfaces in the injection system can all reduce
permeability near the wellbore (153). The consequent reduced
injection rate can result in a lower rate of oil production at
offset wells. Dealing with corrosion and bacterial problems,
compatibility of ions in the injection water and formation fluids,
and filtration can all alleviate formation damage.
    Formation damage can also be caused by chemical and physical
interactions of fluid and rock. Low salinity fluids can cause
swelling of water-expandable clays. The resulting larger clay
dimensions can reduce the fluid carrying capacity of rock flow
channels. The expanded clay particles are more susceptible to the
shear forces of flowing fluids. In addition, clays act as the
cementing medium in many sandstone formations. Swelling weakens
this cementation and can cause the release of mineral fine

particles. Fines migration in Berea sandstone occurs when the
salinity of the flowing phase drops below a critical salt
concentration (CSC) (155,156). The CSC varies for different
monovalent cations in solution and decreases with increasing ion
exchange affinity of the clay for the cation. The CSC of
multivalent cations is very low (157). Flowing fluids can carry
these fine particles to constrictions in the flow channels where
they form a plug.

Inorganic salts: KCl, NH_4Cl, $CaCl_2$, or high concentrations of NaCl
have been used in wellbore fluids, fracturing fluids, and injection
fluids to temporarily reduce formation damage by converting the
more water-expandable smectite and mixed layer clays to less
swelling forms through ion exchange processes. However, the
potassium, ammonium, or calcium ions on the clays are themselves
subject to ion exchange processes and the clays may later be
converted back to the more water-expandable sodium form. Once clay
swelling has occurred, injection of salts will not reverse formation damage. An acidizing treatment to partially dissolve the
clays is required for this.

The addition of potassium hydroxide to injection waters has been
used to stabilize clays and maintain injectivity (158). Some
degree of permanence appears to result from this treatment since
injectivity appeared to be substantially maintained during
subse-quent injection of low salinity water.

More permanent stabilization of water-swelling clays may be
achieved by bonding the clay surface cation exchange sites together
so that simultaneous ion exchange at a large number of sites is
required for desorption of the clay stabilizer. This may be accomplished by injection of hydroxyaluminum, zirconyl chloride, or
certain quaternary ammonium salt polymers. A 6-12 hour well
shut-in period is required to allow polymerization of hydroxyaluminum on formation surfaces to occur (159). Because hydroxyaluminum
is removed from mineral surfaces by fluids at pH<3, it cannot be
used in conjunction with acidizing treatments. However, treated
clays are stable to high temperatures and hydroxyaluminum can be
used in 500°F steam injection wells.

Zirconyl chloride can be used to stabilize swelling clays in
both acidic environments and in the presence of 600°F steam (160).
No well shut-in time is required for polymerization to occur so
zirconyl chloride may be used in conjunction with hydraulic
fracturing treatments (161).

Quaternary ammonium salt polymers are more versatile and have been
used in drilling fluids, completion fluids, acidizing treatments,
and hydraulic fracturing. No well shut-in period is required.
Care in choosing the particular polymer to be employed in a
frac-turing treatment is needed because some polymers can interfere
with the function of the crosslinker. Some of these polymers are
also stable to high temperature steam and have been successfully
used to treat high temperature steam injection wells. Recent
developments in organic polymer formation damage control polymers
are discussed in chapter 10 of this book.

Fine particle migration can occur in the absence of water-swelling
clays. Migrating fines can include the migrating clays kaolinite,
illite, chlorite, and some mixed layer clays and fine silica
particles (162,163). Fine particle migration is promoted when the

flowing phase is the rock wetting phase (162), is affected by
flowing fluid salinity (155-157) and pH (164), and is accelerated
by rapid flow rates (165).   A critical flow velocity exists below
which fines migration is greatly reduced (166).
        Conventional inorganic and cationic organic polymer clay sta-
bilizers have been shown to be effective in substantially preventing permeability damage due to clay swelling and consequent fines
migration (see Table 1, Chapter 10).   However, most of these
polymers are much less effective at preventing fines migration in
the substantial absence of swelling clays.   Recently developed
quaternary ammonium salt polymers have been shown to be effective
in reducing migration of a variety of mineral fine particles in the
absence of swelling clays (see Chapter 10).
Adsorption of corrosion inhibitors or cationic surfactants can
reduce sandstone formation permeability.   Alcohols can be used to
remove corrosion inhibitors from rock surfaces.   Oil-soluble
corrosion inhibitors may be dissolved by organic solvents such as
xylene or toluene containing a mutual solvent, most often ethylene
glycol monobutyl ether, EGMBE (167).   Aqueous fluids containing
5-10% EGMBE can be used to dissolve cationic surfactants.
Paraffin deposits adjacent to production wells can greatly reduce
productivity by plugging fluid flow channels.   Deposition of these
waxy crude oil deposits can also occur in perforations and production tubing.   Scraping has been used to remove deposits from
production tubing.   Hot oil washes have been used to dissolve
paraffins above the perforations.   Washing with organic solvents
such as xylene or toluene has been used to remove paraffins from
perforations and the formation (167).   Addition of amines to these
solvents can aid in solubilizing asphaltene deposits (167).
Emulsion blocks within the formation can form as a result of
various well treatments and are more easily prevented (by using
surfactants in conjunction with well treatments, see above) than
removed.   Aromatic solvents can be used to reduce the viscosity and
mobilize oil-external emulsions (167).   Low molecular weight urea-
formaldehyde resins have been claimed to function in a similar
manner in steam and water injection wells (168,169).   Water-
external emulsion blocks can be mobilized by injection of water to
reduce emulsion viscosity.
Gypsum scaling (calcium sulfate precipitation) can occur as aqueous
formation fluids cool and experience pressure drops near and in the
production wellbore.   In wells producing from more than one
oil-bearing formation, mixing of different formation waters can
also give rise to this scaling problem.   The most common solution
is washing the wellbore with basic solutions of potassium acetate,
potassium glycolate, potassium citrate or potassium hydroxide.
Scale deposits are converted to dispersed particles which can be
circulated out of the wellbore.   A chelating agent such as
ethylenediamine tetraacetic acid can aid in dissolving calcium
sulfate deposits.   Hydrochloric acid following the basic treatment
can also be used to dissolve calcium sulfate (167).
        Injectivity can be reduced by bacterial slime which can grow
on polysaccharides and other polymer deposits left in the wellbore
and adjacent rock.   Strong oxidizing agents such as hydrogen

peroxide, sodium perborate, and occasionally sodium hypochlorite can be used to remove these bacterial deposits (143,170,171).

Rock Wettability

Wettability is defined as "the tendency of one fluid to spread on or adhere to a solid surface in the presence of other immiscible fluids" (145). Rock wettability can strongly affect its relative permeability to water and oil (145,172). Wettability can affect the initial distribution of fluids in a formation and their subsequent flow behavior. When rock is water-wet, water occupies most of the small flow channels and is in contact with most of the rock surfaces. The converse is true in oil-wet rock. When the rock surface does not have a strong preference for either water or oil, it is termed to be of intermediate or neutral wettability. Inadvertent alteration of rock wettability can strong alter its behavior in laboratory core floods (172).
In water-wet reservoirs being waterflooded, oil is displaced ahead of the water. The injection water tends to invade the small and medium-sized flow channels (or pores). As the water front passes, the remaining oil is left in the form of spherical, unconnected droplets in the center of pores or globules of oil extended through interconnected rock pores but completely surrounded by water. This oil is immobile and there is little oil production after injection water breakthrough at the production well (145).
   In a strongly oil-wet rock, water will tend to invade the larger pores as oil is found in the smaller pores or as a film on rock surfaces. Because the water preferentially flows through the larger pores, flow channels to the producing well develop and water only slowly invades the smaller flow channels. This results in a higher produced water:oil ratio and a lower oil production rate than in the water-wet case.
   Care must be taken in all well treatment and injection operations not to alter rock wettability in an undesired manner. Use of carefully selected surfactants in well treatment fluids is a way to accomplish this. Rock wettability can be altered by adsorption of polar materials such as surfactants and corrosion inhibitors, or by the deposition of polar crude oil components (173). Pressure appears to have little influence on rock wettability (174). The two techniques used to study wettability, contact and and relative permeability measurements, show qualitative agreement (175-177). Deposition of polar asphaltenes can be particularly significant in carbon dioxide enhanced oil recovery.

Primary and Secondary Oil Recovery

Primary oil recovery is the production of oil driven to wellbores by the energy of fluids under pressure in the reservoir. As reservoir pressure is reduced by oil production, additional recovery mechanisms may operate. One is natural water drive as water from an adjacent more highly pressured formation is forced into the oil-bearing formation by the pressure differential between the formations. Gas drive, expansion of a gas cap above the oil as

oil pressure declines can also drive additional oil to the
wellbore.   Additional oil may be produced by compaction of the
reservoir rock as this pressure is reduced by oil production.
Generally the additional oil produced by reservoir compaction is
small.   As the natural pressures in the reservoir decrease, oil
production declines.   The oil well may then be placed "on pump" to
maintain production at economic levels; the pump is used to draw
oil to the surface and keep the production well relatively free of
fluid.   (The pressure of a column of fluid can decrease the rate of
fluid entry into the wellbore.)
        Primary production typically recovers 10-25% of the oil
originally in place in the reservoir.   The efficiency of primary
production is related to oil properties, reservoir properties,
geometric placement of oil wells, and the drilling and completion
technology used to drill the wells and prepare them for production.
Waterflooding or secondary oil recovery is a means of adding energy
to the reservoir.   Water injection through some wells results in a
pressure differential across a reservoir resulting in the movement
of oil and injected water to offset production wells.   The
geographic arrangement of production and injection wells is
critical to maximizing oil recovery and can be related to the
geology of the reservoir (145).   Salinity of the available
injection water can have an important effect on the efficiency of
oil recovery.   Scaling, the formation of insoluble precipitates
when the saline injection water contacts a formation brine, is a
particularly common problem and many scale inhibitors have been
developed to reduce precipitation near the injection wells where
permeability reduction can greatly reduce injection rate (see
above).
        Injection rate can have a major effect on the economics of
secondary oil recovery.   Acidizing or carefully designed hydraulic
fracturing treatments can be used in increase injection rates.
More recently well treatments that do not interrupt normal water
injection operations have increased in frequency.   Addition of
surfactant to the injection water (144,146) can displace the oil
remaining near the production well.   The lower oil saturation
results in an increase in the water relative permeability (145).
Consequently a greater water injection rate may be maintained at a
given injection pressure or a lower injection pressure.   Thus
smaller and cheaper injection pumps may be used to maintain a given
injection rate.   While the concentration of surfactant in the
injection water is relatively high, the total amount of surfactant
used is not great since it is necessary only to displace the oil
from a 6-10 foot radius around the injection well.
        Extensive waterflooding began in the 1940's.   Currently
waterflooding accounts for about 40% of domestic oil production.
Waterflooding typically recovers 15-25% of the oil originally in
place.
Organic and inorganic polymers have been used to improve the
results obtained in waterflooding.   Crosslinked polymers (see
below) have been used to reduce the permeability of fractures and
high permeability streaks so that injected water flows through a
larger fraction of the reservoir volume.   The polymer is injected
with a crosslinker or the crosslinker may be injected after the

polymer. In either case, crosslinking occurs in situ. Sodium
silicate gelation has also been used in this application (178,179).
The use of organic polymers in injection wells is discussed below.
Both in situ crosslinking of partially hydrolyzed polyacrylamides
(180) and quaternary ammonium salt polymers with long sidechains
(181) have been used to reduce the permeability of water producing
zones adjacent to production wells and decrease the produced
water:oil ratio.

Enhanced Oil Recovery

Primary and secondary oil recovery together recovery only 25-50% of
the oil originally in place in a reservoir. Cumulative U.S. production of 133 billion barrels of oil and remaining reserves of 27.5
billion barrels account for only 33% of the 488 billion barrels of
oil discovery to date (182). In a more recent reference (183),
recoverable reserves as of Jan. 1, 1989 were estimated to be 26.5
billion barrels. The increasing cost of discovering major new oil
reserves in the U.S. (which most likely exist in frontier regions
of Alaska and deep water offshore) make unrecovered oil in known
fields an attractive target. Its location is known and much of the
infrastructure: wells, storage tanks, pipelines, roads, etc. are
already in place.
      Major disincentives to enhanced oil recovery are the lack of
tax incentives and a substantial decline in the price of oil since
the end of 1981. All the investment in new wells and surface
facilities and injectants must take place before any incremental
oil is produced.
      This decline in the price of oil has resulted in major changes
in the types of enhanced oil recovery (EOR) being studied in the
laboratory and field tested. Steam injection and injection of
miscible gases, primarily CO_2, remain of great interest due to the
relatively low cost of the injectants (although they are quite
expensive as compared to water). More expensive injectants such as
used in micellar polymer flooding can often efficiently recover
oil. However, the large concentration of surfactant (often as much
as 2-5% wt plus the additional cost of the polymer used to provide
mobility control (see below), usually used in concentrations of
100-1000ppm, have made these fluids prohibitively expensive.
      Recent research and field tests have focused on the use of
relatively low concentrations or volumes of chemicals as additives
to other oil recovery processes. Of particular interest is the use
of surfactants as CO_2 (184) and steam mobility control agents
(foam). Also combinations of older EOR processes such as
surfactant enhanced alkaline flooding and alkaline-surfactant-
polymer flooding have been the subjects of recent interest. Older
technologies: polymer flooding (185,186) and micellar flooding
(187-189) have been the subject of recent reviews. In 1988 84
commercial products: polymers, surfactants, and other additives,
were listed as being marketed by 19 companies for various enhanced
oil recovery applications (190).
      Other important issues influencing the economics of oil
recovery include methods of determining fluid movement and behavior
within the reservoir (191); the effect of oil composition on oil

recovery, particularly in miscible flooding (192); and corrosion control (193). Seismic, geotomographic, controlled-source audio magnetotelluric, and pressure analyses as well as tracer chemical injection and analyses are all used to understand fluid movement within the reservoir and monitor EOR processes (191).

Oil Recovery Mechanisms

The amount of oil recovery promoted by an injected fluid is related to its ability to displace the oil it contacts in the reservoir, termed the oil displacement efficiency (ODE), and to the relative amount of the reservoir invaded by the injected fluid, termed the volumetric sweep efficiency (VSE). Total oil recovery may be expressed as:

$$Oil\ Recovery = VSE \times ODE$$

For example, consider a reservoir which has produced 40% of the oil originally in place. If an injection fluid contacts 70% of the reservoir and has an oil displacement efficiency of 70% of the remaining oil (42% of the oil originally in place) then the maximum enhanced oil recovery is 49% of the oil remaining in place or 29% of the oil originally present in the reservoir. (Trapping and other oil loss mechanisms are neglected in this simplified treatment.) Total oil recovery has increased to 69%.

This example illustrates the importance of using chemicals to improve both the volumetric sweep efficiency and the oil displacement efficiency. Although the greatest attention has been given to increasing the oil displacement efficiency, primarily though the use of surfactants, a government study indicated that volumetric sweep efficiency is the greatest obstacle to increasing oil recovery (194).

Improving Volumetric Sweep Efficiency

Volumetric sweep efficiency is determined by the permeability and wettability distribution in the reservoir and by the properties of injected fluids. Waterflooding characteristically exhibits poor volumetric sweep efficiency. The more expensive the injection fluid, the more important it is to have a high volumetric sweep efficiency so that the injected fluid contacts and thus mobilizes a larger volume of oil. High permeability streaks or layers (thief zones) and natural or induced rock fractures can channel the injected fluid through a small portion of the reservoir resulting in a low volumetric sweep efficiency.

Crosslinked polymers have been widely used to substantially seal these thief zones and fractures thus directing subsequently injected fluids to different parts of the reservoir increasing VSE in waterflooding and chemical flooding. The most commonly used polymers are partially hydrolyzed polyacrylamides (195) although field applications of crosslinked xanthan gum have also been reported (196). These are generally injected at concentrations of 1000-5000 ppm and crosslinked in situ. Treatments are restricted to the near-wellbore region due to the kinetics of the crosslinking

process. The most commonly used crosslinkers are Al(III) (196-198)
and Cr(III) compounds(199). The injected fluid preferentially
enters the thief zone. The well is shut in to allow crosslinking
to occur. After 1-7 days depending on the treatment, normal
injection operations are resumed.

Aluminum (III) citrate and sodium aluminate have been used as
crosslinkers. The polymer and crosslinker solutions are injected
as alternate slugs. A layer of adsorbed polymer is built up which
is then crosslinked to subsequently injected polymer (200).
Cationic polyacrylamide may be used in the initial treatment stages
to promote rapid polymer adsorption (201). Adjustment of the pH
may allow deeper penetration of the fluids in an aluminate
crosslinking system prior to gelation (202). A process involving
injection of alternate slugs of stoichiometrically equivalent
amounts of partially hydrolyzed polyacrylamide and $Al_2(SO_4)_3$ has
been evaluated in the laboratory; permeability of sand packs were
reduced by more than 96% (203). Mixtures of Al(III) and Zr(IV)
have also been evaluated as partially hydrolyzed polyacrylamide
crosslinkers (204).

Sodium bisulfite and thiourea have been used to reduce
injected Cr(VI) to the reactive Cr(III) species that promotes
crosslinking (205,206). Kinetic studies suggest that the crosslink
stucture includes two chromium atoms bridged by oxygen (205).
Gradual dissolution of colloidal $Cr(OH)_3$ has been used to delay
crosslinking to permit deeper polymer penetration in the formation
prior to crosslinking (207) as has the use of Cr(III) propionate
(208). Injection of unhydrolyzed polyacrylamide followed by in
situ hydrolysis delays Cr(III) crosslinking (209). The rate of the
hydrolysis reaction is dependent on temperature but not on
injection and formation water salinity (210). Studies suggest that
5-10% hydrolysis is the optimum to produce a crosslinked
polyacrylamide (211).

Occasionally it may be desirable to have a rapid crosslinking
take place. Blends of chromium triacetate and hydrochloric acid
have been used in this situation (212). Gelation time decreases
substantially as applied shear increases (213,214). Thus, static
laboratory gelation time experiments should not be used to predict
gelation time in actual well treatments.

Organic crosslinkers have also been used. These include
glyoxal (215) and formaldehyde. Use of hypohalite salts (216,217)
and epichlorohydrin (218,219) have been found to increase gel
stability.

Copolymers of sodium acrylate with sodium 2-acrylamido-2-
methylpropane sulfonate (220) or N,N-dimethylacrylamide (221) have
been found useful for preparing crosslinked systems that must
function at high temperatures and relatively high salinity.
Chromium crosslinked gels prepared from a 3:1 blend of partially
hydrolyzed polyacrylamide and guar gum have been found to have a
higher strength and stability than gels prepared from the partially
hydrolyzed polyacrylamide alone (222).
Crosslinked xanthan gums have also been used to reduce the permeability of thief zones. Trivalent chromium and aluminum have been
used as crosslinkers (223,224). While crosslinker effectiveness is
reduced at high salinity, Cr(III) has been used in the field at

salinities as great as 166,000 ppm total dissolved solids (224).
Xanthan gum can be precipitated by quite high concentrations of
divalent metal ions to plug thief zones (225). Xanthan gum plus a
partially methylated melamine-formaldehyde resin has been used to
form a gel (226).

Carboxymethylhydroxyethyl cellulose has been gelled by hydrated lime (227). Succinoglycan has been crosslinked by Cr(III),
Al(III), Zr(IV), Ti(IV), and other trivalent metal ions (228,229).

Careful sizing of the treatment and choice of injection rates
is required to prevent inadvertent overtreatment i.e., excessive
treatment of oil-containing rock. The post-treatment fluid
injection rate is usually significantly less than that prior to
treatment. While successful applications of this technology in
waterfloods and in surfactant polymer floods have been reported,
temperature and pH stability limitations of the polymer and the
crosslinking chemistry result in few if any applications in steam
and CO_2 injection wells.

Reactive monomers such as acrylamide in concentrations of 2-5%w and
various additives including a free radical polymerization initiator
may be used to form polymers in situ (230-232). An optional
reactive difunctional monomer such as N,N'-methylenebis(acrylamide)
can be added to the formulation to form a cross-linked polymer in
the high permeability zone. The low viscosity aqueous fluid may be
injected at relatively high rates and preferentially enters the
high permeability zones to a greater extent than do non-Newtonian
polymer solutions (233). Polymerization takes place forming a high
permeability mass that greatly reduces rock permeability. If no
difunctional monomer is used, this mass may be slowly dissolved by
injection water. This process increases injection water viscosity
(see below).

If the two functional groups of the difunctional monomer
differ substantially in reactivity, the difunctional monomer may be
injected without any comonomer. An example is 2-hydroxyethylacry-
late which can polymerize rapidly through the carbon-carbon double
bond and form crosslinks more slowly through hydrogen atom
abstraction from the hydroxyl group. Less reactive monomers such
as 2-hydroxyethyl acrylate and the use of less reactive
polymerization intiators permit the use of this technology in
somewhat higher temperature formations.

Lignosulfonates may be crosslinked in situ using Cr(III) (234,235)
or an acidic gas such as CO_2 (227) to promote crosslinking.
Crosslinked lignosulfonate has been reported to be quite effective
at high formation temperatures. Lignosulfonate concentration is
usually 2-3% by weight. This gelation technology has been
evaluated in field tests in both waterflood and steam injection
wells. An advantage of this technology appears to be the ability
of this system to crosslink at long distances from the injection
well bore. Chemical reactivity at formation temperatures and in
situ dilution effects can limit the effective treatment radius of
the crosslinked polymer systems described above. Blends of
lignosulfonate and sodium silicate have also been evaluated in the
field (236).

Phenol-formaldehyde resins (237, 238), urea-formaldehyde
resins (239,240), melamine-formaldehyde resins (241), furfuryl

alcohol resins (242) and resins of formaldehyde plus sulfonated
tannin extract (243) or alkali Kraft lignin (244) have been
evaluated to seal thief zones near water and steam injection wells.
Polymer gels formed in situ from polyvinyl alcohol and aldehydes
(245) and by Cr(III) crosslinking of alkylene oxide - styrene
block copolymers (246) have also been evaluated.
Surfactant precipitation may be used for in-depth permeability
reduction of thief zones (247). This process is based on the
sequential injection of a slowly propagating ionic surfactant
followed by an aqueous spacer containing no surfactant, and then a
more rapidly propagating ionic surfactant of the opposite charge
type. In a sandstone reservoir, one would initially inject a
cationic surfactant and then an anionic surfactant. The oppositely
charged surfactants gradually mix in the high permeability portions
of the reservoir resulting in the formation of precipitates. These
precipitates plug flow channels and the cumulative effect is to
reduce permeability in the most flooded portions of the reservoir
diverting injectant to rock zones containing higher oil
saturations. The economically limiting factors in the use of this
process would probably be the cost and propagation rate of the
cationic surfactant.

The use of polyethylene glycol ethers in a process in which a
high viscosity emulsion is formed on contact with residual crude
oil has also been tested as a means of plugging thief zones using
surfactants (248-250). Precipitation of sodium pectate when fresh
water solutions contact brine has been proposed as a method of
plugging high permeability zones (251).

Polymer Flooding (186,187,252)

Even in the absence of fractures and thief zones, the volumetric
sweep efficiency of injected fluids can be quite low. The poor
volumetric sweep efficiency exhibited in waterfloods is related to
the mobility ratio, M. This is defined as the mobility of the
injected water in the highly flooded (watered-out) low oil
saturation zone, m_w, divided by the mobility of the oil in
oil-bearing portions of the reservoir, m_o, (253,254). The mobility
ratio is related to the rock permeability to oil and injected water
and to the viscosity of these fluids by the following formula:

$$M = m_w/m_o = (k_{rw}/\mu_w)/(k_{ro}/\mu_o)$$

wherein k_{rw} and k_{ro} represent the relative permeability to water
and oil respectively and μ_w and μ_o represent the viscosity of the
aqueous and oil phases respectively.

The displacing fluid may be steam, supercritical carbon
dioxide, hydrocarbon miscible gases, nitrogen or solutions of
surfactants or polymers instead of water. The VSE increases with
lower mobility ratio values (253). A mobility ratio of 1.0 is
considered optimum. The mobility of water is usually high relative
to that of oil. Steam and oil-miscible gases such as supercritical
carbon dioxide also exhibit even higher mobility ratios and
consequent low volumetric sweep efficiencies.

The first mobility control agents were partially hydrolyzed
polyacrylamides having molecular weights of $1\text{-}5\text{x}10^6$ (254-259) and
xanthan gum (biopolymer) (1,34,260). Virtually all field projects
have used polymers from one of these two classes although variations in polymer molecular weight and structure have been made to
improve performance properties (see below). Relatively low polymer
concentrations (down to ca. 100ppm) can significantly increase
injected water viscosity. The increase in apparent viscosity in
porous media was often sigificantly greater that that exhibited in
conventional laboratory viscosity measurements (259). Another
benefit of both polyacrylamides and xanthan gum is a long-lasting
decrease in rock permeability to aqueous fluids that persists even
during long periods of water injection. This residual resistance
effect has been observed in laboratory tests and some field trials.
After termination of polymer injection, the North Stanley Field and
North Burbank Unit 29 polymer field tests exhibited injected water
permeability reductions attributed to residual resistance effects
that lasted for more than three years and more than seven years
respectively (261,262).
    Each polymer type has important advantages and significant
disadvantages. These are summarized in Table I and discussed below
for polymers representative of those presently used.

Table I. Properties of Polyacrylamide and Xanthan Gum EOR Polymers

Property	Polyacrylamide	Xanthan Gum
Brine tolerance	very limited esp. to Ca^{+2}, Mg^{+2}	good tolerance to mono- and divalent cations
Transition metal cations	easily crosslinked	easily crosslinked
Shear stability	undergoes irreversible shear degradation	reversible shear thinning.
Thermal stability	maximum use T $225^{\circ}\text{-}250^{\circ}F$	maximum use T $160\text{-}170^{\circ}F$
Hydrolytic stability	hydrolysis promoted by acid or base. Partially hydrolyzed product more sensitive to Ca^{+2}, Mg^{+2}	hydrolytic depolymerization promoted by acid or base esp. at high T
Oxidative stability	susceptible	particularly susceptible esp. at high T
Microbial degradation	susceptible to attack by yeasts, fungi, bacteria	very susceptible in aerobic conditions

Compared to partially hydrolyzed polyacrylamide, xanthan gum is more expensive, more susceptible to bacterial degradation, and less stable at elevated temperatures (1). However, xanthan gum is more soluble in saline waters, particularly those containing divalent metal ions; generally adsorbs less on rock surfaces; and is substantially more resistant to shear degradation (1,34). The extensional viscosity of the semi-rigid xanthan molecule is less that that of the flexible polyacrylamide (263).

In addition to the normal problems of completely dissolving particles of water-thickening polymers, xanthan gum contains insoluble residues the cumulative effect of which is to decrease polymer injectivity. Fermentation broths containing 8-15% by weight xanthan can be used to prepare solutions which contain no undissolved polymer particles (264). Flash drying of polymer broths can produce a solid product which dissolves more readily in injection waters (264). Other means of inproving xanthan solution injectivity include brief (30 sec) ultrasonic treatment at 60-80 MHz (265), solid polymer hydration in solutions of metal complexing agents such as sodium citrate (266) or low concentrations of boron species (266), heat treatment (267,268), bentonite treatment followed by filtration through diatomaceous earth (266,269), passing xanthan solutions through a colloid mill (270), treatment with methylenebis(isocyanate) (271), cellulase enzymes (272,273), proteases (272,273), polysaccharide hydrolases (274), or caustic agents plus an enzyme (274). Flow channels adjacent to the well-bore which have been plugged with solid xanthan residues may be reopened by treatment with oxidizing agents such as hydrogen peroxide.

Use of oxygen scavengers (275,276,277) and bactericides (278) is common parctice in field operations. Among the oxidation stabilizers used are thiourea (279), sodium dithionite (280), guanidine acetate (281), and a blend of sodium sulfite, thiourea, and 2-propanol (282). Among the other chemicals studied as xanthan gum and polyacrylamide stabilizers are: sodium bisulfite, sodium 2-mercaptobenzothiazole and benzoimidizol (283), 2-thioimidazoli-done (284), 1-tolylbiguanide (285), 2-thiazoline-2-thiolate (286), dithiocarbamates (287), methionine (288), thiosulfuric acid (289), and phosphonic acid esters (290), and mixtures of isobutanol, sodium 2,4,6-trichlorophenate, and sodium diethylenetriaminepenta-acetic acid (291).

When dissolved in more saline waters, xanthan gum produces a higher apparent viscosity than the same concentration of polyacryl-amide (292). Prehydration of xanthan in fresh water followed by dilution in the saline injection water has been reported to provide higher viscosity than direct polymer dissolution in the same injec-tion water. Optical rotation and intrinsic viscosity dependence on temperature indicate xanthan exists in a more ordered conformation in brine than in fresh water (293).

Although high concentrations of pyruvate ring-opened polymers exhibit increased tolerance to divalent metal ions in high density completion fluids (294), at low polymer concentrations, xanthan containing the intact pyruvate ring exhibits higher brine solution viscosity and better filterability than its ring-opened analog (295). A xanthan gum containing pyruvate rings in most of the

polymer repeat units has been produced by a proprietary strain of
Xanthamonas campestris (296) and evaluated for polymer flooding
applications (295).
Other microbial polysaccharides have also been evaluated for use in
enhanced oil recovery. These include scleroglucan (299-300) which
is thought to exist in solution in a rigid helical conformation,
polymers produced by the bacterium Pseudomonas methanica (301), by
Leuconostoc mesenteroides (302), by Aerobacterium NC1B11883 (303),
the alga Porphyridium aeruginium (304), and nonionic glucose
homopolysaccharides produced by fungi (305). Xanthamonas bacteria
have also been used to produce a polysaccharide comprising glucose
and mannose units in a 2:1 ratio. This polymer has been claimed to
be a better water viscosifier than xanthan gum (306). Saccharide
polymers may be prepared by the polymerization of
3-0-methacryloyl-D-glucose (307).
Most polyacrylamides used as mobility control agents are actually
partially hydrolyzed or are acrylamide - acrylic acid (or sodium
acrylate) copolymers produced by emulsion copolymerization (308).
Emulsion polymers are used to avoid the high shear degradation and
undissolved solid particle problems often associated with solid
polyacrylamide dissolution. Another method of avoiding problems
associated with hydration of solid polymer particles is acrylamide
solution polymerization at the wellhead. The polymerization can be
designed to proceed at adequate rates and in saline injection
waters to provide polymers of adequate viscosity characteristics
(309). Polyacrylamide is usually hydrolyzed in base to produce a
random distribution of acrylate groups. This random distribution
is similar to that obtained in a copolymer having the same acrylate
group content (310). Acid hydrolysis results in a more block-like
distribution of acrylate units (311). ^{60}Co irradiation has been
used to initiate polymerization and prepare particularly high
molecular weight polyacrylamides (312).

Electrostatic repulsion of the anionic carboxylate groups
elongates the polymer chain of partially hydrolyzed polyacrylamides
increasing the hydrodynamic volume and solution viscosity. The
extensional viscosity is responsible for increased resistance to
flow at rapid flow rates in high permeability zones (313). The
screen factor is primarily a measure of the extensional (elonga-
tional) viscosity (314). The solution properties of polyacrylamides have been studied as a function of NaCl concentra-tion and
the parameters of the Mark-Houwink-Sakaruda equation calculated
(315). Maximum freshwater viscosity occurs at ca. 35% hydrolysis
(316) while maximum viscosity in a Ca^{+2}-containing brine occurs at
10-15% hydrolysis. Metal ions interact with carboxylate groups
reducing their mutual repulsion and thus decreasing hydrodynamic
volume and solution viscosity. Divalent metal ions reduce viscosity more than monovalent ones (317). Above 33-35% hydrolysis,
interaction with Ca^{+2} causes polyacrylamide precipitation (318).
The major mode of polyacrylamide decomposition at elevated
temperature (in the absence of oxygen) is hydrolysis (319,320).
Thus, the concentration of divalent metal ions has an effect on
viscosity retention at high temperature. Chelating and
sequestering agents have been used to reduce the adverse effect of

divalent (316) and multivalent metal ions on polyacrylamide
solution viscosity (321,322).

Proper well completion, particularly perforation design, can
reduce polyacrylamide shear degradation during entry into the
formation (313).
Acrylamide copolymers designed to reduce undesired amide group
hydrolysis, increase thermal stability, and improve solubility in
saline media have been synthesized and studied for EOR applications. These polymers still tend to be shear sensitive. Acrylamide comonomers that have been used include 2-acrylamido-2-
methylpropane sulfonate, abbreviated AMPS, (1,321-324), 2-sulfo-
ethylmethacrylate (325,326), diacetone acrylamide (324, 326), and
vinylpyrrolidinone (327,328). Acrylamide terpolymers include those
with sodium acrylate and acrylamido-N-dodecyl-N-butyl sulfonate
(329), with AMPS and N,N-dimethylacrylamide (330), with AMPS and
N-vinylpyrrolidinone (331), and with sodium acrylate and sodium
methacrylate (332). While most copolymers tested have been random
copolymers, block copolymers of acrylamide and AMPS also have
utility in this application (333).

Acrylamide terpolymers having a heterocyclic ring in the
polymer backbone have been shown to exhibit improved viscosity and
shear degradation properties (334,335). A disadvantage of acrylamide copolymers is their greater cost as compared to partially
hydrolyzed polyacrylamides. Acrylamide graft copolymers have been
studied in an effort to reduce copolymer costs. These include
acrylamide graft copolymers with starch (336), dextran (337), and
lignin (338).
Polymer association complexes (38-43, see above) including those
which form micelle structures by association of hydrophobic groups
such as nonylphenoxy polyethylene glycol acrylates (339),
acrylamide terpolymers containing hydrophobic alkylacrylamides
(340-343), and poly(styrene-co-maleic anhydride) vinylbenzylpoly-
glycol ethers (344) substantially increase water viscosity at quite
low polymer concentrations. Similar hydrophobically modified
polysaccharides such as hexadecyloxyhydroxyethyl cellulose (345)
may be suitable for use in enhanced oil recovery. These polymer
association complexes exhibit much higher solution viscosity than
equal concentrations of conventional polymers.

The substantial decrease of polyacrylamide solution viscosity
in mildly saline waters can be utilized to increase injection rates
by adding a quaternary ammonium salt polymer to the polyacrylamide
mixing water (346,347). If the cationic charge is in the polymer
backbone and substantially shielded from the polyacrylamide by
steric hindrance, formation of an insoluble interpolymer complex
can be delayed long enough for polymer injection. Upon contacting
formation surfaces, the quaternary ammonium salt polymer is adsorbed reducing solution salinity and thus increasing viscosity away
from the wellbore where it will not adversely affect injectivity.
By using a clay stabilizing quaternary ammonium salt polymer,
formation damage associated with low salinity polyacrylamide
solvents can be reduced (348).
Propagation of enhanced oil recovery chemicals through rock is
critical to the success of an EOR project. Polymer retention in
permeable media has been the subject of considerable study (349)

and mechanical entrapment as well as adsorption has been identified
as a cause of polymer loss (350,351). Sacrificial adsorption
agents may be used to reduce the adsorption of expensive polymers
and surfactants. Lignosulfonates and their derivatives have been
extensively evaluated for this application (34,352-356). Other
chemicals tested for this application include poly(vinyl alcohol)
(357), sulfonated poly(vinyl alcohol) (358), sulfonated poly(vinylpyrrolidinone) (358), low molecular weight polyacrylates (359), and
sodium carbonate (360).

Surfactants for Mobility Control

Despite its relatively high mobility, water has been used to
decrease the mobility of even higher mobility gases and
supercritical CO_2 used in miscible flooding (361). While water
mobility can be up to ten times that of oil, the mobility of gases
can be 50 times that of oil (362). The following formula is used
to calculate gas:oil mobility ratios (363):

$$M = \left[(k_g/\mu_s) + (k_w/\mu_w)\right] \bigg/ \left[(k_o/\mu_o) + (k_w/\mu_w)\right]$$

wherein k refers to permeability, μ to viscosity, and g, s, o, and
w to gas, miscible solvent, oil, and water respectively. The water
may be injected simultaneously with the gas or in alternate slugs
with the gas (WAG process). X-ray computerized tomography of core
floods has demonstrated the increased volumetric sweep efficiency
attained in the WAG process (364).

The WAG process has been used extensively in the field,
particularly in connection with supercritical CO_2 injection and
some success have been reported.(365-367). However, it would be
desirable to develop a method to further reduce the viscosity of
injected gas, particularly CO_2, the most commonly used gas
(actually injected as a supercritical fluid) in the U.S.. While
limited studies on increasing the viscosity of CO_2 though the use
of supercritical CO_2-soluble polymers and other additives have been
reported (368, see also Chapter 29 and references therein), the
major direction of research has been the use of surfactants to form
low mobility foams or supercritical CO_2 dispersions within the
formation.
The behavior of foam in porous media has been the subject of
entensive study and recently a collection of papers on this subject
(369), a review of foam rheology (370), and an extensive
bibliogra-phy (371) have been published. X-ray computerized
tomographic analysis of core floods indicated that addition of 500
ppm of an alcohol ethoxyglycerylsulfonate increased volumetric
sweep efficiency substantially over that obtained in a WAG process
(364).

The role of various surfactant association structures such as
micelles and lyotropic liquid crystals (372), adsorption-desorption
kinetics at liquid-gas interfaces (373) and interfacial rheology
(373) and capillary pressure (374) on foam lamellae stability has
been studied. Microvisual studies in model porous media indicate

that the predominant mechanisms of in situ foam generation are snap-off at pore constrictions (375,376), lamellae leave-behind (375), and lamellae division (375).

The reason for wide-spread interest in the use of surfactants as gas mobility control agents (369) is their effectiveness at concentrations of 0.1%wt (377) or less (364). This low chemical requirement can significantly improve process economics.

Another advantage is the wide range of surfactant classes and chemical structure variations within each class of surfactant which can be screened to optimize surfactant performance for a given set of reservoir conditions. Any change in surfactant structure should increase the propagation rate and the displacement efficiency. It has been noted that this is possible only by decreasing surface viscosity (378). Among the classes of surfactants studied for this application are alcohol ethoxylates and their sulfate and sulfonate (364,379-384), and carboxylate (385) derivatives, alkylphenol ethoxylates (382), alpha-olefin sulfonates (383), and alkylated diphenylether disulfonates (386). Increased linear hydrophobe carbon chain length, decreased hydrophobe branching, and increased ethoxy group chain length was found to increase foam stability (380). When using mixtures of surfactants or surfactants plus an alcohol, foam stability, injected breakthrough time at the core outlet, and oil recovery were maximized when the two surfactants or the surfactant and the alcohol had the same carbon chain length (387,388). Addition of a water-thickening polymer to the aqueous phase will stabilize the foam (389-391).

In addition to the mobility control characteristics of the surfactants, critical issues in gas mobility control processes are surfactant salinity tolerance, hydrolytic stability under reservoir conditions, and surfactant propagation. Lignosulfonate has been reported to increase foam stability and function as a sacrificial adsorption agent (392). The addition of sodium carbonate or sodium bicarbonate to the surfactant solution reduces surfactant adsorption by increasing the aqueous phase pH (393).

Alcohol ethoxysulfates have been used in field tests as foaming agents for nitrogen (394) and carbon dioxide (395). Application of alcohol ethoxysulfates is restricted due to its limited hydrolytic stability at low pH and elevated temperature (396).

High temperature steam has also been used in enhanced oil recovery, for the recovery of highly viscous crude oils (397). In heavy oil fields, water flooding is often omitted and steam injection begun immediately after primary production. Steam injection temperature, usually 350-450°F in California oil fields, can reach 600°F in Canadian and Venezuelan projects. Heat is transferred from the steam to the rock and crude oil reducing oil viscosity. This increases oil mobility thereby enhancing oil production.

Gravity override is the migration of the steam to the upper portion of the formation and is caused by the low steam density. This results in channeling of the steam through the upper portion of the reservoir and a low volumetric sweep efficiency.

Surfactants have been used as steam mobility control agents in both laboratory and field tests to prevent this gravity override thereby increasing volumetric sweep efficiency. Surfactants that have been

used in field tests include C_{16-18} alpha-olefin sulfonates,
alkyltoluene sulfonates, and neutralized dimerized alpha-olefin
sulfonic acid.

Careful screening procedures are required to evaluate surfactants as steam foaming agents (398,399). Increasing the hydrophobe
carbon number in alpha-olefin sulfonates from 14-16 to 16-18 to >25
has been reported to improve foam strength (400,401). In alkylaromatic sulfonates, longer linear alkyl groups (402) or dialkyl
substitution (403) seem to have the same effect. Other alkylaromatic sulfonates containing benzene, toluene, or xylene rings (402,
404); two fused aromatic groups (405); and the diarylether group
(406) have been favorably evaluated as steam foaming agents. The
neutralized dimer of an alpha-olefin sulfonate has also been used
(407).

The high temperature steam will cool and eventually condense
as it propagates through the oil reservoir. In order to maintain
foam strength as the steam cools, a noncondensible gas, most often
nitrogen or methane, is usually added to the injectant composition
(408). A method of calculating the optimum amount of noncondensible gas to use has been reported (409).

Critical parameters affecting surfactant performance are surfactant propagation rate and surfactant stability at steam temperatures that can reach more than $600^{\circ}F$. Surfactant propagation rate
can be reduced by adsorption, precipitation, and partitioning into
the oil phase. Adsorption increases with increasing salinity and
decreases with increasing temperature (410). A numerical model of
foaming agent transport has been developed which uses the surface
excess variable in describing surfactant adsorption (411).

Surfactant propagation can be improved by the use of
additives. Both surfactant partitioning and precipitation tend to
increase with increasing calcium ion concentration (412) so minimizing divalent metal ion concentration in the surfactant solution is
desirable. Injection of a surfactant preslug containing NaCl will
convert clays to their sodium form reducing later ion exhange
processes that result in the presence of Ca^{+2} ions in the surfactant solution (413,414). The use of a hydrotrope such as sodium
xylene sulfonate has been reported to increase oil recovery in
laboratory steam foam floods of sandpacks (415). The hydrotropes
may function as sacrificial adsorption agents or interact with the
foaming agent to stabilize lamellae and increase foam strength.
Thermal stability of the foaming agent in the presence of high
temperature steam is essential. While alkylaromatic sulfonates
possess superior chemical stability at elevated temperatures
(416-419), alpha-olefin sulfonates possess sufficient chemical
stability to justify their use at steam temperatures characteristic
of most U.S. steamflood operations. Decomposition is a
desulfonation process which is first order in both surfactant and
acid concentrations (417). Since acid is generated in the
decomposition, the process is autocatalytic. However, reservoir
rock has a substantial buffering effect. The addition of high pH
agents such as sodium hydroxide to the surfactant solution has been
reported to increase foam strength and stability (420). The sodium
hydroxide may function by precipitating with calcium ions to
improve surfactant propagation (421). This is also the mechanism

by which sodium carbonate and trona ($Na_2CO_3/NaHCO_3$) function when used as steam foam additives (422). These additives can also maintain the pH at a high enough value to reduce the rate of surfactant decomposition. In addition, the added base may interact with petroleum soaps naturally found in the crude oil to more efficiently displace oil (420); the consequent lower oil saturation can result in a more stable foam.
High sulfur content heavy crude oil may be recovered more efficiently using transition metal ions such as Ti or V and optionally carbon monoxide as steam additives (423).

Improving Oil Displacement Efficiency

The use of relatively large concentrations of surfactants, usually 2-5%w, to substantially increase oil displacement efficiency has been the subject of very extensive study (188-190,424-426). This complex process usually involves the injection of a brine preflush to adjust reservoir salinity followed by injection of a micellar slug comprising the surfactant, a "cosurfactant" (usually a C_{4-6} alcohol) and a hydrocarbon. A polymer solution is usually injected after the micellar slug to reduce viscous fingering of the drive fluid into and through the micellar slug. This viscous fingering causes dilution of the surfactant, less contact of the micellar slug with the crude oil, and trapping of some of the micellar slug in the reservoir. Process effectiveness depends on maintaining an ultralow interfacial tension (<0.01 dynes/cm) between the injected surfactant slug and the crude oil (427). The surfactant-rich microstructures involved in oil recovery include vesicles as well as micelles (428). Interfacial tension behavior is sensitve to the presence of air and is both temperature and pressure dependent, it can be different for stock tank oil and live (containing dissolved gases under pressure) reservoir crude oil 174). Therefore interfacial tension, phase behavior, and core flood tests should be carefully designed.
By about 1980, the emphasis of research had shifted from inexpensive surfactants such as petroleum sulfonates to more expensive but more effective surfactants tailored to reservoir and crude oil properties. Critical issues are: surfactant performance in saline injection waters, surfactant adsorption on reservoir rock, surfactant partitioning into the crude oil, surfactant chemical stability in the reservoir, surfactant interactions with the mobility control polymer, and production problems caused by produced emulsions. Reservoir heterogeneity can also greatly reduce process effectiveness. The decline in oil prices dating from the end of 1981 halted much of this research due to the relatively high cost of micellar processes (also called surfactant polymer flooding). Since 1982 the number of field projects in progress evaluating this technology has dropped from 20 to 9 (429). Only one field test since 1982 has been successful enough that expansion of the project has been considered (430).
The thrust of surfactant flooding work has been to develop surfactants which provide low interfacial tensions in saline media, require less cosurfactant, are effective at low concentrations, and exhibit less adsorption. The optimal salinity concept and the

salinity requirement diagram (431,432) are extremely useful when
screening surfactants. When comparing the performance of different
surfactants, it is important that the comparison be made at the
optimal salinity of each surfactant or, if it is not possible to
adjust injection water salinity, in the actual injection water to
be used in a given field project.

While nonionic surfactants such as alcohol ethoxylates,
alkylphenol ethoxylates (433) and propoxylates (434) and alcohol
propoxylates (434) have been studied, most recent work has been on
anionic surfactants: alcohol propoxysulfates (434), alkylphenol
propoxysulfates (434), alcohol alkoxyalkylsulfonates (435), alkylphenol alkoxyalkylsulfonates (435), secondary alkane sulfonates
(436), alpha-olefin sulfonates (436), calcium or magnesium salts of
alpha-olefin sulfonates (437), internal olefin sulfonates (438),
blends of alpha- and internal olefin sulfonates (439), blends of
branched olefin sulfonates and polyoxyalkylene alkylphenyl ether
sulfates (440) or alkylaryl alkoxysulfate (441), sulfonated
Friedel-Crafts alkylation products of benzene, toluene, and xylene
with alpha-olefins (442), alkylalkoxyphenol sulfonates (sulfonate
group on the benzene ring) (443), styrylaryloxy ether ethylsulfo-
nates (ethylsulfonate group at end of the alkoxy chain) (444), the
ethoxyethylsulfonate salt of dicyclopentadiene (445), carboxymethylated linear alcohol ethoxylates (445-448), carboxymethylated
alkylphenol ethoxylates (448), carboxypropylated alcohol ethoxylates and alkylphenol ethoxylates (449), and branched (twin-tail)
carboxymethylated alcohol ethoxylates (450). Increasing the length
of an ethoxy chain reduces the critical micelle concentration (380,
451). Cosurfactant requirements can be minimized using a surfactant having a short branched hydrophobe or branched (vs. linear)
alkyl substituent on an aromatic group (452,453) and a long ethoxy
group chain (453).

Blends of surfactants optimized for seawater or reservoir
brine salinity include linear alkylxylene sulfonate/alcohol ether
sulfate mixtures (454,455). Alkyl- and alkylarylalkoxymethylene
phosphonates (456), and amphoteric surfactants (457,458) have also
been evaluated for use in surfactant flooding.
High concentrations (1-10%) of lignosulfonate have sufficient
interfacial activity to increase oil recovery from unconsolidated
sands (459) and have been shown to interact synergistically with
petroleum sulfonates to produce an ultralow interfacial tension
(460) and substantially increase oil recovery (461). Paper
industry spent sulfite liquors function in a similar manner if they
are not in the Ca^{+2} or Mg^{+2} forms which precipitate the petroleum
sulfonates (462). Low molecular weight ethoxylated, sulfated, or
sulfonated lignin phenols have been used alone in surfactant floods
and found to recover more than 75% of the oil remaining after
waterflood (463). The use of alkylated oxidized lignins has also
been studied (464).

Of these surfactants, two classes are worthy of further note.
The alpha-olefin sulfonates have been found to possess good salt
tolerance, chemical stability at elevated temperatures, and appear
to exhibit good oil solubilization and low interfacial tension over
a wide range of temperatures (438,465). While being less salt
tolerant, alkylaromatic sulfonates exhibit excellent chemical

stability. The nature of the alkyl group, the aryl group, and the ring isomer distribution produced in the Friedel-Crafts alkylation of the aromatic compound (452,466) can all be adjusted to optimize surfactant performance under a given set of reservoir conditions.

The effect of temperature, pressure, and oil composition have all been the subjects of intensive study and only a leading reference (467) will be cited. Surfactant propagation is a critical factor in determining the economics of an oil recovery process and has been the subject of many investigations (468). Recently liquid flow microcalorimetry has been used to study surfactant adsorption and determine the adsorption isotherm and entropy of adsorption (469,470). The use of high pressure liquid chromatography to analyze low surfactant concentration core flood effluents has aided in determining which components of commercially produced surfactants most rapidly propagate through rock (471). Commercial surfactant synthesis synthesis can then be modified to maximize the content of these rapidly propagating components. Mass spectral analysis could also be applied to this problem to identify chromatographic elution peaks and reduce the need to synthesize model compounds. Surfactant retention due to partitioning into residual crude oil can be significant relative to adsorption and reduce surfactant propagation rate appreciably (472).

The use of sacrificial agents to increase the surfactant propagation rate through reservoir rock has been proposed. Lignosulfonates and chemically modified lignosulfonates (34,352-356) and sodium saccharite wastes from wood pulping (473) have been evaluated as sacrificial adsorption agents. Injection of solutions containing K^+, NH_4^+, and zirconium ions prior to surfactant injection has been found to decrease surfactant adsorption (474). This is believed to occur by clay stabilization which reduces later swelling and fines migration (processes which increase the surface area exposed to the surfactant solution). Alkaline chemicals (422,423), particularly sodium silicate (475), which precipitate in the presence of divalent metal ions can increase the surfactant propagation rate.

Intermixing of the polymer mobility control fluid with the surfactant slug can result in surfactant - polymer interactions which have a significant effect on oil recovery (476). Of course, oil - surfactant interactions have a major effect on interfacial behavior and oil displacement efficiency. The effect of petroleum composition on oil solubilization by surfactants has been the subject of extensive study (477).

Caustic flooding involves the injection of high pH agents such as sodium hydroxide, sodium carbonate or sodium silicate to generate surfactants in situ by interaction with organic acids present in crude oil (478,479). The crude oil acid number (the number of grams of KOH required to neutralize one gram of crude oil) should be >0.5. A number of different oil recovery mechanisms are thought to be operative: lowering of the capillary number (the ratio of viscous to capillary forces) through interfacial tension reduction, altering rock wettability (usually from oil-wet to water-wet), oil emulsification and entrapment of oil which results in a lower water mobility (in turn resulting in a greater injected water volumetric sweep efficiency), oil emulsification and entrainment in the

flowing aqueous phase, and possibly the solubilization of rigid
films that may form at the oil-water interface.

While the injected chemicals are relatively inexpensive, large
quantities must be injected due to reaction of the high pH agents
with reservoir clays (480) and by precipitation by divalent metal
ions present in formation waters. (This precipitation has been
used to reduce adverse surfactant and polymer interactions with
divalent metal ions by injecting a caustic preflush prior to a
micellar polymer flood (481)). The presence of a lignosulfonate
(482) or a polyacrylate (483) in the alkaline injectant has been
reported to reduce this precipitation. Ion exchange processes
promoting solubilization of divalent metal ions (484) limit the
effectiveness of preflushes injected prior to the caustic solution.
Which is the best of the three major alkaline agents used in this
process: sodium hydroxide, sodium carbonate, and sodium orthosilicate, is the subject of some dispute. At equivalent Na_2O levels,
the three alkaline agents gave equivalent recovery of each of nine
different crude oils was obtained in laboratory core floods (485).
The inclusion of surfactant in the caustic formulation (surfactant
enhanced alkaline flooding) can increase optimal salinity of the
saline (NaOH is a salt) alkaline formulation thereby reducing
interfacial tension and increasing oil recovery (481,486,487).
Both nonionic and anionic surfactants have been evaluated in this
application (488,489) including internal olefin sulfonates (487,
490), linear alkylxylene sulfonates (490), petroleum sulfonates
(491), alcohol ethoxysulfates (487,489,492). Ethoxylated alcohols
have been added to some anionic surfactant formulations to improve
interfacial properties (486). The use of water thickening polymers, either xanthan or polyacrylamide to reduce injected fluid
mobility mobility has been proposed for both alkaline flooding
(493) and surfactant enhanced alkaline flooding (492). Crosslinked
polymers have been used to increase volumetric sweep efficiency of
surfactant - polymer - alkaline agent formulations (493).

While this technology appears quite promising and a field
project is in progress (494), field pilot results are unavailable
as yet.

Miscible gas flooding is currently the preferred technology to
increase oil displacement efficiency. Supercritical CO_2 (361,495,
496) and various hydrocarbon injectants (361,497,498) undergo
complicated physical interactions with crude oil that result in
stripping out of the low molecular weight components (which increases oil production). In addition, the rapid or gradual development
of miscibility with the remaining crude oil constituents results in
mobilization of at least some of the oil. Either partial or complete miscibility with the oil may be developed depending on the
nature of the injectant, crude oil properties, and reservoir conditions, particularly temperature. In addition, the interaction of
the injectant with the crude oil can result in changes in rock
wettability which can affect oil recovery and reduce injectivity.
As noted previously, both surfactants and polymers may be used to
reduce the mobility of these low viscosity injectants.

Steam flooding (397,499,500) can greatly increase the recovery of
high viscosity crude oils through heat thinning processes. As
noted previously, surfactants can be used to reduce the mobility of

the high temperature steam. Interfacial tension reduction promoted
by steam foaming agents can also increase oil recovery (see Chapter
18). Since heavy crude oils have relatively high acid numbers, it
is not surprising that addition of alkaline agents to high
temperature steam can increase recovery of these oils (501,502).
The in situ combustion method of enhanced oil recovery through air
injection (397,503,504) is an exceeding complex process chemically.
However, because little work has been done on the effect of
chemical additives to oil recovery efficiency, this process will
not be discussed herein.

Summary and Conclusions

Current and projected oil prices have resulted in oil recovery
processes based on organic chemicals as the primary oil recovery
agent being uneconomic. As a result, both R&D and commercial
activities have been redirected to the use of relatively low
quantities of chemicals to increase the effectiveness of
waterflooding, supercritical CO_2 and gas injection EOR, and steam
flooding. The recent emphasis has been on the use of chemicals to
increase the volumetric sweep efficiency of these EOR processes.
While a number of crosslinked polymer systems have been shown to be
effective in substantially reducing the permeability of thief
zones, methods for achieving greater treatment radii (from the
injection well) are desirable. Experience with surfactant foams or
dispersions as mobility control agents is more limited so the
potential of this technology in economically improving volumetric
sweep efficiency is less clear. Additional field tests are needed.

Literature Cited

1. Chatterji, J.; Borchardt, J.K. J. Pet. Technol., 1981, 32
    2042-2056.
2. Chilingarian, G.V.; Vorabutr, P. Drilling and Drilling Fluids,
    Elsevier Scence Publishing Co., Inc. New York, 1981.
3. Clark, R.K.; Nahm, J.J. in Kirk-Othmer Encylcopedia of
    Chemical Technology, M. Grayson, Ed.; John Wiley & Sons, New
    York, 1982, 3rd Edit., Vol. 17, pp. 143-167.
4. Saunders, G.C. West German Patent 2 949 741, 1980.
5. Bruton, J.R. World Oil, 1979, 189 (7), 71-4.
6. Poole, G. Oil Gas J., 1981, 79 (28), 159, 161.
7. Ezzat, A.M.; Augsburger, J.J.; Tillis, W.J. J. Pet. Technol.,
    1988, 40, 491-8.
8. Doty, P.A. SPE Drilling Eng., 1986, 1, 17-30.
9. Morgenthaler, L.M. SPE Production Eng., 1986, 1, 432-436.
10. Borchardt, J.K.; Rao, S.P. U.S. Patent 4 554 081, 1985.
11. House, R.F.; Hoover, L.D. French Patent 2 488 325, 1982.
12. Socha, G.E. U.S. Patent 4 373 959, 1983.
13. Pelezo, J.A.; Corbett, Jr., G.E. French Patent 2 578 549,
    1986.
14. Walker, T.E.; Strassner, J.E . West German Patent 3 521 309,
    1985.
15. Miura, M.; Harada, E.; Domon, F. Japan Patent 62 153 382,
    1987.

16. Salamone, J.C.; Clough, S.B.; Salamone, A.B.; Reid, K.I.G.;
     Jamison, D.E. Soc. Pet. Eng. J., 1982, 22, 555 (1982).
17. Haack, T.; Shaw, D.A.; Greenley, D.E.   Oil Gas J., 1986, 84
     (1), 82.
18. Vio, L.; Meunier, G.   French Patent 2 552 441, 1985.
19. Meltzer, Y.L.   Water-soluble Polymers: Technology and
     Applications, Chem. Process Rev., Noyes Data Corp., Park
     Ridge, N.J., 1972, pp. 17-19.
20. Allan, T.O.; Roberts, A.P.   Production Operations, 2nd edit.,
     Oil & Gas Consultants, Inc., Tulsa, 1982, Volumes 1 and 2.
21. Smith, D.K.   Cementing, Society of Petroleum Engineers,
     Dallas, 1974, Monograph No. 4
22. Smith, R.C.   Oil Gas J., Nov. 1, 1982, 72-5.
23. Carter, L.G.; Evans, G.W.   J. Pet. Technol., 1964, 16, 157-60.
24. Brice, Jr., J.W.; Holmes, B.C.   J. Pet. Technol., 1964, 16,
     503-8.
25. Keller, S.R.; Crook, R.J.; Haut, R.C.; Kulakofsky, D.S.
     J. Pet. Technol., 1987, 39, 955-60.
26. Crook, R.J.; Keller, J.; Wilson, M.A.   J. Pet. Technol., 1987,
     39. 961-6.
27. Cheung, P.R.; Beirute,, R.M.   J. Pet. Technol., 1985, 37 (6),
     1041-1048.
28. Griffin, T.J.; Spangle, L.B.; Nelson, E.B.   Oil Gas J., June
     25, 1979, 143-151.
29. Harms, W.M.; Febus, J.S.   J. Pet. Technol., 1985, 37 (6),
     1049-1057.
30. Smith, R.C.; Powers, C.A.; Dobkins, C.A.   J. Pet. Technol.,
     1980, 32, 1438-44.
31. Harms, W.M.; Lingenfelter, J.T.   Oil Gas J., Feb. 2, 1981,
     59-66.
32. Bruckdorfer, R.A.; Coleman, S.E. U.S. Patent 4 635 724, 1987.
33. Cole, R.C.; Borchardt, J.K. Drilling, 1985 (4), 44.
34. Borchardt, J.K.   Encyclopedia of Science and Engineering, John
     Wiley & Sons, New York, 1987, Volume 10, 328-369.
35. Mondshine, T.C.   U.S. Patent 4 175 042, 1979.
36. Saucier, R.J.   J. Pet. Technol., 1974, 26, 205-12.
37. Pober, K.W.; Huff, H.M.; Darlington, R.K.   J. Pet. Technol.,
     1983, 35, 2185.
38. Borchardt, J.K.   U.S. Patent 4 524 003, 1985.
39. Norton, C.J.; Falk, D.D.   U.S. Patent 3 919 092, 1975.
40. Borchardt, J.K.   U.S. Patent 4 508 629, 1985.
41. Knight, R.K.   U.S. Patent 4 039 028, 1977.
42. Kovacs, P.   Food Technol., 1973, 27, 26.
43. Rocks, J.K., Food Technol., 1971, 25, 22.
44. Hong, K.C.; Milhone, R.S.   J. Pet. Technol., 1977, 29, 1657.
45. Muecke, T.W.   J. Pet. Technol., 1974, 26, 157.
46. Rensvold, R.F.   Soc. Pet. Eng. J., 1983, 23, 238.
47. Penberthy, Jr., W.L.; Shaughnessy, C.M.; Gruesbeck, C.;
     Salathiel, W.M. J. Pet. Technol., 1978, 30, 845.
48. Young, B.M.; Totty, K.D.   U.S. Patent 3 404 735, 1968.
49. Yano, A.; Yasue, T.   Japan Kokai Tokkyo Koho 63 23 922, 1988;
     Chem. Abs., 1988, 109, 232346p.
50. Burger, J.; Gadelle, C.; De Chou, J.S.   French Patent 2 575
     500, 1986.

51. Suman, G.O.   World Oil, Nov. 1974, 179, 63.
52. Murphey, J.R.   U.S. Patent 4 259 205, 1981.
53. Underdown, D.R.; Das, K.   J. Pet. Technol.,1985, 37, 2006-12.
54. Watkins, D.R.; Kalfayan, L.J.; Watanabe, D.J.; Holm, J.A.   SPE Production Eng., 1986, 1, 471-477.
55. Davies, D.R.; Richardson, E.A.; Van Zanten,   M. Eur. Pat. Appl. 30 753, 1981.
56. Economides, M.J.; Nolte, K.G., Eds. Reservoir Stimulation, Schlumberger Educational Services, Houston, 1987.
57. Githens, C.J.; Burnham, J.W.   Soc. Pet. Eng. J., 1977, 17, 5.
58. Wu., S.N.   Eur. Pat. Appl. 163 271, 1984.
59. Clark, P.E.; Underwood, J.S.; Steiner, T.M.   Canadian Patent 1 090 112, 1977.
60. N.R. Morrow and J.P. Heller in Developments in Petroleum Science; Donaldson, E.C.; Chilingarian, G.V.; Yen, T.F. Eds., Elsevier, Amsterdam, 1985, Volume 17A, pp. 47-74.
61. Barker, S.A.; Stacey, M.; Zweifel, G.   Chem. Ind. London, 1957, 330.
62. Dill, W.R.   U.S. Patent 4 466 893, 1984.
63. Stauffer, K.B. in Handbook of Water-Soluble Gums and Resins; Davidson, R.L., Ed.; McGraw-Hill, New York, 1980, pp. 11/1-11/31.
64. Nierode, D.E.; Kehn, D.M.; Kruk, K.F.   U.S. Patent 3 934 651, 1976.
65. Almond, S.W.; Conway, M.W.   U.S. Patent 4 553 601, 1985.
66. Brode, G.L.; Stanley, J.P.; Partain, III, E.M.   U.S. Patent 4 579 942, 1986.
67. Roll, D.L.; Himes, R.; Ewert, R.; Doerksen, J.   SPE Production Eng., 1987, 2, 291-6.
68. Constein, V.G.; King, M.T.   U.S. Patent 4 541 935, 1985.
69. Prikryl, J.; Oliva, L.; Kubat, V. Czech. Patent 235 624, 1986.
70. Lundberg, R.D.; Peiffer, D.G.; Sedillo, L.P.; Newlove, J.C. U.S. Patent 4 579 671, 1986.
71. Burnham, J.W.; Tiner, R.L.   U.S. Patent 4 200 539 (1980).
72. Griffin, Jr., T.J.   U.S. Patent 4 174 283, 1979.
73. Zigrye, J.L.; Whitfill, D.L.; Sievert, J.A.   J. Pet. Technol., 1985, 37, 315.
74. Penny, G.S.; Conway, M.W.; Lee, W. J. Pet. Technol., 1985, 37, 1071.
75. Gulbis, J. in reference 56, p. 4-11.
76. Anderson, R.W.; Phillips, A.M.   J. Pet. Technol., 1988, 40, 223-8.
77. Needham, R.B.; Thomas, C.P.; Wier, D.R.   U.S. Patent 4 231 428, 1980.
78. Graham, J.W.; Sinclair,R.A.   U.S. Patent 4 585 064, 1986.
79. Rummo, G.J.   Eur. Pat. Appl. 92 756, 1983.
80. Kucera, C.H.   British Patent 2 108 112, 1983.
81. Harris, L.E.   U.S. Patent 4 324 668, 1982.
82. Reference 75, pp. 4-5, 4-6.
83. Gregory, G.J.   Oil Gas J., 1982, 83 (37), 80.
84. Payne, K.L.   U.S. Patent 4 579 670, 1986.
85. Misak, M.D.   U.S. Patent 3 922 173, 1975.
86. Reference 75, p. 4-9.
87. Malone, T.R.; Foster, J., T.D.; Executrix, S.T.   U.S. Patent 4 290 899, 1981.

88. Rygg, R.H.   British Patent 2 090 308, 1982.
89. Reference 75, p. 4-9 and references therein.
90. Millar, S.W.   U.S. Patent 4 552 591, 1985.
91. Sortwell, E.T.; Solovinski, M.; Mikkelsen, A.R.   U.S. Patent
     4 507 450, 1985.
92. Hoover, L.D.; House, R.F.   British Patent 2 070 611, 1981.
93. House, R.F.; Hoover, L.D.   French Patent 2 488 325, 1982.
94. Bayerlein, F.; Habereder, P.P.; Keramaris, N.; Kottmair, N.;
     Kuhn, M.   French Patent 2 513 265, 1983.
95. Mosier, B.; McCrary, J.L.; Guilbeau, K.G.   U.S. Patent 4 530
     601, 1982.
96. Reference 75, p. 4-10.
97. Clark, H.B.; Pike, M.T.; Rengel, G.L.   J. Pet. Technol., 1982,
     34, 1565-9.
98. Penny, G.S.; et. al.   U.S. Patent 4 425 242, 1984.
99. Hirt, D.E.; Prud'homme, R.K., Rebenfeld, L.   J. Dispersion
     Sci. Technol., 1987, 8, 55-73.
100. Reference 75, pp. 4-7, 4-8 and references therein.
101. Black, H.N.; Langsford, R.W.   J. Pet. Technol., 1982, 34,
     135-40.
102. Reidenbach, V.G.; Harris, P.C.; Lee, Y.N.; Lord, D.L.   SPE
     Production Eng., 1986, 1, 31-41.
103. Harris, P.C.   SPE Production Eng., 1987, 2, 89-94.
104. Ward, V.L.   SPE Production Eng., 1986, 1, 275-8.
105. Garbis, S.J.; Taylor III, J.L.   SPE Production Eng., 1986, 1,
     351-358.
106. Williams, R.B.; Gidley, J.L.; Schechter, R.S.   Acidizing
     Fundamentals, Society of Petroleum Engineers, Dallas, SPE
     Monograph No. 6, 1979.
107. Hung, K.M.; Hill, A.D.; Sepehrnoori, K.   J. Pet. Technol.,
     1989, 41, 65-6.
108. Reference 106, p. 97.
109. Thomas, R.L. U.S. Patent 4 151 878, 1979.
110. Thomas, R.L.; Suby, F.A.   U.S. Patent 4 160 483, 1979.
111. Kunze, K.R.; Shaughnessy, C.A.   Soc. Pet. Eng. J., 1983, 23,
     65-72.
112. Blumer, D.J.   U.S. Patent 4 703 803, 1987.
113. Dill, W.R.   West German Patent 2 925 748, 1980.
114. Nehmer, W.L.   British Patent Appl. 2 163 790, 1986.
115. Norman, L.R.; Conway, M.W.; Wilson, J.M.   J. Pet. Technol.,
     1984, 36, 2011.
116. Roper, L.E.; Swanson, B.L.   U.S. Patent 4 205 424, 1980.
117. Burns, L.D.; Stahl, G.A.   U.S. Patent 4 690 219, 1987.
118. Haltmar, W.C.; Lacey, E.S.   U.S. Patent 4 207 946, 1980.
119. Engelhardt, F.; Schmitz, H.; Hax, J.; Gulden, W.   European
     Patent 44 508, 1982.
120. Engelhardt, F.; Piesch, S.; Balzer, J.; Dawson, J.C.   PCT
     Internat. Patent Appl. 82 02 252 (1982).
121. Swanson, B.L.   Eur. Patent Appl. 7 012, 1980.
122. Costanza, J.R.; DeMartino, R.N.; Goldstein, A.M.   U.S. Patent
     4 057 509, 1977.
123. Hoefner, M.L.; Fogler, H.S.; Stenius, P.; Sjoblom, J.   J. Pet.
     Technol., 1987, 39, 203-8.

124. Hoefner, M.L., Fogler, H.S.   Chem. Eng. Prog., 1985, 40.
125. Ford, W.G.F.   J. Pet. Technol., 1981, 33, 1203-1210.
126. Khasaev, A.M.; Sadykhov, M.G.; Kurbanova, Kh. G.   Azerb. Neft. Khoz., 1978, 32.
127. Jones, W.L.   British Patent 2 141 731, 1987.
128. Reference 106, p. 97.
129. Crema, S.C.   U.S. Patent 4 676 916, 1987.
130. Walton, W.B.   U.S. Patent 4 541 483, 1985.
131. Hall, B.E.   J. Pet. Technol., 1975, 27, 1439-42.
132. Walsh, M.P.; Lake, L.W., Schechter, R.S.   J. Pet. Technol., 1982, 34, 2097-2112.
133. Dria, M.A.; Schechter, R.S.; Lake, L.W.   SPE Production Eng., 1988, 3, 52-62.
134. Labrid, J.C.   Soc. Pet. Eng. J., 1975, 15, 117-28.
135. Crowe, C.W.   J. Pet. Technol., 1986, 38, 1234-40.
136. Street, Jr., E.H.   U.S. Patent 4 167 214, 1979.
137. Dill, W.R.; Walker, M.L.; Ford, W.G.F.   U.S. Patent 4 679 631, 1987.
138. Nehmer, W.H.; Coffey, M.D.   British Patent Appl. 2 158 487, 1985.
139. Walker, M.L.; Ford, W.G.F.; Dill, W.R.; Gdanski, R.D.   U.S. Patent 4 683 954, 1987.
140. Reference 106, pp. 101-102.
141. Brannon, D.H.; Netters, C.K.; Grimmer, P.J.   J. Pet. Technol., 1987, 39, 931-42.
142. Reference 106, p. 100.
143. Clementz, D.M.; Patterson, D.E.; Aseltine, R.J.; Young, R.E.   J. Pet. Technol., 1982, 34, 2087-96.
144. Taggart, D.L.; Heffern, E.W.   U.S. Patent 4 690 217, 1987.
145. Craig, Jr., F.F.   The Reservoir Engineering Aspects of Water Flooding, Society of Petroleum Engineers, Dallas, Monograph No.3, 1971, pp. 19-22.
146. Dymond, P.F.; Spurr, P.R.   SPE Reservoir Eng., 1988, 3, 165-74.
147. Scheuerman, R.F.   SPE Production Eng., 1988, 3, 15-21.
148. Reference 106, pp. 92-95.
149. Wu, Y.   British Patent Appl. 2 082 589, 1982.
150. Tuttle, R.N.   J. Pet. Technol., 1987, 39, 756-62.
151. Wilhelm, S.M.; Kane, R.D.   J. Pet. Technol., 1986, 38, 1051-61.
152. Tate, J.F.; Maddox, Jr., J.; Shupe, R.D.   Canadian Patent 1 051 649, 1979.
153. Redmore, D.; Outlaw, B.T.   U.S. Patent 4 315 087, 1982.
154. Patton, C.C.   J. Pet. Technol., 1988, 40, 1123-6.
155. Khilar, K.C.; Fogler, H.S.   Soc. Pet. Eng. J., 1983, 23, 55-64.
156. Khilar, K.C.; Fogler, H.S.   J. Colloid Interface Sci., 1984, 101 (1), 214-24.
157. Kia, S.F.; Fogler, H.S.; Reed, M.G.; Vaidya, R.N.   SPE Production Eng., 1987, 2, 277-83.
158. Sydansk, R.D.   J. Pet. Technol., 1984, 36, 1366-74.
159. Reed, M.G.   J. Pet. Technol., 1972, 24, 860-64.
160. Veeley, C.D.   J. Pet. Technol., 1969, 21, 1111-18.
161. Peters, F.W.; Stout, C.M.   J. Pet. Technol., 1977, 29, 187-94.

162. Muecke, T.W.   J. Pet. Technol., 1979, 31, 144-150.
163. Gruesbeck,C.; Collins, R.E.   Soc. Pet. Eng. J., 1982, 22,
     847-856.
164. Kia, S.F.; Fogler, H.S.; Reed, M.G.   J. Colloid Interface
     Sci., 1987, 118, 158-68.
165. Reference 20, Volume 2, p. 99.
166. Leone, J.A.; Scott, E.M.   SPE Reservoir Eng., 1988, 3,
     1279-86.
167. Broaddus, G.   J. Pet. Technol., 1988, 40, 685-7.
168. Blair, Jr., C.M.   U.S. Patent 4 337 828, 1982.
169. Blair,Jr., C.M.; Stout, C.A.   Oil Gas J., May 20, 1985, 83
     (55). 133.
170. Cusack, F.; Lappin-Scott, H.M.; Costerton, J.W.   Oil Gas J.,
     Nov. 9, 1987, 85, 87.
171. Hensel, Jr., W.N.; Sullivan, R.L.; Stallings, R.H.   Pet. Eng.
     Int., 1981(5), 155.
172. Anderson, W.G., J. Pet. Technol., 1987, 39, 1453-67.
173. Anderson, W.G., J. Pet. Technol., 1986, 38, 1125-44. 173.
     Reference 144, pp. 19-21.
174. Hjelmeland, O.S.; Larrondo, L.E.   SPE Reservoir Engr., 1986,
     1, 321-8.
175. Owens, W.W.; Archer, D.L.   J. Pet. Technol., 1971, 23, 873-8.
176. Morrow, N.R.; Mungan, , N.   Rev. Inst. Franc. Petrole, 1971,
     629-50.
177. Treiber, L.E.; Archer, D.L.; Owens, W.W.   Soc. Pet. Eng. J.,
     1972, 12, 531-40.
178. Elphingstone, E.A.; McLaughlin; H.C.; Smith, C.W.   Canadian
     Patent 1 070 936, 1980.
179. Downs, S.L.; Gohel, M.G.   J. Pet. Technol., 1974, 26, 557-70.
180. Peddycoart, L.R.   Oil Gas J., Feb. 4, 1980, 78, 52.
181. Weaver, J.D.; Harris, L.E.; Harms, W.M.   U.S. Patent 4 460
     627, 1984.
182. Geosciences Research for Oil and Gas Discovery and Recovery,
     U.S. Dept. of Energy, Washington, D.C., March, 1987.
183. Oil Gas J., December 26, 1988, 86 (52), 49.
184. Smith, D.H. Surfactant-Based Mobility Control - Progress in
     Miscible-Flood Enhanced Oil Recovery, ACS Symposium Series No.
     373, American Chemical Society, Washington, D.C., 1988.
185. Neale, R.R.; Doe, P.H.   J. Pet. Technol., 1987, 39, 1503-7.
186. Kessel, D.G.; Volz, H.; Maitin, B.   Proc. World Pet. Congr.,
     1977, 12th (Vol. 3), 355-64.
187. Ling, T.F.; Lee, H.K.; Shah, D.O.   Spec. Publ. - R. Soc.
     Chem., 1987, 59 (Ind. Appl. Surfactants), 126-78.
188. Miller, C.A.; Qutubuddin, S.   Surfactant Sci. Ser., 1987, 21
     (Interfacial Phenom. Apolar Media), 117085.
189. Tomich, J.F.; Laplante, D.L.; Snow, T.M.   Proc. World Pet.
     Congr., 1977, 12th (Vol. 3), 337-46.
190. Pet. Eng. Int., May, 1988. 79-81.
191. Leighton, A.J.; Wayland, J.R.   J. Pet. Technol., 1987, 39,
     129-36.
192. Holm, L.W.; Josendal, V.A.   Soc. Pet. Eng. J., 1982, 22,
     87-98.
193. Martin, R.L.; Braga, T.G.   Mater. Perform., 1987, 26, 16-22.
194. Technical Constraints Limiting Application of Enhanced Oil

Recovery Techniques to Petroleum Production in the United
States, U.S. Dept. of Energy, DOE/BETC/RI-83/9 (DEB4003910),
Jan., 1984.
195. Reference 34, pp. 350-351.
196. Reference 34, p. 344.
197. Norton, C.J.; Fak, D.O.   U.S. Patent 4 343 363, 1982.
198. Sandiford, B.B.; Dovan, H.T.; Hutchins, R.D.   U.S. Patent
4 413 680, 1983.
199. Prud'homme, R.K.; Uhl, J.T.; Poinsatte, J.P.; Halverson, F.
Soc. Pet. Eng. J., 1983, 23, 804.
200. Wilhite, G.P.; Jordan, D.J.   Polym. Prepr. Am. Chem. Soc. Div.
Polym. Chem., 1981, 22 (2), 53.
201. Sloat, B.   Pet. Eng. Int., 1977, 20.
202. Dovan, H.T.; Hutchins, R.D.   SPE Res. Eng., 1987, 2, 177-188.
203. Norton, C.J.   Canadian Patent 1 217 629, 1987.
204. Hanlon, D.J.; Almond, S.W.   European Patent Appl. 161
858,1985.
205. Prud'homme, R.K.; Uhl, J.T. Proc. Fourth Joint Symposium on
Enhanced Oil Recovery, Tulsa, Okla., Apr. 15-18, 1984, Paper
No. SPE/DOE 12640 .
206. Jordan, D.J.; Green, D.W.; Terry, R.E.; Willhite, G.P.   Soc.
Pet. Eng. J., 1982, 22, 463.
207. Routson, W.G.   U.S. Patent 3 687 200, 1972.
208. Mumallah, N.A.   SPE Reservoir Eng,, 1988, 3, 243.
209. Sydansk, R.D.   U.S. Patent 4 744 418, 1988.
210. Moradi-Araghi, A.; Doe, P.H.   SPE Reservoir Eng., 1987, 2,
189-93.
211. Nanda, S.K.; Kumar, R.; Sindhwani, K.L.; Goyl, K.L.   ONGC
Bull., 1986, 23, 175-85.
212. Sydansk, R.D.   U.S. Patent 4 723 605, 1988.
213. Aslam, S.; Vossoughi, S.; Whillhite, G.P.   Chem. Eng. Commun.,
1986, 48, 287-301.
214. Hyang, C-G.; Green, D.W.; Willhite, G.P.   SPE Reservoir
Engr., 1986, 1, 583-592.
215. Vio, L.   U.S. Patent 4 155 405, 1979.
216. Sullivan, E.J.; Jones, G.D.   U.S. Patent 4 125 478, 1978.
217. Pliny, R.J.; Regulski, T.W.   European Patent 5 835, 1979.
218. Sidorov, I.A.; et. al.   USSR Patent 1 040 118, 1983.
219. Ech. E.; Lees, R.D.   U.S. Patent 4 579 667, 1986.
220. Ryles, R.G.; Robustelli, A.G.; Cicchiello, J.V.   European
Patent Appl. 200 062, 1986.
221. Ryles, R.G.; et. al.   European Patent Appl. 273 210, 1988.
222. Falk, D.O.   U.S. Patent 4 688 639, 1987.
223. Hessert, J.E.; Johnston, Jr., CC.   U.S. Patent 4 110 230,
1978.
224. Abdo, M.K.   U.S. Patent 4 574 887, 1986.
225. Hurd, B.G.   U.S. Patent 3 581 824, 1971.
226. Colegrove, G.T.   U.S. Patent 4 157 322, 1979.
227. Clear, E.E.   U.S. Patent 4 321 968, 1982.
228. Dasinger, B.L.; McArthur, H.A.I.   European Patent Appl. 251
638, 1988.
229. Sampath, K.   U.S. Patent 4 640 358, 1987.
230. McLaughlin, H.C.   U.S. Patent 3 334 689, 1967.
231. Borchardt, J.K.   U.S. Patent 4 439 334, 1984.

232. Argabright, P.A.; Rhudy, J.S.   U.S. Patent 4 503 909, 1985.
233. Seright, R.S.   Proc. 1989 SPE International Symposium on
     Oilfield Chemistry, Society of Petroleum Engineers, Dallas,
     1989, pp. 389-402.
234. Felber, B.J.; Dauben, D.L.   Soc. Pet. Eng. J., 1977, 17, 391.
235. Felber, B.J.; Christopher, C.A.   U.S. Patent 4 428 429, 1984.
236. Lawrence, D.D.; Felber, B.J.   U.S. Patent 4 275 789, 1981.
237. Whitworth, A.J.; Tung, S.Y.S.; Hajto, E.A.   U.S. 3 686 372,
     1972.
238. Allan, B.W.   U.S. Patent 4 299 690, 1981.
239. Soreau, M.; Siegel, D.   French Patent 2 551 451, 1985.
240. Navritil, M.; Batycky, J.P.; Sovak, M.; Mitchell, M.S.   Canada
     Patent 1 187 404, 1985.
241. Balitskaya, Z.A.; et. al.   U.S.S.R. Patent 878 904, 1981.
242. Hess, P.M.   J. Pet. Technol., 1980, 32, 1834-42.
243. Navritil, M.; Mitchell, M.S.; Sovak, M.   U.S. Patent 4 663
     367, 1987.
244. Navritil, M.; Mitchell, M.S.; Sovak, M.   Canada Patent 1 217
     932, 1987.
245. Marocco, M.L.   U.S. Patent 4 664 194, 1987.
246. Chung, H.S.; Sampath, K.; Schwab, F.C.   U.S. Patent 4 653 585,
     1987.
247. Harwell, J.H.; Scamehorn, J.F.   U.S Patent 4 745 976, 1988.
248. Schievelbein, V.H.; Kudchadker, M.V.; Varnon, J.E.;
     Whittington, L.E.   U.S. Patent 4 161 982, 1979.
249. Schievelbein, V.H.   U.S. Patent 4 161 983, 1979.
250. Varnon, J.E.; Schievelbein, V.H.; Kudchadker, M.V.;
     Whittington, L.E.   U.S. Patent 4 161 218, 1979.
251. Schievelbein, V.H.; Park, J.H.   U.S. Patent 4 160 480, 1979.
252. Stahl, G.A.; Schulz, D.N.; Eds.   Water Soluble Polymers for
     Petroleum Recovery, Plenum, New York, 1988.
253. Crafts, B.C.; Hawkins, M.F.   Applied Petroleum Reservoir
     Engineering, Prentice-Hall, New York, 1959.
254. Reference 145, pp. 45-7.
255. Sandiford, B.B.; Keller, Jr., H.F.   U.S. Patent 2 827 964,
     1958.
256. Sandiford, B.B.; Keller, Jr., H.F.   U.S. Patent 3 116 791,
     1964.
257. Sandiford, B.B.   U.S. Patent 3 308 885, 1967.
258. Sandiford, B.B.   J. Pet. Technol., 1964, 16, 917.
259. Pye, D.J.   J. Pet. Technol., 1964, 16, 911.
260. Chang, H.I.   J. Pet. Technol., 1978, 30, 1113-1128.
261. Smith, R.V.; Burtch, F.W.   Oil Gas J., Nov. 24, 1980, 78, 127.
262. Lorenz, P.B.; Trantham, J.C.; Zornes, P.R.; Dodd, C.G.   SPE
     Reservoir Eng., 1986, 1, 341-53.
263. Jones, D.M.; Walters, K.; Williams, P.R.   Rheological Acta,
     1986, 26, 20.
264. Cahalan, P.T.; Peterson, J.A.; Arndt, D.A.   U.S. Patent 4 053
     699, 1977.
265. Carter, W.H.; Christopher, C.A.; Jefferson, T.   West German
     Patent 2 809 136, 1979.
266. Sandford, P.A.; Laskin,A. Eds;   Extracellular Microbial
     Polysaccharides, ACS Symposium Series No. 45; American
     Chemical Society: Washington, D.C., 1977.

267. Abdo, M.K., U.S. Patent 3 771 462, 1973.
268. Rhone-Poulenc Specialties Chemique, Japan Kokai Tokkyo Koho 62
     39 643, 1987; Chem. Abstr., 1987, 107, 42858u.
269. Casad, B.M.; Conley, D.; Ferrell, H.H.; Stokke, O.M.   U.S.
     Patent 4 212 748, 1980.
270. Bragg, J.R.   British Patent Application 2 115 430, 1983.
271. Rinaudo, M.; Milas, M.   Int. J. Biol. Macromol., 1980, 2, 45.
272. Holding, T.J.; Pace, J.W.   British Patent 2 065 689, 1981.
273. Kohler, N.; Longchamp, D.; Thery, M.   J. Pet. Technol., 1897,
     39, 835-43.
274. Stokke, O.M.   U.S. Patent 4 165 257, 1979.
275. Knight, B.L.; Jones, S.C.; Parsons, R.W.   Soc. Pet. Eng. J.,
     1974, 14, 643.
276. Grollman, U.; Schnable, W., Polymer Degradation and Stability,
     Applied Science Publishers, London, 1982, pp. 353-362..
277. Wellington, S.L.   Soc. Pet. Eng. J., 1983, 23, 901-12.
278. Cadmus, M.C.; et. al.   Appl. Environ. Microbiol., Aug. 1982,
     5.
279. Lee, S.L.   West German Patent 2 715 026, 1977.
280. Philips, C.J.   European Patent 106 666, 1984.
281. F. Dawans, Binet, D.; Kohler, N.; Quang, D.V.   U.S. Patent
     4 454 620, 1984.
282. Wellington, S.L.   Canadian Patent 1 070 492, 1980.
283. Kanda, S.; Kawamura, G.   U.S. Patent 4 481 316, 1985.
284. Kanda, S.; Kawamura, G.   Japan Kokai Tokkyo Koho   61 275 337,
     1986.
285. Niita, A.; Arai, T.; Funato, R.; Sato, T.   Japan Kokai Tokkyo
     Koho 62 184 048, 1987.
286. Niita, A.; Arai, T.; Funato, R.; Sato, T.   Japan Kokai Tokkyo
     Koho 62 184 047, 1987.
287. Niita, A.; Arai, T.; Funato, R.; Sato, T.   Japan Kokai Tokkyo
     Koho 62 177 052, 1987.
288. Nakanishi, Y. Japan Kokai Tokkyo Koho 62 277 407, 1987.
289. Judson, C.P.   European Patent Application 196 199, 1986.
290. Nitta, A.; Ito, Y.; Nitta, A. Japan Kokai Tokkyo Koho 61 136
     545, 1986.
291. Contat, F.; Boutin, J.   European Patent Application 241 340,
     1987.
292. Chen, C.S.H.; Sheppard, E.W.   J. Macromol. Sci. Chem. Part A,
     1979, 239.
293. Chen, C.S.H.; Sheppard, E.W.   Polym. Prepr. Am. Chem. Soc.,
     Div. Polym. Chem., 1978, 19 (1), 424-9.
294. Wernau, W.C.   West German Patent 2 848 984, 1979.
295. Holzwarth, G.M.   European Patent 103 483, 1984.
296. Phillips, J.C.; Miller, J.W.; Wernau, W.C.; Tate, B.E.;
     Auerbach, M.H.   Soc. Pet. Eng. J., 1985, 25, 594.
297. Donche, A.F.   World Biotech Rep., 1985, 1, 429-36.
298. Meldrum, I.G.   Comm. Eur. Communities, {Rep.} EUR, 1985, Vol.
     2, 776-83.
299. Doster, M.S.; Nute, A.J.; Christopher, C.A.   U.S. Patent 4 457
     372, 1984.
300. Rinaudo, M.; Vincendon, M.   Carbohydr. Polym., 1982, 2, 135.
301. Hitzman, G.O.   U.S. Patent 4 096 073, 1978.
302. Linton, J.D.; Evans, M.W.; Godley, A.R.   European Patent 135
     953, 1985.

303. Thompson, B.G.; Jack, T.R.   U.S. Patent 4 561 500, 1985.
304. Savins, J.G.   U.S. Patent 4 079 544, 1978.
305. Lindoerfu, R.; et. al.   West German Patent 3 643 467, 1988.
306. Vanderslice, R.W.; Shanon, P.   European Patent Application 211 288, 1987.
307. Graaflund, T.   European Patent Application 237 132, 1987.
308. Frank, S.; Coscia, A.T.; Schmitt, J.M.   U.S. Patent 4 439 332, 1984.
309. Borchardt, J.K.   U.S. Patent 4 439 334, 1984.
310. Klein, J.; Heitzman, R.   Makromol. Chem., 1978, 179, 1895.
311. Halverson, F.H.; Lancaster, J.E.   Macromolecules, 1985, 18, 1139.
312. Boutin, J.; Contat, F.   French Patent 2 495 217, 1982.
313. Southwick, J.G.; Manke, C.W.   SPE Reservoir Eng., 1988, 3, 1193-1201.
314. Prud'homme, R.K.   SPE Reservoir Eng., 1986, 1, 272-6.
315. Munk, P.; Aminabhavi, T.M.; Williams, P.; Hoffman, D.E.; Chmelir,M.   Macromolecules, 1980, 13, 871-5.
316. Muller, G.; Laine, J.P.; Fenyo, J.C.   J. Polym. Sci. Polym. Chem. Ed., 1979, 17, 659.
317. Ward, J.S.; Martin, F.D.   Soc. Pet. Eng. J., 1981, 21, 623.
318. Chen, G.S.; Neidlinger, H.H.; McCormick, C.L.   Prepr. Pap. Nat. Meet. Div. Pet. Chem. Am. Chem. Soc., 1984, 29 (4), 1147.
319. Ryles, R.G.   SPE Reservoir Eng., 1988, 3, 23-34.
320. Moradi-Araghi, A.; Doe, P.H.   SPE Reservoir Eng., 1987, 2, 189-98.
321. Szabo, M.   J. Pet. Technol., 1979, 31, 553.
322. Szabo, M.   J. Pet. Technol., 1979, 31, 561.
323. McCormick, C.L.; Johnson, C.B.   Biotechnol. Mar. Polysaccharides, Proc. Annu. MIT Sea Grant Coll. Program Lect. Semin., 3rd, 1984 (Pub. 1985), 213-48.
324. McCormick, C.L.   J. Macromol. Sci. Chem. Part A. 1985, 22, 955.
325. McCormick, C.L., Hester, R.D.; Neidlinger, H.H.; Wildman, G.C. BETC Prog. Rev., 1979, 20, 61.
326. McCormick, C.L., Blackmon, K.R.; Elliott, D.L.   Prepr. Pap. Nat. Meet. Div. Pet. Chem. Am. Chem. Soc., 1984, 29 (4), 1159.
327. Stahl, G.A.   U.S. Patent 4 644 020, 1987.
328. Doe, P.H.; Moradi-Araghi, A.; Shaw, J.E.; Stahl, G.A.   SPE Res. Eng., 1987, 2, 461.
329. Ching, T.Y.   West German Patent 3 627 456, 1987.
330. Ryles, R.G.   European Patent Application 233 533, 1987.
331. Engelhardt, P.; Greiner, U.; Schmitz, H.; Gulden, W.; von Halasz, S.P.   West German Patent 3 220 503, 1983.
332. Ito, Y.; Niuta, A.; Sudo, Y.; Hayashi, K.   Japan Kokai Tokkyo Koho 62 15 279, 1987.
333. Wu, M.M.; Ball, L.E.   U.S. Patent 4 540 498, 1985.
334. Martin, F.D.; Hatch, M.J.; Abouelezz, M.; Oxley, J.C.   Polym. Mater. Sci. Eng., 1984, 51, 688.
335. Ball, L.E.; Griffin, L.M.; Antloger, K.M.; Nemacek, A.L.   U.S. Patent 4 653 584, 1987.
336. Pledger, Jr., H.; Meister, J.J.; Hogen-Esch, T.E.; Butler, G.B.   Polym. Prepr. Am. Chem. Soc. Div. Polym. Chem., 1981, 22 (2), 72.

337. Neidlinger, H.H.; McCormick, C.L.   Polym. Prepr. Am. Chem. Soc. Div. Polym. Chem., 1979, 20 (1), 901.
338. Meister, J.J.; et. al.   Polym. Prepr. Am. Chem. Soc. Div. Polym. Chem., 1984, 25 (1), 266.
339. Schulz, D.N.; Kaladas, D.A.; Maurer, J.J.; Bock, J.; Pace, S.J.; Schulz, W.W.   Polymer, 1987, 28, 2110-15.
340. Bock, J.; Siano, D.B.; Turner, R.   U.S. Patent 4 694 046, 1987.
341. Bock, J.; Valint, P.L.   U.S. Patent 4 730 028, 1988.
342. Bock, J.; Valint, P.L.; Pace, S.J.   U.S. Patent 4 702 314, 1987.
343. Bock, J.; Pace, S.J.; Schulz, D.N.   U.S. Patent 4 709 759, 1987.
344. Evani, S.; Rose, G.D.   Polym. Mat. Sci. Eng., 1987, 57, 477-81.
345. Landoll, L.M.   Netherlands Patent 80 03 241, 1980.
346. Borchardt, J.K.; Brown, D.L.   U.S. Patent 4 409 110, 1983.
347. Borchardt, J.K.; Brown, D.L.   Oil Gas J., Sept. 10, 1984, 89, pp. 150,152,154,156.
348. Shuler, P.J.; Kuehne, D.L.; Uhl, J.T.; Walkup, Jr., G.W.   SPE Reservoir Engr., 1987, 2, 271-80.
349. Willhite, G.P.; Dominguez, J.G. in Improved Oil Recovery by Surfactant and Polymer Flooding, Academic Press, Washington, D.C., 1977, pp. 511-553.
350. Friedmann, F.   SPE Reservoir Eng., 1986, 1, 261-271.
351. Cohen, Y.; Christ, F.R.   SPE Reservoir Eng., 1986, 1, 113-18.
352. Hong, S.A.; Bae, J.H.   SPE Reservoir Eng., 1987, 2, 17-27.
353. Kalfoglou, G.   U.S. Patent 4 627 494, 1986.
354. Howard, J.; Stirling, M.   U.S. Patent 4 713 185, 1987.
355. Bansal, B.B.; Hornof, V.; Neale, G.   Can. J. Chem. Eng., 1979, 57, 203.
356. Novosad, J.   Can. J. Pet. Technol., 1984, 23, 24.
357. Chen, C.S.H.; Sheppard, E.W.   U.S. Patent 4 284 517, 1981.
358. Clint, J.H.; Hodgson, P.K.G.; Tinley, E.J.   British Patent 2 148 356, 1985.
359. Horton, R.L.   U.S. Patent 4 574 885, 1986.
360. Arf, T.G.; et. al.   SPE Reservoir Eng., 1987, 2, 166-76.
361. Stalkup, Jr., F.I.   Miscible Displacement, Monograph Volume 8, Society of Petroleum Engineers, Dallas, 1983, pp. 62-67.
362. Caudle, B.H.; Dye, A.B.   Trans. AIME, 1958, 213, 281.
363. Reference 361, p. 32.
364. Wellington, S.L.; Vinegar, H.J.   J. Petroleum Technol., 1987, 39, 885-98.
365. Reference 361, pp. 147-158.
366. Stalkup, Jr., F.I.   J. Pet. Technol., 1984, 36, 815-826.
367. Desch, J.B.; et. al.   J. Pet. Technol., 1984, 36, 1592-1602.
368. Heller, J.P.; Dandge, D.K.; Card, R.J.; Donaruma, L.G.   Soc. Pet. Eng. J., 1985, 25, 679-86.
369. Smith, D.H.   Surfactant-Based Mobility Control - Progress in Miscible-Flood Enhanced Oil Recovery, ACS Symposium Series No. 373, American Chemical Society, Washington, D.C., 1988.
370. Heller, J.P.; Kuntamukkula, M.S.   Ind. Eng. Chem. Res., 1987, 26, 318.

371. Ali, S.M. Farouq; Selby, R.J.   Oil Gas J., 1986, 84 (5), 57,
     60-3.
372. Friberg, S.E.; Solans, C.   Langmuir, 1986, 2, 121-6.
373. Malhotra, A.K.; Wasan, D.T.   Chem. Eng. Commun., 1987 , 55,
     95-128.
374. Khatib, Z.I.; Hirasaki, G.J.; Falls, A.H.   SPE Reservoir
     Eng., 1988, 3, 919-26.
375. Ransohoff, T.C.; Radke, C.J.   SPE Reservoir Eng., 1988, 3,
     573-85.
376. Owete, O.S.; Brigham, W.E.   SPE Reservoir Eng., 1987, 2,
     315-23.
377. Craig, Jr., F.F.; Lummus, J.L.   U.S. Patent 3 185 634, 1965.
378. Hahn, P.S.; Ramamohan, T.R.; Slattery, J.C.   AIChE J, 1985,
     31, 1029-35.
379. Borchardt,, J.K.; Bright, D.B., Dickson, M.K.; Wellington,
     S.L.   in Reference 338, pp. 163-80.
380. Borchardt, J.K.   in Reference 369, pp. 181-204.
381. Holm, L.W.   U.S. Patent 4 706 752, 1987.
382. Wellington, S.L.   U.S. Patent 4 380 266, 1983.
383. Buckles, J.J.   U.S. Patent 4 706 750, 1987.
384. Wellington, S.L.; Resiberg, J.; Lutz, E.F.; Bright, D.B.   U.S.
     Patent 4 502 538, 1985.
385. Borchardt, J.K.   U.S. Patent 4 799 547, 1989.
386. Settlemeyer, L.A.; McCoy, M.J.   U.S. Patent 4 739 831, 1988.
387. Sharma, M.K.; Shiao, S.Y.; Bansal, V.K.; Shah, D.O. in Macro-
     and Microemulsions, ACS Symposium Series No. 272, American
     Chemical Society, Washington, D.C., 1985, pp. 87-103.
388. Sharma, M.K.; Shah, D.O.; Brigham, W.E.   SPE Reservoir Eng.,
     1986, 1, 253-60.
389. Mitchell, T.O.   U.S. Patent 4 676 316, 1987.
390. Zhukov, I.N.; Polozova, T.L.; Shatova, O.S.   Kolloidn. Zh.,
     1987, 49, 758-62; Chem. Abstr., 1987, 108, 63029a.
391. Hutchins, R.D.; Dovan, H.T.   European Patent Applicaation 212
     671, 1987.
392. Djabbarah, N.F.   Canadian Patent 1 221 305, 1987.
393. Falls, A.H.   U.S. Patent 4 733 727, 1988.
394. Holm, L.W.   J. Pet. Technol., 1970, 32, 1499-1506.
395. Holm, L.W.; Garrison, W.H.   SPE Reservoir Eng., 1988, 3,
     112-118.
396. Talley, L.D.   SPE Reservoir Eng., 1988, 3, 235-243.
397. Prats, M.   Thermal Recovery, Monograph Volume 7, Society of
     Petroleum Engineers, Dallas, 1982.
398. Castanier, L.M.; Brigham, W.E.   Chem. Eng. Prog., 1985, 81,
     37-40.
399. Strycker, A.R.; Madden, M.P.; Sarathi, P.   SPE Reservoir Eng.,
     1987, 2, 543-8.
400. Isaacs, E.E.; McCarthy, F.C.; Maunder, J.D.   SPE Reservoir
     Eng., 1988, 3, 565-72.
401. Muijs, H.M.; Keijzer, P.P.M.   U.S. Patent 4 693 311, 1987.
402. Janssen van Rosmalen, R.; Muijs, H.M.; Keijzer, P.P.M.   West
     German Patent 3 510 765, 1985.
403. Muijs, H.M.; Keijzer, P.P.M., West German Patent 3 734 075,
     1988.
404. Huang, W.S.; Gassmann, Z.Z.; Hawkins, J.T.; Schievelbein,
     V.H.; Hall, W.L.   West German Patent 3 503 532, 1985.

405. Angstadt, H.P.   U.S. Patent 4 699 214, 1987.
406. Lim, T.   PCT International Patent Application 85 05 146, 1985;
     Chem. Abstr., 1985, 104, 171228a.
407. Duerksen, J.H.; Wall, R.G.; Knight, J.D.   U.S. Patent 4 576
     232, 1986.
408. Falls, A.H.; Lawson, J.B., Hirasaki, G.J.   J. Pet. Technol.,
     1988, 40, 95-104.
409. Falls, A.H.; Lawson, J.B.; Hirasaki, G.J.   U.S. Patent 4 570
     711, 1986.
410. Novosad, J.; Maini, B.B.; Huang, A.   J. Can. Pet. Technol.,
     1986, 25, 42-6.
411. Huang, A.Y.; Novosad, J.   Proc. Eng. Found. Conf. Fundam.
     Adsorpt., 2nd, 1986 (Pub. 1987), 265-75.
412. Lau, H.C.; O'Brien, S.M.   SPE Reservoir Eng., 1988, 2,
     1177-1185.
413. Dilgren, R.E.; Lau, H.C.; Hirasaki, G.J.   U.S. Patent 4 597
     442, 1986.
414. Lau, H.C.   U.S. Patent 4 617 996, 1986.
415. Angstadt, H.P.; Rugen, D.F.; Cayias, J.L.   French Patent 2 557
     198, 1985.
416. Maini, B.B.; Ma, V.   J. Can. Pet. Technol., 1986, 25, 65-9.
417. Angstadt, H.P.; Tsao, H.   SPE Reservoir Eng., 1987, 2, 613-8.
418. Isaacs, E.D.; Prowse, D.R.   U.S. Patent 4 458 759, 1984.
419. Duerksen, J.H.   SPE Reservoir Eng., 1986, 2, 44-52.
420. Shen, C.W.   U.S. Patent 4 702 317, 1987.
421. Lau, H.C.   U.S. Patent 4 609 044, 1986.
422. Lau, H.C.   U.S. Patent 4 727 938, 1988.
423. Hyne, J.B.; Clark, P.D.   U.S. Patent 4 506 733, 1985.
424. Shah, D.O.; Schecter, R.S., Eds.   Improved Oil Recovery by
     Surfactant and Polymer Flooding, Academic Press, New York,
     1977.
425. Poettmann, F.H. in Improved Oil Recovery, Interstate Oil
     Compact Commission, Oklahoma City, Okla., 1983, pp. 173-250.
426. van Poolen, H.K.   Fundamentals of Enhanced Oil Recovery, pp.
     114-151.
427. Morgan, J.C.; Schechter, R.S.; Wade, W.H.   in Proc. Sect. 52nd
     Colloid Surf. Sci. Symp.; Mittal, K.L., Ed.; Plenum Press, New
     York, 1979, Vol. 2, 749-75.
428. Puig, J.R.; Scriven, L.E.; Davis, H.T.; Miller, W.G.   Chem.
     Eng. Commun., 1988, 65, 169-85.
429. Aalund, L.R.   Oil Gas J., 1988, 86, 33-73.
430. B. Cox and J. Schubert, Eds.   1986 EOR Project Source
     Book, Pasha Publications, Arlington, Va., 1986.
431. Nelson, R.C.   Soc. Pet. Eng. J., 1982, 22, 259-70.
432. Puerto, M.C.; Gale, W.W.   Soc. Pet. Eng. J., 1977, 17,
     193-200.
433. Graciaa, A.; Fortney, L.M.; Schechter, R.S.; Wade, W.H.; Yiv,
     S.   Soc. Pet. Eng. J., 1982, 20, 743-9.
434. Puerto, M.C.   West German Patent 3 542 063, 1986.
435. Cardenas, R.L.; Harnsberger, B.G.; Maddox, Jr., J.   U.S.
     Patent 4 270 607, 1981.
436. Barakat, Y.; et. al., Soc. Pet. Eng. J., 1983, 23, 913-18.
437. Morita, H.; Kawada, Y.; Yamada, J.; Ukigai, T.   U.S. Patent
     4 512 404, 1985.

438. Morita, H.; Kawada, Y.; Yamada, J.; Ukigai, T.   U.S. Patent
     4 555 351, 1982.
439. Okada, T.; Morita, H.; Hagiwara, M.   Japan Kokai Tokkyo Koho
     60 152 794, 1985.
440. Thaver, R.   British Patent Application 2 160 242, 1985.
441. Thaver, R.   British Patent Application 2 184 763, 1987.
442. Aldrich, H.S.; Ashcraft, T.L.; Puerto, M.C.; Reed, R.L.   West
     German Patent 3 303 894, 1983.
443. Greif, N.; Oppenlaender, K.; Sewe, K.U.   West German Patent
     3 422 613,1984.
444. Schmidt, R.; Rupp, W.; Schneider, G.; Kohn, E.M.   European
     Patent Application 264 867, 1988.
445. McCoy, D.R.; Gipson, R.M.; Naylor, C.G.   U.S. Patent 4 426
     302, 1984.
446. Chiu, Y.C.; Hwang, H.J.   Colloids Surf., 1987, 28, 53-65.
447. Balzer, D. West German Patent 3 523 355, 1987.
448. Shaw, J.E.   J. Am. Oil Chem. Soc., 1984, 61, 1395-9.
449. Liu, K.C.   U.S. Patent 4 692 551, 1987.
450. Abe, M.; Schechter, R.S.; Selliah, R.D.; Sheikh, B.; Wade,
     W.H.   J. Dispersion Sci. Technol., 1987, 8, 157-72.
451. Abe, M.; Schechter, D.; Schechter, R.S.; Wade, W.H.;
     Weerasooriya, U.; Yiv, S.   J. Coll. Interface Sci., 1984, 114,
     342.
452. Puerto, M.C.; Reed, R.L.   Soc. Pet. Eng. J., 1983, 23, 669-82.
453. Lalanne-Cassou, C.; et. al.   J. Dispersion Sci. Technol.,
     1987, 8, 137-56.
454. Fernley, G.W.   West German Patent 3 535 371, 1984.
455. Bolsman, T.A.B.M.; Daane, G.J.R.   SPE Reservoir Eng., 1986, 1,
     53-60.
456. Heiss, L.; Schneider, G.; Trost, A.   West German Patent 3 407
     565, 1985.
457. Baviere, M.; Durif-Varambon, B.; Salle, R.   French Patent 2
     547 860, 1984.
458. Kalpacki, B.; Chan, K.S.   U.S. Patent 4 554 974, 1985.
459. Bansal, B.B.; Hornof, V.; Neale, G.   Can. J. Chem. Eng., 1979,
     57, 203.
460. Manasrah, K., Neale, G.H.; Hornof, V.   Cellul. Chem. Technol.,
     1985, 19, 291.
461. Kalfoglou, G.   West German Patent 2 918 197, 1980.
462. Babu, D.B.; Neale, G.; Hornof, V.   Cellul. Chem. Technol.,
     1986, 20, 663-72.
463. Nase, D.G.; Whittington, L.E.; Ledoux, W.A.; Debons, F.E.
     U.S. Patent 4 739 040, 1988.
464. Morrow, L.R.; Dague, M.G.; Whittington, L.E.   U.S. Patent
     4 739 041, 1986.
465. Baviere, M.; Bazin, B.; Noil, C.   SPE Reservoir Eng., 1988, 3,
     597-603.
466. Aldrich, H.S.; Puerto, M.C.; Reed, R.L.   French Patent 2 589
     858, 1987.
467. Hjelmeland, O.S.; Larrondo, L.E.   SPE Reservoir Eng., 1986, 1,
     321-8.
468. Somasundaran, P.; Shafick, H.H.   Soc. Pet. Eng. J., 1985, 25,
     343-50 and references therein.

469. Van Os, N.M.; Haandrikman, G. Langmuir, 1987, 3, 1051-6.
470. Noll, L.A. Colloids Surf., 1987, 26, 43-54.
471. Hofman, Y.L.; Angstadt, H.P. Chromatographia, 1987, 24, 666-70.
472. Lorenz, P.B.; Trantham, J.C.; Zornes, D.R.; Dodd, C.G. SPE Reservoir Engr., 1986, 1, 341-53.
473. Hohnson, J.S.; Jones, R.R.M. Ann. Chim. (Rome), 1987, 77, 229-44.
474. Doleschall, S.; et. al. Hungarian Patent 42 575, 1987.
475. Krumrine, P.H.; Ailin-Pyzik, I.B.; Falcone, Jr., J.S. Prepr. Am. Chem. Soc., Div. Pet. Chem., 1981, 26 (1), 195-8.
476. Methemitis, C.; Morcellet, M.; Sabbadin, J.; Francois, Eur. Polym. J., 1986, 22, 619-27 and references therein..
477. Bourrrel, M.; Verzaro, F.; Chambu, C. SPE Res. Eng., 1987, 2, 41-53.
478. Reference 426, pp. 104-111.
479. Neil, J.D.; Chang, H.L.; and Gefferm, T.M. in Improved Oil Recovery, Interstate Oil Compact Commission, Oklahoma City, 1983, pp. 52-66.
480. Mohnot, S.M.; Bae, J.H.; Foley, W.L. SPE Reservoir Eng., 1987, 2, 653-63.
481. Holm, L.W.; Robertson, S.D. J. Pet. Technol., 1981, 33, 161-72.
482. Chan, K.S.; Majoros, S.J. U.S. Patent 4 466 892, 1984.
483. Mohnot, S.M.; Chakrabarti, P.M. U.S. Patent 4 714 113, 1987.
484. Bunge, A.L.; Radke, C.J. Soc. Pet. Eng. J., 1983, 23, 657-68.
485. Burk, J.K. SPE Reservoir Eng., 1987, 2, 9-16.
486. Shuler, P.J.; Kuehne, D.L.; Lerner, R.M. J. Pet. Technol., 1989, 41, 80-88 and references therein.
487. Nelson, R.C.; Lawson, J.B.; Thigpen, D.R.; Stegemeier, G.L. Proc. 1984 SPE Enhanced Oil Recovery Symposium, Society of Petroleum Engineers, Dallas, 1984, pp. 417-424.
488. Saleem, S.M.; Faber, M.J. Rev. Tec. INTEVEP, 1986, 6, 133-42; Chem. Abstr., 1986, 106, 159007m.
489. Saleem, S.M.; Hernandez, A. J. Surf. Sci. Technol., 1987, 3, 1-10.
490. Pitts, M.J. Prepr. Am. Chem. Soc., Div. Pet. Chem., 1988, 33 (1), 169-72.
491. Peru, D.A. Report, 1986, NIPER 212; Chem. Abstr., 1986, 104, 140693k.
492. Lawson, J.D.; Thigpen, D.R. U.S. Patent 4 502 541, 1985.
493. Neale, G.H.; Khulbe, K.C.; Hornof, V. Can. J. Chem. Eng., 1987, 65, 700-3.
493. Lin, F.F.J.; Besserer, G.J.; Pitts, M.J. J. Can. Pet. Technol., 1987, 26, 54-65.
494. Enhanced Recovery Week, June 16, 1984, pp. 1,4.
495. Mungan, N. in Improved Oil Recovery, Interstate Oil Compact Commission, Oklahoma City, 1983, pp. 113-72.
496. Reference 426, pp. 132-145.
497. Johansen, R.T. in Improved Oil Recovery, Interstate Oil Compact Commission, Oklahoma City, 1983, pp. 91-112.
498. Reference 426, pp. 114-131.
499. Reference 426, pp. 3-40.
500. Ali, S.M. Farouq; Meldau, R.F. in Improved Oil Recovery,

Interstate Oil Compact Commission, Oklahoma City, 1983, pp. 311-350.
501. Tiab, D.; Okoya, C.U.; Osman, M.M.   J. Pet. Technol., 1982, 34, 1817-27.
502. Babu, D.M.; Hornof, V.; Neale, G.   Can. J. Chem. Eng., 1984, 62, 156-9.
503. Crawford, P.B.; Cju, C.   in Improved Oil Recovery, Interstate Oil Compact Commission, Oklahoma City, 1983, pp. 251-309.
504. Reference 426, pp. 41-55.

RECEIVED February 21, 1989

Chapter 2

Application of Chemistry in Oil and Gas Well Fracturing

Weldon M. Harms

Halliburton Services, P.O. Drawer 11431, Duncan, OK 73536-0428

Hydraulic pressure stimulation (fracturing) of oil and gas wells has
now accumulated 40 years of history and experience. The actual
practice and application of this technique supports a multi-billion
dollar service industry which annually utilizes in excess of 130
million pounds of chemical additives. This chapter will describe
the fracturing fluids that are used and some of the additives, their
purpose, and the principles that make their use effective as well as
necessary. Information presented will update a body of review
literature that covers the prior years of fracturing(1). Chemicals
are added for specific purposes which are identifiable by their
descriptive title. Veatch(2) has compiled a thorough general list
of the additives added to fracturing fluids.

Typical Functions or Types of Additives Available
For Fracturing Fluid Systems

Antifoaming agents
Bacteria control agents
Breakers for reducing viscosity
Buffers
Clay stabilizing agents
Crosslinking or chelating agents (activators)
Demulsifying agents
Dispersing agents
Emulsifying agents
Flow diverting or flow blocking agents
Fluid-loss control agents
Foaming agents
Friction reducing agents
Gypsum inhibitors
pH control agents
Scale inhibitors
Sludge inhibitors
Surfactants
Temperature stabilizing agents
Water-blockage control agents

To better understand the need for chemical additives we first

0097–6156/89/0396–0055$12.50/0

must become familiar with the physical and engineering aspects of
fracturing. The process is ordinarily conducted with viscous fluids
that are prepared by introducing gelling agents or viscosifiers to
the fracturing fluid. Fracturing of gas and oil wells is performed
for one primary reason: to increase the rate of production of fluids
held by a localized zone beneath the earth's surface. Generally the
zones that require treatment to attain acceptable flow rates consist
of hard, consolidated rock. Rock does contain pores and some rock
has cracks along layers (natural fractures). Rock also possesses
interconnective channels (permeability) through which fluid can flow
under pressure. If permeability is low and flow is not substantial
then fracturing must be considered. Low flow rates usually mean a
well is not profitable.

Fluids, both liquids and gas, readily transmit pressure. By
pressuring up fluids (called frac fluids) on the surface with high
pressure pumps and discharging them into the wellbore of a well at
high rates, the parting pressure of a rock zone can be exceeded and
a crack (a fracture) created. After the well is drilled, wellbores
usually are lined with high pressure-resistant steel pipe which is
cemented into place to protect and isolate sensitive formations,
water zones, or strata of little interest. The entry point of the
pressurized fluid is controlled by perforating the steel pipe lining
opposite the zone of interest, usually with explosive shape charges
or by high pressure jetting(3). Perforating creates small holes
that penetrate through the steel casing. The holes generally are
0.25-0.5 inches (6-13mm) in diameter.

These holes (1-4 per foot of pipe) provide access to the
earthen formation of interest but also create a restriction to fluid
flow and subject the frac fluid to high shear forces. This shear is
in addition to laminar or turbulent shear forces experienced in
transit through the pipe. Well pipes are generally 0.3-5 miles
(0.48-8 Km) in length which requires a fluid transit time of 1-20
minutes at the high rates necessary to fracture a well. Therefore
it is important that the chemical additives contained in the
pressurized fluid be tolerant of high shear. Intramolecular bonds
of polymers can be cleaved by high shear which can drastically alter
molecular weight, and subsequently viscosity. Chemical
intermolecular bonds (crosslinkages) also can be broken, destroying
the fluid colligative properties conferred by crosslinkages;
therefore, chemical additive packages must be carefully tested and
designed to derive optimal performance under shear. Once the
fracturing fluid exits through the perforations and enters the rock,
it then encounters much reduced shear in the fracture, except at the
very leading edge where the crack is being propagated. After the
fluid enters the fracture a new problem other than shear is
encountered: fluid leakoff. The rock formation adjacent to the
induced fracture is capable of accepting fluid as a result of the
permeability and the natural fractures that are present. This loss
of fluid can result in insufficient fluid being available to extend
the fracture. Fluid loss control agents therefore are added(4).

The formation may also contain minerals that are chemically or
physically sensitive to the fluid. For that reason clay stabilizing
agents, surfactants(5), etc. are also added to the fluids. Fluids
that will be produced from the well may contain problem-causing

contaminants which require the addition of demulsifiers, mineral
deposition (scale) inhibitors(6), or paraffin deposition preventives
to the fracturing fluid.

Flow in undisturbed rock normally is radial toward a site of
lower pressure (the wellbore). The fracture crack created by high
pressure injection usually forms perpendicular to the least
principle stress that exists in the rock. The induced fracture
intersects and disrupts the radial flow pattern such that flow
becomes linear and more direct to the well. This phenomenon has
been intensively examined and discussed by authors working in the
discipline of rock mechanics as applied to hydrocarbon reservoirs.
Hydraulic fractures created in oil and gas wells grow mainly
vertically, parallel to the wellbore as depicted in Figure 1 and
extend on either side of the perforated wellbore as "wings"(7-11).
It is common for these wings to exceed 100 ft in height and hundreds
of feet in length (Figure 2), but width remains comparatively
narrow. At the wellbore it usually is less than 0.25 in. (6 mm) and
for the vast majority of the fracture area width never exceeds 0.5
in (13 mm) wide (Figure 3).

The shape and exact dimensions of induced fractures remains a
topic of lively interest. Researchers who model fracture behaviors,
or their shapes and dimensions, disagree as to which mathematical
expressions best predict the job results(12). Other than rate of
fluid leakoff, the two most important parameters that a good model
should predict are height and fracture width. The growth of
fracture height usually determines whether the fracture will stay
within the desired zone, or extend down into a water-bearing zone (a
costly and undesired occurrence), or upward away from the
hydrocarbon saturated layer. Prediction of induced fracture
characteristics is important because a single fracturing treatment
requires a considerable monetary investment ($10,000-2,000,000). If
the prediction of the induced fracture is inaccurate, the fracture
too small or in the wrong place, then hydrocarbon production may
remain inadequate; if the fracture is too massive the additional
hydrocarbon production may not offset the greater expense of the
large-scale treatment. If the model is incorrect then the
fracturing treatment may have to be terminated early and additional,
expensive, remedial formation and wellbore cleanout steps taken
before the well can produce hydrocarbons. It may be necessary to
treat the well a second time(13,14).

Width assumes a great importance when one realizes that to
maintain conductivity through the induced fracture it is nearly
mandatory that the crack be filled with a solid spacing agent
(called proppant) which will prevent the crack from closing
completely after the hydraulic pressure on the well is released.
The fracture must be wide enough to permit entry of proppant
to a distance sufficient to stimulate production. Tons of proppant
are normally required to fill this void. Therefore the fracturing
fluid must suspend the proppant long enough for it to be transported
and placed, by flow, throughout the fracture. To preserve the
maximum accessible flow area, the proppant should be uniformly
suspended inside the entire propped fracture area while the fracture
closes. Kaspereit(15), as well as Smith(16), has made the point
that fracture conductivity can be a limiting factor. If the

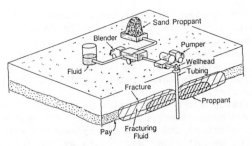

Figure 1. Fracturing process, equipment, and growth of vertical "wings". (Reproduced with permission from ref. 2. Copyright 1983 Society of Petroleum Engineers.)

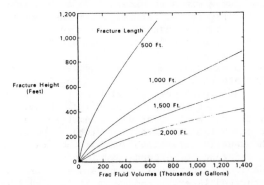

Figure 2. Typical fracture dimensions. (Reproduced with permission from ref. 2. Copyright 1983 Society of Petroleum Engineers.)

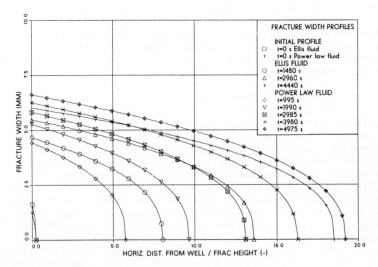

Figure 3. Fracture width vs. length. (Reproduced with permission from ref. 362. Copyright 1983 Society of Petroleum Engineers.)

proppant falls before closure, production increase can be
drastically reduced. Simplistically, viscosity controls the width
of the fracture and to a large degree the time of proppant settling.
Wider fractures result from increased pressure drops within the
fracture. More viscous fluids cause greater pressure drops. Width
can be increased by increasing fluid viscosity and/or fluid
injection rates. These two variables usually work at odds with one
another, since viscous fluids often resist high pumping rates.
    Chemicals must yield fluids with improved viscosity and flow
properties and must provide proppant transport and suspension
properties for the many hours required for a well treatment. This
may be difficult to achieve since temperatures inside the fracture
can range between 60°- 500°F+ according to depth. Without the
correct physical fluid properties it is not possible to place much
proppant before excessive pressure resistance is generated (typical
steel pipe and pumping equipment can operate between 4,000-15,000
psi; special equipment allows treatments to 20,000 psi) and pumping
has to be terminated due to well "screen out" or "pressure out."
Three years (and 10,500 fracture treatments) after fracturing was
introduced, it was clearly recognized by Clark, et al.(9), that
successful fracturing treatments demanded effective viscosifiers as
well as fluid loss control agents.

Fracturing Fluids

Gelled Oils. The first pressurized fluids used to fracture
stimulate wells were hydrocarbon liquids. Credit for conception of
well stimulation by hydraulic fracturing generally has been given to
Clark(7,9,17,18) or alternately to Farris(19), both of whom worked
for Stanolind Oil Co. and filed for patent protection simultaneously
on May 28, 1948. Other authors(20) place the conception and use as
early as 1925. The first recorded application was conducted by
Stanolind Oil (now Amoco Production Co.) in 1947 in Grant County,
Kansas in the giant Hugoton gas field(7). One thousand gallons of
gasoline without any proppant was used. The treatment design
involved multiple stages. Typical results of this type of treatment
were later presented by Clark(9). The gasoline was viscosified by
use of napalm(21,22). Napalm is a specially precipitated, granular
aluminum soap(23) prepared from a mixture of oleic, naphthenic, and
coconut fatty acids. Gelled hydrocarbons remained the preferred
hydraulic fracturing fluids for about 10 years. These fluids could
carry moderate quantities of proppant (1-3 lb/gal) and viscosity
could be adjusted by changing the oil used or the gelling agent
concentration.
    Refinements in napalm chemistry were introduced throughout the
industry (1950-1960) using other fatty acid aluminum salts as well
as sodium or potassium tall oil mixtures(23-26), which are rosin and
fatty acids derived from digested, pulped pine wood. Gelling agent
chemicals were selected and formulated to give best performance with
different hydrocarbon liquids such as kerosene, diesel No. 1, diesel
No. 2, residual oils, "refined oil," and some crude oils according
to customer preference and acceptance. Performance can be achieved
with these metallic soaps but the very viscous gels created are
limited in temperature application (≤150°F) and most are very

sensitive to moisture contamination. The gels can be intentionally broken by the introduction of chelating agents(27), by acids, or by brine water.

In 1970, Monroe and Rooker(28) claimed the use of aluminum salts of acid orthophosphate esters as viscosity builders for use in fracturing fluids. The application of these materials began a new era of hydrocarbon gelling agents. Monroe(29) later claimed the use of Fe_3O_4 as a metal activator of phosphate esters and in 1971 described several other metals(30) that could be used with amine neutralization agents. Numerous metallic ionic derivatives can be used as effective "activators" or crosslinkers to prepare a gel. Most of these metals exist as several species in equilbrium. Aluminum is ordinarily used. Ryles(31) commented on the properties of aluminum species as has Arnson(32), Lauzon(33), Schecher(34), and Baes and Mesmer(35). The distribution of aluminum species is very sensitive to pH (Figure 4).

Phosphate esters can function as friction pressure additaments(36) as well as viscosity builders in hydrocarbon fracturing liquids. A bothersome feature of pumping fluids in narrow steel pipe at the high rates and pressures necessary to fracture rock is that friction is created and a significant pressure drop in the pipe occurs. Friction pressure drop is proportional to the square root of the fluid velocity. A high enough pressure to cause a fracture to open and elongate must be applied by pumps on the surface at all times during the treatment. Additional rate and subsequently more horsepower is required to overcome pressure losses due to friction. More horsepower increases treatment costs and can be a limiting factor in whether a well can be successfully treated. Therefore "friction reducers" are a very important class of additives routinely added to fracturing fluids to lower costs and improve treatment efficiency.

Aluminum crosslinked orthophosphate esters are still the major hydrocarbon gellants used today. Typically 1% by volume acid phosphate ester (ca. 100 moles) plus the stoichiometric equivalent of 33 moles of sodium aluminate or aluminum chloride in solution are added to 1000 gallons of hydrocarbon. Sufficient alkaline solution (caustic or amine) is then added to neutralize the acid phosphate ester and create a viscous gel. Viscosity development depends on the correct degree of neutralization. Excess acid or alkali will result in much lower viscosities. In large industrial applications such as fracturing chemistry it is usually not economically practical to use isolated, highly purified materials. Instead the phosphate ester product mixtures have been adapted. The esters can be prepared from several routes such as phosphorous pentoxide, phosphorous pentachloride, phosphorous oxychloride, or ester exchange. Different product mixtures are obtained from these routes. Zangen(37), et al. and Moule and Greenfield(38) have commented on some of the complexities of phosphate ester products.

Oil Gel Breakers. Many fracturing gels remain too viscous inside the fracture after placement, thereby preventing flow of hydrocarbons from the well. Fracture fluids should eventually lose their viscosity after the fracture job has been completed. Breakers are chemicals intentionally added to fracturing fluids to decrease or

destroy their viscous properties to permit recovery of fluids after
the well has been fractured and propped. Since breakers are
preferably added during treatment it is necessary that their action
be delayed during the job duration, often 1-8 hours long. Hill(39)
described various formate, acetate, benzoate, etc derivatives that
are useful to break gelled oils. McKenzie(40) published data on the
use of selected acids as gelled oil breakers. Daccord, et al.(41)
claimed that addition of a partially neutralized acidic aluminum
salt provides a one stage preparation of oil gels which can be
broken by the addition of 4-nitro-benzoic acid. Since gelled oils
contain little water it takes a period of time before these
additives change the "non-aqueous" pH of the oil gel and break it.
Numerous improvements(42) in orthophosphate gellants have been
reported during the 19 years since their introduction.

Mechanism of Oil Gelation by Orthophosphate Esters

The mechanism of how these viscosifiers alter the properties of
hydrocarbon fluids has not been intensively examined but some
results and proposals have been published. In 1970 Baker, et al.
summarized their findings on the friction reduction properties of
metallic soap derivatives(43) and proposed a colloidal association
configuration of the molecules in micelles as the operative
mechanism. Earlier Rose and Block(44-46) had proposed a bridged
associational model between metal and phosphate molecules to explain
the high intrinsic viscosity development and low-temperature
flexibility imparted by mixed alkyl phosphates with Zn(II) or Co(II)
ions. This association structure is thought to be operative in
poly(phosphonatoalanes)(47) and aluminum orthophosphate ester gels,
as per Burnham, et al.(48) McKenzie(40) proposed a similar,
three-dimensional associated structure involved in a dynamic
chemical equilibrium (Figure 5) that could assimilate all the
modifying agents known to affect these type gels.
     Fendler(49) published regarding his pursuit of "surfactant
vesicles" as a mimic of biological membranes. These are defined as
smectic mesophases of synthetic surfactant bilayers containing
entrapped water(50,51). Dihexadecyl phosphate was the surfactant
used. This molecule bears a resemblence to the dialkyl phosphate
esters used in oil gelation technology. It may be that some of the
oil gels used as fracturing fluids consist of micelles having
associated hydrophobic layers that surround the water contained in
the oil gel. Oil fracturing fluids are never entirely water-free.
Humidity contributes significant moisture and the metallic
"activator" is usually added as an aqueous solution. It is known
that micelles require shear or sonication energy in order to form.
Oil gels must be sheared in order to form. Micelles are dynamic
species possessing interior viscosities. Micelles can maintain
substantial pH gradients between the bulk liquid phase and the
entrapped phase. Oil gels are viscous and can maintain stability
for hours in the presence of acids or alkalis which are added as
breakers. One representation of a vesicle(51) consists of a
three-dimensional equilibrium specie -- similar in some respect to
the suggestions presented by McKenzie.
     Whatever the active mechanism is, from these publications it

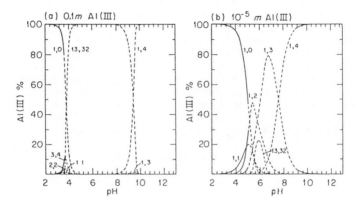

Figure 4. Aluminum cross-linker moiety depends upon pH of gelled fluid. (Reproduced with permission from ref. 35. Copyright 1976 Wiley & Sons.)

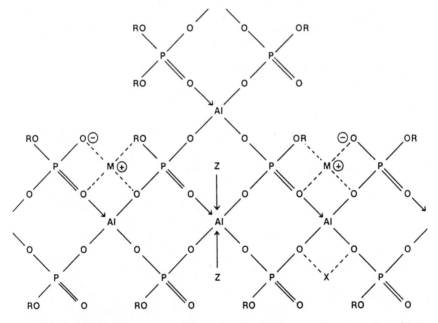

Figure 5. Proposed 3-D associated gel network model. (Reproduced with permission from ref. 40. Copyright 1980 Society of Petroleum Engineers.)

can be concluded that orthophosphate ester gels exist as a dynamic, associated network in equilibrium. Since the network is formed by associative attractions rather than rigid covalent bonds it has the ability to eventually reheal if destroyed by physical shear forces, an attractive property for a fracturing fluid. At the same time equilibrated associative attractions can be altered by numerous polar molecules, much more easily than covalently bonded gels can be altered. Associative oil gels are sensitive to polar contaminants. The gels can be broken by water, alcohols, acids, bases, and surfactants.

Water Gelling Agents

Fracturing effectiveness was quickly proven in 1947 using hydrocarbons as the pressurized fluid(9). Well owners were reluctant to use water as a fluid because of the massive damage to hydrocarbon flow and recovery that can occur when untreated water is contacted with water-sensitive formations(52,53). Certain clays and shales are notorious for sloughing and swelling in fresh waters. Microscopically clays consist of deposited mineral layers which act as ion exchange sites (Figure 6). The tiny clay particles shrink or expand depending on the cation to which they are exposed. Even if the localized swelling at the point of contact doesn't completely close all permeability tiny disturbed clay solids may be released by ionic or mechanical shock and transported by flow as "fines" which can accumulate and plug rock permeability.

   Acids were an early exception to the no water rule. It was recognized that aqueous solutions of acids would inhibit swelling of clays and shales as well as dissolve any acid-soluble minerals contained in a formation. By 1933 commercial well stimulation with hydrochloric acid was of great interest. A whole separate methodology and treatment chemistry has since evolved around acidizing and fracture acidizing(54). Water emulsions, mainly emulsified acids, and gelled acids thickened with polymeric additives were applied early in the history of well treatment.

   By 1953 gelled water fracture treatments were used to improve injectivity in water injection wells in waterflood projects. Frictional pressures with water or brine fluids generally were lower than with hydrocarbon fracturing fluids and could be lessened further by the addition (0.1 to 0.5% W/W) of natural polysaccharide or synthetic polymers. Flammability problems and cost were minimal with water and viscosity thinning due to increasing temperature often affected water gels to a lesser degree than hydrocarbon gels. Yet interest in fracturing oil and gas wells with water-based gels remained subdued until 1957.

Clay Stabilizers

Perhaps the key development that began to tip the balance in favor of water-base fluids was recognition that formation damage by water could be controlled. Control was first provided by inorganic salts dissolved in the water. Operators knew that native brine solutions (usually 6-37% NaCl) caused little or no damage to the formations they were produced from.

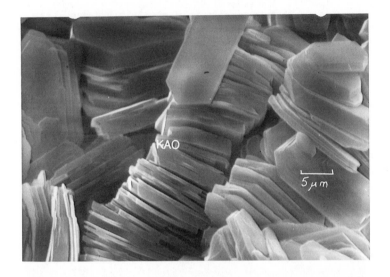

Figure 6. Scanning electron microscope photos of the troublesome clay minerals: kaolinite, chlorite, smectite, and illite. (Reproduced with permission, Halliburton Services.) *Continued on next page.*

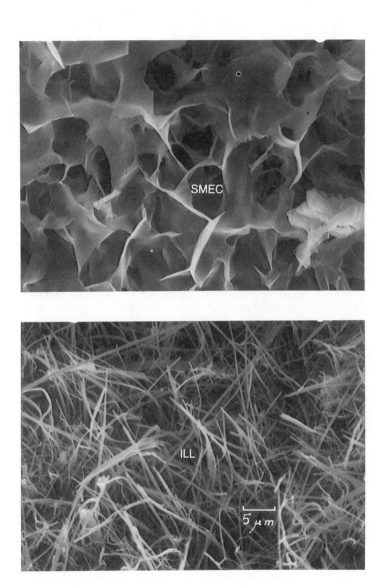

Figure 6. *Continued.*

From 1958 to 1970 inorganic salts such as $CaCl_2$(53,55,56),
$MgCl_2$(57), KCl(55-58), hydroxyaluminum chloride(59-63), and zirconyl
chloride(64-67) were introduced and found acceptance as clay
protection additives. Each salt has its own peculiar set of
advantages and disadvantages. It is now routine to include 2% KCl
by weight of water as a clay treatment chemical in nearly all
fracture waters.
     Each of these cations participates in an ion-exchange reaction
with protons on the surface of clay particles. Aluminum and
zirconium are said to form polynuclear complexes with multiple
positive charge centers that can exchange with more than one proton
site and therefore possibly provide more effective protection than
can simple monovalent or divalent cations. Multiple bonding is
belived to offer some benefit in preventing the migration of fines.
All the benefits derived from treating with metal salts can be
largely nullified by treating the formation with HCl solutions.
This can be a disadvantage since acidization is a common practice
employed to economically increase hydrocarbon production from
depleted wells.
     In 1956 Brown, in a series of patents(68-75), disclosed that
clays could be treated with di-, tri-, or tetra-substituted ammonia
derivatives. Later, McLaughlin, et al.(76,77), introduced cationic
polymers as permanent clay protective chemicals. A series of
published results describing laboratory and field applications soon
became available(78-81). Structural details of the cationic
polymers appeared in patents(82-85). In general the polymers are
polyamine derivatives, mostly quaternary in nature. Theng(86,87)
has discussed how the multiple cationic centers in these polymers
can interact and permanently protect clays. Callaway(88) et al. has
noted that cationic polymers may interfere with the performance of
crosslinked fracturing fluids.
     The effect of pH on both clay swelling and fines production has
been widely discussed(89-95). Little consensus is found in this
literature. Suggested treatments range from application of
fluoboric acid(96) to 15% KOH(92) solutions -- both treatments are
believed to create a protective silicate film that inhibits release
of fines. Polyacrylate polymers can provide protection against
swelling of smectite clays and shales(97-100).
     Several articles with informative bibliographies covering
formation protection additives have appeared recently(97,101,102).
The exact rock formation sample in question, the ionic strength of
the treatment fluid, the preventive additives that are present, the
pH of the fluid, and the test procedure employed all have
significant effects on the test results. However, with careful
experimentation using representative materials a preventive additive
package can be administered as part of a water-based fracture
treatment to allow effective stimulation of most hydrocarbon
reservoirs. Because of this, water-based fracturing fluids are used
in approximately 90% of all fracture treatments performed today.

Gelling Agents (Viscosifiers)

During the 1950-1965 era, a number of water viscosifying agents were
examined and introduced in rapid succession. Anderson and

Baker(103) state that Karaya gum was the first to be used.
Synthetic products, e.g., polyethylene oxides(104), polyacrylates,
polyacrylamides, and polyetherglycols were in competition with
natural polymers like starch, guar, cellulose derivatives,
alignates, carrageenan, and locust bean gum. The basic physical and
structural properties of the various polysaccharide thickeners have
been compiled and reviewed by numerous authors and editors(105-109).
     Cost and performance effectiveness of the natural
polysaccharide thickeners quickly established these materials as
preferred water gelling agents over synthetic polymer viscosifiers.
Synthetic polymers were found to be more cost effective as friction
reducers(110) and remain in routine use as such. Generally these
friction reducers are acrylamide (for cost effectiveness) copolymers
that contain up to 40% of a second monomer (for improved
performance) to impart anionic or cationic properties; such as
acrylamidopropanesulfonate (AMPS), acrylic acid, or
dimethylaminoethylmethacrylate (DMAEM), etc. The greater
effectiveness of synthetic polymers as friction reducers as compared
to polysaccharides is related to high molecular weight as well as
the size of the extended, hydrated synthetic polymer chain. Monomer
identity and ratio alone do not tell the whole story on
effectiveness since other parameters can result in divergent
friction reducing performances(111).

Guar Viscosifiers

Guar gum, a heteropolysaccharide, introduced in 1953(103), quickly
became the most widely used aqueous viscosifier. Its viscosification capability ranks ahead of the other natural gums (Table 1).
Guar is a branched galactomannan found in the interior portion of
the bean seeds of the guar plant (cyamopsis tetragonolobus). The
beans are harvested and then processed by grinding away the outer
seed hull (to remove proteins and trash material) to obtain the
relatively pure interior endosperm that is practically all
water-soluble polygalactomannan. Molecular weight of the polymer
has been estimated as 440,000 to 1.6 million(112,113).
     Derivatized guars are also in common use. Derivatization
usually lowers insoluble residue content, improves rate of viscosity
yield, and increases the high temperature stability of the polymer.
The most prevalent derivatized guar in use is hydroxypropyl guar
(HPG) prepared by caustic treatment of guar with propylene
oxide(114). Field introduction of this polymer was described by
various authors(115-117). Carboxymethylhydroxypropyl guar (CMHPG)
also has found general acceptance(118-120). The presence of a
carboxyl group allows access to crosslinking chemistry not available
with hydroxyl groups alone. The structures of guar and some other
commonly used polysaccharide gelling agents are represented in
Figure 7. Chatterji and Borchardt(121) have assembled a chart of
the general applications of these materials (Table 2).
     Numerous other derivatives of guar have been prepared by
attaching modifying molecules to the guar backbone. Illustrative of
the modifying molecules are chloroacetic acid(122), acrolein(123),
ethylenimine(124), acrylamide(125), aminomethylphosphonic acid(126),
and methyl bromide(127). None of these have achieved widespread use

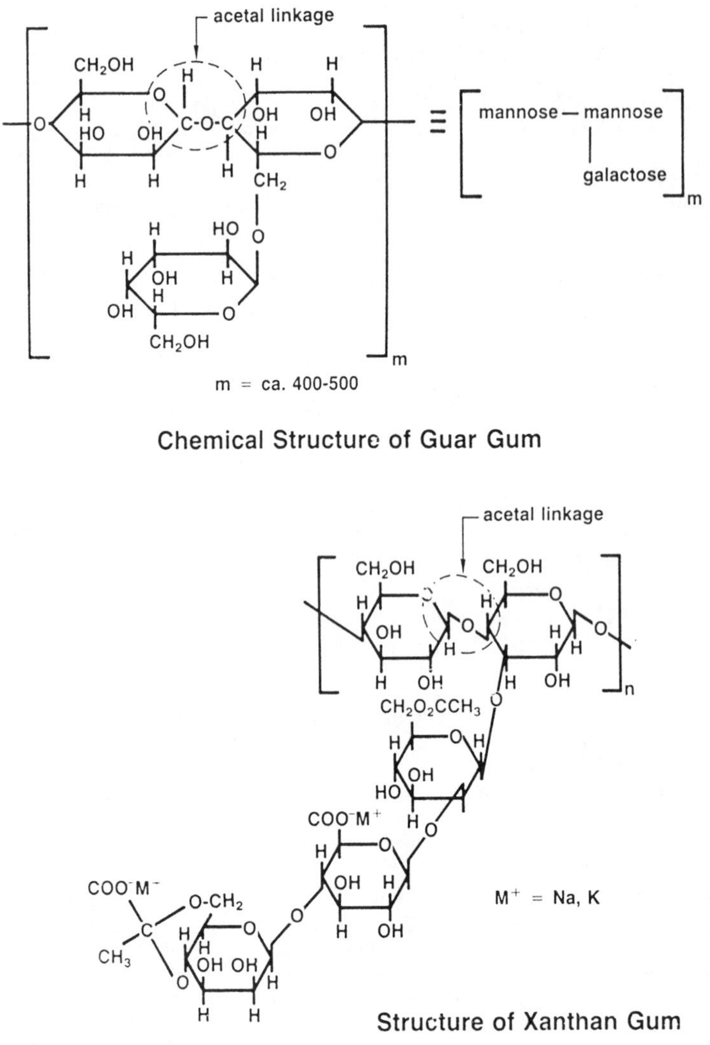

Chemical Structure of Guar Gum

Structure of Xanthan Gum

Figure 7. Structures of several polysaccharide gelling agents. (Reproduced with permission from ref. 121. Copyright 1980 Society of Petroleum Engineers.) *Continued on next page.*

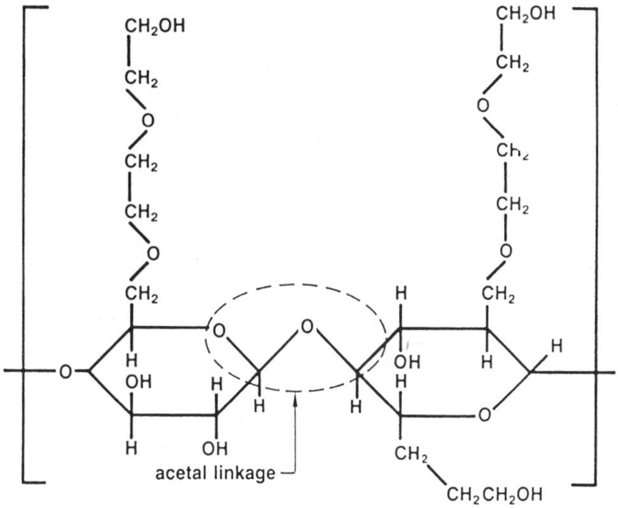

Structure of Hydroxyethyl Cellulose

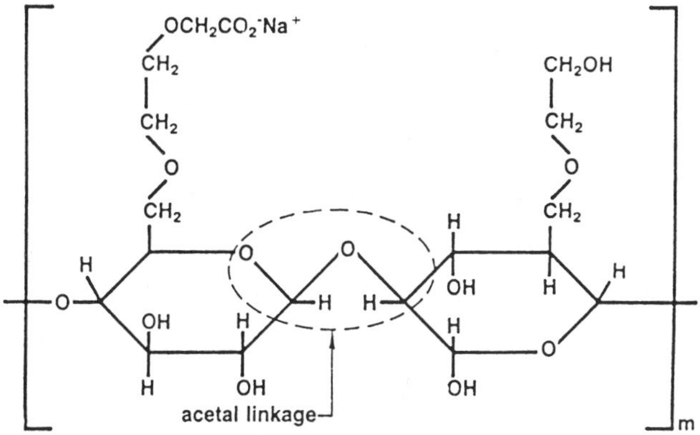

Idealized Structure of Sodium
Carboxymethylhydroxyethyl Cellulose

Figure 7. *Continued.*

Table 1. Relative Viscosities of Natural Polysaccharide Gelling Agents

Gum	cps
Gum arabic (20% by wt)	50
Locust bean gum	100
Methylcellulose	150
Gum tragacanth	200
Carrageenan	300
High-viscosity sodium carboxymethylcellulose	1200
Gum karaya	1500
Sodium alginate	2000
Guar gum	4200

NOTE: Reproduced with permission from ref. 105. Copyright 1973 Academic Press.

Table 2. General Properties of Common Oil Field Gelling Agents

Polymer	Cost[b]	Viscosity[c] (cps) 40 lb/1000 gal at 300 rpm	Shear[d] Stability	Salt[e] Tolerance	Acid[f] Stability	Enzyme[g] Stability	Residue[h] in Broken Gel	Applications
Guar Gum	1	34	3	C	N.S.	N.S.	R	Drilling fluids, spacers, friction reduction, in stimulation, fracturing and lost circulation (crosslinked guar gum gels), fluid loss additive in drilling, spacer and fracturing fluids
Hydroxypropyl Guar	1	36	3	C	N.S.	N.S.	R	Drilling fluids, spacers, completion and workover fluids, friction reducer in fracturing, fracturing and lost circulation (crosslinked HPG gels), fluid loss additive in drilling spacer and fracturing fluids
CMC	2	55	3	IC	N.S.	N.S.	RF	Drilling fluids
HEC	2	37	3	C	N.S.	N.S.	RF	Fluid loss additive for cementing spacers, completion and workover fluids, fracturing fluids, friction reduction in stimulation, enhanced recovery
CMHEC	2	32	3	C	N.S.	N.S.	RF	Fluid loss additive and retarder for cementing, spacers, gelling weak acids, temporary diverting agents in fracturing (crosslinked gels)
Xanthan Gum	4	34	1	C	M.S.	M.S.	RF	Drilling, completion, fracturing and enhanced oil recovery
Polyacrylamide (Partially Hydrolyzed)	3	34	2	IC	S	S	RF	Fluid loss additive in cementing, drilling fluids, friction reduction in fracturing, enhanced oil recovery, scale inhibitor
Copolymer of Polyacrylamide	3	25	2	MC	S	S	RF	Fluid loss additive in cementing, drilling fluids, friction reduction in fracturing, enhanced oil recovery, water-oil ratio reduction, improvement of injection profile (usually crosslinked)

a - These properties were determined using representative polymers of each class. Changing the specific polymers could substantially alter the properties listed here. This table is meant to be a generalized description and not a rigid list of specifications.

b - The order of cost per pound of polymer, 1 being the cheapest and 4 being the costliest. The cost will vary with the supplier and the amount of order.

c - Viscosity data reflect the numbers obtained with a particular polymer in fresh water. Changing the molecular weight, degree of substitution, moles of substitution or the nature of monomers in the copolymer will drastically alter the viscosity.

d - The order of shear stability, 1 being the least shear stable and 3 being the most. Xanthan gum solution viscosity is not permanently reduced by shear.

e - C = compatible, IC = Incompatible.

f - N.S. = Not Stable, M.S.= Moderately Stable, S = Stable.

g - N.S. = Not Stable, M.S.= Moderately Stable, S = Stable.

h - R = Residue present, RF = Residue Free.

NOTE: Reproduced with permission from ref. 121. Copyright 1980 Society of Petroleum Engineers.

in fracturing fluids. Cationic derivatives are prepared by treating
polysaccharides with vinyl monomers such as 3-chloro-2-hydroxypropyl
trimethyl ammonium chloride. Cationic guar was used as the gelling
agent in one high temperature crosslinked fracture fluid(128,129).

Cellulose Viscosifiers

Cellulose polymer has also been studied extensively as has its
derivatives. Cellulose is a linear homopolysaccharide of glucose
units linked via β(1-4) glycoside bonds. It is highly crystalline
in nature and insoluble in water. Derivatization is required to
render the polymer water soluble. The most common cellulose
derivatives used in fracturing fluids are hydroxyethylcellulose (HEC)
prepared by ethoxylation of cellulose, and carboxymethylhydroxyethyl
cellulose (CMHEC) prepared by treatment of cellulose with ethylene
oxide followed by treatment with chloroacetic acid to provide
carboxymethyl substitution on the backbone(130). Similar in many
respects to CMHPG, this anionic, doubly-derivatized cellulose gains
the valuable properties of the carboxymethyl group (water solubility
and complexation ability) while the presence of the hydroxyethyl
group improves salt tolerance and acid resistance. These properties
make the use of HEC or CMHEC preferred over guar when preparing
viscosified acids or brines.
    Carboxymethyl cellulose (CMC) is widely used as an additive to
prepare drilling fluids and certain specialty fluids but failed to
maintain acceptance in fracturing fluids because of its salt
sensitivity and narrow temperature application range(131).
Cellulose derivatives are widely used as viscosifiers but do not
command the volume of usage of guar and guar derivatives. This
preference is due in part to the stereochemical and structural
difference between guar and cellulose and in part due to the higher
cost of cellulose derivatives. Guar has a mannose backbone with
numerous branches of galactose sub-units, which contain cis
3,4-hydroxyl groups. Cellulose is composed entirely of glucose
sub-units, where all adjacent hydroxyl groups are trans to each
other. Branching provides guar with different characteristics than
the more linear cellulose derivatives and the presence of
cis-hydroxyl groups in the galactose moieties promotes interaction
of the guar molecules with many metal cations which do not interact
with the trans-hydroxyl groups of cellulose derivatives.
    Xanthan is a natural, highly-branched polysaccharide produced
by the bacterium Xanthomonas campestris(132). The backbone consists
of a cellulose chain with attached trisaccharidic side chains. The
terminal saccharide of the branch is a pyruvic acid derivatized
mannose unit (Figure 7). As such this terminal unit provides both a
carboxyl group and cis-hydroxyl moieties. These groupings provide
interactive sites that endow xanthan with special properties in
solution as well as providing sites for crosslinker interactions to
occur. Xanthan self-associates to form helical structures. For
this reason an accurate molecular weight is difficult to determine.
Southwick(133) et al. believes the base molecular weight of
individual strands to be 2x10^6. Salamone(134), et al. reviewed the
overall uniqueness of xanthan. Most applications of xanthan seek to
take advantage of the suspending properties of the polymer or its
ability to interact with other polymers(135,136).

Xanthan has several undesirable properties. It ordinarily
forms microgel particles that can plug permeability(137), it is
expensive relative to other natural polysaccharide gel agents, and
it resists degradation by ordinary gel breaker additives. These
features have kept xanthan applications in fracturing gels to a
minimum. Some literature(138) on xanthan has appeared.
Carico(139-140) has published comparative information about the
various gelling agents. Xanthan has been adapted to foam fracturing
gels. Xanthan gels can be broken by the addition of lithium
hypochlorite(141).

Selection of a gelling agent to prepare viscous fluids is
determined by many factors. Cost and performance are primary
properties to be respected. Secondary properties can change
original choices based solely on cost and viscosity performance
alone. Factors such as pH stability, interaction with metal
crosslinkers, response to brines, rate of hydration, temperature
susceptibility, dispersibility, compatibility with other additives,
and residue left after break often(142) influence the ultimate
selection. Derivatization of polysaccharides is done to enhance
certain performance features. Variations in moles of substitution
and degree of substitution can cause very pronounced changes in
behavior(143). These differences can arise because of crosslinking
density as well as changes in the binding sites.

Effective complexation chemistry with metal ion complexing
agents is highly desirable and is a most important feature of the
majority of fracturing fluids used today. Crosslinking ordinarily
increases the viscosity of a polymer solution 5 to 20 fold (Figure
8) and therefore offers a cost effective method of providing highly
viscous polymer gels. Complexation by many metal crosslinking ions
is a function of multiple interactions with hydroxyl groups and is
dependent on hydroxyl spatial orientation. Cis-hydroxyl groupings
appear to provide increased crosslinking abilities. Angyal(144,145)
attributes complexation effectiveness of polyhydroxy materials to
the ability to assume a sequence of axial, equatorial, axial
hydroxyl groups. Cis-diol characteristics can be conferred on
cellulose polymers by derivatization with appropriate dihydroxy
monomers(146,152). Ordinary derivatized cellulose fluids used in
gravel packing or fracturing fluids are useful only to about
220°F(149,150). The presence of the cis-dihydroxy grouping
increases the operating temperature range of cellulosic polymers
beyond 220°F.

Dispersion of Gelling Agents

Dispersion of polymeric viscosifiers is often difficult because the
initial contact of the untreated polymer with water results in very
rapid hydration of the outer layer of particles which creates a
sticky, rubbery exterior layer that prevents the interior particles
from contacting water. The net effect is formation of what are
referred to as "fish eyes" or "gel balls." These hamper efficiency
by lowering the viscosity achieved per pound of gelling agent and by
creating insoluble particles that can restrict flow both into the
formation and back out of it. The normal remedy for this behavior

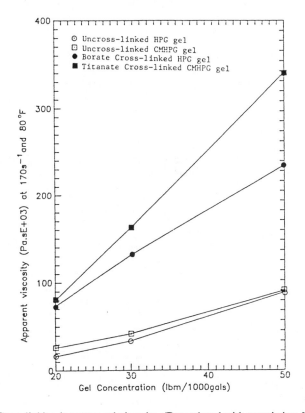

Figure 8. Cross-linking increases gel viscosity. (Reproduced with permission from ref. 330. Copyright 1987 Society of Petroleum Engineers.)

is to control particle size and provide surface treatment
modifications to the polymer. One seeks to delay hydration long
enough for the individual polymer particles to disperse and become
surrounded by water. Then no dry particles are trapped inside a
gelled coating. This can be achieved by slightly reacting or
coating the polymer with borate salts(155,156), glyoxal(157,158),
non-lumping HEC(159), sulfosuccinate(160), metallic soaps(161),
surfactants(162), or materials of opposite surface charge to the
polymer(163). Polymers have also been dispersed into a water-in-oil
emulsion which is then inverted after addition to water(164).
    Another recent innovation to improve the efficiency of polymer
addition to water and derive the maximum yield from hydrophylic
polymers was introduced by Briscoe(165,166). The method involved
the preparation of a stabilized polymer slurry (SPS) to be added to
water. Briscoe used water as the suspension liquid, usually also
containing dissolved KCl as a clay stabilizer, and formulated a
package of inhibitors (borate and caustic) to prevent the polymer
from hydrating until the pH was lowered. These concentrates remain
in routine use today.
    Aqueous concentrates can suspend a limited quantity of polymer
(ca 0.8 lb/gal) due to the physical swelling and viscosification
that occurs in a water-based medium. Higher quantities (up to 5
lb/gal) of solids can be suspended in a diesel fuel carrier fluid.
This fact and a desire for greater efficiency of equipment use led
to the development of diesel-based SPS concentrates. First used in
1985(167-170), use of these SPS concentrates has spread
rapidly(171-173).
    Suspension of water soluble solids in oil can be achieved by a
variety of chemical additives. Chemical suspension additives that
have been suggested include alkyl mercaptophosphonic acids(174),
organophylic clay plus hydroxypropyl cellulose(175), polyols(176),
aluminum stearate(177), organophylic clay plus surfactant(178-181),
aluminum phosphate esters(182), and acrylate copolymers(183-184).
The most commonly cited suspending agent employed to prepare oil
dispersions is organophylic clay. This material is prepared by
treating smectite-type clays with fatty-quaternary amines.
Organophylic clays were first described by Hauser(185-188) and
Hauser(189,190) and have routinely been used to viscosify oil-based
drilling fluids(191). Improvements have been claimed in preparing
these materials(192-195). Since organophylic clays are insoluble
solids their use in preparing SPS concentrates should be minimized
and generally restricted to applications where their presence is not
objectionable. The presence of solid residues can lower fracture
conductivity and therefore production(142,196). SPS concentrates
can be prepared using residue-free synthetic polymers(184) that will
not affect fracture conductivity.

Crosslinked Fracture Fluids

Historically, as wells are drilled deeper and operating temperatures
increased, new methods are sought to improve fluid properties to
suspend proppant at the higher temperatures. The use of very high
loading levels of gelling agent soon reaches a practical economical
and operational limit. High viscosities developed from large

polymer loadings in the surface mixing equipment leads to poor
mixing efficiency of additional polymer solids into an already
thickened gel. Also excessive frictional pressure resistances can
be encountered during placement.

Distinctly different approaches to avoid these problems and
still achieve high viscosities were conceived and applied. The
different approaches can be categorized as 1) preparation of
emulsions or foams and 2) addition of crosslinkers to the polymer.
Two separate processes which utilized crosslinking of polymer
gelling agents were pursued. These are the use of secondary (or
delayed) gelling agents and the use of metallic crosslinkers added
on-site.

All the methods involve preparation of an initial, primary,
hydrated, viscous solution referred to as base gel or linear gel.
The viscosity of this base gel at surface conditions is between
25-100 centipoise (as measured at 511 sec^{-1} shear rate with a Fann
viscometer). Base gel viscosity is then caused to increase even
further for performance at higher temperatures. How this viscosity
increase is achieved distinguishes the two processes that utilize
crosslinking of a gelling agent. Ely(197), employed addition of a
second polysaccharide that had been treated (crosslinked) with a
difunctional aldehyde or other chemical coating agent to render the
polysaccharide very slowly hydratable below 140°F. This crosslinked
polysaccharide can be added to the viscous base gel as a secondary
gelling agent to be activated by solvation at higher temperature.
The use of a secondary gelling agent delays the initial hydration of
a major portion of viscosifier and protects it from degradation for
a period of time. Eventually it hydrates and provides extended
viscosity in the formation. Horton(198) has provided a brief
overview of secondary gelling agents compared to a primary
crosslinked fluid. Field application data and well treatment
results after use of this technique have been reported(199-201).
Refinements have been made to improve the characteristics of the
secondary agents(202,203). This technique can provide viscosity for
long periods of time at high temperatures but suffers from the
requirement of large amounts of total polymer (up to 2%), and higher
amounts of residue(142).

The second type of crosslinked gelled fluid, the one most
widely used today, involves the addition of a metal crosslinking
agent to a fully hydrated base gel to create exceptionally viscous
fluids. These types of gels were being pursued in other industrial
applications(204) besides oilfield. The first use of a crosslinked
fluid as a fracturing gel(205-207) was introduced by Holtmyer, et
al. in 1969. It took advantage of the ability of antimony(V) to
crosslink guar related materials at pH 5. The onset of antimony
crosslinking can be somewhat delayed by control of pH(208). Insight
into the rheological behavior of antimony crosslinked gels has been
provided by Chakrabarti, et al(209). Nearly simultaneously(205-207)
with the introduction of antimony crosslinked guar gels a sodium
dichromate plus sodium sulfite crosslinked guar gel system employing
in situ reduction of Cr(VI) to Cr(III) was also field tested. In
response other groups introduced fluids(153,210-213) based on
Kern's(214) preparation of viscous borate crosslinked fracturing
gels at pH greater than 8.

Boron gels had been known for many years. Boeseken and
coworkers(215) in 1913-1933 had examined in detail borate
crosslinked gels prepared from polysaccharides that contained
cis-diol (galactose) sugar moeities. Deuel and Neukom(216) verified
Boeseken's work and proposed a three-dimensional network structure.
Roy, et al.(217) reviewed the complexation ability of tellurate,
borate, and arsenite ions and found that the complexes formed were
decreasingly stable in that order. Slate(218) determined that
borate-contaminated fluids could be viscosified by guar gum after
treatment of the water with polyols (e.g. sorbitol, mannitol,
glycerol, etc) that had superior complexation ability with borates.
McIrvine(219) noted that the effect of borate crosslinking could be
delayed by addition of MgO as a delayed-action caustic agent.
Similarly Chrisp(220) would later suggest this technique for use
with transition metal crosslinkers. Conner and Bulgrin(221)
published complexation constants for borate and 15 simple polyols
and concluded that the capability to delay crosslinking depended on
the ease of the polyol to attain coplanarity of hydroxyl carbons and
oxygens when entering the complex. This ability was affected by
degree of substitution of the carbon atoms. Savins(222) examined
the rheological behavior of polyvinylalcohol (PVA) -- borate
complexes and commented on several postulated mechanisms to explain
the "bizarre rheological properties of shear thickening complexes."
He preferred the mechanism proposed by Bourgoin(223) as the source
of the rheological effects. Schultz and Myers(224) employed dynamic
oscillatory rheometry to examine PVA-borate gels and found that the
borate crosslink density dropped with increasing temperature, that
the borate bond required 5.8 Kcal/mol to break, and that effective
borate crosslinks create a network four-fold the mass of the
original polymer. The low activation energy probably explains why
borate gels are useful only as relatively low temperature fracturing
fluids, generally used at temperature lower than 150°F. Another
possible cause is the change in boron species(35) created by
increasing temperature (Figure 9).

Gorin and Mazurek(225,226) used 13C-NMR to monitor the effects
of borate complexation on carbon atoms of polyols and categorized
the cyclic complexes into four types. The type of complex observed
was dependent on borate concentration as well as the stereochemical
configuration of the hydroxyl groups on the polyol. Nickerson(227)
also commented on the mechanism of borate crosslinking. Ochiai et
al(228,229). reported ionic strength effects on borate crosslinking
density. The sensitivity of borate crosslinking to polyhydroxyl
stereochemistry has also been extensively studied by Moore, et
al.(230). Maerker and Sinton(231) found borate/PVA gels to be both
shear-thickening and shear-thinning fluids. Only mesohydroxyl dyads
were capable of complexation with boron and hysteresis studies
indicated that shear could change intramolecular and intermolecular
bonding. Sinton(232) extended this examination by ^{11}B NMR to
conclude that two complexed species may exist and that possibly
these represent intrachain and interchain bonds. The gel behavior
under shear would reflect elongation of the bonded polymer
molecules. Knoll and Prud'homme(233) suggested that crosslink
formation was a result of electron sharing through pi-orbital
overlap as had been described by Noll(234) in silicon polymers.

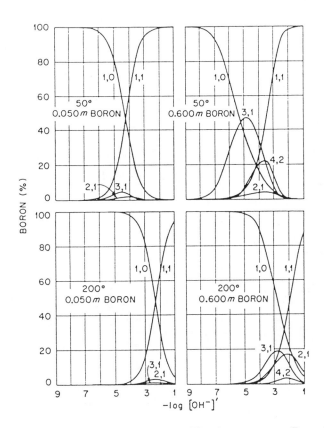

Figure 9. Boron cross-linking moiety depends on pH and temperature. (Reproduced with permission from ref. 35. Copyright 1976 Wiley and Sons.)

In 1971, Rosene and Shumaker(235) described the field
application of borate crosslinked fracturing gels. Free(213,236)
taught an improved "continuous" application of borate crosslinked
fracturing fluids by incorporation of a delayed onset alkaline agent
(MgO). During preparation of borate crosslinked fluids, the
crosslink occurs very rapidly once the pH and borate concentration
becomes high enough(237). The performance of the fracturing gel is
dependent upon borate concentration and pH(263). At pH 8 to 10, a
vigorous thickening of the polymer gel is noticed. Mondshine(234)
has accomplished the design of a delayed onset borate crosslinked
fracturing fluid by selecting appropriate buffers and natural borate
ores to provide a slowly soluble borate source. Application of this
type borate crosslinked fluid allows useful viscosity development
even at temperatures as high as 220°F.

Use of boron crosslinked fracturing gels in low temperature
applications remains commonplace as they have the reputation of
being inexpensive and also can re-heal after being sheared. This
re-healing and their high pH nature can also lead to difficulty in
obtaining a break in the gel after it has been placed in the
fracture. The enzyme breakers normally used to break
polysaccharides do not perform very well at elevated pH. If
oxidative breakers are applied they provide limited effectiveness
since borate gels are used mainly for lower temperature applications
where the oxidative breakers have difficulty with thermal
initiation, unless special activaters are included(239-241). These
activators require extra attention to detail to achieve the correct
break times.

The use of other crosslinking metals developed simultaneously
with the use of antimony, chromium, and boron(borate). Tiner, et
al.(242) introduced titanium (IV) crosslinkers in 1975 as ammonium
tetralactonate or bis(triethanolamine)bis(isopropyl)titanium(IV).
Upon contact with water soluble titanium (IV) derivatives ordinarily
form orthotitanic acid, $Ti(OH)4$, which rapidly forms oligimeric
metatitanic acid, $[Ti(OH)2]_x$ and titanium dioxide. Electron donors
such as the hydroxyl groups of polysaccharides, if properly
oriented, can participate in the sequence of titania reactions and a
crosslinked gel network results. Various titanium metal
crosslinkers remain in common use today. More will be said about
titanium crosslinked gels later.

McDaniel, et al. in 1975 published(243) a description of the
crosslinking chemistry of Cr(III), reduced in situ from Cr(VI),
which apparently had developed from the work of Clampitt(244-247).
This system, similar in many respects to the previously mentioned
Cr(III)-guar gel, used sodium dichromate and sodium hydrosulfite as
a redox couple activated by HCl to complex CMC or CMHEC.
Davidson(248) has also adapted chromium crosslinked gels.
Significant effort has been recently expended to understand the
nature of in situ Cr(III) formation with the ultimate goal being to
control the time of crosslinking. Willhite and
collaborators(249-252) have published (1981-84) a series of
findings. Prud'homme, et al.,(249,250) recently examined the
rheological behavior of this type crosslinking with polyacrylamide.
Interpretation of storage modulus (G') rheology behavior indicated a
sequential increase in crosslink sites by an olation mechanism. The

use of chromium crosslinkers in fracturing gels has declined because
of environmental concerns.

Aluminum(III) continues to be used in crosslinked aqueous
fracturing fluids as well as EOR fluids(255-259). Al(III) is more
environmentally acceptable than Cr(III)(259). Crosslinking polymers
with aluminum can be achieved at pH up to 4.5 or in the transition
region near 9.5(255). Aluminum chemistry is very pH dependent since
the distribution of aluminum species changes from solubilized Al(3+)
cations at low pH (e.g. aluminum citrate) to insoluble Al(OH)$_3$ in
the neutral regions, to soluble Al(OH)$_4$ at higher pH regions (Figure
4). The use of partially hydrolyzed polyacrylamides crosslinked by
Al(III) in citric acid to prepare viscous fluids was claimed by Gall
and Johnston(260). Aluminum crosslinked fracturing gels are
normally restricted to temperatures less than 225°F(153,261).

The antimonate, borate, and titanate crosslinked gels as
already described have remained in general use. Modifications of
buffers, ligands, and accessory additives were made as necessary to
improve and regulate performance. By the late 1970 the industry
became aware that delay of the crosslinking reaction was a much
desired behavior. Fully crosslinked gels can generate significantly
more frictional resistance to flow than uncrosslinked base gel.
Frictional resistance is costly in injection pump horsepower
requirements and represents an undesirable property that needs to be
minimized whenever possible. One can lessen frictional resistance
if the fluid being pumped does not begin crosslinking until it nears
the perforations of the well pipe. Even more undesirable is the
fact that crosslinked gels are sensitive to high shear and most do
not recover from the destructive effects of shear. To determine
shear susceptibility one can run a rheogram, where shear stress vs
shear rate is determined as shear is methodically increased and then
methodically decreased. Any change in fluid behavior between the
increasing shear regime vs decreasing shear response is called
hysteresis. Hysteresis is a reflection of the shear sensitivity of
a crosslinked fluid. If a fluid has not been altered or damaged by
shear it will give identical performance in the downward regime as
it did in the upward shear regime. A practical way of minimizing
the effects of high shear on a crosslinked fluid during the
fracturing treatment is to delay the onset of crosslinking until the
fluid reaches a region of lower shear (i.e., beyond the well pipe
and inside the fracture). Turbulence in the pipe and shear
encountered when the fluid passes through the perforations are the
primary sources of high shear. Once the fluid exits the pipe into
the formation, shear forces drop to a lower level, usually 10-100x
less.

Early methods of delaying the metal crosslinker additive
centered on mixing an additional chemical with the crosslinker or in
the fracture mixing water to cause a delayed activation. For example,
undiluted titanium triethanolamine crosslinks aqueous polymer base
gels nearly instantaneously upon addition. The reaction can be
delayed by up to 2 minutes by diluting the titanate with water(262).
This approach provides a workable system but not an ideal one since
the actual length of delay is affected by numerous factors such as
temperature, concentration of water, time of contact with water,
alcohols, and any chelating agents present(263-269). This type

"delayed" crosslinker also suffers from the lack of long term
stability. Titanium triethanolamine water solutions have to be
discarded after aging a few hours.

Small quantities of 2,4-pentanedione (acetylacetone) added to
the polymer gel solution can regulate the crosslink onset(263-269)
from organo-Ti(IV) and organo-Zr(IV) crosslinking agents.
Elphingstone and Dees(269) achieved delay from titanium bis
acetylacetone enolate dihydroxide by virtue of the presence of the
complexing ligand as well as adding the crosslinker as a solid. The
use of secondary complexing agents suffers from unpredictable
response behaviors because of the same variables that affect the
"water-delayed" crosslinkers. Also, increasing the concentration of
chelating agents can quickly maximize the delay threshold and
completely prevent the metal from creating a crosslinked gel.

In 1980 a new concept in the application of delayed metal
crosslinkers appeared--the concept of using a storage-stable,
delayed onset crosslinking agent to avoid the undesirable effects of
instantaneous or early crosslinking on the surface was introduced.
Conway(270,271) prepared these agents by chemically reacting equal
volumes of titanium triethanolamine, water, and polyhydroxyl
compound (such as glycerol, mannitol, xylitol). The length of delay
was regulated by the selection of polyhydroxyl compound. These
delaying effects of various polyols upon titanium triethanolamine
paralleled the much earlier results of Lagally and Lagally(272),
Mills(273), and Russell(274).

Zirconium Crosslinkers. The history of the discovery of new oil and
gas reserves shows a clear trend toward deeper drilling, where
temperatures are very hot. In order to provide better fluids to
withstand these temperatures more effective crosslinkers and gelling
agents have gradually been introduced. Thus, antimony and boron
complexes gave way to chromium and titanium crosslinked gels. Use
of zirconium(IV) crosslinkers for preparation of improved high
temperature fracturing gels began about 1981. Rummo
reported(263-265) on acetylacetonate delayed titanium(IV) and
zirconium(IV) complexes. Baumgartner, et al.(275), compared their
Zr(IV) high temperature fracturing gel to a Ti(IV) gel(1983).
Additional details were later made available (1985) by LaGrone, et
al.(276) and Williams(277,278). This crosslinker apparently is a
zirconium(IV) triethanolamine mixture (1:8.9)(277). Kucera(279)
claimed the usefulness of a "water-activated" mixture of
Zr(IV):triethanolamine chelate (1:3). Field application of this
type system was described(280) by Walser (1985). Baranet, et al.,
taught that these type gels had better temperature stability if the
triethanolamine ratio was increased; also the crosslink time became
more delayed(281). Payne(282) described techniques whereby control
of crosslinking onset of Zr(IV) and Ti(IV) metals could be
accelerated by polyamines or delayed by glyoxal or triethanolamine.
Horton formulated a "one-bag" gelling mixture(283) that contained
guar derivatives along with solid Zr(IV) acetylacetonate and a
regulating amount of buffering agent. A similar "one-bag"
formulation for acid fracturing gels was developed by Githens(284).
Almond (1984) was issued a series of patents claiming Zr(IV) delayed
crosslinking agents with α-hydroxyacid ligands which were found

useful to crosslink CO_2-containing fluids(285–287). Applications of
these type fluids were described by Almond and Garvin(288), and
Sandy, et al.(120) Zhao(289) used $ZrOCl_2$ to prepare inexpensive
crosslinked polyacryalmide or guar gels. He also claimed simple
Ti(IV) salts for prepartion of fracturing fluids used in the Peoples
Republic of China. Hodge(290,291) claimed hydroxyacetic acid
mixtures of Zr(IV) for use with CMHPG gelling agents, with or
without CO_2 being present. Anderson and Paktinat(199) preferred a
mixed metal crosslinker containing Zr(IV) lactate and Al(III)
chlorohydrate for use with CMHPG.

Titanium Crosslinkers. Improved Ti(IV) crosslinkers for high
temperature application were also introduced during this period.
Hollenbeak and Githens(292) prepared delayed Ti(IV) complexors by
adding sorbitol or other saccharide agents. Putzig claimed Ti(IV)
amine chelates treated with monosaccharide delaying agents could be
prepared(293) as solids which were useful for fracturing gel
crosslinkers as well as creating thixotropic properties to cementing
slurries. Wadhwa found that mixed titanate crosslinkers that
contained boric acid(294) could give early crosslinked gel strength
as well as provide high temperature performance as the titanate(s)
activated. Putzig(295) similarly claimed that mixtures of several
Ti(IV) chelates could be used to achieve regulated crosslink onset.
Smeltz(296) examined mixed ligand Ti(IV) derivatives by
incorporating α-hydroxyacids and polyols.
     As more stable crosslinked gels for application at high
temperatures have been developed efforts have also been made to find
additives that extended the life of gels at high temperatures. The
incorporation of alcohols (5–30%) helps but suffers several
disadvantages, such as cost, hazardous handling, and behavorial
interferences. Thiosulfate and thiourea salts have found routine
application(297–299). Use of 0.5% triethanolamine(300) is claimed
to provide protection as has been isothiazolones(301),
dolomite,(302) borohydride,(303) aminoacetamides,(304)
dithiocarbamides,(305) iodates,(306) benzoquinone(307–308), and
alcohol plus thiourea(309). Researchers such as Shupe(310),
Wellington(311), Thomas(312), Glass(313), Braga(314), and
Ryles(31,315) have examined the high temperature stability of
polymers.
     The level of interest in crosslinked gels continued at an
accelerated pace. Conway et al.(316–318) discussed (1980) the
instrumentation and evaluation procedures commonly in use at that
time and compared crosslinked gel systems at high temperature. They
discussed and ranked the performance capabilities of various metal
crosslinkers Ti(IV), Sb(V), Cr(III), B(III), Sb(III). A chemical
model for crosslinked fracturing gels was proposed
(transesterification) and it was mentioned that crosslinked fluid
rheology performance was dependent upon early shear history. The
model represented a covalent metal ester intramolecular bridge
between strands of polymer. Rummo(263) invoked a similar
description. Clark(319) introduced the use of a pressure rheometer
to examine the behavior of crosslinked fracturing fluids. This
instrument employed parallel plate instrumentation to measure fluid
viscosity as opposed to a Couette geometry rheometer in general use

throughout the oil field laboratories(320). The use of the parallel
plate device in a dynamic oscillatory mode allows one the ability to
measure fluid stress response to an applied in-phase shear and an
out-of-phase strain. It also allows characterization of viscous
fluids at very high shear(321) The in-phase stress measurement
provides information about the elasticity of the fracturing gel and
is defined as the storage modulus, G'. The stress component that is
out-of-phase with the applied strain (but in-phase with the velocity
gradient) is designated the loss modulus, G''. This parameter
offers a measure of the viscosity provided by the gel system. Gel
performance can thus be tested and expressed as two fundamentally
different parameters, elasticity and viscosity. Since viscosity of
a gel can be characterized by a variety of methods including
determination of G'' most researchers have since focused their
attention on the changes observed in the storage modulus G'. An
increase in G' indicates formation of a greater gel network with
more three-dimensional structure, and hence represents a direct
correlation to the extent of crosslinking. Careful analysis of G'
and G'' behavior offers clues to the molecular interactions that
occur. Prud'homme(322-324) and several
collaborators(143,233,253-255,325-329), Acharaya(330),
Menjivar(331), and Jiang(332) have probed the mechanism of how
crosslinkers interact with polymer gels.
     The mechanistic description provided by Conway(317,318) was
modified(262,323) after observing the behaviors of three
crosslinkers [boron (sodium tetraborate, pH ∿9), titanium
acetylacetonate (1:9 with isopropanol, pH ∿5) and titanium
triethanolamine (1:9 with water, aged 30 min, pH ∿9)] with two model
sugars; methyl-β-D-galactopyronose and methyl-α-D-mannropyranoside.
The borate solution interacted with the sugars as expected.
Unequivocal ^{13}C NMR shifts attributed to formation of complexes were
observed. Neither titanium crosslinker solution caused an
observable ^{13}C NMR shift with the model sugars(262). Still it is
well known that these crosslinkers cause a dramatic change in the
properties of polysaccharide base gels. It was concluded(262,323)
that the vast bulk of both of the organotitanate crosslinkers
rapidly formed unreactive, collodial TiO_2. It was reasoned that
this left only a few reactive TiOH sites exposed on the outer
surface of the colloidial TiO_2 spheres. These sites were
insufficient in number to significantly interact with the model
sugars to cause an observable ^{13}C NMR shift; below the level of
detection of the instrument, but are effective enough to crosslink
polysaccharide polymers. Kramer and Prud'homme concluded(262) that
crosslinking of polysaccharide gels by addition of organotitanate
complexes is by TiO_2 sols, not by a monomeric Ti(IV) cation.
     Dynamic light scattering (DLS) experiments(327) with water aged
titanates indicated several important features of these colloidial
TiO_2 sols; A) their size is important to their crosslinking
behavior, B) their size is regulated by the amount of water added
and the length of time aged after mixing with water, C) the
colloidial TiO_2 particles are in equilibrium with the
organo-titanate precursor, and D) pH affects the equilibrium.
Because pH, ionic strength, time-of-aging, and water content are
widely variable parameters in a field preparation of polysaccharide

fracturing gels it is not surprising that the final fracturing gel
properties can sometimes be difficult to control(268,333,334).
Buffering agents are commonly added to regulate pH and 2% KCl is
ordinarily added to prevent clay swelling but its presence also
provides a more uniform ionic strength. Unfortunately the addition
of KCl cannot overcome the undesired effects exerted on crosslinking
chemistry by the natural contaminants of many field waters. The
effectiveness of buffering agents is more than a pH effect alone.
Factors that can affect crosslinking rates and fracturing gel
performance include surface temperature, polymer concentration, salt
concentrations, fluid loss additives, surfactants, mutual solvents,
alcohols, and also the anionic identity of buffering agents. For
this reason, quality control procedures have been established and
are routinely followed to help ensure optimum fracturing gel
performance(335).

Zasadzinski, Chu, and Prud'homme, by use of transmission
electron microscopy (TEM) of freeze-fractured samples of gels,
observed ionic strength effects on the microscopic physical gel
structure(336). Increased salt content resulted in a higher degree
of association of polymer molecules, condensation occured, and the
polymer chains appeared much thicker. The presence of salt is
believed to strengthen polymer hydrogen bonding at the expense of
hydroxyl group water of hydration. The effect makes the polymer
more hydrophobic and likely accounts for their observation that the
presence of salt made the guar gels more susceptible to mechanical
shear degradation. Shearing resulted in the appearance of ruptures
in the crosslinked gel network along weak zones. The gel network
bounded inside by the rupture lines appeared unaffected. The TEM
procedure also showed that the colloidial TiO_2 crosslinks appeared
to weld together polymer strands already associated by hydrogen
bonding.

Kramer, et al.(262,327) and Knoll(233), on the basis of G', G''
response concluded that gels kept in continuous motion during the
crosslink formation period have completely different 3-D networks
than gels allowed to stand in a quiescent state prior to analysis on
the rheometer. This observation confirmed the early comments of
Conway(317) and has had important implications for fracturing gel
research and modeling since crosslinked fracturing gels as applied
in the field ordinarily do not experience a quiescent period, yet
nearly all test data accumulated prior to 1986 was collected from
samples that had experienced at least a momentary quiescent period.

Emulsions. Emulsion fluids and foams came into routine use in
competition with crosslinked fluids during 1970-80. Simple, barely
stable emulsions had been used early in fracturing. These were
mainly emulsified acids that "broke" when the acid spent on the
formation surfaces. In the late 1960's Kiel became a proponent of
very high viscosity oil fluids as a method to place exceptional (at
the time) amounts of proppant(337,338). To avoid the frictional
resistance typical of gelled oils he advanced the concept of
preparing a very viscous oil-external emulsion with one part fresh
water, 0.1% sodium tallate surfactant, and two parts oil. The
viscous emulsion had to be pumped simultaneously with a water stream
to minimize frictional pressure. This process was clumsy and still

created frictional problems. A better emulsion fluid using a
quaternary amine surfactant and brine water was developed. Also the
inclusion of a polymer in the aqueous phase caused a significant
increase in emulsion viscosity(339-341). These emulsions reportedly
had very low fluid loss properties, and the field data published
suggested that wells treated with the emulsion cleaned up with
greater production than from wells treated with other contemporary
fluids. A variety(342) of emulsion fluids (usually 70% water)
continue in use today(338) and are preferred by some on the basis of
causing minimal damage to conductivity(343-346). Even so, emulsion
treatments usually are more costly than crosslinked aqueous
fracturing fluids.

Foamed Frac Fluids. Foams have properties very similar to
emulsions. Both fluids consist of a dispersed phase carried inside
a continuous phase(347). In foams the dispersed phase is a gas.
Emulsions are liquid-in-liquid dispersions. Both types of
dispersions provide large increases in viscosity when higher
percentages of dispersed phase are introduced. This viscosity is
useful to create fracture width and suspend proppant. Figure 10
illustrates the viscosity behavior of a typical foam. Operationally
foams and emulsions are ordinarily applied with 55-90% internal
dispersed phase, where the enhanced viscosity properties are
unmistakeable. The placement of a pressurized gas into a
hydrocarbon bearing formation offers a significant advantage over
other fracturing fluids. When pressure is released on the surface the
gas expands and rises, carrying liquids, fines, and debris with it.
Foams generally are excellent well clean-out fluids. In some
instances, foam fluids cause much less conductivity damage to
fracture channels than other stimulation fluids. Several recent
articles (which include useful bibliography references) on oil-field
foam applications are available(348-352).
     Chemically, the preparation of a "stable" foam or emulsion
requires the use of a surfactant to aid in dispersion of the
internal phase and prevent the collapse of the foam (or emulsion)
into separate bulk phases. The selection of a surfactant is made on
the basis of severity of conditions to be encountered, the gas to be
entrained (N_2, CO_2, LPG, CH_4, or air), the continuous phase liquid
(water, alcohol, or oil), and half-life of foam stability desired.
Foam viscosity and stability can be enhanced viscosifying the
continuous phase with thickening agents. These are mostly the same
thickening agents used to prepare viscous fracturing base gels.
Public information about the specific chemical identity of the
surfactants and stabilizers in use is scant(353-355) (Figure 11).
Performance of foamed fluids is heavily dependent upon the size and
distribution of the individual foam cells that are present,
therefore the generator, testing apparatus, pressure and procedures
employed are critical parts of the evaluation and the observed
results. Contaminants (salts, acids, alkalies, etc) in the liquid
phase also can cause drastic changes in foam performance.
     Aqueous fracture foams are the least expensive not only because
of the low cost of water but also due to the availability of
inexpensive surfactants that foam water and brines easily and
effectively. Oil and alcohol foams are much more costly due to the

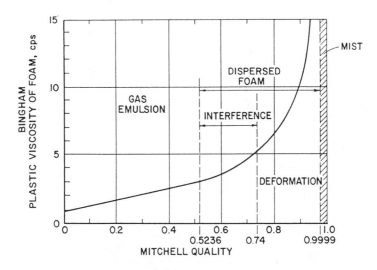

Figure 10. Foam viscosity depends on percent internally dispersed gas (Mitchell quality). (Reproduced from ref. 363. U.S. Patent 3 937 283, 1976.)

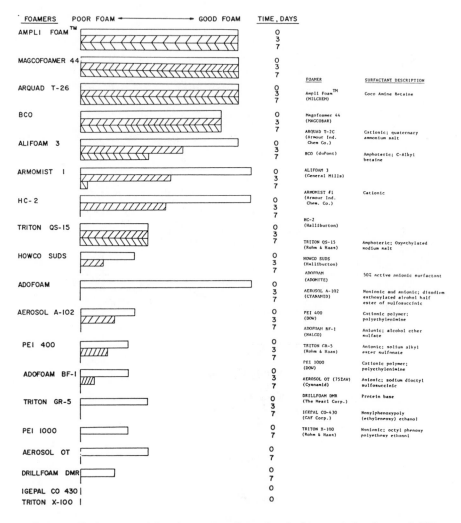

Figure 11. Performance of foaming agents. (Reproduced with permission from ref. 353. Copyright 1978 Society of Petroleum Engineers.)

expense of the hydrocarbon or alcohol and the requirement for high
performance (and expensive) surfactants. Fluorinated surfactants
are normally required(356-359). These materials are noted for their
ability to drastically lower the surface tension between phases
which makes them effective foamers for oil or alcohol.

Summary

   This chapter has presented an overview of some major areas of
fracturing chemistry from a historical, functional, mechanistic, as
well as bibliographical viewpoint. Space constraints dictated that
other very important fracturing chemistry topics be only lightly
addressed. Notably absent is the chemistry of numerous surface
active fracturing additives. Surface active chemicals play major
roles in antifoaming agents, clay stabilization, de-mulsifying
additives, dispersing agents, emulsifying agents, foaming agents,
sludge inhibitors, and water-blockage control additives(5,360,361).
Some of these have been discussed but many were not. A limited
bibliography for de-mulsifying (non-emulsifiers) agents, sludge
inhibitors, and water blockage control chemicals is provided in the
Appendix(361).

Acknowledgment
I wish to thank Halliburton Services for the opportunity to assemble
this manuscript and S. K. McCulloch; only with her assistance was
the project completed.

Literature Cited

1.   See Appendix Reviews of Oil & Gas Fracturing
2.   Veatch, R.W., Jr. J. Pet. Technol. 1983, 35(4) 677-864; ibid
     Part 2 1983, 35(5) 853-864.
3.   Cramer, D.D. Oil Gas J. 1987, 85(50), 40-45.
4.   See Appendix Fracturing Fluid Leakoff
5.   Anderson, W.G.: "Wettability Literature Survey;" Part 1, J.
     Pet. Technol.1986, 38(10), 1125-1144; ibid, Part 2, 38(11),
     1246-1262; ibid, Part 3, 38(12), 1371-1378; ibid, Part 4, 1987,
     39(10), 1283-1300; ibid, Part 5, 39(11), 1453-1468; ibid, Part
     6, 39(12), 1605-1622.
6.   Rogers, L.A.; Tomson, M.B.; Matty, J.M.; Durrett, L.R. Oil Gas
     J. 1985, 83(13), 97-108.
7.   Howard, G.C.; Fast, C.R. Hydraulic Fracturing, Monograph
     Series, Society of Petroleum Engineers of the American
     Institute of Mining, Metallurgical and Petroleum Engineers:
     Dallas, TX., 1970, 2.
8.   Dickey, P.A.; Andersen, K.H. "The Behavior of Water-Input
     Wells;" Amer. Pet. Institute, Secondary Recovery of Oil in the
     United States, 2nd Ed. 1950, 332-40.
9.   Clark, J.B.; Fast, C.R.; Howard, G.C. "A Multiple Fracturing
     Process for Increasing the Productivity of Wells," 1952 Spring
     Meeting of the Midcontinent District API, Wichita, KS, March
     19-21.
10. Scott, P.P., Jr.; Bearden, W.G.; Howard, G.C. "Rock Rupture as
     Affected by Fluid Properties;" Soc. Pet. Eng. J. 1953, 111-124;
     Trans., AIME, 198.

11. Hubbert, M.K.; Willis, D.G.: "Mechanics of Hydraulic
    Fracturing;" Soc. Pet. Eng. J. 1957, 153,166,167; Trans., AIME,
    210.
12. See Appendix Modelling of Fracture Behavior
13. Conway, M.W.; McMechan, D.E.; McGowen, J.M.; Brown, D.;
    Chisholm, P.T.; Venditto, J.J. "Expanding Recoverable Reserves
    Through Refracturing"; SPE paper 14376, 1985 SPE Annual
    Technical Conference and Exhibition, Las Vegas, September
    22-25.
14. Venditto, J.J.; Chisholm, P.; Wiggins, M.; Conway, M.
    "Refracturing Completed Wells Can Be Cost Effective", PETROLEUM
    ENGINEER International August 1986, 26-34.
15. Kaspereit, D.H. "Fracture Design for Suspended Sand Packs"; SPE
    paper 1843, 1979 SPE Annual Technical Conference and
    Exhibition, Las Vegas, September 23-26.
16. Smith, J.E. "Effect of Incomplete Fracture Fill-Up at the
    Wellbore on Productivity Ratio"; Proc. Annu. Southwest. Pet.
    Short Course, April 1975, 135-144.
17. Clark, J.B. US Patent 2 596 844, 1952.
18. Clark, J.B. US Patent 2 596 845, 1952.
19. Farris, R.F. US Patent 23 733, reissued 1953.
20. Prusick, J.H.; Morgan, Z.V "The Use of Emulsions and Related
    Techniques in the Treatment of Oil and Gas Wells"; Pet.
    Engineer (May 1954) B54-B56.
21. Fieser, L.F.; Harris, G.C.; Hershberg, E.B.; Morgana, M.;
    Novello, F.C.; Putnam, S.T. Ind. Eng. Chem. 1946, 38(8),
    768-773.
22. Fieser, L.F. US Patent 2 606 107, 1952.
23. Kirk-Othmer Encyclopedia of Chemical Technology; Wiley:New
    York, 1978, 3rd Ed, Vol 2, 202 "Aluminum Carboxylates" and Vol
    8, 34 "Driers and Metallic Soaps".
24. Self, E.S. US Patent 2 461 483, 1949.
25. Dawson, R.D. US Patent 2 350 154, 1944.
26. Anderson, F.M. US Patent 2 430 039, 1947.
27. Magram, S.J. US Patent 2 774 740, 1956.
28. Monroe, R.F.; Rooker, B.E. US Patent 3 494 949, 1970.
29. Monroe, R.F. US Patent 3 505 374, 1970.
30. Monroe, R.F. US Patent 3 575 859, 1971.
31. Ryles, R.G. "Elevated Temperature Testing of Mobility Control
    Reagents", SPE paper 12008, 1983 SPE Annual Technical
    Conference and Exhibition, San Francisco, CA, October 5-8.
32. Arnson, T.R. Tappi, 1982, 65(3), 125-129.
33. Lauzon, R.V. Oil Gas J. 1982, 80(36), 124,126.
34. Schecher, W.D.; Driscoll, C.T. Water Resources Res., 1987,
    23(4), 525-534.
35. Baes, C.F. Jr.; Mesmer, R.E. "The Hydrolysis of Cations",
    Chaper on Boron, Aluminum, Scandium; Wiley:New York, 1976.
36. Crawford, D.L.; Earl, R.B.; Monroe, R.F. US Patent 3 757 864,
    1973.
37. Zangen, M.; Marcus, Y.; Bergmann, E.D. Sep. Sci. 1967 2(2),
    187-197.
38. Moule, H.A.; Greenfield, S. J. Chromatog. 1963, 11, 77-83.
39. Hill, D.G. US Patent 3 990 978, 1976.

40. McKenzie, L.F. "Hydrocarbon Gels of Alumino Alkyl Acid Orthophosphates", SPE paper 9007, 1980 Fifth International Symposium on Oilfield and Geothermal Chemistry, Standford May 28-30.
41. Daccord, G.; Lemanczyk, R.; Vercaemer, C. US Patent 4 507 213, 1985.
42. See Appendix Chemistry of Gelled Oils
43. Baker, H.R.; Bolster, R.N.; Leach, P.B.; Little, R.C. Ind.Eng.Chem.Prod.Res.Develop. 1970, 9(4), 541-548.
44. Rose, S.H.; Black, B.P. J. Am. Chem. Soc. 1965, 87(9), 2076-77.
45. Rose, S.H.; Black, B.P. J. Polym. Sci., 1966 4(A-1), 573.
46. Rose, S.H.; Black, B.P. J. Polym. Sci., 1966 4(A-2), 583.
47. Schmidt, D.L.; Flagg, E.E. J. Polym. Sci., 1968, 6(A-1), 3235-3244.
48. Burnham J.W.; Harris, L.E.; McDaniel, B.W. "Developments in Hydrocarbon Fluids for High Temperature Fracturing", SPE paper 7564, 1978 SPE Annual Technical Conference and Exhibition, Houston, October 1-3. See also J. Pet. Technol., 1980, 32(2), 217-220.
49. Fendler, J.H. Acc. Chem. Res., 1976, 9, 153-161.
50. Fendler, J.H. Acc. Chem. Res., 1980, 13, 7-13.
51. Fendler, J.H. J. Phys. Chem., 1980, 84, 1485-1491.
52. Hewitt, C.H. J. Pet. Technol., 1963,15(8), 813-818.
53. Gray, D.H.; Rex, R.W. "Formation Damage in Sandstones Caused by Clay Dispersion and Migration", 1966 Fourteenth National Conference on Clays and Clay Minerals, 355-366.
54. See Appendix Acidizing, Fracture Acidizing
55. Monaghan, P.H.; Salathiel, R.E.; Morgan, B.E. J. Pet. Technol. 1959, 11, 209-215.
56. Reed, M.G. Production Monthly, 1968, 32, 18.
57. Jones, F.O. Jr. J. Pet. Technol., 1964, 16(4), 441-446.
58. Black, H.N.; Hower, W.F. "Advantageious Use of Potassium Chloride Water for Fracturing Water Sensitive Formations", 1965 Spring Meeting of the Mid-Continent District Division of Production, Wichita, KS, March 31-April 2.
59. Reed, M.G. "Stabilization of Formation Clays with Hydroxy-Aluminum Solutions", SPE paper 3694, 1981 SPE Annual California Regional Meeting, Los Angeles, November 4-5.
60. Reed, M.G.; Coppel, C.P. "Sand Stabilization with Hydroxy Aluminum Solutions," SPE Paper 4168, 1972 SPE Annual California Regional Meeting, Bakersfield, November 8-10.
61. Coppel, C.P.; Jennings, H.Y. Jr.; Reed, M.G. "Field Results from Wells Treated with Hydroxy-Aluminum", SPE paper 3998, 1972 SPE Annual Meeting, San Antonio, October 8-11.
62. Haskin, C.A.; Reed, M.G.; Coppel, C.P. Oil Gas J. 1975, 73(45), 217-220.
63. Haskin, C.A. "A Review of Hydroxy-Aluminum Treatments"; SPE Paper 5692 1976 SPE Symposium for Formation Damage Control Meeting of AIME, Houston, January 29-30.
64. Veley, C.D. "How Hydrolyzable Metal Ions Stabilize Clays to Prevent Permeability Reduction"; SPE paper 2188, 1968 SPE Annual Meeting of AIME, Houston, September 29-October 2.
65. Christenson, R.M.; McBane, B.N. US Patent 3 382 924, 1968.

66. Veley, C.; Casey, G., Jr. "In-Place Clay Stabilization Cuts Formation Sand Entry"; World Oil(June 1973)52-54.
67. Peters, F.W.; Stout, C.M. J. Pet. Technol. 1977, 29(2), 187-194.
68. Brown, W.E. US Patent 2 761 835, 1956.
69. ibid. 2 761 836.
70. ibid. 2 761 837.
71. ibid. 2 761 838.
72. ibid. 2 761 839.
73. ibid. 2 761 840.
74. ibid. 2 761 841.
75. ibid. 2 761 842.
76. McLaughlin, H.C.; Elphingstone, E.A.; Hall, B. "Aqueous Polymers for Treating Clays in Oil and Gas producing Formations"; SPE paper 6008, 1976 SPE Annual Technical Conference and Exhibition, New Orleans, October 3-6.
77. McLaughlin, H.O., Sr.; Elphingstone, E.A.; Remington, R.E., II; Coates, S. "Clay Stabilizing Agent Can Correct Formation Damage," World Oil(May 1977) 58-60.
78. Black, H.N.; Ripley, H.E.; Beecroft, W.H.; Pamplin, L.O. "Drilling and Fracturing Fluid Improvements for Low-Permeability Gas Wells in Canada"; SPE paper 7926, 1979 SPE Symposium Low Permeability Gas Reservoirs, Denver, May 20-22.
79. Young, B.M.; McLaughlin, H.C.; and Borchardt, J.K. J. Pet. Technol., 1980, 32(12), 2121-2130.
80. Hall, B.E. Oil Gas J. 1978, 76(30), 68-70.
81. Borchardt, J.K.; Roll, D.L.; Rayne, L.M. "Use of a Mineral Fines Stabilizer in Well Completions"; SPE paper 12757, 1984 SPE California Regional Meeting, Long Beach, April 11-13.
82. McLaughlin, H.C.; Weaver, J.D. US Patent 4 366 071, 1982.
83. ibid. 4 366 072.
84. ibid. 4 366 073.
85. McLaughlin, H.C.; Weaver, J.D. US Patent 4 374 739, 1983.
86. Theng, B.K.G.: "Clay-Polymer Interactions: Summary and Perspectives", Clays and Clay Minerals, 1982, 30(1), 1-10.
87. ibid. "The Chemistry of Clay-Organic Reactions", Wiley:New York, (1974).
88. Callaway, R.E. "Clay Protection Chemicals: A Practical Evaluation and Application Technique for Their Use in Stimulation Fluids", SPE paper 10663, 1982 SPE Formation Damage Control Symposium, Lafayette, March 24-25.
89. Coulter, G.R.; Hower, W. "The Effect of Fluid pH on Clays and Resulting Formation Permeability", Proc. Annu. Southwest Pet. Short Course, April 1975.
90. Simon, D.E.; McDaniel, B.W.; Coon, R.M. "Evaluation of Fluid pH Effects on Low Permeability Sandstones"; SPE paper 6010, 1976 SPE Annual Technical Conference and Exhibition, New Orleans, October 3-6.
91. Sydansk, R.D. "Stabilizing Clays With Potassium Hydroxide", SPE paper 11721, 1983 California Regional Meeting, Ventura, March 23-25.
92. Coulter, A.W., Jr.; Frick, E.K.; Samuelson, M.L. "Effect of Fracturing Fluid pH on Formation Permeability", SPE paper 12150, 1983 SPE Annual Technical Conference and Exhibition, San Francisco, October 5-8.

93. Kia, S.F.; Fogler, H.S.; Reed, M.G. J. Colloid Interface Sci., 1987, 118(1), 158-168.
94. Burk, J.H. "Comparison of Sodium Carbonate, Sodium Hydroxide, and Sodium Orthosilicate for EOR", SPE Reservoir Engineering, February 1987, 10-15.
95. Radenti, G.; Palumbo, S.; Zucca, G. "Potassium Carbonate Fluid Inhibits Highly Reactive Clays", PETROLEUM ENGINEER International, September 1987, 32,34,36,40.
96. Thomas, R.L.; Crowe, C.W. "New Chemical Treatment Provides Stimulation and Clay Control in Sandstone Formations," SPE Paper 7012, 1978 SPE Third Symposium on Fomation Damage Control, Lafayette, February 15-16.
97. Clark, R.K. "Polymer Effectiveness in Stabilizing Shales of Various Compositions," Polym. Mater. Sci. Eng. 1984, 51, 1-5.
98. Hawkins, R.R. US Patent 3 122 203, 1964.
99. Chesser, B.G. "Design Considerations for an Inhibitive, Stable Water-Based Mud System", SPE Drilling Engineering, December 1987, 331-336.
100. Wilcox, R.D.; Jarrett, M.A. "Polymer Deflocculants: Chemistry and Application", IADC/SPE paper 17201, 1988 IADC/SPE Drilling Conference, Dallas, February 28-March 2.
101. Krueger, R.F. J. Pet. Technol., 1986, 38(2), 131-152.
102. Leone, J.A.; Scott, E.M. "Characterization and Control of Formation Damage During Waterflooding of a High-Clay-Content Reservoir", SPE paper 16234, 1987 SPE Production Operations Symposium, Oklahoma City, March 8-10.
103. Anderson, R.W.; Baker, J.R. "Use of Guar Gum and Synthetic Cellulose in Oilfield Stimulation Fluids", SPE paper 5005, 1974 SPE Annual Meeting, Houston, October 6-9.
104. Davis, J.A., Jr.; Rhudy, J.S. US Patent 3 747 681, 1973.
105. Whistler, R.L.; BeMiller, J.N., Eds "Industrial Gums; Polysaccharides and Their Derivatives", 2nd Ed, Academic, New York, 1973.
106. Rowell, R.M.; Young, R.A., Eds, "Modified Cellulosics"; Academic, New York, 1978.
107. Davidson, R.L., Ed, "Handbook of Water-Soluble Gums and Resins," McGraw-Hill, Inc, 1980.
108. Carraher, C.E.; Moore, J., Eds, "Modification of Polymers"; Plenum Press, New York, 1983.
109. Glass, J.E. "Water-Soluble Polymers: Beauty with Performance", Advances in Chemistry Series 213, Amer. Chem. Soc., Washington, D.C., 1986.
110. See Appendix Friction and Drag Reducers
111. Malachosky, E.: "Status Report: Oil Field Copolymers," PETROLEUM ENGINEER, International, April 1987, 48-53.
112. Barth, H.G.; Smith, D.A. J. Chromatog., 1981, 205, 410-415.
113. Robinson, G.; Ross-Murphy, S.B.; Morris, E.R. Carbohydr. Res., 1982, 107, 17-31.
114. Jordan, W.A. US Patent 3 483 121, 1969.
115. White, J.L.; Means, J.O. J. Pet. Technol. 1975, 27(9), 1067-1073.
116. Holcomb, D.L.; Smith, M.O. "The Use of a Low-Concentration Crosslinked Hydroxyalkyl Polymer System as a Highly Efficient

Fracturing Fluid", 1975 Proc. Annu. Southwest. Pet. Short
Course, April, 129-134.

117. Githens, C.J.; Burnham, J.W. Society of Petroleum Engineers
Journal(February 1977) 5-10.

118. Nordgren, R.; Jones, D.A.; Wittcoff, H.A. US Patent 3 723 408,
1973.

119. Anderson, J.L.; Paktinat, J. US Patent 4 635 727, 1987.

120. Sandy, J.M.; Wiggins, M.; Venditto, J.J. Oil Gas J., 1986
84(35), 52-54.

121. Chatterji, J.; Borchardt, J.K. J. Pet. Technol., 1980, 32(11),
2042-2056.

122. Moe, O.A. US Patent 2 520 161, 1950.

123. Nordgren, R. US Patent 3 225 028, 1965.

124. Nordgren, R. US Patent 3 303 184, 1967.

125. Nordgren, R. US Patent 3 346 555, 1967.

126. Tessler, M.M. US Patent 4 276 414, 1981.

127. DeMartino, R.N.; Wayne, N.J. US Patent 4 169 798, 1979.

128. Harms, S.D.; Goss, M.L.; Payne, K.L.: "New Generation
Fracturing Fluid for Ultrahigh-Temperature Application", SPE
paper 12484, 1984 SPE Formation Damage Control Symposium,
Bakersfield, February 13-14.

129. Payne, K.L. US Patent 4 579 670, 1986.

130. Chrinsenson, R.M. US Patent 2 618 632, 1968.

131. Thomas, D.C.: "Thermal Stability of Starch and Carboxymethyl
Cellulose Polymers Used in Drilling Fluids," SPE Paper 8463,
1979 SPE Annual Technical Conference and Exhibition, Las Vegas,
September 23-26.

132. Patton, J.T. US Patent 3 729 460, 1970.

133. Southwick, J.G.;Lee, H.; Jamieson, A.M.; Blackwell, J.
Carbohydr. Res., 1980 84, 287-295.

134. Salamone, J.C.; Clough, S.B.; Salamone, A.B.; Reid, K.I.G.;
Jamison, D.E. Soc. Pet. Eng. J, 1982, 22(8), 555-56.

135. Rees, D.A. Biochem. J., 1972, 126(1), No. 2, 257-273.

136. Clark, P.E.; Halvaci, M.; Ghaeli, H. "Proppant Transport by
Xanthan and Xanthan-Hydroxypropyl Guar Solutions: Alternatives
to Crosslinked Fluids," SPE/DOE paper 13907, 1985 SPE/DOE Low
Permeability Gas Reservoirs, Denver, May 19-22.

137. Chauveteau, G.; Kohler, N. "Influence of Microgels in Xanthan
Polysaccharide Solutions on Their Flow Through Various Porous
Media," SPE paper 9295, 1980 SPE Annual Technical Conference
and Exhibition, Dallas, September 21-24.

138. See Appendix Xanthan Gum

139. Carico, R.D.; Bagshaw, F.R. "Description and Use of Polymers
Used in Drilling, Workovers, and Completions," SPE paper 7747,
1978 SPE of AIME Production Technology Symposium, Hobbs,
October 30-31.

140. Carico, R.D. "Suspension Properties of Polymer Fluids Used in
Drilling, Workover, and Completion Operations," SPE paper 5870,
1976 SPE of AIME Annual California Regional Meeting, Long
Beach, April 8-9.

141. Gall, B.L.; Maloney, D.R.; Raible, C.J. "Permeability Damage to
Artificially Fractured Cores," NIPER- Final Report, National
Institute for Petroleum and Energy Research, Bartlesville, OK,
May 1988.

142. See Appendix <u>Frac Gel Residue and Formation Damage</u>
143. Kramer, J.; Prud'homme, R.K.; Norman, L.R.; Sandy, J.M.
     "Characteristics of Metal-Polymer Interaction in Fracturing
     Fluid Systems," SPE paper 61914, 1987 SPE Annual Technical
     Conference and Exhibition, Dallas, September 27–30.
144. Angyal, S.J.; Greeres, D.; Mills, J.A. <u>Aust. J. Chem.</u>, 1974 <u>27</u>,
     1447–56.
145. Angyal, S.J. <u>Aust. J. Chem.</u>, 1972, <u>25</u>, 1957–66.
146. Engelskirchen, K.; Galinke, J. US Patent 4 001 210, 1977.
147. Engelskirchen, K.; Galinke, J. US Patent 4 013 821, 1977.
148. Reid, A.R. US Patent 4 096 326, 1978.
149. Lukach, C.; Majewicz, G.; Reid, A.R. US Patent 4 523 010, 1985.
150. Almond, S.W.; Himes, R.E. US Patent 4 552 215, 1985.
151. Almond, S.W.; Conway, M.W. US Patent 4 553 601, 1985.
152. Brode, G.L.; Stanley, J.P.; Partain, E.M. III. US Patent 4 579
     942, 1984.
153. White, J.L.; Rosene, R.B.; Hendrickson, A.R.: "New Generation
     of Frac Fluids", 1973 SPE Annual Meeting, Edmonton, May 8–12.
154. Majewicz, T.G. US Patent 4 400 502, 1983.
155. Goldstein, A.M. US Patent 2 868 664, 1959.
156. Shelso, G.J.; Seaman, B. US Patent 3 808 195, 1974.
157. Engelskirchen, K.; Galinke, J. US Patent 3 350 386, 1967.
158. Boudreaux, J.R. US Patent 3 475 334, 1969.
159. Bishop, R.G.; Desmarais, A.J.; Reid, A.R. US Patent 3 455 714,
     1969.
160. Whelan, K. US Patent 3 503 895, 1970.
161. McClain, D.M. US Patent 4 148 766, 1979.
162. Majewicz, T.G. US Patent 4 400 502, 1983.
163. Almond, S.W. US Patent 4 487 866, 1984.
164. Anderson, D.R.; Frisque, A.J. US Patent Re. 28 474, 1974.
165. Briscoe, J.E. US Patent 4 336 145, 1982.
166. ibid.: US Patent No. 4,466,890(1984).
167. Harris, P.C.; Harms, W.M.; Norman, L.R.: "Study of Continuously
     Mixed Crosslinked Fracturing Fluids With a Recirculating Flow
     Loop Viscometer", SPE paper 17044, 1987 SPE Eastern Regional
     Meeting, Pittsburgh, October 21–23.
168. Harms, W.M.; Yeager, R. <u>Oil Gas J.</u>, 1987, <u>85</u>(44), 37–39.
169. Harms, W.M.; Watts, M.; Venditto, J.J.; Chisholm, P.
     "Diesel-Based HPG Concentrate is Product of Evolution",
     <u>PETROLEUM ENGINEER International</u>(April 1988), 51,53.
170. Yeager, R.R.; Bailey, D.E. "Diesel-Based Gel Concentrate
     Improves Rocky Mountain Region Fracture Treatments", SPE paper
     17535, 1988 SPE Rocky Mountain Regional Meeting, Casper, May
     11–13.
171. Constien, V.G.; Brannon, H.D.; Bannister, C.E. <u>Oil Gas J.</u>,
     1988, <u>86</u>(23), 49–54.
172. Brown, E.; Hoover, M. "An Improved System for Making
     Predictable, High Quality Fracturing Fluids Under Field
     Conditions", 1988 Proc. Annu. Southwest. Pet. Short Course,
     Lubbock, 49–61.
173. Brannon, H.D.: "Fracturing FLuid Slurry Concentrate and Method
     of Use," Europ. Pat. Appl., 0 280 341 A1, 1988.
174. Fuller, G.; Toombs, A.J.L. US Patent 3 779 723, 1973.

175. Pickens, P.A.; Lindroth, T.A.; Carico, R.D. US Patent 4 312 675, 1982.
176. Dawans, F.; Binet, D.; Kohler, N.; Vu, Q.D. US Patent 4 393 151, 1983.
177. Hoover, L.D. US Patent 4 330 414, 1982.
178. House, R.F. US Patent 4 435 217, 1984.
179. Hatfield, J.C. US Patent 4 566 977, 1986.
180. Pelezo, J.A.; Corbett, G.E., Jr.; Siems, D.R. US Patent 4 615 740, 1986.
181. Burkhalter, J.F.; Weigand, W.A. US Patent 4 687 516, 1987.
182. Watson, K.E.; Sharp, K.W. US Patent 4 622 153, 1986.
183. Dymond, B; Langley, J; Howe, M. US Patent 4 670 501, 1987.
184. Harms, W.M.; Norman, L.R. US Patent 4 772 646, 1988.
185. Jordan, J.W. J. Phys. Chem. 1949 53, 294-306.
186. Jordan, J.W.; Hook, B.J.; Finlayson, C.M. J. Phys. Chem. 1950, 54 1196-1208.
187. Jordan, J.W. US Patent 2 966 506, 1960.
188. Jordan, J.W.; Nevins, M.H.; Stearns, R.O.; Cowan, J.C.; Beasley, A.E., Jr. US Patent 3 168 475, 1965.
189. Hauser, E.A. US Patent 2 531 427, 1950.
190. Hauser, E.A. US Patent 2 531 812, 1950.
191. Schmidt, D.D.; Roos, A.F.; Cline,J.T., "Interaction of Water with Organophylic Clay in Base Oils to Build Viscosity," SPE Paper 16683, 1987 SPE Annual Technical Conference and Exhibition, Dallas, 311-326.
192. Finlayson, C.M.; Jordan, J.W. US Patent 4 105 578, 1978.
193. Finlayson, C.M. US Patent 4 081 496, 1978.
194. Goodman, H. US Patent 4 631 019, 1986.
195. Knudson, M.I., Jr.; Jones, T.R. US Patent 4 664 842, 1987.
196. Pye, D.S.; Smith, W.A.: "Fluid Loss Additive Seriously Reduces Fracture Proppant Conductivity and Formation Permeability," 1973 SPE Annual Meeting, Las Vegas, Sept. 30-Oct. 3.
197. Ely, J.W.; Chatterji, J.; Holtmyer, M.D.; Tinsley, J.M. US Patent 3 768 566, 1973.
198. Horton, R.L.: "Fracturing Fluids for High-Temperature Reservoirs", Drilling(December 1982) 72-78.
199. Seidel, W.R.; Stahl, E.J. Jr. J. Pet. Technol. 1972, 24(11), 1385-1390.
200. Hsu, C.H.; Conway, M.W. J. Pet. Technol. 1981, 33(11), 2213-2218.
201. Holditch, S.A.; Ely, J. "Successful Deep Well Stimulation Utilizing High Proppant Concentration", SPE paper 4118, 1972 SPE Annual Meeting, San Antonio, October 8-11.
202. Ely, J.W.; Chatterji, J.; Holtmyer, M.D.; Tinsley, J.M. US Patent 3 898 165, 1975.
203. Ely, J.W.,; Tinsley, J.M US Patent 4 210 206, 1980.
204. See Appendix Crosslinked Gels from Other Industries
205. Holtmyer, M.D.; Githens, C.J. "Field Performance of a New High-Viscosity Water Base Fracturing Fluid", SPE paper 875-24-E, 1970 Spring Meeting of the Rocky Mountain District Division of Production, Denver, April 27-29.
206. Holtmyer, M.D.; Githens, C.J.; Tinsley, J.M. US Patent 4 021 355, 1977.

207. Holtmyer, M.D.; Githens, C.J.; Tinsley, J.M. US Patent 4 033 415, 1977.
208. Harris, W.F., Jr. US Patent 4 568 481, 1986.
209. Chakrabarti, S.; Guillot, D.; Rondelez, F.: "Gelation of Hydroxypropylguar Under Simple Shear", Polymer Preprints 27(1), 247(April, 1986).
210. Alderman, E.N. "Super Thick Fluids Provide New Answers to Old Fracturing Problems"; SPE paper 2852, 1970 SPE Spring Symposium, Fort Worth, March 8-10.
211. Dysart, G.R.; Richardson, D.W.; Kannenberg, B.G. "Second Generation Fracturing Fluids," paper 906-15-H, 1970 Spring Meeting of the Southwestern District Division of Production, Odessa, March 18-20.
212. Holcomb, D.L.; Smith, M.O. "The Use of a Low-Concentration Crosslinked Hydroxyalkyl Polymer System as a Highly Efficient Fracturing Fluid," 1975 Proc. Annu. Southwest. Pet. Short Course, April, 129-134.
213. Free, D.L.; Frederick, A.F.; Thompson, J.E.: "Fracturing with a High-Viscosity, Crosslinked Gel- Continuous Fracturing Technique", J. Pet. Technol. 1978, 30(1), 119-122.
214. Kern, L.R. US Patent 3 058 909, 1962.
215. Boeseken, J., Ber., 46,(1913)2612; Bl.,(4),53(1933)1332.
216. Deuel, von H. and Neukom, H.: "Uber die Reaktion von Borsaure und Borax mit Polysacchariden und anderen hochmolekularen Polyoxy-Verbindungen", Nature 1948, 161, 96.
217. Roy, G.L.; Laferriere, A.L.; Edwards, J.O. J. Inorg. Nucl. Chem. 1957, 4, 106-114.
218. Slate, R.L. US Patent 3 096 284, 1963.
219. McIrvine, J.D. US Patent 3 108 917, 1963.
220. Chrisp, J.D. US Patent 3 301 723, 1967.
221. Conner, J.M.; Bulgrin, V.C. J. Inorg. Nucl. Chem. 1967, 29, 1953-1961.
222. Savins, J.G. Rheol. Acta, 1968, 7(1), 87-93.
223. Bourgoin, D. J. Chemie. Phys., 1963, 59, 923.
224. Schultz, R.K.; Myers, R.R. "The Chemorheology of Poly(vinyl Alcohol)-Borate Gels", Macromolecules 1969, 2.
225. Gorin, P.A.J.; Mazurek, M. Can. J. Chem. 1973, 51, 3277-3286.
226. Gorin, P.A.J.; Mazurek, M. Carbohydr. Res. 1973 27, 325-339.
227. Nickerson, R.F. J. Polym Sci. 1971, 15, 111-116.
228. Ochiai, H.; Fujino, Y.; Tadakoro, Y.; Murakami, I. Polymer J. 1982, 14(5), 423-426.
229. Ochiai, H.; Kurita, Y.; Murakami, I. Makromol. Chem. 1984, 185, 167-172.
230. Moore, R.E.; Barchi, J.J., Jr.; Bartolini, G. J. Org. Chem. 1985, 50, 374-379.
231. Maerker, J.M.; Sinton, S.W. J. Rheol. 1986, 30(1), 77-99.
232. Sinton, S.W. Macromolecules, 1987, 20, 2430-2441.
233. Knoll, S.K.; Prud'homme, R.K.: "Interpretation of Dynamic Oscillatory Measurements for Characterization of Well Completion Fluids", SPE paper 16283, 1987 SPE International Symposium on Oilfield Chemistry, San Antonio, February 4-6.
234. Noll, W.: "Chemistry and Technology of Silicons, Academic, 1968, 368.

235. Rosene, R.B.; Shumaker, E.F. "Viscous Fluids Provide Improved Results from Hydraulic Fracturing Treatments," SPE paper 3347, 1971 SPE Rocky Mountain Regional Meeting, Billings, June 2-4.
236. Free, D.L. US Patent 3 974 077, 1976.
237. Shah, S.N.; Harris, P.C.; Tan, H.C. "Rheological Characterization of Borate Crosslinked Fracturing Fluids Employing A Simulated Field Procedure"; SPE Paper 18589, 1988 SPE Production Technology Symposium, Hobbs, N.M.
238. Mondshine, T.C. US Patent 4 619 776, 1986.
239. Chatterji, J. US Patent 4 144 179, 1979.
240. Brown, R.A.; Norris, R.D. US Patent 4 552 675, 1985.
241. Hinkel, J.J. US Patent 4 560 486, 1985.
242. Tiner, R.L.; Holtmyer, M.D.; King, B.J.; Gatlin, R. US Patent 3 888 312, 1975.
243. McDaniel, R.R.; Houx, M.R.; Barringer, D.K. "A New Generation of Solid-Free Fracturing Fluids", SPE paper 5641, 1975 SPE Annual Meeting, Dallas, September 28-October 1.
244. Clampitt, R.L. US Patent 3 727 689, 1973.
245. Clampitt, R.L. US Patent 3 727 688, 1973.
246. Clampitt, R.L.; Hessert, J.E. US Patent 3 727 687, 1973.
247. Clampitt, R.L. "Gelled PROD Fluid for High Temperature Fracturing," 1975 Proc. Annu. Southwest. Pet. Short Course, April, 109-114.
248. Davidson, C.J. Eur Pat Appl 0 142 407, 1984.
249. Southard, M.Z.; Green, D.W.; Willhite, G.P. "Kinetics of the Chromium (VI)/Thiourea Reaction in the Presence of Polyacrylamide", SPE/DOE paper 12715, 1984 SPE/DOE Fourth Symposium on Enhanced Oil Recovery, Tulsa, April 15-18.
250. Aslam, S.; Vossoughi, S.; Willhite, G.P. "Viscometric Measurement of Chromium(III)-Polyacrylamide Gels by Weissenberg Rheogoniometer", SPE/DOE paper 12639, 1984 SPE/DOE Fourth Symposium on Enhanced Oil Recovery, Tulsa, April 15-18.
251. Huang, C-G.; Green, D.W.; Willhite, G.P. "An Experimental Study of the In-Situ Gelation of Chromium(+3)-Polyacrylamide Polymer in Porous Media", SPE/DOE paper 12638, 1984 SPE/DOE Fourth Symposium on Enhanced Oil Recovery, Tulsa, April 15-18.
252. Terry, R.E.; Huang, C.G.; Green, D.W.; Michnick, M.J.; Willhite, G.P. Soc. Pet. Eng. J. 1981, 22(4), 229-235.
253. Prud'homme, R.K.; Uhl, J.T. "Kinetics of Polymer/Metal-Ion Gelation", SPE/DOE paper 12640, 1984 SPE/DOE Fourth Symposium on Enhanced Oil Recovery, Tulsa, April 15-18.
254. Prud'homme, R.K.; Uhl, J.T.; Poinsatte, J.P.; Halverson, F. Soc. Pet. Eng. J., 1983, 24(10), 804-808.
255. Ghazali, H.A.; Willhite, G.P. "Permeability Modification Using Aluminum Citrate/Polymer Treatments: Mechanisms of Permeability Reduction in Sandpacks," SPE Paper 13583, International Symposium on Oilfield and Geothermal Chemistry, Phoenix, April 9-11.
256. Green, P.C.; Block, J.C. US Patent 4 486 318, 1984.
257. Abdo, M.K.; Chung, H.S.; Phelps, C.H.; Klaric, T.M. "Field Experience with Floodwater Diversion by Complexed Biopolymers"; SPE/DOE paper 12642, 1984 SPE/DOE Symposium on Enhanced Oil Recovery, Tulsa, April 15-18.
258. Abdo, M.K. US Patent 4 141 842, 1979.

259. Dovan, H.T.; Hutchins, R.D.: "Development of a New
     Aluminum-Polymer Gel System for Permeability Adjustment,"
     SPE/DOE Paper 12641, 1984 SPE/DOE Fourth Symposium on Enhanced
     Oil Recovery, Tulsa, April 15-18.
260. Gall, J.W.; Johnston, E.L. US Patent 4 018 286, 1977.
261. Majewicz, T.G. US Patent 4 486 335, 1984.
262. Kramer, J.; Prud'homme, R.K.; Wiltzius, P.; Miran, P.; Knoxll,
     S. Colloid & Polymer Science, 1988 266:1-11.
263. Rummo, G.J. Oil Gas J., 1982, 80(37), 84,89.
264. Chemical Abstracts 102(18):141960h
265. Chemical Abstracts 100(8):54290w
266. Conway, M.W. US Patent 4 470 915, 1984.
267. Crowe,C.W. US Patent 4 317 735, 1982.
268. Payne, K.L.; Harms, S.D. "Recent Developments in Polymer
     Fracture Fluid Technology"; 1984 AIChE National Meeting,
     Anaheim, May 20-24.
269. Elphingstone, E.A.; Dees, J.M. US Patent 4 369 124, 1983.
270. Conway, M.W. US Patent 4 462 917, 1984.
271. ibid. 4 502 967.
272. Lagally, P.; Lagally, H. TAPPI, 1956, 39(11), 747-754.
273. Mills, J.A. Biochem Biophys Res Commun, 1961/62, 6(6),
     418-421.
274. Russell, C.A. US Patent 2 894 966, 1959.
275. Baumgartner, S.A.; Parker, C.D.; Williams, D.A.; Woodruff,
     R.A., Jr. "High Efficiency Fracturing Fluids for
     High-Temperature, Low-Permeability Reservoirs", SPE/DOE paper
     11615, 1983 SPE/DOE Symposium on Low Permeability, Denver,
     March 14-16.
276. LaGrone, C.C; Baumgartner, S.A.; Woodroof, R.A., Jr. Soc. Pet.
     Eng. J. 1985, 25(9), 623-628.
277. Williams, D.A. US Patent 4 534 870, 1985.
278. Williams, D.A.; Woodroof, R.A., Jr.; Box, P.C. Oil Gas J.,
     1982, 80(13), 141-145.
279. Kucera, C.H. US Patent 4 683 068, 1987.
280. Walser, D.W. "Field Study of a New High-Temperature Fracturing
     Fluid in South Texas", SPE Production Engineering(May 1988)
     187-191.
281. Baranet, S.E.; Hodge, R.M.; Kucera, C.H. US Patent 4 686 052,
     1987.
282. Payne, K.L. US Patent 4 579 670, 1986.
283. Horton, R.L. US Patent 4 505 826, 1985.
284. Githens, C.J. US Patent 4 566 979, 1986.
285. Almond, S.W. US Patent 4 448 975, 1984.
286. Almond, S.W. US Patent 4 477 360, 1984.
287. Hanlon, D.J.; Almond, S.W. US Patent 4 460 751, 1984.
288. Almond, S.W. and Garvin, T.R.: "High Efficiency Fracturing
     Fluids for Low-Temperature Reservoirs," 1984 Proc. Ann.
     Southwest Pet. Short Course, 76-80.
289. Zhao, F. Chemical Abstract 106(24); 199015A, Chemical Abstract
     105(12), 100219h.
290. Hodge, R.M. US Patent 4 657 080, 1987.
291. Hodge, R.M. US Patent 4 657 081, 1987.
292. Hollenbeak, K.H.; Githens, C.J. US Patent 4 464 270, 1984.

293. Putzig, D.E. Eur Pat Appl 0 138 522 A2, 1984. Chemical
     Abstracts 103(6):39588t.
294. Wadhwa, S.K. US Patent 4 514 309, 1985.
295. Putzig, D.E. Eur Pat Appl 0 195 531 A2, 1986. Chemical
     Abstracts 106(2):7301q
296. Smeltz, K.C. US Patent 4 609 479, 1986.
297. Goldstein, A.M. US Patent 3 146 200, 1964.
298. Jordan, W.A. US Patent 3 084 057, 1963.
299. Elbel, J.L.; Thomas, R.L.: "The Use of Viscosity Stabilizers in
     High Temperature Fracturing," SPE Paper 9036, 1980 SPE Rocky
     Mountain Regional Meeting, Casper, May 14-16.
300. Chemical Abstracts 102(10):81253j and Chemical Abstracts
     101(24):213581f
301. Haack, T.K.; Shaw, D.A.; Greenley, D.E. Oil Gas J. 1986, 84(1),
     81-83.
302. Malone, T.R.; Foster, T.D., Jr.; Foster, S.T. US Patent 4 290
     899, 1981.
303. Phillips, J.C.; Tate, B.E. US Patent 4 458 753, 1984.
304. Lai, J.T. US Patent 4 310 429, 1982.
305. Vio, L.; Meunier, G. US Patent 4 599 180, 1986.
306. Sandell, L.S. US Patent 4 486 317, 1984.
307. Podlas, T.J. US Patent 4 183 765, 1980.
308. Kohn, R.S. US Patent 4 452 639, 1984.
309. Wellington, S.L. US Patent 4 218 327, 1980.
310. Shupe, R.D. J. Pet. Technol., 1981, 33(8), 1513-1529.
311. Wellington, S.L. Soc. Pet. Eng. J., 1983, 23(12), 901-912.
312. Thomas, D.C. Soc. Pet. Eng. J., 1982, 22(4), 171-180.
313. Glass, J.E.; Soules, D.A.; Ahmed, H. "Viscosity Stability of
     Aqueous Polysaccharide Solutions", SPE paper 11691, 1983
     California Regional Meeting, Ventura, March 23-25.
314. Braga, T.G. "Effects of Commonly Used Oilfield Chemicals on the
     Rate of Oxygen Scavenging by Sulfite/Bisulfite," SPE Production
     Engineering, May 1987, 137-142.
315. Ryles, R.G. "Chemical Stability Limits of Water-Soluble
     Polymers Used in Oil Recovery Processes," SPE Paper 13585, 1985
     International Symposium on Oilfield and Geothermal Chemistry,
     Phoenix, April 9-11.
316. Conway, M.W.; Pauls, R.W.; Harris, L.E.: "Evaluation of
     Procedures and Instrumentation Available for Time-Temperature
     Stability Studies of Crosslinked Fluids," SPE paper 9333, 1980
     SPE Annual Technical Conference and Exhibition, Dallas,
     September 21-24.
317. Conway, M.W.; Almond, S.W.; Briscoe, J.E.; Harris, L.E.:
     "Chemical Model for the Rheological Behavior of Crosslinked
     Fluid Systems," SPE Paper 9334, 1980 SPE Annual Technical
     Conference and Exhibition, Dallas, September 21-24.
318. Conway, M.W.; Almond, S.W.; Shah, S.N. Polym Mater Science Eng
     1984, 51, 7-12.
319. Clark, P.E. "Stimulation Fluid Rheology - A New Approach," SPE
     Paper 8300, 1979 SPE Annual Technical Conference and
     Exhibition, Las Vegas, September 23-26.
320. Cameron, J.R.; Gardner, P.C.; Veatch, R.W., Jr. "New Insights
     on the Rheological Behavior of Delayed Crosslinked Fracturing
     Fluids", SPE paper 18209, 1988 SPE Annual Technical Conference
     and Exhibition, Houston, October 2-5.

321. Kramer, J.; Uhl, J.T.; Prud'homme, R.K. Polym Eng Sci, 1987, 27(8), 598-602.
322. Prud'homme, R.K. "Rheological Characterization of Fracturing Fluids," Final Report API PRAC Project 84-45, American Petroleum Institute, Dallas, TX (1985).
323. Prud'homme, R.K. "Rheological Characterization of Fracturing Fluids," Final Report API PRAC Project 85-45, American Petroleum Institute, Dallas, TX (1986).
326. Prud'homme, R.K. Polym Mat Sci Eng, 1986, 55, 798.
327. Kramer, J.; Prud'homme, R.K.; Wiltzius, P. J. Colloid and Interface Sci., 1987.
328. Yoshimura, A.S.; Prud'homme, R.K.: "Viscosity Measurements in the Presence of Wall Slip in Capillary, Couette, and Parallel-Disk Geometries," SPE Reservoir Engineering, May 1988, 735-742.
329. Prud'homme, R.K.; Ellis, S.; Constien, V.G. "Reproducible Rheological Measurements on Crosslinked Fracturing Fluids," SPE paper 18210, 1988 SPE Annual Technical Conference and Exhibition, Houston, October 2-5.
330. Acharya, A.; Deysarkar, A.K. "Rheology of Fracturing Fluids at Low-Shear Conditions," SPE paper 16917, 1987 SPE Annual Technical Conference and Exhibition, Dallas, September 27-30.
331. Menjivar, J.A.: "On the Use of Gelation Theory to Characterize Metal Crosslinked Polymer Gels," Chapter 13, Advances in Chemistry Series 213, American Chemical Society, Washington, DC 1986.
332. Jiang, T.Q., Young, A.C., and Metzner, A.B. "The Rheological Characterization of HPG Gels: Measurement of Slip Velocities in Capillary Tubes," Rheol Acta 25:397-404 (1986).
333. Freck, J.;Gottschling, J. "A Field and Laboratory Study of Polysaccharides in Fracturing Treatments," 1984 Proc. Annu. Southwest. Pet. Short Course, April, 141-156.
334. Hodge, R.M.; Baranet, S.E. "Evaluation of Field Methods to Determine Crosslink Times of Fracturing Fluids," SPE paper 16249, 1987 SPE International Symposium on Oilfield Chemistry, San Antonio, February 4-6.
335. See Appendix Quality Control of Fracturing Fluids
336. Zasadzinski, J.A.N.; Chu, A.; Prud'homme, R.K. Macromolecules, 1986, 19 2960.
337. Matthews, T.M. "Field Use of 'Superfrac' - A New Hydraulic Fracturing Technique," SPE paper 2625, 1969 SPE Annual Meeting of AIME, Denver, September 28-October 1.
338. Kiel, O.M. J. Pet. Technol. 1970 22(1), 89-96.
339. Kiel, O.M. US Patent 3 710 865, 1973.
340. Sinclair, A.R. J. Pet. Technol. 1970, 22(6), 711-719.
341. Sinclair, A.R.; Terry, W.M.; Kiel, O.M. J. Pet. Technol. 1974, 26(7), 731-738.
342. See Appendix Emulsion Frac Fluids
343. Gidley, J.L.; Mutti, D.H.; Nierode, D.E.; Kehn, D.M.; Muecke, T.W. J. Pet. Technol. 1979, 31(4), 525-531.
344. Roodhart, L.; Kuiper, T.O.; Davies, D.R. "Proppant Rock Impairment During Hydraulic Fracturing," SPE paper 15629, 1986 SPE Annual Technical Conference and Exhibition, New Orleans, October 5-8.

345. Roodhart, L.P.; Davies, D.R. "Polymer Emulsion: The Revival of a Fracturing Fluid", SPE/DOE paper 16413, 1987 SPE/DOE Low Permeability Reservoirs Symposium, Denver, May 18-19.

346. Davies, D.R.; Kulper, T.O.H. J. Pet. Technol. 1988, 40(5), 550-552.

347. Schwartz, L.W.; Princen, H.M. J. Colloid Interfacial Sci. 1987 118(1), 201-211.

348. King, G.E. "Foam and Nitrified Fluid Treatments - Stimulation Techniques and More", SPE paper 14477, Distinguished Lecturer Program, 1985-86.

349. Ali, S.M.F.; Selby, R.J. Oil Gas J. 1986, 84(5), 57-63.

350. Ely, J.: "Recent Mechanical and Chemical Improvements in Foam Fracturing", 1985 Proc. Annu. Southwest. Pet. Short Course, April 23-25.

351. Reidenbach, V.G.; Harris, P.C.; Lee, Y.N.; Lord, D.L. "Rheological Study of Foam Fracturing Fluids Using Nitrogen and Carbon Dioxide", SPE paper 12026, 1983 SPE Annual Technical Conference and Exhibition, San Francisco, October 5-8.

352. Grundmann, S.R.; Lord, D.L. J. Pet. Technol., 1983, 35(3), 597-601.

353. Elson, T.D.; Marsden, S.S., Jr. "The Effectiveness of Foaming Agents at Elevated Temperatures Over Extended Periods of Time", SPE paper 7116, 1978 SPE California Regional Meeting of AIME, San Francisco, April 12-14.

354. Eakin, J.L.; Taliaferro, R.W. Oil Gas J., 1962, 60(49), 131-134.

355. Harms, S.D.; Payne, K.L. "Factors Affecting the Selection of Foaming Agents for Foam Stimulation", 1983 Proc. Annu. Southwest. Pet. Short Course, November.

356. Holcomb, D.L.; Callaway, R.E.; Curry, L.L. "Foamed Hydrocarbon Stimulation Water Sensitive Formations", SPE paper 9033, 1980 SPE Rocky Mountain Regional Meeting, Casper, May 14-16.

357. Crema, S.C.; Alm, R.R. "Foaming of Anhydrous Methanol for Well Stimulation", SPE paper 13565, 1985 International Symposium on Oilfield and Geothermal Chemistry, Phoenix, April 9-11.

358. Crema, S.C. US Patent 4 609 477, 1986.

359. Clark, H.B. J. Petro. Technol. 1980, 32(10), 1695-1697.

360. Hall, B.E. "Workover Fluids. Part I - Surfactants have differing chemical properties that should be understood to ensure proper application," World Oil(May 1986) 111-114; ibid "Part 2 - How the various types of surfactants are used to improve well productivity," World Oil(June 1986) 64-67; ibid "Part 3 - Use of alcohols and mutual solvents in oil and gas wells," World Oil(July 1986) 65-67; ibid "Part 4 - Use of Clays and fines stabilizers and treaters," World Oil(October 1986) 61-63; ibid "Part 5 - How certain chemicals react to stabilize clays and fines in the formation," World Oil(December 1986), 49,50.

361. See Appendix Non-Emulsifiers, Water Blockage Additives

362. Guillot, D; Dunand, A. "Rheological Characterization of Fracturing Fluids Using Laser Anemometry" SPE paper 12030, 1983 58th Annual Technical Conference and Exhibition, San Francisco, October 5-8.

363. Blauer, R.E.; Durborow, C.J. US Patent 3 937 283, 1976.

RECEIVED February 21, 1989

POLYMER PROPERTIES AND DESIGN

Chapter 3

Effect of Anionic Comonomers on the Hydrolytic Stability of Polyacrylamides at High Temperatures in Alkaline Solution

R. W. Dexter and R. G. Ryles

American Cyanamid Company, Chemical Research Division, Stamford, CT 06904-0060

Hydrolysis of amide groups to carboxylate is a major
cause of instability in acrylamide-based polymers,
especially at alkaline pH and high temperatures. The
performance of oil-recovery polymers may be adversely
affected by excessive hydrolysis, which can promote
precipitation from sea water solution. This work has
studied the effects of the sodium salts of acrylic acid
and AMPS*, 2-acrylamido-2-methylpropanesulfonic acid,
as comonomers, on the rate of hydrolysis of polyacrylamides in alkaline solution at high temperatures.
Copolymers were prepared containing from 0-53 mole % of
the anionic comonomers, and hydrolyzed in aqueous
solution at pH 8.5 at 90°C, 108°C and 120°C. The
extent of hydrolysis was measured by a conductometric
method, analyzing for the total carboxylate content.
It was found that the rate of hydrolysis decreased as
the mole ratio of the anionic comonomers increased, and
that AMPS was more effective in preventing hydrolysis
at all of the temperatures studied.

Polymers designed for the enhancement of oil recovery must remain
in solution throughout the predicted life of the flood to provide
the required viscosity for oil displacement. During polymer
flooding using brines or sea water as solvents, the hydrolysis of
amide groups present in polyacrylamides to form carboxylates
limits the useful lifetime of the polymer due to the formation of
complexes with magnesium and calcium ions [2,3]. These may be
precipitates or gels. The objective of this work was to investigate copolymers of acrylamide having greater resistance to hydrolysis at high temperatures, and with a lower tendency to form
insoluble precipitates in sea water.

AMPS* is a registered trademark of the Lubrizol Corporation

0097-6156/89/0396-0102$06.00/0
© 1989 American Chemical Society

The hydrolysis of polyacrylamide and acrylamide/sodium acrylate copolymers has been extensively studied [1,2,3,5,6,7,8,-9,10], in relatively strongly alkaline conditions, above pH 12. These studies demonstrated that the hydrolysis of the amide groups is hydroxide-catalyzed and that neighboring ionized carboxyl groups in the polymer inhibit the hydrolysis by electrostatic repulsion of the hydroxide ions. Senju et al. [6] showed that at temperatures up to 100°C, there is an apparent limit to the extent of hydrolysis of polacrylamide when approximately 60% of the amide groups are hydrolyzed.

The hydrolysis of acrylamide copolymers in dilute alkaline conditions for long periods at high temperatures up to 120°C, as found in harsh reservoirs, has not been extensively studied. Metzner et al. [1] have considered the chemical stability of acrylamide copolymers with various proportions of sodium acrylate in sea water at 90°C. They showed that precipitation of the polymers occurred when approximately 40% of the polymer was in the carboxylate form, and attributed this to the formation of magnesium and calcium complexes. Nonionic polyacrylamides are unsatisfactory in polymer floods because of their low viscosity and high adsorption. Anionic polyacrylamides containing 20-30% acrylic acid can be obtained with higher viscosity and lower adsorption, and are generally quite effective. However, even moderate increases in the hydrolysis of these polymers raises the carboxylate content to a level at which precipitation of calcium and magnesium complexes occurs, thereby shortening the effective lifetime of the polymer. (The separate and independent instability due to chain scission of polyacrylamides by oxidation has been shown to be minimal in reservoirs in the absence of molecular oxygen [4]).

EXPERIMENTAL

Materials. Monomers used in the preparation of the copolymers were as follows: acrylamide as a 50% solution in water, stablized with cupric ion, supplied by American Cyanamid Company; acrylic acid supplied by BASF; and AMPS*, 2-acrylamido-2-methylpropanesulfonic acid, (recrystallized grade) obtained from Lubrizol. The sodium salts of acrylic acid and AMPS were prepared by gradual neutralization of the monomers with sodium hydroxide solution, maintaining a temperature of 0 to 5°C, to give a final concentration of 50%.

All copolymers were prepared by solution polymerization, under adiabatic conditions, giving at least 99.9% conversions. The polymer gels were granulated and then dried at 90°C to a residual water content of 10 to 12%. The active polymer content of each sample was calculated from the initial weight of the comonomers and the weight of the dried gel. Hydrolysis of the polymers was determined by conductometric titration to be less than 0.2% of the acrylamide charge. The molecular weight of the polymers was 8-10 million as determined by intrinsic viscosity measurements.

The composition and concentration of polymers in the test solutions for hydrolysis are shown in Table 1. The concentration of the sodium acrylate and sodium AMPS copolymers with acrylamide were calculated to provide 0.025 molar solutions of amide units, to simplify the kinetics.

Table 1. Composition and Concentration of Polymers

Mole % Anionic Comonomer in Copolymers	% w/w Active Polymer in Solution	
	Sodium acrylate Copolymers	Sodium AMPS Copolymers
0.00	0.178	0.178
7.70	0.197	0.226
15.40	0.222	0.286
22.40	0.254	0.349
33.50	0.296	0.467
43.00	0.355	0.609
53.30	0.444	0.803

Solutions of each copolymer were prepared by dissolving the appropriate amount of copolymer in deionized water and rolling the solutions for 16 hours.

ANIONIC ACRYLAMIDE COPOLYMERS

SODIUM ACRYLATE

SODIUM AMPS®

HYDROLYSIS EXPERIMENTS

Conditions for hydrolysis experiments were selected to simulate harsh reservoir environments. Moderately alkaline (pH 8.5) solutions, high temperatures, and long reaction times up to 120 days were used. The pH of the solutions remained at 8.5 or slightly higher, ensuring that all anionic groups were fully ionized.

A portion of each solution was retained for analysis of carboxylate content at zero time. Samples of the polymer solutions were weighed into glass jars, the pH adjusted to 8.5 and the jars were sealed with tightly fitting screw caps. The jars were placed in thermostatted ovens at 90°, 108°, and 120°C. After the appropriate time, the jars were removed, cooled and weighed to ensure no loss of contents, prior to analysis for hydrolysis.

The extent of hydrolysis of the copolymers was determined by conductometric titration. The increase in carboxylate content was determined by difference, before and after hydrolysis. (The AMPS content of the polymers, where measured, was determined by colloid titration with poly [diallyl dimethyl ammonium chloride].)

KINETICS

The rate of hydrolysis of acrylamide is assumed to be equal to the
rate of formation of carboxylate groups in the early stages of
reaction, for both sodium acrylate and AMPS copolymers.

$$\frac{d[COO^-]}{dt} = - \frac{d[AMD]}{dt} = K'[AMD][OH]$$

$$\frac{d[COO^-]}{dt} = K[AMD] \qquad \text{at zero time and constant pH}$$

$$K = \frac{\frac{d[COO^-]}{dt}}{0.025} \qquad [AMD] = 0.025 \text{ at zero time}$$

Plots of the concentration of carboxylate formed vs. time were
drawn for each copolymer, and the initial rates of hydrolysis were
determined by measurement of the slope of the tangent to the curve
at zero time. The pseudo-unimolecular rate constant (K) is given
by:

$$K = \frac{\text{initial slope}}{0.025}$$

Confirmatory values of K were obtained from plots of $\log_e$
$\frac{0.025}{0.025-[COO^-]}$ vs. time in the early stages of the reaction,

although deviations from a straight line occurred in later stages.

RESULTS

The rate of hydrolysis of acrylamide in copolymers with sodium
acrylate or AMPS, 2-acrylamido-2-methylpropanesulfonic acid,
decreased as the proportion of the anionic comonomers was
increased. This effect was much more marked with AMPS than with
sodium acrylate, and occurred at 90°, 108°, and 120°C. Typical
results at 108°C [Figs. 1 and 2] show the increase in carboxylate
content of acrylamide copolymers containing sodium acrylate and
AMPS respectively.
    The calculated pseudo-unimolecular rate constants (k) for the
hydrolysis reaction [Fig. 3], clearly show the inhibiting effect of
AMPS, relative to sodium acrylate at all three temperatures.
    The total carboxylate content in a range of sodium acrylate
copolymers is shown in Fig. 4, calculated as the sum of the initial
carboxylate and carboxylate formed during hydrolysis. These values
may be compared with the total carboxylate content in AMPS
copolymers [Fig. 2].

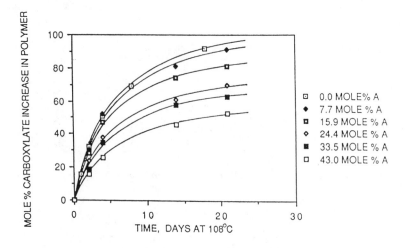

Figure 1. Hydrolysis of amide in sodium acrylate (A) copolymers at 108 ° C, pH 8.5, 0.025 M amide.

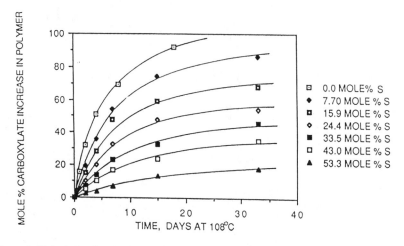

Figure 2. Hydrolysis of amide in AMPS (S) copolymers at 108 ° C, pH 8.5, 0.025 M amide.

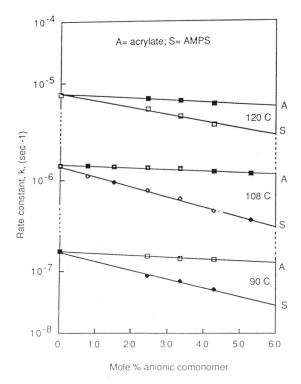

Figure 3. Initial rate constants for hydrolysis of acrylamide copolymers at 90, 108, and 120 °C.

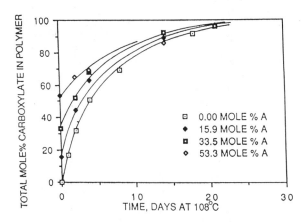

Figure 4. Total mole percent carboxylate in sodium acrylate (A) copolymers at 108 °C, pH 8.5, 0.025 M amide.

DISCUSSION

The mechanism of base catalyzed hydrolysis of either homopoly-
acrylamide or of copolymers of acrylamide and acrylic acid has been
studied extensively. It is well known that the rate of hydrolysis
of amide groups in such copolymers decreases significantly as the
charge on the polymer is increased [2,3,5,7]. This phenomenon has
been mainly attributed to electrostatic effects, repulsion between
charges on the macroion and on the approaching hydroxide anion. It
is generally believed that specific neighboring group effects,
inhibition by adjacent pendant carboxyl groups, dominate [7].
However, Morowetz [5] has proposed that the total charge on the
polymer does play an important role.
    The data presented here confirms the work of these previous
authors showing that the rate of amide group hydrolysis decreases
as the level of anionicity is increased. This was found to be true
for both carboxylated and sulfonated copolymers. However, the rate
of amide group hydrolysis in the AMPS copolymers was found to be
further inhibited for any given level of anionicity, e.g., at 30
mole % anionicity the rate of hydrolysis of amide groups in the
AMPS copolymers was found to be ca. 1/2 that of the corresponding
acrylate copolymer at all of the temperatures studied. Since the
total charge on these copolymers was the same and all groups were
fully ionized under these reaction conditions, this difference
cannot be attributed to a macroion charge effect.
    As far as neighboring group inhibition is concerned, sequence
distribution can play an important role, e.g. Morowetz [5] showed
that partially hydrolyzed polyacrylamide hydrolyzes more slowly
than an acrylic acid copolymer of the same charge. The former has
a more even distribution of groups leading to a greater proportion
of the least reactive BAB triad (where A are the acrylamide and B
are the acrylic acid moieties).
    Also, Higuchi and Senju [7] have proposed that the overall
rate constant is composed of three distinct rate constants corresponding to the hydrolysis of the three possible triad configurations, AAA, AAB, and BAB, and have found that the relative reactivity is 1:0.25:0.005. Thus, the overall rate is determined by
the relative proportions of these configurations and a relative
composite rate constant K can be derived as follows:

KINETIC SCHEME FOR THE HYDROLYSIS OF ACRYLAMIDE COPOLYMERS

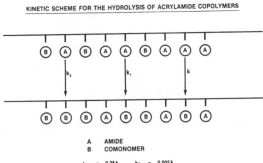

A   AMIDE
B   COMONOMER

k_1 = 0.25 k k_2 = 0.005 k

$$K = \frac{k\,[AAA] + 0.25\,k\,[AAB] + 0.005\,k\,[BAB]}{[AAA] + [AAB] + [BAB]}$$

where k is the rate for AAA hydrolysis

The relative proportions of triads is determined by the synthetic conditions chosen as described above for acrylic acid copolymers of acrylamide derived by either direct copolymerization or by hydrolysis. Also, the polymerization pH has a considerable effect on the reactivity in acrylamide/acrylic acid copolymerization. Table 2 shows the triad distribution, integrated to full conversion, for 30 mole % anionic copolymers of acrylamide using the reactivity ratios taken from Ponratnam [11] for acrylic acid and from McCormick [12] for AMPS. From these data (the acrylamide/acrylic acid copolymer prepared at pH = 4 is shown for comparison) composite relative rate constants K were obtained assuming equal reactivity for the acrylic and AMPS based triads of the same sequence. These data show that K for the sodium acrylate copolymer should be ca. 17% lower than for the sodium AMPS copolymer. Since our experimental data show a significant reduction, ca. 50%, for the sodium AMPS copolymer, we can only conclude that sequence distribution effects on neighboring group inhibition cannot be the dominant controlling factor in the hydrolysis of these copolymers. However, the pendant group of the AMPS monomer does possess a geminal dimethyl group which may associate more strongly with the hydrophobic polymer backbone. Such a configurational arrangement may place the negatively charged sulfonate group in very close proximity to any neighboring amide group resulting in increased repulsion of hydroxide anion. The carboxyl groups in acrylic acid copolymers are bonded directly to the polymer chain and are therefore, unlikely to form associations over several bond lengths.

Table 2. Triad Distributions and Composite Rate Constants (K) for 30 Mole % Anionic Acrylamide Copolymers

Monomer B	Polym. pH	r_B	r_A	Triad Distribution						K
				AAA	AAB	BAB	ABA	ABB	BBB	
Acrylic Acid	4.0	0.57	0.32	0.30	0.28	0.12	0.22	0.07	0.01	0.53k
Sodium Acrylate	8.0	0.12	0.63	0.25	0.34	0.11	0.27	0.03	0	0.47k
AMPS Na	9.0	0.49	0 98	0.31	0.31	0.08	0.18	0.10	0.02	0.55k

LITERATURE CITED

1. P. Davidson and E. Mentzer, SPE 9300, presented at the 55th Annual Technical Conference, Dallas, TX, 1980.
2. R.G. Ryles, SPE 13585, presented at the International Symposium on Oilfield and Geothermal Chemistry, Phoenix, AZ, April, 1985.
3. A. Moradi-Araghi and P.H. Doe, SPE 13033, presented at the 59th Annual Technical Conference, Houston, TX, Spet. 1984.

4. R.G. Ryles, SPE 12008, presented at the 58th Annual Technical
    Conference, San Francisco, CA, 1983.
5. S. Sawant and H. Morowetz, Macromolecules, 17, 2427, (1984).
6. K. Nagase and K. Sakaguchi, J. Polym. Sci. (A), 3, 2475,
    (1965).
7. M. Higuchi and R. Senju, J. Polym. Sci., (3), 3, 370, (1972).
8. G. Smets and A.M. Hesbain, J. Polym. Sci., 40, 217, (1959).
9. J. Moens and G. Smets, J. Polym. Sci., 23, 931 (1957).
10. S. Mukhopadhyay, B. Ch. Mitra, and S.R. Pailt, Indian J. Chem.,
    7, 903 (1963).
11. S. Ponratnam and S.L. Kapur, Makromol. Chem., 178, 1029,
    (1977).
12. C.L. McCormick and G.S. Chen, J. Polym. Sci., 20, 817, (1982).

RECEIVED September 7, 1988

Chapter 4

Predictions of the Evolution with Time of the Viscosity of Acrylamine–Acrylic Acid Copolymer Solutions

Houchang Kheradmand and Jeanne François

Institut Charles Sadron, CRM–EAHP, CNRS–ULP, 6 rue Roussingault, 67083 Strasbourg-Cedex, France

This work deals with an attempt to predict the evolution at long term of the thickening properties of acrylamide-acrylic acid copolymer solutions, from their hydrolysis and degradation kinetics. A Monte-Carlo method is proposed to simulate hydrolysis process. By introducing at each step of this calculation the molecular weight deduced from degradation equations and using semi-empirical laws for the molecular weight and charge density dependences of the intrinsic viscosity, we have obtained some tendencies for the variations of the thickening power with time, under various conditions of temperature, pH and salinity.

One of the main problem encountered when hydrosoluble polymers are used in chemical tertiary process of oil recovery is the prediction of the evolution of the thickening properties of their solutions.

In the case of acrylamide-acrylic acide copolymers, such a prediction requires a good knowledge and understanding on the three following aspects:

i) the hydrolysis of amide groups which leads to the enhancing of the polyelectrolyte character of the polymer. Different experimental works have dealt with the dependence of this kinetics on pH ,temperature and initial polymer composition(1-6). More recently a Monte-Carlo simulation method has been proposed in order to predict the variation of the hydrolysis degree under different conditions(7-8).

ii) the chemical degradation of the chain which can be due to various mechanims according to the pureness of the samples,the method used for its synthesis,the nature of the ions present in the brine (oxidizing or reducing ions),the oxygen content of the brine and the temperature.

0097–6156/89/0396–0111$06.00/0
© 1989 American Chemical Society

iii) these two chemical processes leading to changes in the polymer charge and molecular weight respectively are expected to strongly modify the solution viscosity. Then the relation between viscosity and these two parameters must be known for the given conditions of application.

In recent works, we have studied the kinetics of both hydrolysis and degradation of a acrylamide-acrylic acid copolymer containing 17% of acrylate groups. The purpose of this paper is to give some predictions of the thickening properties evolution based upon semi-empirical viscosity laws.

Hydrolysis kinetics

It is well known that the base hydrolysis of polyacrylamide is catalyzed by OH^- ions (first order reaction) and obeys autoretarded kinetics due to the electrostatic repulsion between the anionic reagent and the polymeric substrate(3-5). In the range of slightly acid pH (3 < pH < 5), Smets and Hesbain(6) have demonstrated a mechanism of intramolecular catalysis by undissociated neighbouring carboxylate groups analogous to that observed in low molecular weight compounds such as phtalimic acid(10). By assuming that Hydrolysis simply results from these two mechanisms no reaction was expected in the range of pH near neutrality. However, Muller(2) has shown that the modification of already partially hydrolyzed polyacrylamide cannot be neglected if one considers reaction times of several months. A more recent systematical study(7,8) of the reaction at different pH ,temperatures and initial carboxylate contents led us to propose the simple following model ,for 3 < pH < 9. We have considered two types of reacting monomer units:

- the units which have an undissociated neighbouring group which catalyses the reaction with a rate constant k_h independent on pH (units X)
- the other units (units Y) whose hydrolysis rate k_a is not simply proportional to (OH^-) concentration (as in the range of high pH) but varies with pH according to an empirical rule:

$$\log k_a = \log k_{a0} + pH \ (\ C_A - C_B \ \alpha \ r - C_C \ (\ \alpha \ r)^2) \qquad (1)$$

where r is the fraction of carboxylate groups in the polymer, α is their ionization degree and C_A, C_B, C_C are constants. This expression corresponds to the following experimental observations:

i) for the unhydrolyzed polyacrylamide ,only units of type Y must be considered in the initial step of the reaction and intramolecular catalysis has not to be taken into account. In relation (1), we have obtained a rate constant k_a varying as :

$$\log k_a = \log k_{a0} + C_A \ pH \qquad (2)$$

This shows that pH increase favors hydrolysis reaction of Y units.

ii) for partially hydrolyzed polymers (or polyacrylamide in a second step), the increase of k_a with pH is lowered when the charge density $(\alpha \ r)$ increases (see fig.10 of ref.8). This retardating effect is expressed by the terms $(C_B \ \alpha \ r)$ and $(C_C \ (\ \alpha \ r \ )^2)$.

At each reaction time , and for a given polymerization degree N, the number of units X is:

$$N_x = N_{AB} (1 - \alpha) \qquad (3)$$

where N_{AB} is the number of diades AB (acrylamide :A; acrylic acid :B). The number of Y units is $N-N_x$.

Then the modelization of the hydrolysis kinetics requires at each time the knowledge of α and N_{AB}. α can be calculated by writing the different relations of dissociation equilibria of water,polyacid and NH_3 (produced by the hydrolysis reaction). We have proposed to determine N_{AB} at each reaction step and simulate the whole kinetics by using a Monte-Carlo method .(see ref.8).

In Figure 1 we compare the calculated and measured variations of τ for a copolymer of initial τ = 17%. Let us remark that in these experiments, the pH has not been adjusted at a constant value and the hydrolysis process induces a change of pH in the solution. We have taken into account this effect in our calculations. In fig.2, we give the predicted hydrolysis kinetics for the same polymer sample but in the case where pH remains constant.

Degradation kinetics

We have previously performed a systematical study of the degradation of a acrylamide-acrylic acid copolymer called sample C(of acrylate content τ = 17% and molecular weight M_w = $6*10^6$) prepared by photocopolymerization by using benzyl methyl ketal as catalyst(11). The observed behaviors have been compared with those of a sample obtained by base hydrolysis of polyacrylamide called sample H (approximately same M_w and τ = 30%), in some particular cases.

In the practical application, the chemical stability of these polymers at high temperature will essentially depend on the content of oxygen and ions of transition metals in the brine. It is the reason why we have investigated their behaviors for several months (at least 3) ,at 80°C and under three main conditions (see fig.3).

-without oxygen and salts of transition metals

Under such conditions , sample C is not degradated ,M_w measured by light scattering remaining constant for 6 months. A slight increase of reduced viscosity (η_{red}) will be later explained by the hydrolysis of amide groups.

- in the presence of oxygen and without salts of transition metals

After 3 months of ageing, η_{red} is reduced by a factor 6 while M_w is 10 times lower than initial $M_{w(0)}$. We have observed that the degradation follows a random process of link breaking since the linear expression[12,13]:

$$\ln (1 - M_{w(0)}) - \ln (1 - M_{w(t)}) = k\, t \qquad (4)$$

has been obtained. Moreover, the degradation leads to the formation of shorter linear chains and at each time the intrinsic viscosity (η) can be obtained from relation (4) and the classical Mark-Houwink law:

$$(\eta)= K\, M_w{}^a \qquad (5)$$

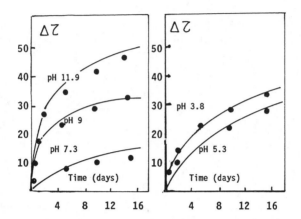

Figure 1.   Calculated and experimental (●) hydrolysis kinetics
            Copolymer C, T=80°C , different initial pH

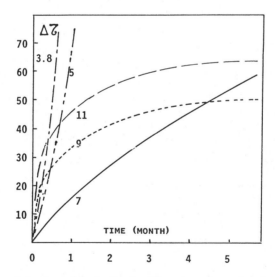

Figure 2.   Calculated hydrolysis kinetics for Copolymer C at
            80°C at different constant pH

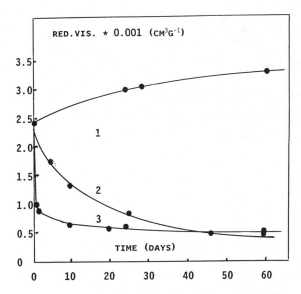

Figure 3.   Degradation kinetics of Copolymer C at 80°C
1: without oxygen ; 2 : with oxygen ;
3: without oxygen and with 5 ppm of Fe^{++}

Under these conditions, the degradation kinetics of sample H is similar,with a different value of k in relation (4). However, the origin of the phenomenon is different: for sample H, the degradation is due to the decomposition of chain hydroperoxides by traces of Fe^{III}. The sample C does not contain peroxides nor Fe^{III} and we have explained its instability by the presence of catalyst residues.

The purification by precipitation in methanol of both samples allows to obtain a very good stability for 6 months.

-in absence of oxygen and with salts of transition metals

In fig. 3, we compare the evolutions of η_{red} of free oxygen solutions of copolymer C with and without 5 ppm of Fe^{++}. As confirmed by light scattering measurements, the presence of ferrous ions induce a quasi instantaneous degradation of this polymer followed by a much slower reaction.

The purified sample C is not degradated under such conditions while degradation is measured before and after purification of sample H. This results confirms the difference of degradation mechanism according to the method of preparation of the polymer.

Oxidizing ions (Cu^{++} or Fe^{+++}) as well as reductants ions (Cu^{+} or Fe^{++}) induce the degradation of these polymers.

Dependence of the viscosity on molecular weight and charge density of the polymer

If the polymer concentration c_p is lower than c_p^* ,the critical concentration of chain overlapping , one can express the viscosity η of the solution by:

$$\eta = \eta_0 + \eta_0 (\eta) c_p + K' \eta_0 (\eta)^2 c_p^2 \qquad (6)$$

where η_0 is the solvent viscosity

K' is the Huggins constant which varies with the thermodynamical quality of the solvent (salinity and temperature)

For the calculation of η of a solution of copolymer of given r ,α and M_w under given conditions of salinity and temperature , one must know the variation laws of (η) and K' with these parameters.

- intrinsic viscosity (η)

In the classical theories of polyelectrolytes , the chain expansion is characterized by the electrostatic excluded volume parameter ,z_{el} with:

$$z_{el} = i^2 M_w^{1/2} / c_s \qquad (7)$$

where i is the ionization degree and c_s the concentration of added electrolyte(14). A dependence of the intrinsic viscosity with $1/C_s$ is then predicted while experiments show that (η) varies as $1/c_s^{1/2}$. Fixmann et Al.(15) have proposed an expression which gives a better account for experimental data ((α'^3-1) varying as $1/c_s^{1/2}$, α' being the chain expansion). Nevertheless, in spite of the numerous expressions proposed, they generally do not take into account the strong interactions between counter ions and polyions for the high values of the charge parameter ξ

$$\xi = e^2 / D k T b \qquad (8)$$

where e is the proton charge, D the dielectric constant, k the Boltzamm constant, T the absolute temperature and b the average distance between two charged groups along the chain, varying as $1/\alpha r$.

The Manning theory(16,17) predicts a critical value ξ_c above which counterions are condensated on polyion. In the case of monovalent ions , $\xi_c = 1$ and for $\xi > \xi_c$, the chain expansion predicted by the classsical theories is overestimated if the ionization i is not corrected by a fixation term. Koblansky et Al have shown(18) that ion condensation occurs for values of ξ much lower than ξ_c in the case of partially hydrolyzed polyacrylamide. They have found for Na^+ that the activity coefficient γ does not vary with ξ as predicted by the Manning theory (16,17) but according to a empirical law:

$$\gamma = 0.96 - 0.42 \, \xi^{1/2} \qquad (9)$$

in the absence of simple electrolytes.

Kowblansky et Al (19) have also measured the intrinsic viscosity of hydrolyzed polyacrylamide of a given molecular weight as a function of r for $\alpha = 1$ and their results can be fitted by the expression:

$$(\eta) = (\eta)_0 + k_s * \xi * \gamma * M_w \qquad (10)$$

where $(\eta)_0$ is the intrinsic viscosity of the uncharged polymer

$$(\eta)_{0(25)} = 9.3*10^{-3} * M_w^{0.75} \; (cm^3 g^{-1}) \quad \text{at } 25°C \; {}^{(19)} \qquad (11)$$

and k_s is a coefficient which depends on the salinity:

$$k_s = 4.7*10^{-4}/c_s^{1/2} - 1.33*10^{-4} \qquad (12)$$

c_s is the molar concentration of monovalent salt (NaCl)

Such empirical expressions have been also verified by Kulkarni et Al.(21) and Kheradmand(7) but they have been established for room temperature. In fact, we have to know the temperature dependence of the two terms of relation (10):

-For the first non electrostatic term, such a dependence can be calculated from the classical Flory theory and the value of the theta temperature of unhydrolyzed polyacrylamide ($\theta = 265°K$ (22))

$$(\eta)_{0(T)} = (\eta)_{0(25)} * (1-\theta/T)/(1-\theta/298))^{3/5} \qquad (13)$$

-For the electrostatic term, there is a lack of reliable data and predictions dealing with the variation of the ions -polyion interactions with temperature. If one considers the relation (8), the product D * T is a decreasing function of temperature but the term $\xi * \gamma$ detreases for the high values of ξ and increases for the low values of ξ. Nervertheless, these variations are negligeable and we have considered , in a first approximation, that the electrostatic term remains constant for $20 < T < 80°C$.

The final expression for (η) is:

$$(\eta) = 9.33*10^{-3} * M^{0.75} * ((1 - 265/T)/0.11)^{3/5}$$

$$+ (4.7*10^{-4}/c_s^{1/2} - 1.3*10^{-4}) * \xi * \gamma * M_w \tag{14}$$

(cm^3/g^{-1})(see examples of results in Figure 4)

   - Huggins constant K'

Let us remark that relation (6) is given for polymer concentration c lower than the critical overlapping concentration c^* above which higher terms in c must be considered. In fact, the concentration practically used (around 10^{-3} g/cm^3) corresponds to the semi-dilute regim for which the behavior is not well known in the case of polyelectrolytes. We have however kept relation (6) by introducing for K' a mean apparent value determined from our experiments (K' = 1)

By combining expressions (6) and (14), one can obtain an approximative value of the solution viscosity if only monovalent ions are present in the solution.

In the presence of alkalno-earth cations, one must take into account not only viscosity decrease due to electrostatic screening and condensation which theoritically occurs at lower values of ξ (ξ_c= 0.5) but also phase separation (23-25). We will take the case of Ca^{2+} as example.

   - Kowblansky et Al.(26) have obtained for activity coefficient of Ca^{2+} :

$$\gamma = 0.43 - 0.26 \; \xi^{1/2} \tag{15}$$

We have recently performed systematical measurements of the intrinsic viscosity of acrylamide-acrylic acid copolymers for large ranges of r and α ,in the presence of $CaCl_2$(26). Our results show that the empirical relation (14) can be extended to the case of divalent cations by using the value of γ given in relation (15). It should then possible to predict the variation of intrinsic viscosity at infinite dilution ,but at finite concentration the formation of aggregates makes difficult the determination of the Huggins constant.

   - It is now well known that the addition of Ca^{2+} in aqueous solutions of these copolymers induces phase separation with precipitation of a phase constituted by a polymer-Ca complex. From turbidity measurements, it is possible to define a critical concentration c_a^* above which this phenomenon occurs(23-25).

c_a^* decreases with c_p for $c_p < c_p^*$ and increases for $c_p > c_p^*$

c_a^* is an decreasing function of temperature ,the demixing is obtained by heating

c_a^* increases by addition of NaCl

c_a^* decreases by increasing r for $r < 50\%$

Some aspects of this behavior are given in Figure 5 and more details can be found in ref. 24 and 25;

<u>Viscosity evolution</u>

   - in absence of degradation

We have seen that such polymers are not degradated in absence of oxygen. On an other hand, it is possible to avoid degradation

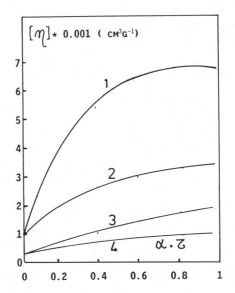

Figure 4.   Calculated variations of (η) as a function of $\alpha * \tau$
1 and 2 : $M_w = 510^6$ $c_s = 0.1$ and 0.5N ; 3 and 4 : $M_w = 10^6$
same c_s

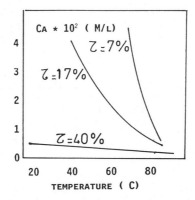

Figure 5. Solubility limit of different copolymers (1000 ppm)
in the presence of Ca^{2+} as a function of temperature

either by purification or by addition of stabilizing compounds
(7,9,27) under all the practical conditions. We have then calculated
the variations of viscosity with time by combining the hydrolysis
kinetics and viscosity expressions (at constant M_w) under different
conditions of constant pH and salinity.

The Figure 6 corresponds to the predictions for the coplymer C
of 17% of acrylic content at 80°C. For pH>7 , an increase of
viscosity can be expected but the form of the curves changes with
pH: at high pH , the initial hydrolysis rate is higher than at pH 7
and this leads to a higher initial increase of viscosity. But in the
second step of the hydrolysis, the kinetics is strongly autoretarded
at ph>9 and the slow variation of τ corresponds to a slight
increase of viscosity while at pH 7 , complete hydrolysis can be
reached and a higher viscosity is expected for a long aging time.

For pH<7 , the hydrolysis is very fast (see figure 2) and a
limiting value of viscosity can be reached after 1 month but the
polymer is slightly ionized : the product $\alpha\tau$ passes through a
maximum which also induces a maximum in the viscosity curves.

Finally, all these variations are strongly reduced by
increasing salt concentration.

We have reported in Figure 7 the expected variations of
viscosity at five temperatures for the same polymer in the presence
of O.1NaCl at pH 7. Two opposite effects explain these curves: the
initial viscosity is a decreasing function of T while the hydrolysis
rate increases with increasing T.

Some remarks can be made about the evolutions in the
presence of Ca^{2+}, although it is difficult to think that some brine
could contain only this type of cations without monovalent cations.
Before any aging, a copolymer of 17% of acrylate content begins to
demix for $CaCl_2 > 2.5*10^{-2}$ M/l and this limit is lower for higher
initial values of τ. If $CaCl_2 < 5\ 10^{-3}$ M/l , no precipitation will
occur even if τ reachs the value 1. For an intermediate
concentration of $CaCl_2$, for instance 10^{-2} M/l, phase separation can
be expected after 20 months and 3.5 months at 60°C and 80°C
respectively.

-when degradation occurs
    i) in the presence of oxygen
                    In this case, we must take into account the
variation of M_w through relation (4). An example of results is given
in Figure 8. It is obvious that the degradation is the main
phenomenon and a high loss of viscosity can be predicted. Since
hydrolysis has an increasing effect on the viscosity, it was
interesting to determine the M_w variation which could lead to a
constant value of viscosity (see Figure 5b). We can observe that M_w
must remain higher than $3.7*10^6$
    ii) in absence of oxygen and with ions of transition metals
                    In this case, the very fast degradation induces the high
initial loss of viscosity as already measured and predictions for a
long time do not present interest.

Conclusion

In this work, we propose a method based on different combined semi-
empirical laws to predict the evolution of the viscosity of a
solution of acrylamide - acrylic acid copolymer. In fact, it appears

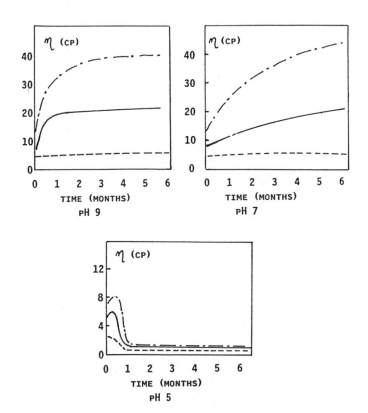

Figure 6. Predictions of viscosity evolution for Copolymer C
without degradation at 80°C and different pH; (—·—)
c_s=0N; (———) c_s = 0.1N ; (-----) c_s = 0.5N

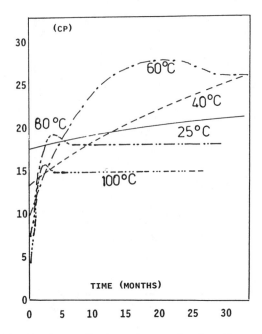

Figure 7. Predictions of viscosity evolution for copolymer C
at pH 7, 0.1 NACl at different temperatures

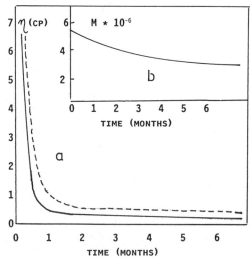

Figure 8. a:Predictions of viscosity evolution for Copolymer C
with degradation in the presence of oxygen at 80°C
and pH7 ; (-----) c_s = 0.05N ; (——————) c_s =
0.1N

b : calculated decrease of M_w able to compensate
hydrolysis effect

that in absence of degradation, the behaviors must be rather independant on the origin of the polymer and only depend on its initial properties and on the pH , salinity and temperature of the brine. When degradation occurs, it is generally due to impurities,then the kinetics is not universal and must be the object of a specific study for each sample.

This work has benefited from grants from Compagnie Française des Pétroles. The authors are indebted for fruitful discussions to Doctor V.Plazanet.

Literature Cited
1. Muller G.; Fenyo J.C.; Selegny E.J. J.Appl.Polym.Sci. 1980, 25 627
2. Muller G. Polymer Bulletin 1981, 5 , 31
3. Higuchi M.; Senju R. Polymer 1972, 3, 370
4. Sawant S.; Morawetz H. Macromolecules,1984,17, 2427
5. Truong N.D.;,Galin J.C.; François J.; Pham Q.T. Polymer, 1986, 27, 459
6. Smets G.; Hesbain A.M., J.Polym.Sci. 1959, 23, 217
7. Kheradmand H. thesis L.Pasteur University Strasbourg 1987
8. Kheradmand H.; François J.; Plazanet V. Polymer 1988, 29, 860
9. Kheradmand H.; François J.; Plazanet V. J.Appl.Polym.Sci under press
10. Bender M.L.; Chow Y.L.; Chloupek F. J.Amer.Chem.Soc. 1958, 80, 5380
11. Boutin J.; Contat S. French Patent N 249 217 issued to Rhone Poulenc Ind.
12. Jellinek H.H.G.; Degradation of Vinyl Polymers Academic press, New-york 1955
13. Vink H. Makromol.Chem. 1963, 67, 105
14. Yamakawa H. Modern theory of Polymer Solutions Harper and Row New-York 1971
15. Fixman M.; Solnick J. Macromolecules, 1978, 11, 863
16. Manning G. J.Chem.Phys. 1969, 51, 924
17. Manning G. Acc.Chem.Res. 1979, 12, 442
18. Kowblansky M.; Zema P. Macromolecules, 1981, 14,166
19. Kowblansky M.; Zema P. Macromolecules, 1981, 14, 1451
20. François J.; Sarazin D.; Schwartz T.; Weill G.Polymer, 1979 20, 969
21. Kulkarni R.A.; Gundiah S. Makromol.Chem., 1984, 185, 957
22. Kanda A.; Sarazin D.; Duval M.; François J. Polymer, 26, 406
23. Ikegami A.; Imai N. J.Polym.Sci., 1962, 56, 133
24. Truong N.D.; Galin J.C.; François J.; Pham Q.T. Polymer Communications, 1984, 25, 208
25. Truong D.N.; FRançois J. in Solid-Liquid Interactions in Porous Media, Ed.Technip, Paris , 1982, p.251
26. Medjadhdi G.; Sarazin D.; François J. Unpublished results
27. Muller G.; Kohler N. in 2nd European Symposium on Enhanced Oil Recovery, Ed. Technip, Paris, 1982 p. 87

RECEIVED February 2, 1989

Chapter 5

Interactions Between Acrylamide–Acrylic Acid Copolymers and Aluminum Ions in Aqueous Solutions

Ramine Rahbari, Dominique Sarazin, and Jeanne François

Institut Charles Sadron, CRM–EAHP, CNRS–ULP, 6 rue Boussingault, 67083 Strasbourg-Cedex, France

Phase separation, gelation and viscosity of acrylamide-acrylic acid copolymer solutions containing aluminium chloride have been studied as a function of pH, salinity and composition of the polymer. In the range of polymer concentration investigated, the most common behavior is phase separation with loss of viscosity and formation of large agregates. At pH 5 gelation phenomenon can occur due to the presence of polynuclear ions of aluminium and the conditions of gel formation are studied as a function of salinity. At pH 7, phase separation is due to flocculation of $Al(OH)_3$ particles by polymer bridges. These behaviors are discussed from ^{27}Al NMR data giving the fraction of Al ions bound on the polymer and from of a model of electrostatic interactions.

The strong interaction of polyvalent cations with polyions is well known to strongly alter the rheological properties of hydrolyzed polyacrylamide used in the tertiairy oil recovery process (1-4). The influence of divalent cations have already been studied(5-7) but the rôle played by the presence of small quantities of aluminium ions has never been investigated.

In a first part of this paper, we will discuss results of a ^{27}Al NMR study of the binding of Al ions on acrylamide-acrylic acid copolymers as a function of pH, at the light of a simple electrostatic model. The second part deals with the phase diagrams , physical gelation and precipitation phenomenon, for different copolymer compositions and under various conditions of concentrations , pH and salinity.

Experimental

in this study AD10,AD17,AD27 and AD37 (Rhône Poulenc Industries(8) of acrylate content r (mole %) equal to 1.5, 7, 17 and 27 , and

0097–6156/89/0396–0124$06.00/0

of weight average molecular weight M_w approximately $5*10^6$. An other sample (PAMNH) ($\tau=0.3\%$ and $M_w=10^6$) has been prepared in our laboratory by photopolymerization.

The details concerning the experimental methods used in this work can be found in ref. 9 and 10.

Solutions of Aluminium Chloride

The pure $AlCl_3$ solutions neutralized by NaOH contain ions of general formula $Al_i(OH)_p^{(3i-p)+}$, which can be mononuclear (Al^{+++}, $Al(OH)^{++}$, $Al(OH)_2^+$ and $AL(OH)_4^-$) or polynuclear ($Al_2(OH)_2^{4+}$ and $Al_{13}(OH)_p^{(39-p)+}$) and also non ionic species ($Al(OH)_3$). Their amounts depend on the neutralization ratio R ($R= (NaOH/AlCl_3)$) according to six diferent equilibrium laws of constants $K_{i,p}$

$$(Al_i(OH)_p^{(3i-p)+}) = \frac{(Al^{+++})*K_{i,p}*f_{i,p}}{f_{i,p} * (H^+)^i} \qquad (1)$$

where $f_{i,p}$ is the activity coefficient of $Al_i(OH)_p^{(3i-p)+}$ which can be obtained from Debye-huckel expression (11-13). The solubility of Al^{+++} is limited by:

$$(Al^{+++})*(OH^-)^3 < K_s \qquad (2)$$

By using a set of $K_{i,p}$ and $f_{i,p}$ values found in literature, we have calculated the compositions of $AlCl_3$ solutions (without polymer) at two different concentrations (see figure 1).

If one considers the pH range of major interest in application ($4.5 < pH < 8$) , it appears:

-around pH = 5, for the low $AlCl_3$ concentrations only mononuclear species are present while at higher concentrations ,polynuclear ions and non ionic species are preponderant. The time is also an important parameter since polyions (Al_{13}^{3+}) progressively disappear and after 6 months aging ,only non ionic species are in equilibrium with monovalent ions $Al(OH)_2^+$. 15 days aged solutions have been used in this study ; The composition of the solutions given in Figure 1 have been confirmed by NMR and correspond to this aging time (9,11-12).

- around pH = 7, $Al(OH)_3$ is present at more than 90% and it has been shown that the size of aggregates depend on concentration and aging time. We have recently shown (14) by electrophoresis measurements that the $Al(OH)_3$ particles are positively charged and become neutral only after a long time (6 months) of heating at 80°C. Elastic and quasi-elastic light scattering measurements have revealed that their size scales as $C^{1.3}$: at 1 ppm of Al (or $3*10^{-5}$ M/1) , the radius of gyration R_{ga} is 500 Å while at 12 ppm $R_{ga}=15000$ Å. Moreover these particles have a form of plaquets as already described(15,16)

Model of electrostatic interactions

It is well known that the concentration of ionized groups COO^- on the polymer chain obeys the classical law:

$$pK_a = (COO^-)*(H^+) / (C_{ip} - COO^-) \qquad (3)$$

where C_{ip} is the molar concentration of carboxylic groups.

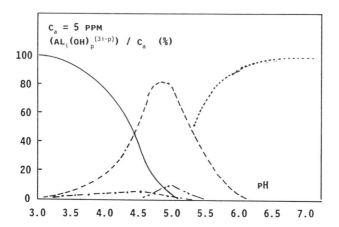

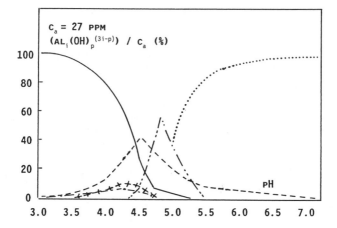

Figure 1 : AlCl$_3$ solutions :Calculated fractions of Al^{+++} (————),
Al(OH)$^{2+}$ (—·—·),Al(OH)$_2^+$ (————), Al$_2$(OH)$_2^{4+}$ (+++++)
Al$_{13}$(OH)$_{27}^{3+}$ (—··—) and Al(OH)$_3$ (·······)

In order to obtain a rough evaluation of the fraction of aluminium bound on polymer, we have assumed in a first approach:

i) the polyelectrolyte effects due to the polymer are negligible and the distribution of ionic species in the bulk can be considered as uniform. Such an assumption can be considered as correct if the polymer concentration c_p is higher than c_p^*, the critical concentration of chain overlapping.

ii) among the different possible interactions between polymer and aluminium species, the most important are the electrostatic interactions between carboxylate groups COO^- and the Al trivalent ions which are oppositively charged. Then we have taken into account the equations corresponding to the following equilibria, with constants K_i and K_0:

$$(Al^{+++}) + 3 \,----\,(COO^-) \;=\; COO_3Al \qquad K_i \qquad (4)$$

$$(Al_{13}^{\;+++}) + 3 \,----\,(COO^-) \;=\; COO_3Al_{13} \qquad K_0 \qquad (5)$$

By using the expressions (1) to (5), it is possible to calculate the concentrations of the different ionic $(Al_i(OH)_p^{(3i-p)+}$ and $COO-$) and non ionic $(Al(OH)_3$, $COOH$, COO_3Al and $COO_3Al_{13})$ species in the mixed $AlCl_3$- polymer solutions as a function of pH and composition.

As shown in Figures 2 and 3, we have found two maxima in the amount of the "bound" monomeric ions or more precisely the difference between the total monomeric ions concentration in absence and presence of polymer (NMR experiments give this quantity): for pH = 4 and pH = 4.6. Comparison with figure 1 indicates that these maxima correspond to the formation of COO_3Al and COO_3Al_{13} respectively. For very aged or very dilute $AlCl_3$ solutions where polynuclear ionic species are absent, only one maximum is found, around pH = 4.5.

^{27}Al NMR

Bottero et Al.(11,12) have shown that ^{27}Al NMR allows to distinguish the mononuclear from the polynuclear species of Aluminium ions in $AlCl_3$ solutions. We have performed the same type of studies by adding different amounts of copolymers and we have measured the decrease of monomeric Al ions concentration by increasing the polymer content. Figures 2 and 3 give some examples of results; "Al bound" as measured by NMR is plotted as a function of pH, for two different systems and can be compared with the values calculated for $K_i=10^{14}$ and $K_0=10^{16}$.

As expected from the model, we observe:

i) Two reproducible maxima at pH slightly shifted with respect to predictions ,ii) the amount of Al ions bound onto polymer increases by increasiong r and the same couple of K_i and K_0 values gives a rather good account of experimental results for $1.5\% < r < 30\%$. This result shows that is reasonable to neglect other type of interactions between aluminium ions and the polymer: for instance coordination binding with amide groups.

This first study is very important: the good agreement between calculation and experiments justifies the previous

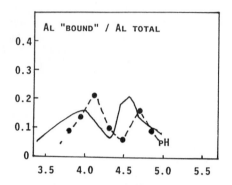

Figure 2 : Interactions polymer-aluminium : concentration of
bound aluminium ions for AlCl$_3$ (27 ppm) and AD27
(0.25 g/l) Full line : Calculated curve ; Dotted
line :NMR results

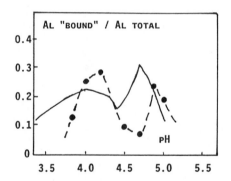

Figure 3: Interactions polymer-aluminium : concentration of
bound aluminium ions for AlCl$_3$ (27 ppm) and AD37
(0.25 g/l) Full line : calculated curve; dotted line
:NMR results

hypothesis. Moreover it clearly demonstrates that polynuclear ions of Al strongly interact with polymers and will play an important role in their stability ,particularly around pH 5. However, we must point out that, by hypothesis, the interactions between non ionic species of aluminium $(Al(OH)_3$ are not taken into account in our calculation and in NMR studies these Al species do not give any signal. Thus , our model is rather good for pH<6 where non ionic species are negligible and is predicted to fail for ph>6.

Phase diagrams and viscosity

This part deals with the stability of the polymer in the presence of the different species of aluminium. For strongly charged polyelectrolytes such as polyphosphate or sodium polyacrylate in the presence of divalent cations $(Ca^{++}, Ba^{++}....)$, it has been shown that precipitation occurs when a given fraction of charged groups is "neutralized" by the binding of the cations and this situation is realized for a molar concentration of cations of the same order of magnitude than that of charged groups(17,18). In the case of acrylamide-acrylic acid copolymers, the problem becomes more complex since the distance between the charged groups is much higher depending on r. Truong(5,7) has recently shown that the relative probability to form intra or inter molecular bridges must be taken into account. If intramolecular fixation is preponderant, precipitation can be expected while in the case where the intermolecular bridgings are favoured physical gelation should occur with or without syneresis effects.

From these qualitative considerations and taking into account only electrostatic interactions according to our model, we could expect the following features for the behavior of acrylamide-acrylic acid copolymer in the presence of aluminium:

- no interaction with unhydrolyzed polymer (PAMNH) since only electrostatic interactions are considered in the model

-for $r > 0$, the maximum of instability should be observed at the pH values where the interaction with ionic Al species has found to be maximum in NMR experiments and no effect is expected at pH 7 if really only uncharged species of Aluminum are present. Moreover, the high valency and great size of Al_{13} ions could lead to gelation.

We will summarize the results of a systematical study of these systems by phase titration, turbidimetry and viscosimetry. The concentration ranges were $0<c_a<20$ppm for Al and $0<c_p< 1000$ppm for the polymer, pH being ranging between 4 and 7.

Phase titration

In a first series of experiments, we have studied the phase separation of a same composition (C_a and c_p constant) ,for varying pH. The fraction of precipitated polymer was determined by potentiometric titration of the supernatant phase and from the ratio of its viscosity η with the viscosity η_0 of free aluminium polymer solution. An example of results is given in Figure 4 for a copolymer with $r=17\%$, $c_a= 5$ ppm, $c_p= 500$ ppm. . The experiments show that instability is maximum at pH 4.7. The calculated fraction of carboxylate groups neutralized by the formation of a specie $Al(COO)_3$ has also a maximum around the same pH. In fact, the good agreemnt between calculation and experiment is limited to low Al concentration, as we will see later.

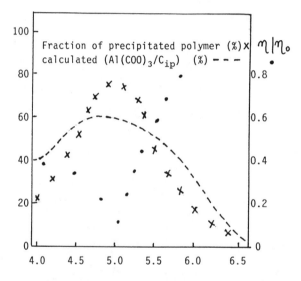

Figure 4: Phase titration of system AD27 (5 g/1) - Al (5 ppm)
         (see test)

Turbidimetry and viscosimetry
In a second series of experiments, we have maintained the pH at two
values 5 and 7 and changed the concentrations of Al and polymer.
    i)" Unhydrolyzed" polyacrylamide (r=0.3%)
No turbidity has been detected in the concentration ranges
investigated. However the reduced viscosity η_{red} is a decreasing
fonction of C_a at pH 5 and pH 7. This result shows that this
polymer interacts with aluminium although no phase separation has
been observed.
In fact this "unhydrolyzed" polyacrylamide sample is slightly
charged and its low polyectrolyte character is confirmed by a
slight difference of η_{red} values at pH 7 and 5, for salt free
solutions. A really neutral polymer should be necessary to
differentiate low effects of electrostatic interactions from non
ionic interactions ,coordination binding at low pH and hydrogen
bonds at pH 7. Nevertheless, at this pH, the adsorption of the
chain on Al(OH)$_3$ aggregates can probably be considered as the main
origin of the loss of viscosity.
    ii) Copolymer of low acrylate content : AD10
The behaviors are quite different at pH 7 and pH 5
    pH 5: we have observed the phase diagram represented in Figure
5 where three domains must be distinguished:
            - domain A for $c_a < 7*c_{ip}$: transparent solutions (both
concentrations expressed in Mole/l)
- domain B for $c_p>c_p^*$ (c_p^* being the critical polymer
concentration of chain overlapping) and $7*c_{ip}<c_a<10*c_{ip}$:
transparent gels
- domain C for $c_a>10*c_{ip}$: phase separation with dense microgels in
equilibrium with solution.
In the domain A, η_{red} increases by increasing c_a and in the domain
B the gel formation has been confirmed by measurements of elastic
modulus. This phenomenon is due to the predominant presence of
polynuclear ions of Al (see Figure 1). The pH range where gelation
occurs is very narrow and at pH 4 and 6 only phase separations are
observed.
            Gelation does not occur for high content of NaCl (20
g/l): only phase precipitation occurs at higher values of c_a and
can be attributed to the interactions with the Al(OH)$_3$ species.
    pH 7: the turbidity of the solutions increases for aluminium
concentrations higher than a limit ,c_a^*, indicating phase
separation and this limit slightly increases with c_p. In the range
of homogeneous solutions η_{red} decreases by increasing c_a. This
polymer is then less stable then PAMNH with respect to Al(OH)$_3$
particles and this phase separation was not predicted by our
electrostatic model.
    iii) copolymers of $r > 7\%$
    pH 5 in pure water
        In the absence of added salts (NaCl) and in the whole ranges
of pH and composition investigated, gelation has never been
observed with copolymers of $r > 7\%$.
        There is simply a domain ($c_a<c_a^*$) of transparent solutions
and a domain ($c_a>c_a^*$) of phase separation. As shown in Figure 6 c_a^*

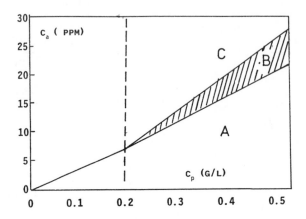

Figure 5 : Phase diagram of AD10-ALCl$_3$ at pH 5 in pure water

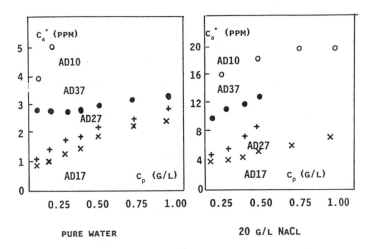

PURE WATER   20 G/L NaCL

Figure 6 : Variations of c_a^* as a function of c_p at pH 5

slightly increases with c_p and also depend on r. The polymer which appears to be the less stable is AD17. The values of c_a^* are very low (< 3ppm) and the phase separation is clearly due to the interactions with mononuclear trivalent ions Al^{+++} because they are preponderant at this Al conentration range.

Our calculation made for $c_a=c_a^*$ shows that phase separation begins when only 20 to 30% of the totality of carboxylate groups are neutralized but one can deduce from phase titration that , at c_a^*, only a low fraction of polymer is precipitated containing the major part of Al. These different observations reveal for the stability with respect to Al features very different from that defined with alkalino-earth cations. For instance, in the presence of Ca^{2+}, the order of stability is inverse: AD17>AD27>AD37>AD60 and the values of c_a^* are much higher(7).

The solution viscosity decreases in the presence of very low number of aluminium species (see examples on Figure 7). This decrease is observed even for $c_a<c_a^*$ when the solutions are still transparent and for all the values of c_p. The variation of η_{red} is found to be faster for the lower polymer concentrations.

pH-5 with added NaCl

The phase diagrams of AD17,AD27 and AD37 have been studied in the presence of 20g/l of NaCl. The same behavior as in pure water was found again for AD17, with higher values of c_a^* (see Figure 6). On the contrary, AD27 and AD37 are able to form gels for $c_p>c_p^*$ with phase diagrams quite identical to that presented in Figure 5 for AD10 in pure water. The gel domain corresponds to the same Al concentration range and gelation must also be attributed to polyions Al_{13}^{3+}.

In the case of AD17, only decrease of viscosity is observed while for $c_p>c_p^*$, the viscosity of AD27 and AD37 solutions measured at the newtonian plateau is an increasing function of c_a.

pH 7 in pure water

The stability of these copolymers is slightly better at pH 7 than at pH 5 (see Figure 8). For instance for AD37, at $c_p>0.5$ g/l, separation begins for $c_a = 7$ ppm to be compared with 3 ppm at pH 5. Nervertheless, they are less stable then PAMNH and AD10 and the same order of stability is found : AD17<AD27<AD37. This confirms the role played by the elestrostatic interactions.

Phase titrations have well confirmed that the precipitated phase contains polymer ,then the turbidity cannot be due to a flocculation of $Al(OH)_3$ particles without polymer adsorption.

pH 7 with added NaCl

As expected from electrostatic screening hypothesis, the addition of NaCl has an increasing effect on the stability (see Figure 8). Moreover, while c_a^* increases with c_p, in pure water, the reverse is observed at 20g/l of NaCl.

Discussion

pH 5

The comparison between phase diagrams and composition of $AlCl_3$ solutions shows that gelation occurs:
- if $c_p > c_p^*$
-when Al_{13}^{3+} polyions are present. It seems that the size of these ions (diameter 20 Å) favors intermolecular bridges: for the same

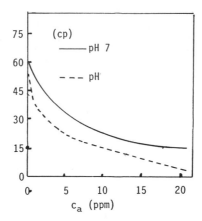

Figure 7 : Variations of the viscosity of the AD27 solution
(0.5 g/l) as a function of c_a

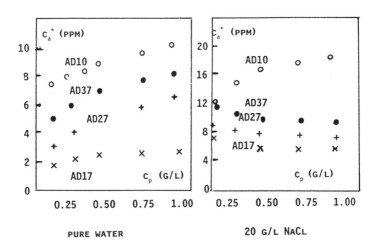

Figure 8 : Variations of c_a^* as a function of c_p at pH 7

calculated fraction of neutralized carboxylate groups, at lower pH, only phase separation is observed with mononuclear ions Al^{3+}.
-for a narrow range of hydrolysis degree depending on the ionic strength
-if the carboxylate groups are completely neutralized (results of our calculation and of electrophoresis measurements)(14)
     We can then deduce that gelation requires well defined conditions corresponding to suitable respective probabilities of intra and intermolecular bridges:
-for $c_p < c_p^*$, in all cases only intramolecular bridges can be expected and then phase separation occurs when a fraction of carboxylate groups is neutralized at an Al concentration increasing with increasing salinity and effects of screening out of the electrostatic counter ions-polymer interactions. The same phenomenon was observed with Ca^{2+} cations(7).
-for $c_p < c_p^*$, , the formation of intermolecular bridges becomes possible but it can lead to a tridimensional network only if the probability of intramolecular bridges remains lower than that of intermolecular ones: this is the case for $\tau > (1.5 + x)\%$, in pure water or $(7 + x)\% < \tau < (27 + x)\%$ for 20 g/l of NaCl (our studies do not allow us to precise the values of x and the exact limits). This means that the average distance between the dissociated carboxylate groups inside a chain must be of the same order of magnitude as the average intermolecular distance. This is realized only for slightly charged polymers in pure water, where all carboxylate groups are effective for fixation (case of AD10). This become possible in large excess of salt which acts by decreasing the number of groups able to bind AL counter ions : then gelation occurs for polymers of higher hydrolysis degree but disappears for AD10 because the binding is too weak and the number of bridges too low.
On the other hand ,a stable gel can only be formed if all the polymer charges are neutralized so that electrostatic repulsions of polymer segments between reticulation points are weak enough.
pH 7
     The problem is slightly different since the size of $Al(OH)_3$ particles are much higher than that of Al_{13}^{3+} polyions. This is a problem of flocculation of charged particles by polymer of opposite charge. Nervertheless, it is well known that hydrogen bonds can be formed between hydroxyl groups of $Al(OH)_3$ and amide groups of the polymer(19).
     If only electrostatic effects are responsible for polymer adsorption and flocculation, our results can be explained according to the same scheme as that used by Furusawa et Al.(20) to interprete the destabilization of negatively charged latex by a cationic polymer. In a first step, the adsorption of the polymer leads to the neutralization of the particles which are no more stabilized by electrostatic repulsions and there is flocculation (we have not studied this step since in our experiments polymer was always in large excess with respect to $Al(OH)_3$). In a second step the adsorption inverses the charge and (we have indeed measured by

electrophoresis a negative charge of the aggregates) the
electrostatic   repulsion is established again leading to the
restabilization (c_a increases with c_p in pure water but there is
screening out of this repulsion at high ionic strength). This could
qualitatively explain the variations of c_a^* with c_p,r and ionic
strength.
    In a second hypothesis, one can consider that the adsorption
is mainly due to hydrogen bonds and that electrostatic attraction
between polymer and particles only brings them together. In such
case, the influence of c_p and r on the stability should be related
to the number of amide groups available for hydrogen bounds and the
increase of salinity should lead to the collapse of the chain and
reduce the probability of interparticles bridging.
Only measurements of adsorption isotherms and determination of the
conformation of the adsorbed chains could help the interpretation.
One could indeed expect large loops in the first hypothesis and
more flatted conformation if hydrogen bonds are predominant.

This work has benefited from grants from Institut Français du
Pétrole and we thank Dr.J. Lecourtier and Dr. G.Chauveteau for
fruitful discussions

Literature Cited
1.   Mungan N. Soc.Pet.Eng.J. 1972 , 469
2.   Nouri H.H.;   Root, P.J. Paper SPE 3523 ,1971, SPE 46th Meeting
3.   Martin F.D.; Sherwood N.S.Paper SPE 5339,1975, SPE Meeting
4.   Muller G.; Lainé J.P.; Fenyo J.C. J.Polymer.Sc.1979, 17,659
5.   Truong N. Thésis 1984 University Louis Pasteur Strasbourg
6.   Schwartz T.; François, J.Makromol.Chem., 1981,182,2757
7.   Truong N.; François, J. in "Solid-Liquid Interactions in
     Porous Media Ed Technip Nancy 1984 p.251
8.   Boutin J.; Contat S. French Patent,1980 n°249 217
9.   Rabhari R.; François J. Polymer, 1988, 29, 845
10. Rabhari R.; François J.Polymer, 1988 , 29, 851
11. Bottero J.Y.; Cases J.M.; Fiessinger F. J.Phys.Chem.,
     1980,84, 2933
12. Bottero J.Y.; Marchal J.P.; Cases J.M.; Poirier J.E.
     Bull.Soc.Chim. de France, 1982,11,1439
13. Akitt J.W.;   Farthing A. J.of Magn.Res.,1978,32,345
14. R.Rahbari Thesis 1988 University Louis Pasteur Strasbourg
15. Bottero J.Y.; Tchoubar D.; Cases J.M.; Fiessinger F.
     J.Phys.Chem. 1984,86,3667
16. Bale M.D.; Schmidt P.W. J.Phys.Chem., 1958,11,1179
17. Strauss U.P.; Siegel A. J.Phys.Chem. 1963, 67,2683
18. Ikegami A.; Imai, N. J.Polym.Sci., 1962 ,56, 133
19. Pefferkorn E.; Nabzar L.; Carroy A. J.Coll.Interface.Sci.,
     1985,106,96
20. Furusawa K.,; Kanesawa M.; Yamashita S. J.Coll.Interface.
     Sci., 1984, 99,341

RECEIVED February 2, 1989

Chapter 6

Gelation Mechanism of Chromium(III)

Mobil Research and Development Corporation, Central Research
Laboratory, Princeton, NJ 08540

The chromium(III) ion is a common crosslinker of many
polymer gels used in reservoir permeability profile
control. It has been generally recognized that olated
chromium(III) species are involved in the crosslinking,
but the detailed mechanism is not fully understood.
Also, Cr(III) salts and redox generated Cr(III) species
do not always yield the olates. Furthermore, Cr(III)
olates can vary greatly in their crosslinking reactivity
and therefore result in gels of different properties.
After studying the gelation of xanthan gum with various
Cr(III) species ranging from simple salts to Cr olates
of varying degrees of hydrolysis, we propose that the
formation of Cr olates is the rate-determining step in
a gelation reaction and the simple binuclear Cr olate is
the most reactive species for crosslinking.

Chromium(III) is a commonly-used crosslinker for preparing profile
control gels with polymers having carboxylate and amide
functionalities (1a,b). Cr(III) is applied in many forms. For
example, it can be used in the form of simple chromic salts of
chloride and sulfate, or as complexed Cr(III) used in leather tanning
(2), or as in situ generated Cr(III) from the redox reaction of
dichromate and bisulfite or thiourea. The gelation rate and gel
quality depend on which form of Cr(III) is used.
    We have found that the Cr olates produced by hydrolysis of
Cr(III) ions are the reactive crosslinking species. The different
gelation rates are due to the different degrees of olation. Furthermore, by controlling the degree of hydrolysis, Cr(III) derived from
various sources mentioned above can exhibit the same gelation rate.

Hydrolysis of Cr(III)

Due to the high charge-to-radius ratio, a hexaaqua Cr(III) cation
loses protons to form olates (3a,b) in this hydrolysis process. One,
two and three protons can be lost from Cr-coordinated H_2O to yield
the mono-, di- and tri- hydroxides of hydrous Cr species,

0097–6156/89/0396–0137$06.00/0
© 1989 American Chemical Society

respectively. These hydroxides then dimerize or polymerize to form Cr olates (Equation 1) through OH or "ol" bridges. Isolation and identification of dimer, trimer, and tetramer were reported by Stunzi and Marty (4a) and higher oligomers by Marty and Spiccia (4b).

$$Cr(H_2O)_6^{3+} \overset{-H^+}{\underset{+H^+}{\rightleftharpoons}} [Cr(H_2O)_5OH]^{2+} \overset{-H^+}{\underset{+H^+}{\rightleftharpoons}} [Cr(H_2O)_4(OH)]_2^+ \overset{-H^+}{\underset{+H^+}{\rightleftharpoons}} Cr(OH)_3(H_2O)_3$$

$$\downarrow \text{Olation} \qquad\qquad \downarrow \text{Olation} \qquad\qquad \text{Olation} \downarrow \qquad (1)$$

$$[(H_2O)_4Cr\overset{OH}{\underset{OH}{<>}}Cr(H_2O)_4]^{4+} \qquad (H_2O)_4Cr\left[\overset{OH}{\underset{OH}{<>}}Cr(H_2O)_2\right]_n^{n+}OH \sim OH \qquad [Cr(OH)_3]_n \text{ gel}$$

Dimer   Linear Polymer   3-Dimensional Polymer

The hydrolysis reaction is very slow at ambient temperatures and is accelerated by boiling chromium salt solutions (5). The hydrolysis reaction is characterized by the transformation of the deep blue colored $Cr(H_2O)_6^{3+}$ to green colored hydrolyzed olates. Another indication is that an aged or boiled Cr(III) salt solution has a higher neutralization equivalent than a fresh one due to the hydrolytically produced protons. One way to establish hydrolytic equilibria quickly is to add appropriate equivalents of bases such as NaOH to Cr(III) salt solutions.

Olated Cr(III) reagents were prepared according to Equation 2 by reacting $Cr(NO_3)_3$ with a calculated equivalent of NaOH. Chromic nitrate was used because the freshly prepared solution affords the hexaaqua Cr(III) cations.

$$Cr(H_2O)_6^{3+} + nNaOH \rightarrow Cr(OH)_n^{3-n}; \quad n = 0\text{-}3 \qquad (2)$$

The pH and UV-VIS spectral data are listed in Table I. For n=3, the product $Cr(OH)_3$ is a precipitate. Therefore, the UV-VIS spectrum of $Cr(OH)_3$ was not obtained.

Table I. UV-VIS of Olated Cr(III)*

$(Cr(NO_3)_3^{3+} + n \text{ NaOH} \rightarrow \text{Olates})$

	λ1	A1	λ2	A2	pH
n=0	406	1.367	574	1.159	2.48
1/3	410	1.551	576	1.256	2.66
2/3	414	1.811	578	1.372	2.78
1	418	2.080	580	1.498	2.87
2	420	2.75	584	1.837	3.25
3	NA	NA	NA	NA	4.89

*0.088M

A gradual shift of absorption maxima to longer wavelength and an increased absorbance were observed when Cr^{3+} reacted with more and more NaOH. The shift of peak position and the change in absorbance were also found by Ardon and Stein (6). The spectra of olated Cr prepared by us agree with the literature.

As increasing amounts of NaOH are added to the $Cr(NO_3)_3$ solution, the hydrolyzed Cr forms dimeric, polymeric and three-dimensional species. Gelled, amorphous and colloidal $Cr(OH)_3$ is eventually formed. E. Matijevic reported the preparation of a monodispersed $Cr(OH)_3$ sol by forced hydrolysis of Cr(III) salt at 90°C (7). Because of the differences in structural features, each olated species (n=1,2,3) should react differently with polymers and form gels of different properties.

Gelation Mechanism and Rate of Cr(III) Olates

The gel time of a 2000 ppm Flocon 4800 (a Pfizer xanthan polymer) in 2% NaCl solution was measured with various Cr(III) crosslinkers at room temperature (Table II). In this series of experiments Cr(III) concentration was 90 ppm. The most reactive Cr(III) species were olates derived from $Cr(NO_3)_3$ with one and two equivalents of NaOH. Gels formed within 5 minutes and the reaction rate appeared to be diffusion-controlled. $Cr(NO_3)_3$ without NaOH required 48 hours to gel the polymer solution. This reflects the time needed to hydrolyze $Cr(NO_3)_3$ in Equation 3.

$$Cr(NO_3)_3 \xrightarrow{H_2O} Cr(H_2O)_6^{3+} \xrightarrow[Slow]{-H^+} [Cr_2(OH)_2(H_2O)_4]^{4+} \xrightarrow[Fast]{Xanthan} Gel \qquad (3)$$

Table II. Variation of Gelation Time by Different Cr(III) Sources

Polymer = 2000 ppm Pfizer Flocon 4800 in 2% NaCl Cr = 90 ppm	
Cr Source	Gel Time
Simple Cr olate	
$OH^-/Cr = 1,2$	<5 Minutes
Cr(III) salts*	24-48 Hours
Redox Cr, $Cr_2O_7^= + S_2O_5^{=*}$	1-2 Weeks
Hydrous $Cr(OH)_3$ colloid	3 Weeks

*Without pH control.

Since low pH suppresses this hydrolysis process, xanthan/NaCl mixtures with a pH lower than 3 did not gel at room temperatures for over a month. The $Cr(OH)_3$ colloid formed by forced hydrolysis at 90°C (7) takes 3 weeks to gel the xanthan solution. This could be due to the slow dissolution of three dimensional networks of $Cr(OH)_3$ by polymer ligands. These results strongly suggest that olates are the reactive crosslinking species and the hexaaqua Cr(III) is not. The crosslinking reactivities are about the same for the olates of various degrees of oligomerization. Because the rate of gelation decreases due to heterogeneous reactions and to the reduction of effective Cr concentration at higher degrees of polymerization, the dimeric olate should be the most effective crosslinker.

Prud'homme observed a second-order rate dependency of Cr in the gelation of polyacrylamide and redox-produced Cr(III) (9). He suggested that the binuclear Cr olate is involved in the crosslinking reactions. It was not determined whether the olation reaction of Cr occurred before or after the polymer ligands attached to Cr ions (Equations 4A and 4B). To determine this, we used $Cr(NO_3)_3$ and NaOH in 1:1 molar ratio to gel Flocon polymer with different mixing sequences. The mixing sequence markedly affected gel time. Based on the following observations, we strongly favor the "B" gelation scheme. Adding 90 ppm Cr olate prepared by pre-mixing $Cr(NO_3)_3$ and NaOH to 2000 ppm Flocon in 2% NaCl led to rapid gelation (<5 min). Adding $Cr(NO_3)_3$ to the polymer solution first and followed by adding an equivalent amount of NaOH resulted in a much slower gel time of 6 hours. Adding an equivalent amount of NaOH to the polymer solution and then $Cr(NO_3)_3$ resulted in a gel time of 1 hour. These results suggest that the reaction between the polymer ligand and Cr olate is much faster than that of forming Cr olate. Therefore the rate determining step is the formation of olate which involves deprotonation and dimerization. Since the gelation time of $Cr(NO_3)_3$ + polymer without NaOH is the longest (1-2 days) among the reaction schemes, the critical step of olation must be the deprotonation.

(A)

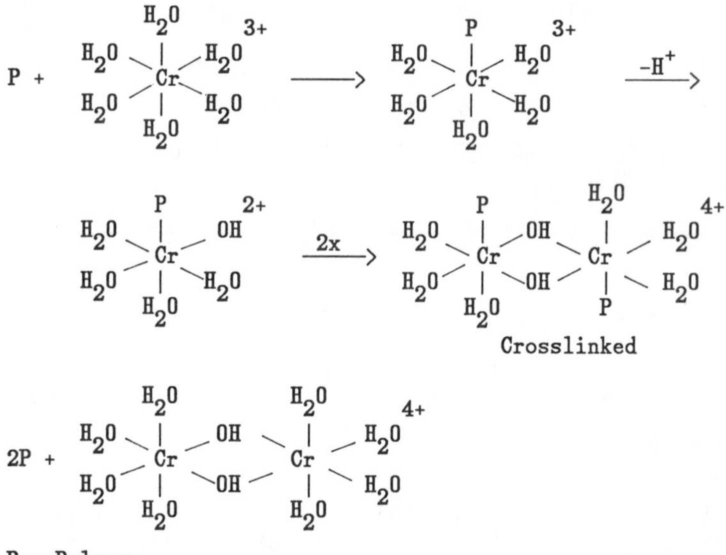

(4)

Crosslinked

(B)

P = Polymer

Gelation Reaction of Redox Generated Cr(III)

$$2Cr_2O_7^{2-} + 3S_2O_5^{2-} + H_2O \rightarrow 2 \text{ }"Cr_2O_3" + 2H^+ + 6SO_4^{2-} \qquad (5)$$

Gelation time of a 2000 ppm Flocon 2% NaCl solution with 90 ppm Cr(III) according to Equation 5 was 2 weeks (Table II), which is in the range of the Cr colloid gelation discussed earlier. Based on the earlier discussion, the gelation reaction of redox generated Cr(III) can also be accounted for with the olation mechanism. However it is very important to have a better understanding of Cr(VI) → Cr(III) reduction, since the reaction rate and the reaction product are highly dependent on the amount of acid present in the redox reaction.

The pH Dependence of Cr(VI) Reduction and Its Products

Chromate is a strong oxidizing agent in acidic media and produces Cr(III) ions in the hydrate form, Equation 6. In neutral and basic media, chromate is a rather weak oxidizer, as evidenced by its negative oxidation potential (Equation 7), and the product is chromic hydroxide (8).

$$Cr_2O_7^{2-} + 14\ H^+ + 6e^- \rightarrow 2\ Cr^{3+}(aq) + 7\ H_2O\quad E^\circ=1.33V \tag{6}$$

$$CrO_4^{2-} + 4\ H_2O + 3e^- \rightarrow Cr(OH)_3(s) + 5OH^-\quad E^\circ=-0.13V \tag{7}$$

None of the Cr(III) products from Equations 6 or 7 are effective crosslinkers since a chromic aqua ion must be hydrolyzed first to form olated Cr to become reactive. Colloidal and solid chromium hydroxides react very slowly with ligands. In many gelation studies, this critical condition was not controlled. Therefore, both slow gelation times and low Cr(VI) → Cr(III) conversion at high chromate and reductant concentrations were reported (9,10).

By adjusting the reaction pH, one can achieve a thermaldynamically favorable redox reaction and produce reactive Cr olates in the dimeric or linear polymer forms for crosslinking.

Gelation Reactions by Acidity-Controlled Redox Reactions

The approach is based on proper control of the acidity of the redox reaction mentioned earlier. A general equation of the dichromate-disulfite reaction as a function of acidity is expressed in Equation 8 where $Cr(OH)_3$ and H^+ are the hypothetical products.

$$2\ Cr_2O_7^{2-} + 3\ S_2O_5^{2-} + n\ H^+ + 7\ H_2O \rightarrow 4\ Cr(OH)_3 + (n+2)\ H^+ + 6\ SO_4^{2-} \tag{8}$$

The reaction of $Cr(OH)_3$ and H^+ is the reverse of the hydrolysis of $Cr(H_2O)_6^{3+}$ (Equation 1). Therefore by adjusting the acidity of the redox reaction (Equation 8), Cr olates of all oligomerizations can be prepared.

At HCl stoichiometries of n=0, 2, 6 and 10 (Equation 8), the reduction product showed wide variations in gelation reactivity (Table III). At n=2 and 6, the reaction products were very effective in gelling a 2000 ppm Flocon 4800 xanthan polymer because dimeric and linear polymeric Cr olates are formed. On the other hand, at n=0 and 10, gelation was very slow, because highly hydrolyzed material similar to $Cr(OH)_3$ is the product when n=0, and aqua ion of Cr(III) with blue color is the product when n=10. Furthermore, in the n=10 case, where no gelation occurred after 24 hrs, a gel formed in one hour after enough NaOH was added to yield the suggested dimeric Cr olate. All the above results have shown the importance of acidity in determining the reactivity of Cr(III) produced in a redox reaction.

Table III.   Gelation by Acidity-Adjusted Redox Reactions

Polymer = Flocon 2000 ppm in 2% NaCl ; Cr(III) = 90 ppm

$$2Cr_2O_7^{2-} + 3S_2O_5^{2-} + nH^+ + 7H_2O \rightarrow 4Cr(OH)_3 + (n+2)H^+ + 6SO_4^{2-}$$

nH^+	Cr (III) Product	Degree of Cr Polymerization	Gelation Time
0	$Cr(OH)_3$	3-d Polymer	No gel in 2 weeks
2	$Cr(OH)_2^+$	2-d Polymer	5 Minutes
6	$Cr(OH)^{2+}$	Dimer	30 Minutes
10	$Cr(H_2O)_6^{3+}$	Monomer	2-3 Days

Similar species are formed from both the acid-adjusted redox and the Cr(III) salt - NaOH reactions. A comparison is given in Table IV. The pH of each corresponding pair at the same Cr concentration is very close, futher supporting this theory. The UV absorption at ~400 nm of the redox products shifted to shorter wavelength when the starting redox mixture was made more acidic, suggesting that less hydrolyzed Cr was formed at higher acidity. This trend was observed in the preparation of Cr olates by the $nNaOH + Cr(NO_3)_3$ reaction (Table I).

Table IV.   Similar Cr Olates Derived From Redox
and From $Cr(NO_3)_3 + xNaOH$
Cr conc: 90 ppm

Redox Reaction		Cr(III) Product	$Cr(NO_3)_3 + xNaOH$	
nH^+	pH		pH	xNaOH
0	5.5	$Cr(OH)_3$	4.9	3
2	3.6	$Cr(OH)_2^+$	3.3	2
6	2.8	$Cr(OH)^{2+}$	2.9	1
10	1.5	$Cr(H_2O)_6^{3+}$	2.5	0

Further Evidence on Acidity Influence of the Cr Redox Reaction

We noticed that the gelation of polymers by the redox method is promoted if 2-3 times the calculated molar ratio of thiosulfite is used (see half Equation 9 below). The gelation rate was very slow when x=1 or x=4 (Table V).

$$2 \; Cr_2O_7^{2-} + x(3S_2O_5)^{2-} + H_2O + Polymer \rightarrow Gel \qquad (9)$$

Table V.   Gelation Rate as a Function of $S_2O_5^{2-}$ Concentration

Polymer = 1500 ppm Flocon in 2% NaCl
Cr=90 ppm

$$2\ Cr_2O_7^{2-} + x\ (3\ S_2O_5)^{2-} + Polymer$$

x	Gel Time, hr.	Color of Reaction Mixture
1	No gel in 2 weeks	Yellow
2	4	Green
3	2	Green
4	24	Blue

Here, disulfite is functioning as a latent acid, releasing protons and bisulfite upon hydrolysis (Equation 10). At the proper proton concentrations, (x=2, 3), rapid Cr(VI) reduction and fast gelation take place. Therefore at x=2 to 3, the redox reaction should be the same as if acid were added at n=2 to 6 (Equation 9). The gelation reactivity of the two are comparable under these conditions.

$$S_2O_5^{2-} + H_2O \rightarrow 2H^+ + 2SO_3^{2-} \qquad (10)$$

At x=1, the redox reaction was very slow and the UV-VIS absorption showed no change with time. The orange - yellowish color persisted for weeks, indicating that there was little or no reduction (i.e., poor conversion) of Cr(VI). At x=4 or more, development of blue color occurred instantly, which is evidence of $Cr(H_2O)_6^{3+}$ production via the acidic redox mechanism. At x=2 to 3, the green color of olated Cr(III) developed in minutes, followed by gelation of the polymer.

Conclusions

• The reactive Cr(III) species in polymer crosslinking are the olates derived from the hydrolysis of hydrated Cr(III) cations.

• The rate-determing step is the deprotonation in the hydrolysis of the Cr^{3+} hydrate.

• In the presence of NaOH or other basic materials, dimerization to form olates becomes the rate-determining step.

• Various Cr(III) olates can be generated from the reduction of Cr(VI) by controlling the amount of the acid.

• The gelation mechanism of redox-Cr(III) follows the same pathway as Cr(III) salt gelation.

Acknowledgments

The author wishes to thank the management of Mobil Research and Development Corporation for permission to present this work. The diligent work of Marie J. Wszolek is greatly appreciated.

Literature Cited

1. (a) Abdo, M. K.; Chung, H. S.; Phelps, C.H.; Klaric, T.M. SPE/DOE
   Paper 12642, 1984.
   (b) Hessert, J. E.; Fleming, P. D.   1979 Tertiary Oil Recovery
   Conf., Wichita, KS, Apr. 25-26, 1979; p 58-63.
2. Udy, M. J.   Chemistry of Chromium and Its Compounds; Vol. 1, ACS
   Monograph Series No. 132, 1956; p 302.
3. (a) Bailar, J. C., Jr., Ed.   The Chemistry of Coordination
   Compounds; Reihold Publishing Co.: New York, 1986; Chapter 13, p
   448-471.
   (b) Colton, R.   Coordination Chemistry Reviews 1985, 62, p
   85-130.
4. (a) Stunzi H.; Marty, W.   Inorg. Chem. 1983, 22, p 2145.
   (b) Spiccia, L.; Marty, W.   Inorg. Chem. 1986, 25, p 266.
5. Hall, H. T.; Eyring, H.   J. Am. Chem. Soc. 1950, 72, p 782.
6. Ardon, M.; Stein, G.   J. Chem. Soc. 1956 78, p 2095.
7. Popey, C. G.; Matijevic, E.; Patel, R. C.   J. Colloid and
   Interface Science 1981, 80, No. 1, p 74.
8. Cotton, F. A.; Wilkinson, G.   F.R.S., Advanced Inorganic
   Chemistry, Interscience Publishers, 1972; 3rd Edition, p 841.
9. Prud'homme, R. K.; Uhl, J. T.; Poinsatte, J. P.; Halverson, F.
   Soc. Pet. Eng. J. 1983, p 804.
10. Southard, M. Z.; Green, D. W.,; Willhite, G. P.   Paper 12638
    SPE/DOE 4th Symposium on Enhanced Oil Recovery, Apr. 16-18,
    1984, Tulsa, OK.

RECEIVED January 27, 1989

Chapter 7

Electron Microscopy of Xanthan

Topology and Strandedness
of the Ordered and Disordered Conformation

Bjørn T. Stokke, Arnljot Elgsaeter, and Olav Smidsrød

Division of Biophysics and Division of Biotechnology, Norwegian Institute of Technology, University of Trondheim, N–7034 Trondheim, Norway

Xanthan is a polyelectrolytic exopolysaccharide of potential interest for polymer flooding in high salinity and high temperature reservoirs. Optical rotation dispersion measurements reveal that xanthan undergoes a temperature- or salt-driven cooperative order-disorder conformational transition. Here we present electron micrographs of vacuum-dried and heavy-metal replicated dilute aqueous xanthan solutions. In the ordered conformation xanthan appears as highly elongated molecules with uniform thickness. Observed weight average contour length combined with experimentally determined molecular weight yield a mass per unit length of (1900±200) Dalton/nm for xanthan from several different sources. This finding is consistent with xanthan being double-stranded in the ordered conformation. Electron micrographs of xanthan under disordering conditions show a mixture of species ranging from purely single- and perfectly matched double-stranded species, to double-stranded chains branching into their two subchains as well as different degrees of mismatched chains.

The usefulness of xanthan in polymer flooding for enhanced oil recovery is based on its ability to yield large increase in viscosity at low polymer concentrations under high-temperature and high salinity conditions. This important property of xanthan is determined both by its molecular weight and by the conformation adopted in solution (1).

Xanthan is reported to undergo a chiroptically detected temperature or salt-driven conformational change from an ordered conformation at high salt and low temperature to a disordered conformation either associated with lowering the salt concentration, or with increasing the temperature (2-5). The primary structure of xanthan has been known for about a decade (6,7), but different structures have been suggested both for the ordered and disordered conformation. Some workers (8-13) conclude that the ordered conformation is double-stranded or double-helix, whereas others (14-17) claim that a single stranded description can account for the observed data under

0097–6156/89/0396–0145$06.00/0
© 1989 American Chemical Society

conditions where the ordered conformation is prevailing. For highly purified, low molecular weight xanthan (Mw = 2 10^5 Dalton), it is reported using light scattering and intrinsic viscosity, that there are no major changes neither in Mw nor in hydrodynamic volume on going through the conformational transition in aqueous solution (18-20). For higher molecular weight species, the radius of gyration decreased while Mw remained essentially constant on passing through the conformational transition (19). Use of cadoxen as solvent yields disordering conditions reported (11,21) to give complete strand separation into single strands.

In this study we use electron microscopy (EM) to study xanthan strandedness and topology both in the ordered and disordered conformation. Correlation of data obtained from electron micrographs to physical properties of dilute aqueous solution on the same sample will be used to provide a working hypothesis of the solution configuration of xanthan. Electron micrographs obtained from xanthan of different origins will be compared to assess similarities and differences in secondary structure at the level of resolution in the used EM technique.

Materials and methods.

Xanthans from several different sources were used in this study: Xanthan samples A, B and C were kindly provided as freeze dried powder of ultrasonic degraded xanthan by Dr. B. Tinland, CERMAV, Grenoble, France. The molecular weights of these samples were determined experimentally in dilute solution by Dr. B. Tinland. Xanthan D was kindly provided as pasteurized, ultrafiltrated fermentation broth by Dr. G. Chauveteau, Institut Francais du Petrole, France. Xanthan E was kindly provided as a freeze dried sample from Dr. I. W. Sutherland, Edinburgh, Scotland. Xanthan F was obtained as a commercial, powdered material (Kelzan, Kelco Inc., a Division of Merck, San Diego CA.). Xanthan G was obtained as a commercial concentrated suspension (Flocon 4800, Pfizer, New York, NY)

Preparation for electron microscopy was carried out by first vacuum-drying dilute glycerol containing aqueous xanthan solutions. A volatile buffer, ammonium acetate, was used as supporting electrolyte to keep xanthan in the desired conformation and to avoid major changes in salt concentration during the vacuum-drying step. The dried preparations were then rotary replicated with Platinuum at an angle of 6°. The contours of the visualized molecules in the electron micrographs were digitized and analyzed as described earlier (22,23).

Results

Ordered conformation. Figure 1. shows representative electron micrographs of samples A, B, C, and D vacuum dried from xanthan solutions under ordering conditions as specified in the legend. Xanthan samples A -D appear as unbranched, uniformly thick, convoluted chains. The contour length varies from molecule to molecule as expected for a polydisperse polymer. The electron micrographic

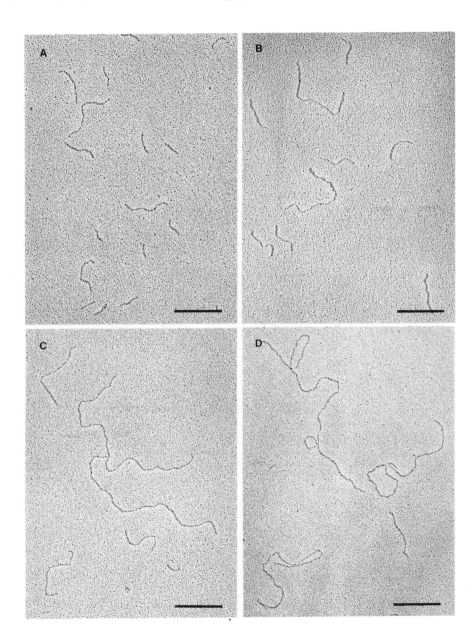

Figure 1. Electron micrographs of xanthan sample A (A), sample B (B) sample C (C) and sample D (D). The electron micrographs were obtained from replicas of vacuum dried solutions containing 100 mM NH$_4$Ac, 50% glycerol and 3 -10 µg /ml polymer. Scale bar = 200 nm.

"snapshot" of the molecules reveals a curvature that varies along each polymer chain and is different from chain to chain. This is as expected for random-coil and worm-like polymers undergoing internal thermal fluctuations. Quantitative analysis of these electron micrographs show that the contour length distribution depends on weight average molecular weight determined by light scattering in dilute aqueous solution. Figure 2 shows the contour length distributions for samples A -D. Each contour length distribution is shifted on the ordinate according to its weight average molecular weight determined in dilute aqueous solution. Figure 2 indicates that the contour length distribution can roughly be described as a log-normal distribution. There is no indication of a bimodal distribution. The weight average molecular weight determined by light scattering divided by the weight average contour lengths yields estimates of the mass per unit length in the range 1750 - 2100 Dalton /nm (Table I) for all three xanthan samples (A-C). The estimated ML for the three samples are in the same range as observed for xanthan of other origins (Table II) using the same approach.

Table I. Molecular weights, contour lengths, and linear mass densities

Sample	Mw/10^5	Ln (nm)	Lw (nm)	n	M_L = Mw/Lw (D/nm)	Ip = Lw/Ln
A	2.6	118	150	197	1730	1.3
B	6.5	235	337	184	1930	1.4
C	11.	352	558	160	1971	1.6

Mw is the experimetally determined weight average molecular weight in dilute aqueous solution.
Ln is the number average contour length from the EM data
Lw is the weight average contour length from the EM data
n   is the number of molecules in the distribution
M_L is the apparent linear mass density
Ip is the polydispersity index.

Figure 3 shows an electron micrograph of a representative replica region of "native" xanthan E. This native xanthan (Fig. 3) contain a large number of aggregates, whereas after exposure to 80 °C for 2 months in synthetic brine (27) this sample reveals the general features of a polydisperse, uniformly thick worm-like polymer.

Figure 4 shows average end-to-end distance $\langle r^2 \rangle^{1/2}$ versus contour distance L for xanthan D, F and G obtained by quantitative analysis of electron micrographs. Electron micrographs of xanthan samples F and G are reported previously (22,25). The limiting slopes α, log ($\langle r^2 \rangle^{1/2}$) ∼ α log L for L> 300.0 nm are observed to α = 0.69 for xanthan D, α = 0.57 for xanthan F, and α = 0.73 for xanthan G.

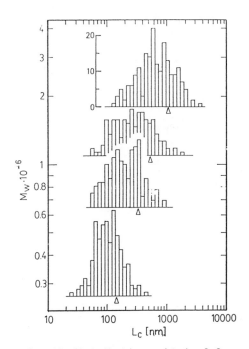

Figure 2. Contour length distributions obtained from electron micrographs of samples A, B, C and D (Fig. 1). The displacement along the y -axis of the contour length distributions corresponds to the reported Mw for each sample. The weight average contour length for each distribution is marked with an arrow.

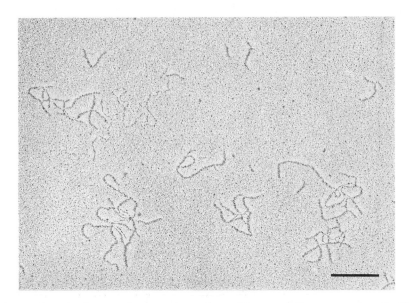

Figure 3. Electron micrographs of native xanthan sample E. The electron micrographs were obtained from replicas prepared as described in Figure 1. Scale bar = 200 nm.

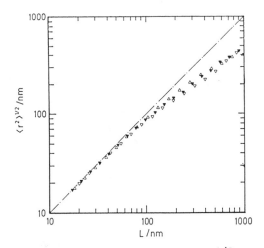

Figure 4. Average end-to-end distance $\langle r^2 \rangle^{1/2}$ versus contour distance L for xanthan from samples D ($\bigtriangledown$) , F ($\triangle$) and G ($\blacktriangledown$) calculated from electron micrographs obtained as described in Figure 1. $\langle r^2 \rangle$ was averaged over a total contour distance of 123 μm , 162 μm, and 124 μm for samples D, F and G respectively.

Table II. Molecular weights, contour lengths, and linear mass densities of xanthan reported elsewhere

Sample	Mw/10^5	Ln (nm)	Lw (nm)	M_L = Mw/Lw (D/nm)
D	18.	732	1050	1714
(ref. 26)	21.	671	1079	1946
(ref. 24)	20.3	578	1017	2000
(ref. 24)	5.7	157	238	2390
(ref. 24)	3.1	98.6	165	1880

Disordered conformation. Figure 5 shows electron micrographs of xanthan D and F obtained from xanthan vacuum-dried from solutions yielding the disordered conformation. The various molecular assemblies are assigned as follow: I = single -stranded, II = perfectly matched double stranded, III = branched from double - to single stranded. This assignment will be discussed below.

Discussion

Ordered conformation. The picture emerging from the electron micrographs of xanthan A, B, C and D is that of a highly elongated, uniformly thick, polydisperse polymer. The relation between solution configuration and the model proposed for xanthan D has been dis-cussed in detail elsewhere (26). The electron micrographs further suggest a much higher degree of association in xanthan E before exposure to 80 °C than after (27). The electron micrographs cannot alone be used to determine whether xanthan of samples A, B, and C is double-stranded or single stranded because of potential decoration effects. However, contour length data from electron micrographs in combination with molecular weight of the same samples in solution yields estimated mass per unit contour length, ML, which for xanthan A - D range from 1700 Dalton/nm to 1970 Dalton/nm (Table I). X-ray fiber diffraction (28) yields axial translation of 0.94 nm per pentasaccharide repeat unit for xanthan corresponding to 950 Dalton/nm for a single- stranded molecule, slightly dependent on acetyl and pyruvyl substitution. The obtained ML therefore suggests a double-stranded configuration for xanthans A - D under ordering conditions. We use the notation double-strand rather than double--helix merely to point out that the resolution in the electron micrographs using the present preparation procedure can not distin-guish between these two possibilities. Our working hypothesis that xanthans A , B and C are double-stranded is contrary to earlier interpretations of solution data of parallell preparations (29).

The native sample E (Fig. 3) contains a relatively large amount of aggregated structures. Most of them appear to rearrange into perfectly matched double-stranded chains after incubation at 80 °C for 2 months (27). This rearrangement is reported to result in a fivefold increase in the apparent viscosity at a shear rate of 1 s-1 (27). A similar rearrangement of xanthan assemblies is also observed in a unpasteurized fermentation broth after exposure of the sample

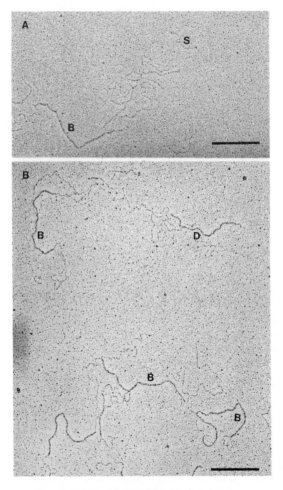

Figure 5. Electron micrographs of xanthan sample F (A), and D (B) obtained from replicas of xanthan vacuum dried from solutions containing 0.1 mM ammonium acetate, 50 % glycerol and 3 - 10 µg /ml polymer. Specie designation: I = single-stranded, II = perfectly matched doublestranded, III = branching from double- to single-stranded. Scale bar = 200 nm.

to disordering conditions (10-5 M NaCl, T = 30 °C) at a low polymer
concentration (26). The association of xanthan single-chains may
result in a double-stranded configuration, or even higher molecular
weight aggregates as exemplified in Fig. 3. There might be several
factors affecting what structure will be the preferred one: polymer
concentration, concentration of divalent ions, and kinetics in the
order - disorder conformational transition. Based on model calculations, it can be shown that the perfectly matched double- stranded
configuration corresponds to the lowest free energy (30). The
aggregates observed in the native material can therefore at the low
polymer concentration used in the preparation for electron microscopy be regarded as metastable state, kinetically trapped from the
perfectly matched double-stranded configuration.

There appears to be close similarity between the observed
scaling behavior of xanthans D, F and G (Fig. 4) indicating that the
adsorption of all these xanthan types results in a comparable
dimensionality change. However, some atypical chains with highly
convoluted segments are occasionally observed in xanthan G. Because
of difficulties in unique assignment of a polymer contour to these
highly convoluted chains they were excluded from the quantitative
analysis. The exclusion of the relative small number of atypical
chains affects the scaling behavior of $\langle r^2 \rangle^{1/2}$ versus L only to a
minor degree. Xanthan G did not appear to be as densely aggregated
as xanthan E possibly because xanthan G was supplied as a concentrated fermentation broth and not as a dried material where more
intermolecular association could have occurred.

Disordered conformation. The difference between Fig. 1 and Fig. 5
indicates that during the last stage of the vacuum drying from 10^{-4}
M NH$_4$Ac there is not enough time for xanthan to rearrange to the
state observed in Fig. 1. Although we use a volatile buffer, an
increase in the salt concentration may occur during the vacuum-drying process resulting in a shift towards the ordered condition from
the starting solution which corresponds to disordering conditions.
The presence of glycerol slows down any tendency to conformational
ordering because of increased solution viscosity. It is therefore
likely that Fig. 5 reflects some aspects of the disordered state of
xanthan, and below we will discuss molecular models suggested by the
electron micrographs of the disordered form.

The electron micrographs in Fig. 5 reveal a mixture of molecular assemblies. These assemblies are assigned with a letter based
on their thickness and on the following interpretation: Because the
configuration under ordering conditions is shown to be double
stranded, the observed branching into two thinner strands most
likely corresponds to the two single-stranded subchains. The
electron micrographs also show that the thin strands are more
flexible than the thick ones. The thick strands have close similarity to those observed in electron micrographs prepared from
solutions where xanthan is under ordering conditions. This is true
both with respect to thickness and apparent turtoisity. Correlation
between observed solution properties and the molecular models
suggested by this type of electron micrographs is detailed elsewhere
(26). Note that similar branching behavior is observed in the
disordered conformation (Fig. 5) for xanthan D and G although
xanthan D have been reported to be single- stranded (16) whereas

xanthan of the same origin as sample G has been reported to be double-stranded (9). The similar behavior of xanthan D and G in the disordered conformation is also consistent with xanthan D and G having similar ordered conformation.

Similarities and differences between xanthan of different origins. The linear mass densities of the xanthan samples A-C (Table I) are comparable with those obtained using the same method, for xanthan of two alternative sources (Table II). For these three origins, the linear mass densities of highly purified samples are all within experimental uncertainty of that of a double-stranded structure. Thus, the EM method in combination with a molecular weight moment (i.e. Mn from osmometry or Mw from light scattering) yields consistent results as far as the linear mass density concerns irrespective of the origin for these xanthans. This is contrary to interpretations of data obtained from solution properties, in which xanthans of origins A - C and D have been suggested to adopt a single-stranded conformation in the ordered state (16,28). The estimates of ML from the EM method is based on the ratio between a experimentally determined molecular weight average and the corresponding contour length average. Although the number of molecules are at least 10 orders of magnitude smaller in the data from the electron micrographs than the data from dilute aqueous solution, it is in principle possible to estimate contour length averages to be used in combination with any molecular weight average because the distribution is obtained.

In conclusion, we observe only minor differences in configurational properties of highly purified and post-fermentation processed xanthans from various sources. However, other workers have reported differences in physical properties among products of various origins (16,31). Our electron micrograph data on the native material and the observed effect of exposing native xanthan samples to high temperature suggest that differences among various products may reside in differences in post-fermentation processing rather than basic configurational features of xanthan. The forces stabilizing a double-stranded structure could well result in formation of more aggregated structures on double-strand formation at sufficiently high polymer concentration. At temperatures well below the transition temperature such aggregated structures would not dissolve when diluted. Increasing the temperature towards the transition temperature would lower the kinetic barriers needed to overcome to reach the most stable (double-stranded) conformation. The observation of aggregate hydration as a result of high temperature exposure is also supported by the work of Holzwarth and coworkers (33), who patented a method for improving filterability based on temperature treatment. This improved filterability probably arise from dissolution from aggregated molecules to dispersed ones under the temperature treatment.

Acknowledgments

Xanthan samples A, B and C was kindly provided by Dr. B. Tinland, Grenoble, France. Xanthan D was kindly provided Dr. G. Chauveteau,

Paris, France, Xanthan E was kindly provided by Dr. I. W. Sutherland, Edinburgh, Scotland. This work was partly supported by grant
V6617 from The Norwegian Academy of Science and Letters.

Literature cited

1.   Holzwarth, G.; Dev. Ind. Microbiol. 1985, 26, 271.
2.   Rees, D.A.; Biochem. J. 1972, 126, 257.
3.   Holzwarth, G.; Biochemistry 1976, 15, 4333.
4.   Morris, E.R.; Rees, D.A.; Young, G.; Walkinshaw, M.D.; Darke,
     A.; J.Mol. Biol. 1977, 110, 1.
5.   Norton, I.T.; Goodall, D.M.; Frangou, S.A.; Morris, E.R.;
     Rees, D.A.; J. Mol. Biol. 1984, 175, 371.
6.   Jansson, P.E.; Kenne, L.; Lindberg, B.; Carbohydr. Res. 1975,
     45, 275.
7.   Melton, L.D.; Mindt, L.; Rees, D.A.; Sanderson, G.R.;
     Carbohydr. Res. 1975, 45, 245.
8.   Holzwarth, G.; Carbohydr. Res. 1978, 66, 173.
9.   Paradossi, G.; Brant, D.A.; Macromolecules 1982, 15, 874.
10. Sato, T.; Norisyue, T.; Fujita, H.; Polymer J. 1984, 16, 341.
11. Sato, T.; Kojima, S.; Norsiyue, T.; Fujita, H.; Polymer J.
     1984, 16, 423.
12. Sato, T.; Norisyue, T.; Fujita, H.; Macromolecules 1984, 17,
     2696.
13. Coviello, T.; Kajiwara, K.; Burchard, W.; Dentini, M.;
     Crescenzi, V.; Macromolecules, 1986, 19, 2826.
14. Muller, G.; Lecourtier, J.; Chauveteau, G.; Allain, C.;
     Makromol. Chem., Rapid Commum., 1984, 5, 203.
15. Lambert, F.; Milas, M.; Rinaudo, M.; Int. J. Biol. Macromol.
     1985, 7, 49.
16. Muller, G.; Anrhourrache, M.; Lecourtier, J.; Chauveteau, G.;
     Int. J. Biol. Macromol. 1986, 8, 167.
17. Lecourtier, J.; Chauveteau, G.; Muller, G.; Int. J. Biol.
     Macromol. 1986, 8, 306.
18. Liu, W.; Sato, T.; Norisyue, T.; Fujita, H.; Carbohydr. Res.,
     1987, 160, 267.
19. Hacche, L.S.; Washington, G.E.; Brant, D.A.; Macromolecules,
     1987, 20, 2179.
20. Liu, W.; Norisuye, T.; Int. J. Biol. Macromol., 1988, 10, 44.
21. Kitagawa, H.; Sato, T.; Norisuye, T.; Fujita, H.; Carbohydr.
     Polymers, 1985, 5, 407.
22. Stokke, B.T.; Elgsaeter, A.; Smidsrød, O.; Int. J. Biol.
     Macromol., 1986, 8, 217.
23. Stokke, B.T.; Brant, D.A.; in preparation.
24. Kitamura, S.; Kuge, T.; Stokke, B.T.; in preparation.
25. Stokke, B.T.; Smidsrød, O.; Marthinsen, A.B.L.; Elgsaeter,
     A.; In Water-Soluble Polymers for Petroleum Recovery, Stahl,
     G.A., Schulz, D.N., Eds.; Plenum Press, New York, 1988; p.
     243.
26. Stokke, B.T.; Smidsrød, O.; Elgsaeter, A.; Biopolymers 1988
     (in press)
27. Stokke, B.T.; Kierulf, C.; Foss, P.; Christensen, B.E.;
     Sutherland, I.W.; in preparation.

28.   Moorhouse, R.;   Walkinshaw, M.D.; Arnott, S.; Am. Chem. Soc.
       Symp. Series 1977, 45, 90.
29.   Milas, M.; Rinaudo, M.;   Tinland, B.;   Carbohydr. Polymers,
       1986, 6, 95.
30.   Washington, G.E.; Brant, D.A.; personal communication.
31.   Morris, V.J.;   Franklin, D.;   l'Anson, K.; Carbohydr. Res.
       1983, 121, 13.
32.   Lange, E.A.; In Water-Soluble Polymers for Petroleum Recove-
       ry, Stahl, G.A., Schulz, D.N., Eds.; Plenum Press, New York,
       1988; p. 231.
33.   Holzwarth, G.M.; Naslund, L.A.; Sandvik, E.I.; U. S. Patent
       4 425 246, 1984.

RECEIVED December 12, 1988

Chapter 8

Succinoglycan

A New Biopolymer for the Oil Field

Anthony J. Clarke-Sturman,[1] Dirk den Ottelander, and Phillip L. Sturla

Shell Research Ltd., Sittingbourne Research Centre, Sittingbourne, Kent, ME9 8AG, United Kingdom

Succinoglycan, a microbially produced polysaccharide with an eight sugar repeating unit, has similar properties to xanthan, and has been successfully used in well completion fluids in the North Sea, where an apparently unique property -- partially reversible viscosity collapse, at a temperature, T_m, determined by the brine composition, was considered advantageous. This viscosity collapse is associated with a structural order-disorder transition, which in sea water occurs at around $75°C$. (A similar transition, normally at a higher temperature, occurs in xanthan.) In calcium bromide brines, the T_m values of both succinoglycan and xanthan fall, leading to viscosity loss as both biopolymers are more readily degraded in the disordered state. Stability of both biopolymers can be improved by using brines based on potassium formate rather than calcium halides.

The oil price rises in the 1970s stimulated interest in Enhanced Oil Recovery (EOR), and fairly rapidly the biopolymer xanthan, the extracellular polysaccharide from the bacterium Xanthomonas campestris, an organism which normally resides on cabbage leaves, was identified as a leading contender as a viscosifier for polymer enhanced water flooding.

Xanthan, used in EOR trials in the USA, and still being considered elsewhere, has found a niche in drilling fluids, which, together with other oilfield uses, accounts for some 2000 tons per year. Xanthan solutions have several useful properties; they display a highly pseudoplastic rheology, are tolerant to salt, and have good thermal stability. There was we felt, however, some scope for improvement.

----

[1]Current address: Shell International Petroleum Company, PAC/31, Shell Centre, London, SE1 7NA, United Kingdom

0097–6156/89/0396–0157$06.00/0
© 1989 American Chemical Society

"Shell" and Biotechnology. Xanthan is manufactured by fermentation, a biotechnological process. How could "Shell", an oil company, be interested in such processes? The Royal Dutch/Shell Group is, however, no newcomer to biotechnology. The Milstead Laboratory of Chemical Enzymology was set up in 1962 and was headed by Professor John Cornforth, who went on to win the 1975 Nobel prize for Chemistry shortly after he retired. In 1970 a fermentation laboratory was built on the same site.

The first large scale process studied was the conversion of natural gas to 'single cell protein' for animal feed. Although technically successful, the project was finally defeated by a combination of high oil, (and gas), prices and cheap soya beans. It did, however, leave us with a strong background in fermentation technology and microbial physiology.

Work on the fermentation of microbial polysaccharides started in the mid 1970's, with the aim of producing improved polymers. Many thousands of samples were screened for microorganisms which produced viscous polymers. Out of over 2000 such 'slime producing' organisms isolated, only one, identified as a Pseudomonas species, now NCIB 11592, seemed to produce a polymer with interesting new properties.

Succinoglycan. Initially identified advantages of this polymer, succinoglycan, were that aqueous solutions of it were more viscous than solutions containing an equal concentrations of xanthan, and that the polymer tolerated higher concentrations of salt, in the sense that solutions passed more readily through microporous filters. These properties made the polymer of potential interest for EOR.

An interesting characteristic, however, was that at a particular temperature, often around $70^{\circ}C$ in sea water for example, the viscosity would collapse, to be partially recovered on cooling.

Experimental

Polysaccharides. Many strains of bacteria produce succinoglycan (1). The Rhizobia, particularly, grow very slowly, and the rate of polymer production is low. Much effort was spent obtaining a strain which produced succinoglycan at a high rate and of good quality (2,3). An organism was selected and a fermentation process developed at laboratory scale. The process has been scaled up successfully and operated at 220 cubic metre scale.

Succinoglycan is now available commercially, in Europe, under the trademark Shellflo-S and as it is a concentrated solution rather than a powder, it readily disperses in brines commonly used in the oilfield.

Shellflo-XA, a proprietary grade of xanthan and cellobond X-100 a hydroxyethylcellulose (HEC) were used for comparative purposes.

Chemical analysis. Succinoglycan, purified by micro-filtration and dialysis, was hydrolysed in 0.5M sulphuric acid at a concentration of approximately 5mg/ml for 16 hours at $95^{\circ}C$. Sugars and acids were determined by HPLC using Biorad HPX-87 columns. No pretreatment was required for acids analysis - detection was by measurement of UV absorption at 206nm. Sugars analysis required the neutralization of the solution with barium carbonate - detection was by refractive index measurement. Typically over 95% of the carbon in the original sample, which was determined by microanalysis, was recovered in the sugars and acids.

Hakomori methylation analysis, and preparation and GLC of the PAAN derivatives (4), were used to determine the linkages. ^{13}C NMR of partially hydrolysed samples confirmed the octasaccharide repeat unit, the beta linkages and the presence of the carboxylic acids.

Rheological measurements. Routine viscosity measurements were made with a Wells-Brookfield micro-cone and plate viscometer, or a Brookfield LVT(D) viscometer with UL adapter. Viscosity-temperature profiles were obtained using the latter coupled via an insulated heating jacket to a Haake F3C circulator and PG100 temperature programmer or microcomputer and suitable interface. Signals from the viscometer and a suitably placed thermocouple were recorded on an X-Y recorder, or captured directly by an HP laboratory data system.

A number of other viscometers were also used, including Haake CV100 and RV3 models. The latter was coupled with a D40/300 measuring head and oil bath circulator for measurements above 100°C. Back pressures up to 4 bar were used and measurements made up to about 160°C.

Hydrolysis rate measurements. Hydrolysis rates were examined by mixing polymer solutions with hydrochloric acid, in apparatus previously described (5). Solutions of polymer and acid are mixed rapidly, and the torque on a rotating PTFE coated fork, attached to a Brookfield LVTD viscometer, recorded as a function of time. Decreases in viscosity were approximated to first-order, and half-lives for viscosity loss calculated.

Results and Discussion

Composition and Structure. Chemical analysis of the polymer from our first strain (NCIB 11592) indicated that it was a polysaccharide containing the sugars, glucose and galactose, and the carboxylic acids -- succinic and pyruvic -- in the approximate ratios 7:1:1:1. A similar polymer had been discovered some years earlier by Harada and his coworkers in Japan (6), produced by an organism which grew on ethylene glycol as sole carbon source. Almost simultaneous with our discoveries (2), the structure of succinoglycan, as the polymer was called, was published (7) (Figure 1).

Like all polymers, succinoglycan, is not a single polymer but a family. Succinoglycan is the extracellular water-soluble polysaccharide produced by a number of Agrobacterium, Rhizobium, and related species of soil bacteria. All have octasaccharide repeating units (Figure 1) with tetra-glucose units in side chains attached to tetrasaccharide main-chain repeating units, each containing a single galactose sugar. Pyruvic acid is bound as pyruvate ketal on the terminus of each side chain, and succinic acid is bound as mono-ester, at an unknown position or positions, as are acetate groups which are found in some strains.

We have examined succinoglycans from a number of bacteria. The polymers from different organisms have different proportions of acid substituents (Table 1). They also have slightly different physical properties, particularly viscosity (Figure 2). In some cases it may be useful to select a polymer with particular characteristics.

Rheology. The viscosity of dilute solutions rises rapidly with concentration (Figure 3). The rheology is, of course, pseudoplastic: the apparent viscosity falls with increasing shear rate. The viscosity of concentrated solutions, the commercial product for example, is, however, lower than might be expected from extrapolation of low concentration data. This is because, in common with many stiff or rigid molecules such as surfactants, a liquid crystalline phase is present. In succinoglycan the mesophase appears at concentrations greater than about 0.5 per cent w/v.

Comparison of Xanthan and Succinoglycan. The physical properties of succinoglycan and its solutions are similar to those of xanthan. Both polysaccharide molecules are relatively stiff, stiffer even than simple cellulosics such as HEC, and have a molecular masses in excess of two million. Recent work by Rinaudo and coworkers (Personal communication) and Crecenzi and colleagues (Int. J. Biol. Macromol., submitted) has shown that succinoglycan molecules are also stiffer than those of xanthan.

This greater stiffness is one reason why solutions of succinoglycan are more viscous than xanthan solutions of equal concentration. The stiffer the molecule, for a given molecular length, the larger the volume of solution that is swept out as the molecules rotate, and the greater the interaction with neighbouring polymer molecules. Such interactions begin to occur at quite low concentrations, much less than 1 g/l. If the interactions are 'sticky', that is, there is a long contact time, entanglement can occur leading to higher viscosities and ultimately a gel.

Entanglement and weak gel formation are characteristic of some oilfield polysaccharides such as guar and starch, but are present only weakly, if at all, in both xanthan and succinoglycan solutions. Solutions of xanthan and succinoglycan are thus able to pass through porous media such as rock, while guar and starch cannot because of their gel-like nature. Hence the different uses of these polymers in the oilfield.

Order-disorder Transition. In common with xanthan, succinoglycan exhibits an order-disorder transition, which in xanthan has been characterized by a number of techniques, including optical rotation and birefringence measurements (8). Viscosity measurements, as a function of temperature, can show the transition in xanthan (9), although the viscosity change is often small as the xanthan backbone is cellulose and still quite stiff. In contrast, however, the viscosity change in succinoglycan solutions is dramatic (Figure 4).

Effect of Temperature. The fall in viscosity usually occurs over a small temperature range, and to a level close to that of the solvent itself. The recovery of viscosity on cooling is partial (Figure 5)

Table 1. Composition of Succinoglycan from Different Strains

	Glucose	Galactose	Pyruvate	Acetate	Succinate
Pseudomonas sp. NCIB 11592	7	1.02	0.96	<0.3	1.08
Pseudomonas sp. NCIB 11264	7	0.94	0.91	<0.3	0.67
Rhizobium meliloti K24	7	1.00	1.00	1.20	1.65
Rhizobium meliloti DSM 30136	7	1.05	1.02	1.07	1.98
Agrobacterium radiobacter NCIB 8149	7	1.05	0.98	0.35	0.74
Agrobacterium radiobacter NCIB 9042	7	0.94	1.00	<0.3	0.57
Agrobacterium tumefaciens DSM 30208	7	0.97	0.89	<0.3	0.51

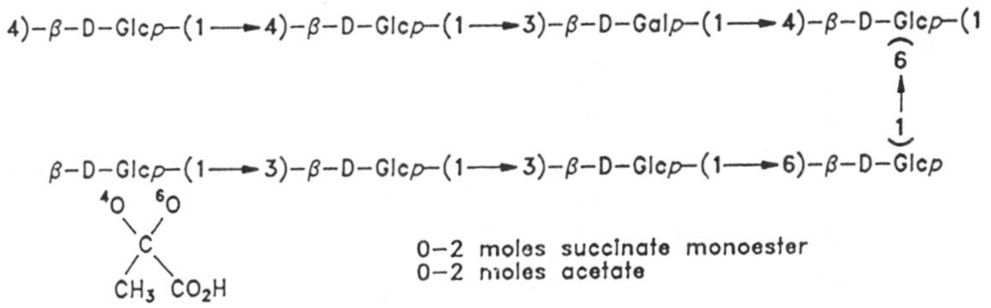

4)−β−D−Glc*p*−(1⟶4)−β−D−Glc*p*−(1⟶3)−β−D−Gal*p*−(1⟶4)−β−D−Glc*p*−(1

β−D−Glc*p*−(1⟶3)−β−D−Glc*p*−(1⟶3)−β−D−Glc*p*−(1⟶6)−β−D−Glc*p*

0−2 moles succinate monoester
0−2 moles acetate

Figure 1.   Structure of succinoglycan.

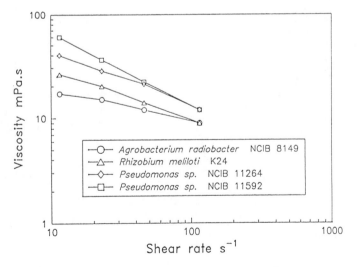

Figure 2.   Viscosities of some succinoglycans.

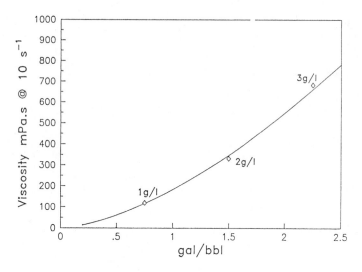

Figure 3.   Viscosity as a function of concentration.

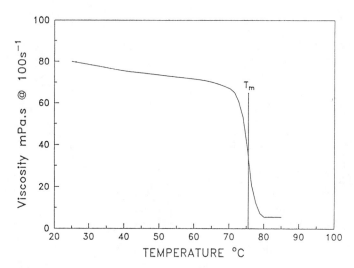

Figure 4.   Viscosity of succinoglycan as a function of
temperature.

and depends on the structure of the succinoglycan and the nature of the environment.

Transition Temperature. The order-disorder transition temperature mid-point T_m is also determined both by the environment - salinity, pH value, and the nature of the ions in solution, and the structure of the polymer - charge, pyruvate, succinate and acetate content. Most published work describes the low salinity behaviour of such polymers, and has shown that for xanthan, the transition temperature exceeds 100°C as the salinity exceeds about 1 per cent sodium chloride (10).

Salt Effects. In the low salinity region, the charge on the polymer determines the slope (Figure 6), and the acetate content changes the T_m by about 15°C per mole/repeat unit. We have obtained data for solutions of higher salinity. Not only have we looked at sodium chloride, but also salts such as calcium chloride and bromide which are used in heavy brines for drilling and workover operations.

The results (Figure 7), in this case for succinoglycan, are rather surprising. The transition temperature does not always increase or decrease as a function of the salt concentration, but rather, in some brines, a maximum value is reached after which the transition temperature falls with increasing salinity. This is particularly apparent for the two calcium salts, calcium bromide and chloride, both of which are used extensively in heavy brine drilling fluids.

One line, however, that of compound 'X', does increase steadily. It is significant, and we will return to it later.

We have examined a number of different salts and have shown that the transition temperature can almost be altered at will (12), in many cases in line with the Hofmeister or lyotropic series (13-15), a relationship found for many other aqueous systems such as hydrocarbon solubility and protein stability.

The viscosity of xanthan solutions in calcium bromide brines was also measured as a function of temperature. We found that the transition temperature fell rapidly as the concentration of salt was increased, so much so that above 2 molar it fell below that of succinoglycan.

Significance. What is the significance of these observations? For succinoglycan solutions the answer is obvious, above the transition temperature they have little or no viscosity, which may be undesirable. Such polymers are usually used as viscosifiers or for particle suspension. On the other hand, a drop in viscosity may be an advantage if fluid penetrates a formation hotter than the well as there could be little or no subsequent formation damage.

Degradation. There is, however, a second, more important, feature. Above the transition temperature we have shown that both xanthan and succinoglycan are more susceptible to degradation (5). This may be because the polymer side chains become dissociated from the main chain, in a similar manner to the unfolding of proteins. Viscosity changes, however, indicate that main chain linkages have become more

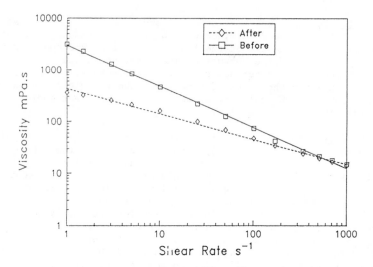

Figure 5.   Viscosity of succinoglycan before and after heating.

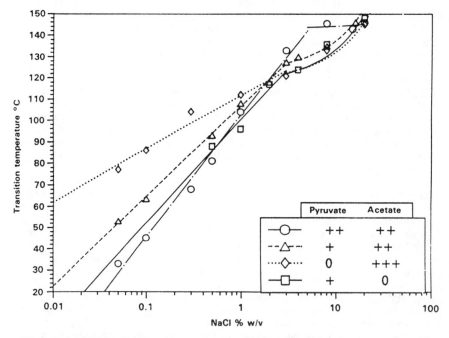

Figure 6.   Transition temperature/salinity profile for a number of
different xanthans.

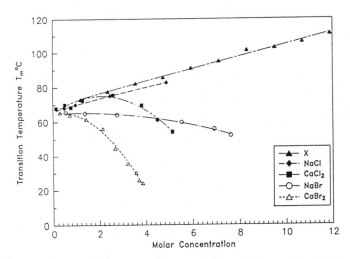

Figure 7. Transition temperature of succinoglycan in different brines.

mobile, perhaps leading to a lower activation energy for hydrolysis, and hence decreased stability. The main chain linkages also become more exposed.

Under many conditions, the viscosity of HEC solutions falls about 100 times faster than that of xanthan solutions. The rate of viscosity loss of Shellflo-S solutions, on the other hand is comparable with that of xanthan solutions at temperatures below the transition temperature, and comparable with that of HEC solutions above.

The relative stabilities of xanthan and Shellflo-S can be reversed. We have made measurements in a number of brines, including calcium bromide (5), and found that xanthan solutions were less stable than those of succinoglycan above about 400g/l (c. 2M).

Transition temperature and stability are, therefore, closely linked. The rate of viscosity loss of both xanthan and succinoglycan solutions increases about 100 fold as the molecules become disordered above T_m. This may be a problem, as some users of xanthan in heavy brines have discovered, but it can be used to advantage.

<u>Well Completion</u>. In well completions it is essential that clean fluids are used, otherwise the production zone can become blocked and oil and gas flows reduced. This is particularly important if the structure is tight. It is also important that the polymer degrades rapidly and leaves little or no residue.

The Troll field, in the Norwegian sector of the North Sea, is the tenth largest gas field in the world, and the biggest offshore in Europe. During the 1985 and 1986 drilling seasons, a formulation containing Shellflo-S was used in some 20 well completion operations. Mixing in the polymer was straightforward and the results were extremely favourable.

It seems probable that in this case the conditions were ideal for Shellflo-S. The water depth is in excess of 300 metres and the reservoir relatively shallow, with a bottom hole temperature of around $70^{\circ}C$, close to the transition temperature.

This combination of long riser and relatively shallow hole probably favours a polymer with a high pseudoplastic index - high viscosity at the low shear rate in the slow-moving fluid in the riser - and rapid, clean breaking at the well bottom temperature.

Such favourable conditions may not occur in many, or any other places. So, is the polymer a one field wonder? Probably not. It was mentioned earlier that it is possible to manipulate the transition temperatures of both Shellflo-S and xanthan by altering the brine. By this means the breaking of the polymer could be adjusted to the required value. The problem is that conventional heavy brines, such as calcium chloride and bromide, decrease the stability of these polymers, when just the reverse is often required, as heavy brines are generally used in deeper, hotter wells.

<u>Increased Stability</u>. Returning to Figure 7, we see that compound 'X' is able to increase the transition temperature substantially. This compound is potassium formate, which can be used at specific gravities up to 1.6. It is possible that, by using a brine based on formate, or perhaps some other salt, biopolymers such as Shellflo-S

and xanthan might be used more widely. We have recently shown (<u>12</u>) that most of the viscosity of a xanthan solution can be retained for up to 40 hours at $153^{\circ}C$ in a formate based brine which has the same density as a calcium chloride brine in which xanthan is unstable.

Conclusions

Shellflo-S, succinoglycan is an interesting new polymer for use in the oilfield, in many ways complementary to xanthan. It is particularly appropriate for use in well completion fluids at moderate temperatures, where a more pseudoplastic rheology than that provided by HEC, and a more rapidly breaking polymer than xanthan is required.

The order-disorder transition temperature is a crucial parameter for biopolymers like Shellflo-S and xanthan, as it controls both rheology and breaking sensitivity.

Brine composition and polymer type should be considered together, and could lead to improved control of the viscosity of polysaccharide based drilling fluids.

<u>Shellflo-S</u>. In one area at least, biotechnology has something to offer the oil industry. Many of the polymers used in drilling fluids are based on natural products. They can be cheap like starch or guar or offer performance no synthetic can match, such as xanthan or succinoglycan. Shellflo-S is for use now in such areas as well completion, but more widespread use could follow if its particular range of properties are seen to be useful, and the cost of production is reduced.

Literature Cited

1. Zevenhuizen, L. P. T. M. In <u>Industrial Polysaccharides</u>; Stivala, S. S.; Crecenzi, V.; Dea, I. C. M., Eds.; Gordon and Breach: New York, 1987; pp 45-68.
2. Cripps, R. E.; Ruffell, R. N.; Sturman, A.J. European Patent 40 445 B, 1981.
3. Linton, J. D.; Evans, M. W.; Godley, A.R. European Patent 138 255 A, 1985.
4. Seymour, F. R.; Chen, E. C. M.; Bishop, S. H. <u>Carbohydr. Res</u>. 1979 , <u>73</u>, 19.
5. Clarke-Sturman, A. J.; Pedley, J.B.; Sturla, P.L. <u>Int. J. Biol. Macromol</u>. 1986, <u>8</u>, 355.
6. Harada, T. <u>Arch. Biochem. Biophys</u>. 1965, <u>112</u>, 65.
7. Harada, T.; Amemura, A.; Jansson, P. E.; Lindberg, B. <u>Carbohydr. Res</u>. 1979, <u>77</u>, 285.
8. Milas, M.; Rinaudo, M. <u>Carbohydr. Res</u>. 1979, <u>76</u>, 186.
9. Morris, E.R.; Rees, D.A.; Young, G.; Walkinshaw, M. D.; Darke, A. <u>J. Mol. Biol</u>. 1977, <u>110</u>, 1.
10. Holzwarth, G. <u>Biochemistry</u> 1976, <u>15</u>, 4333.
11. Holzwarth, G.; Ogletree, J. <u>Carbohydr. Res</u>. 1979, <u>76</u>, 277.
12. Clarke-Sturman, A. J.; Sturla, P. L. European Patent 259 939 A, 1988.
13. Hofmeister, F. <u>Arch. Exp. Path. Pharmakol</u>. 1888, <u>24</u>, 247.

14. von Hippel, P. H.; Schleich, T. In <u>Structure and Stability of Biological Macromolecules</u>; Timasheff, S. N.; Fasman, G. D. Eds.; Dekker: New York, 1969; Chapter 6, p 417.
15. Friedman, H. L.; Krishnan, C. V. In <u>Water: a comprehensive treatise</u>; Franks, F. Ed.; Plenum: New York and London, 1973; Vol. 3, p 1.

RECEIVED January 18, 1989

Chapter 9

Complex Copolymers for Mobility Control, Water Purification, and Surface Activity

John J. Meister and Chin Tia Li

Department of Chemistry, University of Detroit, Detroit, MI 48221–9987

Complex polymers containing different functional groups
and structures with sharply different chemical and
physical properties are needed to meet the varied and
multifaceted demands of resource recovery.
Simultaneously, the supply of oil on our planet is
dwindling while the cost of this increasingly scarce
resource must eventually go up. The price to use oil to
recover oil will then become economically unjustified and
additives or process chemicals that are produced from
renewable resources will be the materials of choice.
The polymers that can be converted to industrial process
chemicals are described here and the chemistry that can
convert two of these natural products, starch and lignin,
to such chemicals is detailed. Synthesis,
characterization, and testing of nonionic and anionic
graft copolymers of starch show the products to be
viscosifiers, drag reducing agents, and surface active
agents. Synthesis, characterization, and testing of
nonionic, cationic, and anionic graft copolymers of lignin
show the products to be viscosifiers, thinning agents,
drilling mud additives, and surface active agents.

Most polymers used in oil field operations and resource recovery are
synthetic. The man-made materials in common use are:
poly(1-amidoethylene) (= polyacrylamide),
poly(1-amidoethylene-r-(sodium 1-carboxylatoethylene) (=
partially hydrolyzed polyacrylamide), poly(1-amidoethylene-r-
(sodium 1-(2-methylprop-1N-yl-1-sulfonate)amidoethylene) (AMPS-
acrylamide copolymer), and xanthan gum. Xanthan gum is a synthetic
because no one finds a pool or river contaminated with Xanthomonas
compestris that experiences the right sequence of solute to naturally
produce the exocellular gum polymer. A fermenter is a man made
object, a tree is not.
The natural polymers in common use are: cellulose, lignin,
chitin, starch and guar gum. The natural products had complete

0097–6156/89/0396–0169$09.75/0
© 1989 American Chemical Society

control of the polymer market prior to 1930 but the greater effectiveness of synthetics produced at low cost from petroleum have sharply
cut the market share of the natural polymers. This market shift can
not be permanent since the growing scarcity of oil will sooner or
later affect the price of petroleum-derived chemicals. While oil
prices have dropped from 1982 to 1988, this has not added one drop of
oil to the world's oil reserves. The price of all petrochemicals
must rise in the future and, at that time, we must be ready to make
greater use of the renewable sources of carbon that nature provides.
     Energy, entropy, and economic considerations make it obvious that
the easiest way to replace a synthetic polymer is to start with a
natural polymer. The five natural polymers listed above can be used
in one of four ways. These are:
1. Use the polymer directly;
2. Decompose the natural polymers into constituent chemicals and
react these constituent chemicals into solvents, reagents, and
monomers;
3. React the natural polymer to add small substituents to it,
thereby forming a useful derivative; and
4. Form a graft, block, or interpenetrating network polymer from
the natural polymer.
While the decomposition process of utilization method 2 will become
more common in the future, most applications of natural polymers
will be based on methods 1, 3, and 4 to make use of the chemical
energy and physical structure stored in the molecule by the plant or
animal. Current utilization, problems in utilization, and prospects
or methods for the future use of the above, five natural products can
be assessed on the basis of current technology. After discussing the
nature, benefits, and features of all five natural polymers, this
analysis will focus an developments in the formation of complex
polymers from starch and lignin.

NATURAL POLYMERS.

CELLULOSE AND CELLULOSE DERIVATIVES. Cellulose is the most abundant
of all the natural polymers and comprises at least 30 weight percent
of the dry mass of all the vegetable matter in the world. The larger
plants contain higher fractions of cellulose, so, on a dry basis,
wood contains 40 to 50% cellulose[1]. This is one of the few natural
products that is currently being degraded into chemicals and feedstocks (method 2). Cellulose is a linear molecule formed by the
polymerization of a simple sugar, glucose. Due to very strong
intermolecular and intramolecular hydrogen bonding, cellulose is insoluble in water. Water is unable to break these hydrogen bonds and
enter into association with cellulose to dissolve it. Thus, to
utilize cellulose in resource recovery, water-soluble cellulose
derivatives are required. The water-soluble cellulose derivatives
available are: sodium carboxymethyl cellulose (CMC,2), hydroxyethyl
cellulose (HEC), and sodium carboxymethylhydroxyethyl cellulose
(CMHEC) [1,2].

STARCH.   Starches are used as components and/or processing aids in
the production of resources such as aluminum, paper, copper, water,
and oil. The use of this natural polymeric material is based on its

thickening, gelling, adhesive, and film-forming properties, as well as its low cost, controlled quality, and ready availability.

The characteristics of a starch can be modified by chemical, physical, and/or enzyme treatment to enhance or repress its intrinsic properties, or to impart new ones. This capability for modification has been a necessary factor in developing new uses for starch and in maintaining old markets.

There are five prime factors that determine the properties of starches: 1. starch is a polymer of glucose (dextrose); 2. the starch polymer is of two types: linear and branched; 3. the linear polymeric molecules can associate with each other giving insolubility in water; 4. the polymeric molecules are organized and packed into granules which are insoluble in water; and 5. disruption of the granule structure is required to render the starch polymer dispersible in water. The modification of starch takes into account these factors.

Starch is widely distributed as the reserve carbohydrate in the leaves, stems, roots, and fruits of most land plants. It is currently used directly after being separated from the plant as a thickener and dispersant (method 1). It is also used after being phosphorilated or digested (method 3) with several uses under investigation for graft copolymers of starch (method 4). The commercial sources are the seeds of cereal grains (corn, sorghum, wheat, rice), certain roots (potato, tapioca or cassava, arrowroot), and the pith of the sago palm. Since the growth conditions are different in each plant, the starch from each plant source will vary somewhat in appearance, composition, and properties. Consequently, the starch is described by its plant source as corn starch, potato starch, tapioca starch, rice starch, wheat starch, etc(3).

Chemical Composition.   Starch is composed of carbon, hydrogen, and oxygen in the ratio of 6:10:5 ($C_6H_{10}O_5$), placing it in the class of carbohydrate organic compounds. It can be considered to be a condensation polymer of glucose and yields glucose when subjected to hydrolysis by acids and/or certain enzymes. The glucose units in the starch polymer are present as anhydroglucose units, the linkage between the glucose units being formed as if a molecule of water is removed during a condensation polymerization. Since the starch is formed in the plant by a biosynthetic process, the polymerization process is a complex one involving enzymes. The glucose units are connected through an oxygen atom attaching carbon atom 1 of one glucose unit to carbon atom 4 of the next glucose unit, forming a long chain or polymer of interconnected glucose units.

Molecular Structure.   Most starches consist of a mixture of two polysaccharide types: amylose, an essentially linear polymer, and amylopectin, a highly branched polymer. The relative amounts of these starch fractions in a particular starch are a major factor in determining the properties of that starch.

Amylose. This linear polymer consists of a chain of glucose units connected to each other by 1-4 linkages. The glucose units are in the "alpha-D-glucopyranose" form. This means that the glucose is arranged in the form of a six-membered ring in such a way that the hydroxyl group on carbon atom 1 is on the same side of the ring as the hydroxyl group on carbon atom 2 . This spatial configuration of the alpha-glucose units in amylose differs from that of beta-glucose

units in cellulose where the beta configuration indicates that the
hydroxyl group on carbon atom 1 is on the opposite side of the
glucopyranose ring from the hydroxyl group on carbon atom 2. Thus,
the link connecting the glucosidic oxygen atom to carbon atom 1 of
the anhydroglucose units in the amylose chain has a different spatial
relationship to the glucopyranose ring than the analogous link in
cellulose.

Amylopectin. Amylopectin has a highly branched structure consisting
of short linear amylose chains, with a segment chain length ranging
from 12 to 50 anhydroglucose units and an average chain length of
about 20 anhydroglucose units, connected to each other by alpha-1,6-
linkages. These alpha-1,6-linked, anhydroglucose units make up about
4 to 5% of the total repeat units in amylopectin and are the branch
points of the molecule. There is no single definite structure for
amylopectin. The structure is therefore described in the statistical
picture of its structural features and details(4-7).

LIGNIN.   Lignin [8068-00-6] is a natural product produced by
all woody plants. It is second only to cellulose in mass of polymer
formed per annum(8). Lignin constitutes between 15 and 40 percent of
the dry weight of wood with variation in lignin content being caused
by growing conditions, species type, the parts of the plant tested,
and numerous other factors(9). Plants use lignin to 1. control fluid
flow, 2. add strength, and 3. protect against attack by
microorganisms(10). Each cell of the plant grows its own lignin.
The cell undergoes "lignification" in response to an internally-
orchestrated series of reactions which take place all during cell
differentiation (10). Lignin appears first in the primary
(exterior) wall of the cell "corners". As the cell grows, lignin
deposits throughout the primary wall and then appears in the
secondary, interior wall of the cell. During this growth period,
lignin deposits develop in the intercellular region, also. Lignin
appears to be attached to the crystalline microfibrils of cellulose
by phenylpropane linkages to carboxyl groups. Such a bond structure
would be a uronic acid ester linkage (10).

Recent work by Atalla(11) supports the idea that lignin is at
least a semi-ordered substance in wood with the plane of the aromatic
ring parallel to the cell wall surface. Woody plants synthesize
lignin from trans-coniferyl alcohol 1 (pines), trans-sinapyl alcohol
2 (deciduous), and trans-4-coumaryl alcohol 3 by free radical
crosslinking initiated by enzymatic dehydrogenation(12). Structures
of these alcohols are given in Figure 1.

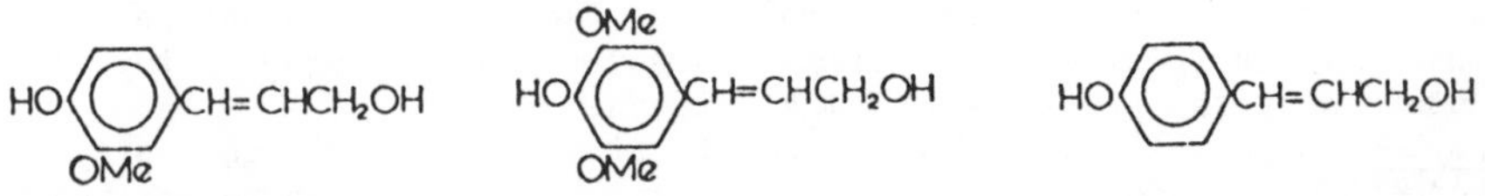

| 1 | 2 | 3 |

Figure 1. Structure of alcohols **1**, **2**, and **3**.

Different ratios of these alcohols are used by different species
of plants to form lignin, with the result that lignins from different
sources will have different elemental and functional-group
compositions. This alone would give lignin an extensive chemical
diversity. Lignin recovery processes which extract lignin from wood,
change the chemical composition of lignin and make this material
extremely heterogeneous.

Methods for recovering lignin are the alkali process, the sulfite
process, ball milling, enzymatic release, hydrochloric acid
digestion, and organic solvent extraction. Alkali lignins are
produced by the kraft and soda methods for wood pulping. They have
low sulfur content (< 1.6 wt. %), sulfur contamination present as
thioether linkages, and are water-insoluble, nonionic polymers of low
(2,000 to 15,000) molecular weight. Approximately 20 million tons of
kraft lignin are produced in the United States each year.

Recovery Methods.   The sulfite process for separating lignin from
plant biomass produces a class of lignin derivatives called
lignosulfonates. Lignosulfonates contain approximately 6.5 weight
percent sulfur present as ionic sulfonate groups. These materials
have molecular weights up to 150,000 and are very water- soluble.
Lignosulfonates are used in resource recovery as cement grouting
agents, sacrificial agents in EOR, and thinning agents in drilling
muds. The material is therefore directly utilized in energy recovery
(method 1).

Milled wood lignin (MWL) is produced by grinding wood in a rotary
or vibratory ball mill. Lignin can be extracted from the resulting
powder using solvents such as methylbenzene or
1,4-dioxacyclohexane(13). Milling only releases 60 weight percent or
less of the lignin in wood, disrupts the morphology of lignin in
wood, and may cause the formation of some functional groups on the
produced lignin(14). Despite these limitations, milling appears to
be an effective way of recovering lignin from plants with only slight
alteration. Enzymes which hydrolyze polysaccharides can be used to
digest plant fiber and release lignin. After digestion, the lignin
is solubilized in ethanol(15). Extensive analytical studies support
the idea that enzymatically produced lignin has undergone no major
modification in removal from plant material(16-20). Milling and
enzyme release are not commercial methods to recover lignin at
present but the commercialization of ethanol from biomass processes
may make enzyme lignin available in large quantities.

Acid hydrolysis of the polysaccharide portion of wood will release
lignin but also causes major condensation reactions in the
product(21). These reactions can be minimized by using 41 wt.
percent hydrochloric acid in place of other mineral acids but some
condensation reactions still occur(22). This is not an effective
method by which to obtain unaltered lignin. On the other hand,
lignin can be solvent extracted from wood at temperature of 175°C
using solvent mixtures such as 50/50 by volume water/1,4-
dioxacyclohexane(23). Changes in lignin under these conditions
appear to be minor.

Once lignin is separated from other plant products, it is not now
widely used in resource recovery. Extensive studies on the
modification of lignin have been made(24) because of the enormous
mass of kraft lignin produced each year by the pulp and paper

industry. A major method of forming derivatives in the formation of graft copolymers of lignin, molecules in which a sidechain of synthetic polymer has been grown off of a lignin molecule. Graft copolymerization sharply changes the properties of lignin and allows useful products to be made from this waste biomass(25.)

CHITIN.   Chitin is the exoskeletal polymer used by most anthropods for structural and protective body parts. It is available in large quantities but is difficult to use. The limited solubility of chitin in all but a few very special solvents and the very limited chemistry and technology available to alter chitin into useful materials has hindered the application of this material to articles of commerce.

GUAR GUM AND ITS DERIVATIVES.(26,27)   This polymer is currently utilized by methods one (directly) and three (adding small functional groups to make a derivative) and this will most likely continue to be the case. Guar gum is derived from the seed of the guar plant, Cyanopsis tetragonolobus L. Grinding the endosperm of the guar bean produces relatively pure guar gum. The structure of guar gum is that of a branched polymer with the backbone of the polymer composed of mannose units and every other mannose unit having a galactose branch bonded to it. The two monomers - mannose and galactose - are simple sugars, and guar gum is termed a "polysaccharide" or, more specifically, a "galactomannan." There are no ions present in the polymer structure; thus, the polymer is termed "nonionic."
        Treatment of guar with ethylene oxide, propylene oxide, and 2-chloroethanoic acid in alkaline medium results in formation of hydroxyethyl guar, hydroxypropyl guar, and carboxymethyl guar, respectively(26). Sequential treatment of guar with two of these chemical reagents can result in the formation of a "double derivative" such as carboxymethyl-hydroxypropyl guar.
        Hydroxypropyl guar is the most widely available guar derivative. By controlling the moles of propylene oxide substitution per polymer chain, significant improvement of guar gum properties are observed. Broken hydroxypropyl guar gels contain no more than 2% insoluble residue(28). Hydroxypropyl guar hydrates more rapidly than guar gum in cold water and is more soluble in water-miscible solvents such as methanol, ethanol, and ethylene glycol. Hydroxypropyl guar has a somewhat greater high-temperature stability than guar and is more resistant to enzymatic degradation.

APPLICATIONS TO RESOURCE RECOVERY

·Many processes that are basic to the extraction of natural resources are facilitated by addition of polymers. To be useful, the polymers must meet an interrelated list of chemical and physical properties as well as economic criteria. The chemical and physical properties demanded of the polymers are:
Solubility. Solubility in the primary process solvent is a mandatory criteria which is easily met in most extraction processes conducted in water.
Rheology. Polymers are often added to change solvent or process flow properties. The addition of polymers almost always causes nonnewtonian flow behavior in the resulting fluid.

Stability. Polymers must be chemically stable, mechanically stable, and thermally stable for long-term applications. Chemical stability to polymer chain scission can be built into the molecule by 1. increasing backbone bond strength, 2. attaching sterically hindering groups to the backbone to protect it, and 3. making a "ladder-backbone" polymer such as is shown in Figure 2. Typical reactions with the polymer which must be inhibited are illustrated by the redox degradation of a polymer by an iron/oxygen couple (29).

$$\text{Poly} + \text{Fe}^{3+} \;\rightleftharpoons\; \text{Poly}_{ox} + \text{Poly}^{\bullet} + \text{H}^{+} + \text{Fe}^{2+} \quad (1)$$

$$\text{Poly}^{\bullet} + \text{R}^{\bullet} \;\rightleftharpoons\; \text{Poly-R} \quad (2)$$

$$\text{Fe}^{2+} + \text{O}_2 \;\rightleftharpoons\; \text{Fe}^{3+} + \text{O}_2^{-} \quad (3)$$

Stability of the polymer to mechanical forces is required because extension and shear forces exerted on the polymer tend to be concentrated, by slippage of the polymer chain, in the center of the molecule(30).

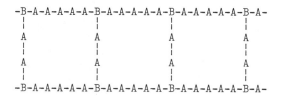

Figure 2. Ladder backbone copolymer.

A requirement that the polymer be thermally stable over a given range of temperature is actually just a demand for chemical stability as a function of temperature. As temperature increases, there is an exponential increase in the rate of any reaction. The reaction-specific temperature (T) at which a given reaction increases sharply in rate is given by

$$T = \Delta E_a / R$$

where ΔE_a is the energy of activation of the reaction and R is the gas constant (26,27).

Surface Behavior. Most extraction processes deal with several phases. At the boundaries between these phases, an interface exists which can be populated with or depopulated of polymer. Situations in which the polymer should accumulate at the surface of one phase are 1. the flocculation of clays and fines or 2. the formation of foams, while situations in which the polymer should depopulate the surface of the phase boundary are 3. minimizing adsorption in mineral acid leaching or 4. minimizing surface tension with surfactants in oil recovery by miscible flooding.

Methods of controlling surface behavior are to: 1. create polar and nonpolar regions in the molecule thus producing a hydrophilic-lipophilic balance in the molecule, 2. charge the

molecule by introducing ionic sites with the same or opposite charge
as the boundary, or 3. introduce or remove functional groups in the
molecule which produce binding reactions, such as a. hydrogen bond
creation or b. nitrogen lone-pair donation, with the surface.

This multitude of properties the polymer must possess dictate that
better polymer performance will be obtained from materials with
complicated structures. Such polymers are complex polymers:
1) random copolymers, 2) block copolymers, 3) graft copolymers, 4)
micellizing copolymers, and 5) network copolymers. There has been a
dramatic increase in the past decade in the number and complexity of
these copolymers and a sizable number of these new products have been
made from natural products. The synthesis, analysis, and testing of
lignin and starch, natural product copolymers, with particular
emphasis on graft copolymers designed for enhanced oil recovery, will
be presented.

COMPLEX POLYMERS FOR RESOURCE RECOVERY

Early applications of polymers to extraction processes were based on
the ability of polymers to alter solution rheology. Pye(31) and
Sandiford(32) showed that fluid mobility could be effectively reduced
by the addition of small amounts of polymer to the solvent.
Similarly, Thompson(33) showed that the addition of trace quantities
of high molecular weight polymer to solvent reduced friction loss
during turbulent flow of the solution. As knowledge of the
characteristics needed to do more than change solution rheology grew,
complex polymers were prepared and applied.

GRAFT COPOLYMERS.   Goggarty was one of the first applications
specialists to propose, in 1978, that graft copolymers by used for
resource recovery(34). A number of such graft copolymers have now
been made and tested and the knowledge gained about products based on
lignin and starch will be summarized below.

STARCH GRAFT COPOLYMERS.

Several groups have conducted graft copolymerizations on starch or
its purified components, amylose and amylopectin. The syntheses are
based on attack of the anhydroglucose unit of starch by cerium (+4)
ion. The free radical produced from this attack is then immersed in
a monomer solution polymerizable by free radical, chain
polymerization and a graft copolymer is formed. This reaction
apparently occurs by the formation of a chelate between the
cerium (+4) ion and the hydroxyl groups on the number 2 and 3 carbon
atoms of the anhydroglucose ring(35-40). This mechanism is not
completely confirmed for starch because the investigations on cerium
(+4)-polysaccharide reactions were done on cellulose, not starch.
Cellulose has a axial-equatorial positioning of the hydroxyl groups
on the number 2 and 3 carbon atoms of the repeat unit while starch
has these hydroxyl groups both equatorial. These structures are shown
in Figure 3. This means that it is possible that the space between
the two hydroxyl groups in starch and cellulose may differ enough to
cause a different initiation mechanism in the two products. If the
same chemistry applies to both molecules, the complexed cerium (+4)

ion executes an oxidation-reduction reaction to convert one hydroxyl group to a carbonyl group and break the 2-3 carbon-carbon bond. This is shown in Figure 4. The free radical may then react with monomer or engage in termination of crosslinking reactions.

Synthesis.   Graft copolymer was formed in aqueous solution by ceric-ion-initiated, radical polymerization of monomer on starch. Polymerization was conducted in an inert, N_2 atmosphere. Details of the synthesis procedure may be found in references 41 to 43. In recovering the polymer product, freeze drying was used with care since freeze drying produces a more dissolvable and useful product but can degrade polymers with molecular weights of 1 million or more. Poly(starch-g-(1-amidoethylene))   Poly(starch-g-(1-amidoethylene)) copolymers can be made by ceric-ion-initiated, free radical polymerization of 2-propenamide on starch(41-43). At monomer concentrations in the synthesis mixture of 1.5 m or more, this reaction produces a water-soluble viscosifier with thickening capacity above that of poly(1-amidoethylene), the homopolymer of 2-propenamide. This graft copolymer is readily made. Yields of product are 90 wt % or above(44) and high-molecular-weight samples are made by using molal concentrations of 1.5 or above to insure gelling of the sample during polymerization. The copolymers are good drag reducing agents for promoting pipeline flow(45). Data demonstrating drag reduction by parts-per-million concentrations of these polymers is given in Table 1. Drag reduction tests were run by connecting a 50 mL buret to 2 meters of 0.318 cm o.d. nylon tubing and flowing water by gravity pressure from the vertically held buret through the horizontally held tubing. Passage time measured in this experiment was the time taken for 20 mL of fluid to pass from the buret to the tubing as measured by the movement of the fluid meniscus between the 1 mL and 21 mL marks on the buret. Passage time measurements were then used to calculate percent drag reduction from the equation

$$\% \text{ Drag Reduction} = 100*(t_w - t_p)/t_w \qquad (4)$$

where t_w = average efflux time for water, and t_p = average efflux time for polymer solution. Drag reductions of more than 17 percent were achieved with concentrations of polymer of 3 to 4 ppm by weight. Some unpublished work has been performed in industrial laboratories to determine if these copolymers are effective beneficiating agents for bauxite ore.

TABLE 1. Reduction of Drag in the Flow of Water Caused by the Addition of Starch-1-amidoethylene copolymer

Sample Number	Fluid	Graft Copolymer Concentration (wppm)	Passage Time[1] (second)	Percent of Drag Reduction (%)
32	water	0	35.63 ± 1.11	0
33	Copolymer Solution	3.72	29.3 ± 0.49	17.75
34	Copolymer Solution	6.10	33.5 ± 3.35	6.02

[1] Passage times given are the means of 4 determinations.

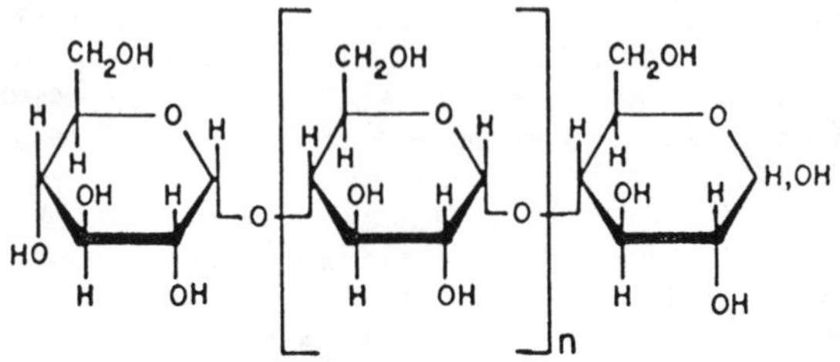

Figure 3. Positioning of hydroxyl groups.

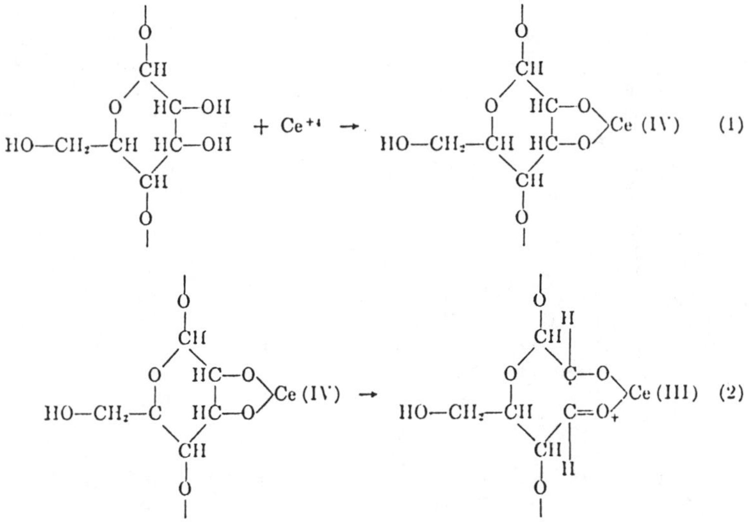

Figure 4. Initiation mechanisms in starch and cellulose.

Poly(starch-g-((1-amidoethylene)-co-(sodium 1-carboxylatoethylene))).
Poly(1-amidoethylene) is, however, rarely used as a viscosifier.
Instead, the homopolymer is reacted with base (hydrolyzed with NaOH)
to convert some of the amide units of the polymer to carboxylic acid
units. The acid units on the hydrolyzed polymer dissociate in water
and produce a polyanionic polymer. This polyelectrolyte expands in
water because of ion-ion repulsion and, as an enlarged molecule, is
a better viscosifier.

Poly(starch-g-(1-amidoethylene)) copolymer is not a
polyelectrolyte and will be a smaller molecule in water than an equal
molecular weight, partially hydrolyzed poly(1-amidoethylene).
Polyelectrolyte effect should, however, cause the graft copolymer to
expand in solution in the same way it causes poly(1-amidoethylene) to
expand, so a series of hydrolyzed graft copolymers were prepared from
poly(starch-g-(1-amidoethylene))(41-43) and these derivatives were
tested to determine the effect of hydrolysis on copolymer properties
in solution.

In the following sections, synthesis of the anionic polymers,
copolymer molecular weight, limiting viscosity number, electrolyte
effects, solution shear thinning, screen factor, polymer radius of
gyration, and solution aging will be discussed and data on the
copolymers presented.

Materials. Deionized-distilled water was used for all syntheses and
distilled water was used in all solutions. All salts were reagent
grade and were used as received. Dialysis membrane was Spectrapor
no. 2, 12,000 upper molecular-weight-cutoff membrane from Spectrum
Medical Industries, Los Angeles, CA.

Hydrolysis. The starting material for the hydrolysis reaction is a
2.0 g sample of previously prepared, graft copolymer(41,42). The
copolymer is hydrolyzed in a basic, saline, aqueous solution under
anaerobic conditions. Sufficient copolymer is dissolved in
sufficient sodium chloride brine to form a combined, final reaction
mixture of 2 g/dL copolymer in 1.0 M sodium chloride. Sufficient
sodium hydroxide is dissolved in water to yield a final concentration
in the combined reaction mixture of 0.5 M. The solutions are
saturated with nitrogen, warmed to 40°C, combined, and allowed to
react with stirring for 10 minutes under a nitrogen blanket.

$$-(CH_2 - \underset{\underset{\underset{NH_2}{/}}{\overset{|}{C=O}}}{CH-})_n - CH_2 - \underset{\underset{\underset{NH_2}{/}}{\overset{|}{C=O}}}{CH} - (CH_2 - \underset{\underset{\underset{NH_2}{/}}{\overset{|}{C=O}}}{CH-})_m + Na^+ + OH^- \rightleftharpoons$$

$$-(CH_2 - \underset{\underset{\underset{NH_2}{/}}{\overset{|}{C=O}}}{CH-})_n - CH_2 - \underset{\underset{\underset{ONa}{/}}{\overset{|}{C=O}}}{CH} - (CH_2 - \underset{\underset{\underset{NH_2}{/}}{\overset{|}{C=O}}}{CH-})_m + NH_3 \qquad (5)$$

The reaction vessel is then packed in ice and the reaction mixture is
neutralized to pH 6.5 to 7.0 with aqueous hydrochloric acid. The
neutral solution was dialyzed against nitrogen-saturated, distilled
water for two days. The polymer solution is precipitated in a 5- to

10-fold excess of vigorously stirred 2-propanone. The precipitated
copolymer is filtered from the 2- propanone, slurried in 2-propanone,
filtered, and vacuum dried to constant weight. Alternatively, the
dialysis solution can be freeze dried to constant weight. Freeze
drying produces a more dissolvable and useful product but can degrade
polymers with molecular weights of 1 million or more(46). During
hydrolysis, exposure of the copolymer to oxygen or acid will result
in immediate cleavage of the polyglucoside backbone.

Assays.   Nitrogen assays to determine 1-amidoethylene unit content
were done by Kjeldahl method. Limiting viscosity numbers were
determined from 4 or more viscosity measurements made on a
Cannon-Fenske capillary viscometer at 30°C. Data was extrapolated to
0 g/dL polymer concentration using the Huggins equation(44) for
nonionic polymers and the Fuoss equation(45) for polyelectrolytes.
Equipment.   Viscosities were measured using Cannon-Fenske capillary
viscometers and a Brookfield LV Microvis, cone and plate viscometer
with a CP-40, 0.8° cone. Capillary viscometers received 10 mL of a
sample for testing while the cone and plate viscometer received
0.50 mL.

Results and Discussion. Of the 12 samples of starch graft copolymer
synthesized, half were hydrolyzed to anionic polyelectrolytes.
Synthesis data on these 6 samples are given in Table 2. These
particular samples were chosen for hydrolysis because the samples can
be intercompared to see the effect of synthesis variables on ultimate
product properties. Samples 5, 8, and 11 have the same mole ratio of
cerium ion to starch backbone, N_g, in their reaction mixture. Samples
7, 8, and 9 all have the same reactable mass per starch molecule,
M_{cal}, in their reaction mixtures. In general, as the mass fraction
of hydrolyzable sidechain in the graft copolymer increases, the
degree of hydrolysis produced by treatment with base increases.
Sample 11 is an exception to this rule.

TABLE 2. Hydrolysis of Graft Copolymers to Anionic Terpolymers

Sample[a] Number	Nitrogen Content (wt.%)		Degree of Hydrolysis (number %)	1-amidoethylene Repeat Units (wt. %)
	Before Hydrolysis	After Hydrolysis		
5*	12.24	11.10	9.31	62.04
7	14.82	12.82	13.50	75.11
8	14.85	12.93	12.93	75.27
9*	14.24	12.20	14.33	72.17
10*	15.17	9.64	36.45	76.89
11	14.71	14.14	3.87	74.56

a) Sample numbers are identical to those in previous publications on
these compounds (41,42,43,48).
*Hydrolysis data determined from repeated hydrolysis reaction.

Limiting Viscosity Number. Limiting viscosity numbers for the
polymers in distilled water are given in Table 3. Limiting viscosity
number increases with increasing copolymer molecular weight.
Further, after 12 to 14 percent hydrolysis, limiting viscosity number
of the derived, partially-hydrolyzed copolymer is 3 to 10 times
larger than that of its nonionic precursor. The ratio of $[\eta]$ after

hydrolysis to $[\eta]$ before hydrolysis decreases with increasing weight percent sidechain in the copolymer. Limiting viscosity number ratio is not clearly correlated with molecular weight or degree of hydrolysis, however.

TABLE 3. Limiting Viscosity Numbers for Complex Copolymers in Water and Electrolyte Solution

Sample Number	Limiting Viscosity Number in Distilled Water		Weigh Avg. Molecular Weight $M_w * 10^{-6}$	Limiting Viscosity Number in 1.0 Molar Sodium Nitrate	
	$[\eta]$ Graft Copolymer	$[\eta]$ Hydrolyzed Graft Copolymer		$[\eta]$ Graft Copolymer	$[\eta]$ Hydrolyzed Graft Copolymer
5	3.99	39.48	0.701	3.17	5.00
7	4.12	27.00	2.06	3.32	6.85
8	5.84	25.96	1.19	4.69	6.08
9	6.95	23.29	1.74	5.68	5.58
10	11.83	48.98	1.00	9.94	7.90[a]
11	12.18	28.64	3.71	10.03	3.69[a]

[a]Partially degraded during hydrolysis.

Screen Factor. Screen Factor, the ratio of passage time of a solution to that of a solvent in a screen viscometer(45),

$$\text{Screen Factor} = \frac{(\textbf{Solution} \text{ Passage Time})}{(\textbf{Solvent} \text{ Passage Time})} \qquad (6)$$

measures the relaxation time, ϕ, and viscosity, η, of the solution. A screen viscometer   produces a flow controlled by the viscoelastic properties of the fluid. The following equation,

$$[f_1(\rho, S, \phi) + f_2(\eta, \phi)] * (dh/dt)^2 + f_3(S, \phi) * (dh/dt) + h = 0 \quad (7)$$

where $f_i(x)$ and constants are defined in reference 43, shows the influence of each fluid property on the change of height with respect to time (t). The derivative, (dh/dt), is the differential of the fluid passage time. Since the quadratic term dominates this equation, the screen factor allows fluids to be compared on the basis of relaxation time.

The screen factors for solutions of the hydrolyzed copolymers (terpolymers) are given in Figure 5. These data show that solution relaxation time increases with increasing polymer concentration and increasing molecular weight. Thus, viscoelastic nature of the solution and polymer entanglements increase with increasing hydrolyzed copolymer concentration and increasing molecular weight. Comparison of screen factor data for equal molecular weight co- and terpolymers at equal concentration shows the terpolymers give more viscoelastic solutions.

Electrolyte Effect on Polymer Solution Rheology. As salt concentration in an aqueous poly(1-amidoethylene) solution increases, the resulting brine becomes a more Theta-solvent for the polymer and the polymer coil compresses(47). This effect is particularly pronounced for partially hydrolyzed poly(1-amidoethylene). The effect of salt on solution rheology was tested in two ways. First, limiting viscosity numbers in 1.0 molar sodium nitrate were determined. These data are given in Table 3. Second, a series of 0.15 g/dL polymer solutions were made in solutions containing between 0 and 0.35 molar sodium chloride or between 0 and 2.5×10^{-2} molar calcium chloride. Solution viscosity at a shear rate of 45 s^{-1} for these solutions is given in Figures 6, 7, and 8.

Comparison of the limiting viscosity numbers determined in deionized water with those determined in 1 molar sodium nitrate shows a 20 per cent decrease in copolymer intrinsic viscosity in the saline solution. These results are consistent with previous studies using aqueous saline solutions as theta solvents for 2-propenamide polymers[47]. Degree of hydrolysis controls the value of limiting viscosity number for the hydrolyzed copolymers in distilled water.

Limiting viscosity numbers of the hydrolyzed graft copolymers in 1.0 molar sodium nitrate are proximate to the values for the unhydrolyzed copolymer in the same solvent. This shows the increase in $[\eta]$ in distilled water upon hydrolysis was due to ion-ion repulsion in the hydrolyzed side chains. Lower values of $[\eta]$ in 1.0 molar sodium nitrate after hydrolysis for the two highest molecular weight copolymers hydrolyzed, numbers 10 and 11, show that the starch backbone of these polymers was partially destroyed during the hydrolysis reaction. This backbone scission is probably due to small amounts of oxygen in the hydrolysis solution. Limiting viscosity number data in Table 3 shows that, in the absence of oxygen, this hydrolysis process sharply increases the size of the polymer in solution without degrading the molecule[48]. These graft copolymers possess all the same benefits and blemishes as simple carboxylic-acid containing polymers when applied to resource recovery. The copolymers will 1. adsorb less on negatively charge surfaces, 2. adsorb more on positively charged surfaces, 3. produce higher viscosity at a given concentration in aqueous solution, 4. lose solution viscosity when electrolyte concentration increases, 5. precipitate in the presence of di- or tri-valent cations, and 6. gel in the presence of low solubility, di- or tri-valent cations, more than the nonionic precursor.

Solutions containing 0.15 g/dL polymer and between 0 and 0.342 molar sodium chloride or between 0 and 2.49×10^{-2} molar calcium chloride show declines in viscosity as salt content increases. Solution viscosity of nonionic copolymers declines, at most, 3 percent in the range of electrolyte concentrations tested. Solutions of hydrolyzed copolymer lose viscosity exponentially as electrolyte concentration in the solution increases.

The data show that calcium chloride is 20 times more effective on a concentration basis in compacting hydrolyzed copolymers than is sodium chloride. Radii of gyration shown in Table 4 are calculated with the Flory equation,

$$[\eta] = \frac{\Phi < \bar{s}^2 >^{3/2}}{M_w} \tag{8}$$

where the value for Φ, the Flory constant, is that appropriate for a broad- molecular-weight distribution in the polymer, 3.09×10^{22}/mole when limiting viscosity number is expressed in g/dL. These data show that the copolymers are high radius of gyration viscosifiers which

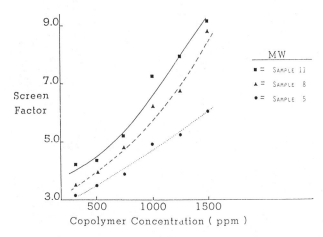

Figure 5. Screen factor of solutions of hydrolyzed poly(starch-*g*-(2-propenamide)) copolymer.

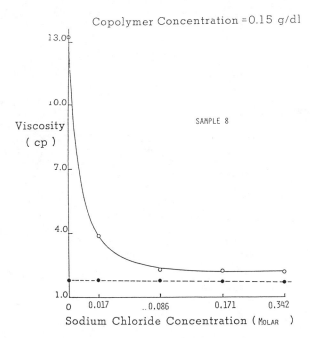

Figure 6. Viscosity of poly(starch-*g*-(2-propenamide)) in sodium chloride brine.

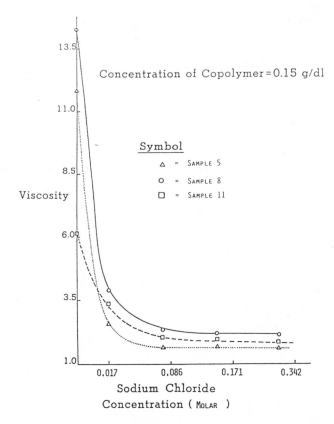

Figure 7. Effect of sodium chloride concentration on the viscosity of hydrolyzed poly(starch-*g*-(2-propenamide)) copolymer solutions.

significantly expand in size in aqueous solution when hydrolyzed. These data further show that addition of 2- propenamide to starch produces a larger molecule than would be obtained by homopolymerization of the 2-propenamide (see reference 49). The radius of gyration in microns of poly(1-amidoethylene) is

$$< \bar{s} \; 2 \; >^{1/2} \; = \; [\; 1.286 \; x \; 10^{-2*(\bar{M}_w x 10^{-6})1.8} \;]^{1/3}$$

TABLE 4.   Radius of Gyration in Microns, $<\bar{s}^2>^{1/2}$

Copolymer Number	Copolymer (unhydrolyzed) H_2O	1 M HNO_3	Hydrolyzed Copolymer H_2O	1 M HNO_3	$<\bar{s}^2>^{1/2}$ for an equal Molecular Wt Poly(1-amidoethylene)[a]
5	.208	.193	4.47	.225	.189
7	.302	.281	.564	.357	.361
8	.282	.262	.464	.286	.260
9	.339	.317	.508	.315	.327
10	.337	.317	b	b	.234
11	.527	1.06	b	b	.514

a) This radius of gyration calculated from the Mark-Houwink equation, $[\eta] = 6.31 \; x \; 10^{-5} \; (\bar{M}_w)^{0.8}$, for poly(1-amidoethylene)(50).

b) $< \bar{s}^2 >^{1/2}$ can not be calculated for these degraded polymers.

Viscosity Loss With Time. Poly(1-amidoethylene) solutions lose viscosity with time(51). Several authors have attributed this viscosity loss to oxygen or radical degradation of the polymer(51), but Francois(52) has shown that changes in viscosity only occur in solutions made from broad-molecular-weight-distribution poly(1-amidoethylene). Since very narrow-molecular-weight-distribution poly(1-amidoethylene) produces a stable solution viscosity and since Narkis(53) has shown that the original solution viscosity can be obtained by precipitating and redissolving the polymer, it would appear that solution viscosity loss is caused by slow disentangling of a broad- molecular-weight polymer mixture.

To determine if poly(starch-g-(1-amidoethylene)) also exhibited this artifact, a series of polymer solutions were prepared and aged for 2 months. Representative data are presented in Figure 9. These data show that graft copolymers also age and, hence, the copolymers and hydrolyzed copolymers slowly disentangle in solution. This behavior is consistent with the broad molecular weight distribution found for these compounds in reference 2. The data in Figure 9 also show that the lowest molecular weight hydrolyzed copolymer, polymer 5, is biodegraded in solution. Hydrolyzed copolymer 5 showed behavior similar to that of hydrolyzed copolymer 8 and 11 when 1% methanal (formaldehyde) was added to the solution.
Pseudoplasticity.   Low-concentration solutions of water viscosifiers are usually nonnewtonian fluids(54) and therefore fail to follow the pressure and flow behavior predicted by newtonian models of flow.   To

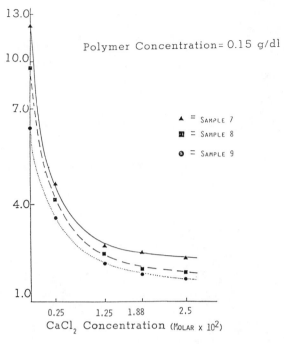

Figure 8. Solution viscosity as a function of calcium chloride concentration for hydrolyzed poly(starch-g-(2-propenamide)) terpolymer.

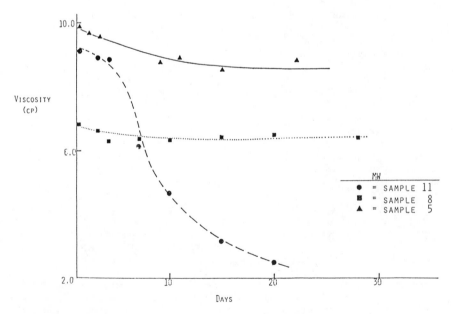

Figure 9. Change with time of viscosity of solutions of hydrolyzed poly(starch-g-(2-propenamide)) copolymer.

determine if the newtonian diffusivity equation or the non-newtonian
Ikoku equation(55) is needed to predict response to shear of a graft
terpolymer solution, the effective viscosity at 5 shear rates was
determined for a series of copolymer and hydrolyzed copolymer
solutions and the Ostwald-DeWaele exponent, n, was obtained from a
match of equation 10 to the data.

$$n_e = H * \overset{\cdot}{\gamma}^{(n-1)} \qquad\qquad (10)$$

In equation 10, H is consistency, n_e is effective viscosity, and $\overset{\cdot}{\gamma}$
is shear rate. Viscosity as a function of concentration in distilled
water and shear rate for hydrolyzed copolymers 5, 8 and 11 are shown
in Figures 10, 11, and 12, respectively. The Ostwald-DeWaele
exponent from the data of Figure 11 is plotted in Figure 13. Data of
Figures 10 through 13 show that graft terpolymer solutions are
pseudoplastic and become more so (decreasing n) with increasing
concentration.

Figure 14 gives limiting viscosity numbers for hydrolyzed
copolymer 11 as a function of shear rate. Since limiting viscosity
number is a function of molecular size, these data show that solution
pseudoplasticity occurs because of compaction of the solvated polymer
with increasing shear.

Effective viscosity as a function of shear rate for 0.15 g/dL of
copolymer 5 in distilled water is given in Figure 15. The
Ostwald-DeWaele exponent for copolymer solutions is greater than that
of matching hydrolyzed copolymer solutions at a given concentration.
Thus, copolymer molecules are less compactable in solution than are
their hydrolyzed derivatives, and pseudoplasticity of polymer
solutions increases upon hydrolysis.

Poly(starch-g-((1-amidoethylene)-co-(sodium 1-(2-methylprop-2N-yl-1-
sulfonate)amidoethylene))).

Strongly anionic, highly water-soluble, graft copolymers of
starch can be made by adding 2-propenamide and sodium
2,2-dimethyl-3-imino-4-oxohex-5-ene-1-sulfonate (Na DMIH) to the
polymerization reactions. See references 43 and 56 for a discussion
of these polymers.

LIGNIN GRAFT COPOLYMERS

Lignin [8068-00-6] is a natural product produced by all woody plants.
It is second only to cellulose in mass of polymer formed per
annum(57). Approximately 20 million tons of kraft lignin are produced
in the United States each year(58). This enormous production of
cheap biomass has induced a significant effort, stretching over 40
years, to alter lignin into industrial or commercial products. A
series of lignin graft(59,60) copolymers have been made which
function effectively as drilling mud additives, flocculating agents,
and thickening agents. Free radical, graft copolymerization of
water-soluble monomers onto kraft or other lignin produces a natural
backbone, graft copolymer which functions as a thickening or
dispersing agent in water-based, bentonite drilling muds. The
complex polymers formed by reacting lignin, calcium chloride, a
hydroperoxide, and ethene monomers in anaerobic solvent have the

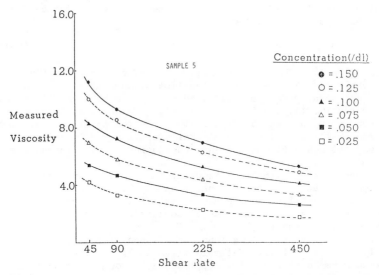

Figure 10. Viscosity as a function of concentration in distilled water and shear rate for hydrolyzed poly(starch-*g*-(2-propenamide)) copolymer, sample 5.

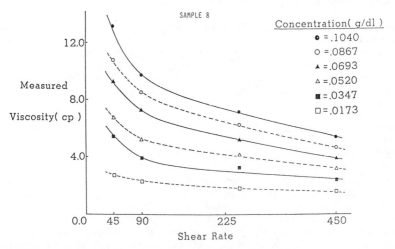

Figure 11. Viscosity as a function of concentration in distilled water and shear rate for hydrolyzed poly(starch-*g*-(2-propenamide)) copolymer, sample 8.

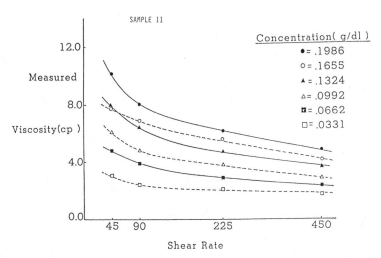

Figure 12. Viscosity as a function of concentration in distilled water and shear rate for hydrolyzed poly(starch-*g*-(2-propenamide)) copolymer, sample 11.

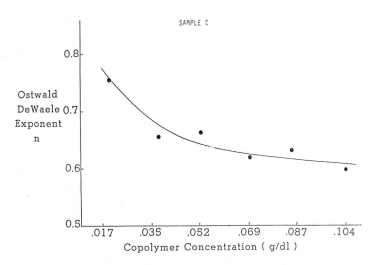

Figure 13. Plot of the Ostwald–DeWaele exponent, sample 8.

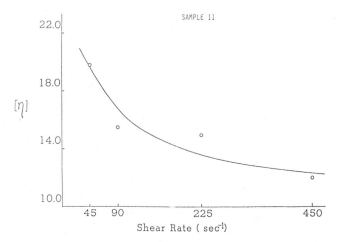

Figure 14. Limiting viscosity numbers for hydrolyzed poly(starch-*g*-(2-propenamide)), sample 11.

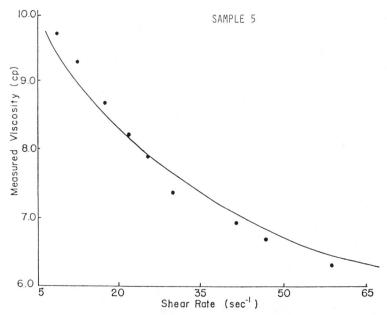

Figure 15. Effective viscosity as a function of shear rate for hydrolyzed poly(starch-*g*-(2-propenamide)), sample 5.

structures given in Figure 16. Synthesis methods, possible synthesis
mechanism insights, characterization, properties, and drilling mud
tests for these samples are presented below.

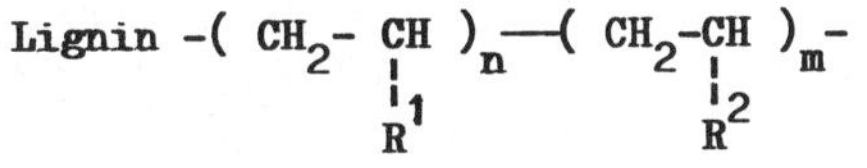

$$\text{Lignin} -\!\!\left(\; CH_2-\underset{\underset{R^1}{|}}{CH}\;\right)_{\!\!n}\!\!-\!\!\left(\; CH_2-\underset{\underset{R^2}{|}}{CH}\;\right)_{\!\!m}\!\!-$$

Figure 16. Structure of lignin polymer formed in anaerobic solvent.

SYNTHESIS: The polymerization can be run in any one of several
solvents, including dimethylsulfoxide, 1,4-dioxacyclohexane,
dimethylformamide, and dimethylacetamide. Dimethylsulfoxide or
mixtures based on dimethylsulfoxide have been used as the solvent for
all reactions reported here. In other solvents, the product often
precipitates as the reaction proceeds. This reaction can be
successfully run with mole ratios of the reactants in the following
ranges: 1. hydroperoxide to calcium chloride: <u>0.25 to 32</u>, and 2.
hydroperoxide to lignin (M_n): <u>21 to 113</u>.
Procedure: To a dry Erlenmeyer flask of appropriate size, add one
half of the reaction solvent. All reactants, including the dry mass
of the hydroperoxide, should not constitute more than 23 weight
percent of the reaction mixture or an insoluble product may be
produced. Add dry lignin and dry calcium chloride to the reaction
vessel and cap with a septum or rubber stopper. In a separate
vessel, dissolve 2-propenamide in about one quarter of the DMSO
solvent and, in a third vessel, dissolve the sulfonated monomer in
the final one quarter of the solvent. Saturate both monomer
solutions with N_2 by bubbling with the gas for 10 minutes. Saturate
the lignin solution with N_2 for 10 minutes. Add the hydroperoxide to
the mixture, bubble with N_2 for 5 minutes, cap, and stir for 10
minutes. While stirring the lignin reaction solution, further
saturate the monomer solutions with N_2. Add the 2-propenamide
solution to the lignin solution with stirring and under an N_2
blanket. Wait 1 minute. Add the second monomer solution to the
reaction vessel in the same way. Place the reaction vessel in a
$30^\circ C$ bath for 48 hr. The reaction is terminated with a small volume
of aqueous, 1% hydroquinone solution and a volume of water equal to
1/3 of the reaction solution volume is added to the product.
This solution is added to 10 times its volume of 2-propanone and
the polymer is recovered by filtration. The solids are redissolved
in water. To remove calcium ion from the product, an amount of
$Na_2C_2O_4$ equal to the moles of $CaCl_2$ added to the reaction is placed
in the solution. The CaC_2O_4 precipitate is removed by filtration.
The filtrate is dialyzed against distilled water for 3 to 5 days
using #6 Spectrapor dialysis tubing. The dilute, aqueous solution is
then freeze dried to recover the product.

POLY(LIGNIN-G-(1-AMIDOETHYLENE)). Data for a series of graft
copolymers made using 2-propenamide are given in Table 5 for
copolymers synthesized in dimethylsulfoxide. These data show that
maximum yield is obtained when chloride ion to lignin* mole ratio is
492. These nonionic molecules are small in size, readily adsorbed on
silica surfaces, and prone to complex di- and tri-valent metal ions

TABLE 5: Lignin-co-(1-amidoethylene) samples. DMSO Data[a]

Sample Number	Reaction Contents					Yield[e]		$[\eta]$ (dL/g)	Product Composition (wt.%)			
	2-propen-amide (g)	Lignin (g)	$CaCl_2$ (g)	RO_2H[b] (g)	$Ce^{(+IV)}$ (mL,.05N)	g	wt.%		N	1-amido	Lignin	Ca
1	3.20	0.50	0.50	0.40/.3532	0.15	4.71/2.799	75.68	0.322	15.21	76.73	6.50	2.789
2	3.20	0.50	0.10	0.40/.3532	0.15	3.30*/2.56	69.32	0.56	13.34	67.38	5.38	0.73
3	3.20	0.50	0.0503	0.40/.3532	0.15	2.72*/2.14	57.83	0.69	13.45	67.91	4.94	0.415
4	3.20	0.50	0.0102	0.40/.3532	0.15	3.16/2.58	70.07	0.77	15.05	76.04	4.60	0.137
5	3.20	0.50	0.50	0.25/.2208	0.15	4.31/3.306	89.36	0.35	13.28	66.95	7.00	2.89
6	3.20	0.50	0.50	0.80/.7064	0.15	4.40/3.34	90.35	0.44	13.03	65.79	6.03	2.86
7@	3.20	0.50	0.50	0.152/.1342	0.15	4.73/3.283	88.73	0.306	12.37	62.31	7.21	2.48
@	3.20	0.50	0.50	0.416	0.15							
@	3.20	0.50	0.50	0.416	0.15							
@	3.20	0.50	0.50	0.416	0.15							
8	3.20	0.50	0.1	0.15/.1325	0.15	4.07/3.307	89.37	0.615	13.55	68.41	5.48	0.757
9	3.20	0.50	0.0515	0.15/.1325	0.15	3.90/3.128	84.53	0.666	13.641	68.89	4.94	0.395
10	3.20	0.50	0.010	0.15/.1325	0.15	3.46/2.863	77.38	0.801	13.734	69.42	4.16	0.0949
11	2.00	0.50	0.50	0.15/.1325	0.15	3.10/2.119	84.77	0.372	11.74	59.12	6.88	3.54
12	2.00	0.50	0.10	0.15/.1325	0.15	3.04/2.091	83.64	0.395	11.67	58.81	6.26	1.007
13	2.00	0.50	0.0516	0.15/.1325	0.15	2.68/2.109	84.36	0.478	13.197	66.43	8.13	0.558
14	2.00	0.50	0.0107	0.15/.1325	0.15	1.84/1.44	57.58	0.565	14.162	71.43	6.54	0.33
15	1.00	0.50	0.50	0.15/.1325	0.15	2.14/1.105	73.64	0.192	7.950	39.53	12.24	5.45
16	1.00	0.50	0.10	0.15/.1325	0.15	1.12/1.063	70.87	0.288	11.515	57.52	13.74	1.65
17	1.00	0.50	0.0512	0.15/.1325	0.15	1.49/.857	57.14	0.275	10.63	52.95	14.95	0.88
18	1.00	0.50	0.0113	0.15/.1325	0.15	1.25/.919	61.29	0.276	11.516	57.41	15.52	0.325
19	3.20	0.50	0.5	0.4/.3532	0.15	3.27/	88.38	0.523	15.802	79.43	11.41	0.2
20	3.20	0.50	0.1	0.15	0.15	4.08/2.896	78.26	0.56	14.062	70.98	5.62	0.679

a All reactions run in 20.0 mL of dimethylsulfoxide save for the 4 marked samples (@) run in 1,4-dioxacyclohexane.

b The results given as #1/#2 are #1=g of crude 2-hydroperoxy-1,4-dioxacyclohexane added to the reaction and #2=g of pure peroxide added to the reaction.

e The yields listed, #1/#2, are: #1 = crude product recovered and #2 = pure product recovered. Weight percent yield is based on pure product recovered.

f 1-amido = 1-amidoethylene repeat units in the polymer. * = Some product lost during recovery.

from aqueous solution(61,62). Synthesis results for several samples
of poly(lignin-g-(1-amidoethylene)) are given in Table 6.

**Table 6. Yield and Limiting Viscosity Number for
Lignin-(2-propenamide) Reactions** [a]

sample no.	anhyd $CaCl_2$ in reaction mixture, wt %	yield g/wt%	limiting[2] viscos no., dL/g	wt % polymerized lignin	2-propen- amide	Ca after ashing
1	4.0	3.36/90.8	0.59	7.5	67.6	4.91
2	2.0	3.64/98.4	0.46	6.95	73.4	2.25
3[c]	2.2	1.5/100.0	0.21	12.4	48.3	6.94

[a]All reaction mixtures contain 20.0 mL of oxygen-bubbled, irradiated
dioxane, 0.5 g of lignin, and 0.15 mL of ceric sulfate solution.
Reactions run in a Pyrex flask and contained 1,4-dioxane irradiated
for 3 h and 0.045 mol (3.2 g) of 2-propenamide.
[b]Determined in distilled water at 30°C.
[c]Reaction run with 0.014 mol (1.0 g) of 2-propenamide.

Other lignins can be reacted with this chemistry. Table 7 shows
synthesis data for the preparation of poly(lignin-g-(1-amido-
ethylene)) from several different lignins. Sample 1 is a kraft pine
lignin grafted in a reaction coinitiated with sodium chloride.
Lignin used in these studies is a commercial product. The material
is a kraft pine lignin prepared in "free acid" form with a number-
average molecular weight of 9,600, a weight-average molecular weight
of 22,000, and a polydispersity index of 2.29. Ash content of the
lignin is 1.0 weight percent or less. The material was used as
received. Elemental analysis is C=61.66, N=0.89, H=5.73, S=1.57,
Ca=0.08, and Fe= 0.014 weight percent.

**Table 7. Poly(lignin-g-(1-amidoethylene)) formed from Various
Lignins and Coinitiators**

Sample	Composition of Reaction(g)					Yield (g/wt %) Solvent
	Lignin	2-Propen amide	Chloride Salt[a]	Hydroper oxide		
1	0.50	3.21	0.68	0.482 mL	21.28	3.46/93.3
2	0.50	3.20	0.62	0.482 mL	21.28	3.48/94.05
3	0.51	3.21	0.62	0.482 mL	21.30	2.48/66.67
4	0.50	3.20	0.62	0.482 mL	21.33	3.50/86.48
5	0.50	3.27	0.64	0.482 mL	21.39	3.20/84.88
6	0.50	3.22	0.63	0.482 mL	21.29	3.26/87.63

a. The same number of moles of chloride ion is used in sample 1 and
samples 2 to 4. Sample 1 received sodium chloride while samples 2
to 4 received calcium chloride.

Sample 2 is run with a steam-exploded, solvent-extracted, aspen
lignin. This backbone, provide by the Solar Energy Research
Institute of Golden, Colorado as DJLX13, is an I-O-TECH process,

wood extract. After steam decompression to disrupt the wood fiber, the wood was extracted with tetrachloromethane at approximately room temperature and reduced pressure. The wood was then extracted with methanol at 60°C and reduced pressure. The lignin sample used was recovered as the methanol extract. Samples 3 and 4 are results on a yellow poplar lignin. The material was produced by BioRegional Energy Associates of Floyd, Virginia. It is produced by steam exploding the wood, washing with water, extracting with alkali, and precipitating with mineral acid. The lignin has a high carboxylic acid content and a high level of phenolic hydroxyl groups. Molecular weight of the product is 1,000 to 1,200.

To see if the phenol hydroxyl group is involved in the grafting reaction, a series of model reactions were run. In these reactions, lignin was replaced by phenol derivatives. The concentration of lignin model compound in each reaction was chosen so that concentration of equal structures in the model reaction and lignin-containing reaction were the same(63). Results from reactions run with phenol, 1-(hydroxymethyl)benzene, 1-(hydroxymethyl)-phen-4-ol, and 3-(4-hydroxyphenyl)-propan-1-ol are given in Table 8. Phenol worked well as a lignin model and allowed 50 percent conversion of monomer to polymer. The two (hydroxymethyl)benzene compounds worked poorly and 3-(4-hydroxyphenyl)- propan-1-ol didn't work at all as a lignin model. These results imply that hydroxyl groups are a probable site for grafting initiation, but that structure controls the capacity of the hydroxyl group to initiate this polymerization.

TABLE 8. Results of Reactions Containing Lignin Models [a]

Lignin Test Compound	g of Model Used	Yield g	Yield %
phenol	0.085	1.64	49.9
1-(hydroxymethyl)benzene	0.088	0.49	14.9
1-(hydroxymethyl)phen-4-ol	0.1023	0.76	23.0
3-(4-hydroxyphenyl))-propan-1-ol	.1254	0	0

[a] All reactions were run using 0.5 g of calcium chloride, 0.15 mL of 0.05 M Ce(+4) solution, and 3.2 g of 2-propenamide in 20 mL of dioxane that had been photolyzed for 3 hr.

POLY(LIGNIN-G-((1-AMIDOETHYLENE)-CO-(SODIUM 1-CARBOXYLATOETHYLENE))).

The copolymer just described can be converted to an anionic polyelectrolyte by hydrolysis of the amide units with strong base. Synthesis data for a sample of poly(lignin-g-(1-amidoethylene)) hydrolyzed to a polyanion are given in Table 9. Hydrolyzed samples are prepared from aqueous solutions containing between 1.0 and 2.5 g/dL of reaction product. A stoichiometric amount of sodium hydroxide that will produce the desired percent hydrolysis is added to the solution, the solution is heated to 50°C with stirring, and is maintained there for 1 hr. Product is recovered by 1. precipitation and drying or 2. dialysis and freeze drying(64). Data from the sample of Table 9, hydrolyzed to four different degrees of hydrolysis, are given in Table 10.

TABLE 9.   Synthesis Data on Graft Copolymers

batch no.   2

reaction mixture composition
1,4-dioxane, mL	20.0	irradiation time after oxygen
lignin, g	0.50	purge = 3.0 hr.
2-propenamide, g	1.00	
0.05 M $Ce(SO_4)_2$ solution, mL	0.15	
$CaCl_2$, g	0.50	
yield		
grams	1.50	limiting viscosity number in
weight percent	100.	water at 30 $^{\circ}$C = 0.21 dL/g
product composition wt %		
lignin	12.44	
1-amidoethylene repeat	48.3	
units		% hydrolysis = 0

[a]These values are averages for four repetitions of the reaction.

Proof of copolymerization is obtained by real time spectra acquisition of absorbance for size exclusion chromatography effluent(65).

TABLE 10.   Elemental Assay, Percent Hydrolysis, and Limiting Viscosity Number for Partially Hydrolyzed Lignin Copolymer.

Sample	\multicolumn{6}{c}{elemental assay, wt %}	degree of hydrolysis from assay[a]	limiting viscosity number (dL/g)
	C	H	N	S	O	Na		
\multicolumn{9}{c}{Batch 2}
e	48.49	7.51	12.76	0.43	28.40	0.02	0	0.13
f	45.69	6.07	8.63	0.57	33.04	7.85	26.6	1.31
g	43.17	5.60	7.14	0.67	33.33	9.72	38.0	4.09
h	46.05	5.94	5.69	0.28	31.47	10.13	50.4	3.84

[a]Based on nitrogen loss from sample e.

Hydrolysis increases the terpolymer limiting viscosity number in water by a factor of up to 45. Other previously published data(66-68) show that these terpolymers are nonnewtonian viscosifiers, metal-ion complexing agents, and effective flocculators. These materials are still "small" molecules in aqueous solution, however, and do not function as effectively when used 1. as nonnewtonian viscosifiers or 2. drag reducing agents as do poly(1-amidoethylene-co-(sodium 1-carboxylatoethylene)) copolymers.

APPLICATIONS TO DRILLING MUDS.

In drilling fluids, water-soluble polymers can perform many functions, some of which are:

-Fluid loss control	-Controlling shale hydration
-Increasing viscosity	-Flocculating drill solids
-Decreasing viscosity	-Reducing friction and torque

Use of polymers of simpler structure has already been reviewed(69). The potential for application of complex polymers to the rheological, interfacial, and suspending job demanded in drilling muds is very large. To determine if the lignin graft copolymers made above would function as drilling mud additives, several nonionic and anionic copolymers were tested in a water based mud. Data for the sample prepared for drilling mud tests are given in Table 9. Batch 2 was broken into parts and hydrolyzed with sodium hydroxide to make samples of differing degrees of hydrolysis. Data on these batch fractions are given in Table 10. Mud test data for mud samples containing the lower molecular size samples of batch 2, Table 9, are given in Table 11. These data show that the reaction product produces the desirable effect of lowering yield point, lowering gel strength, and lowering API filtrate volume as degree of hydrolysis of the reaction product increases. After hot-rolling, the reductions in the above three variables are only significant for the reaction product that is 50% hydrolyzed, sample 2h of Table 10.

Though the data are limited, batch 2 samples, which probably have lower molecular weights, appear to give greater yield point lowering and lower gel strength than do samples synthesized with more 2-propenamide in the reaction mixture.

TABLE 11. Properties of Test Muds Before and After Hot Rolling. Sample: Batch 2, Table 9

property	base mud bef.	aft.	e bef.	aft.	f bef.	aft.	g bef.	aft.	h bef.	aft.
viscosity in cP at a shear rate of										
$1020 \ s^{-1}$	79	80	76	103	78	75	67	68	67	63
$510 \ s^{-1}$	58	50	57	77	52	46	44	41	43	38
$340 \ s^{-1}$	50	38.5	48	66	42	35	35	32	33	28
$170 \ s^{-1}$	39	24.5	38	51	29	22	23	19	22	18
gel strength in $lb/100 \ ft^2$ after mud has set for										
10 s	18	4	9	20	7	4	5	3	3	3
10 min	38	6	44	70	35	5	25	4	12	4
apparent viscosity, cP	40	40	38	52	39	38	34	34	34	32
plastic viscosity, cP	21	30	19	36	26	29	23	27	24	25
yield point, $lb/100 \ ft^2$	37	20	38	41	26	17	21	14	19	13
API filtrate vol, mL	12.4	8.0	12.8	12.6	10.0	10.4	10.0	10.0	9.2	9.7

The data from these first two batch tests indicated that poly(lignin-g-(1- amidoethylene)) has properties which make it a potentially effective drilling mud additive. The tests show that poly(lignin-g-(1-amidoethylene)) and its hydrolyzed derivatives act as a high temperature thinner and as a filtrate control agent. The tests also show (batch 2, Table 11) that reaction products prepared from reactions containing less 2- propenamide acts as thinners for the base mud. Products from low 2-propenamide content reactions have smaller molecular size, as shown by their smaller limiting, viscosity number (Table 9), and should have lower molecular weights than the products from other reactions.

Poly(lignin-g-((1-amidoethylene)-co-(sodium 1-(2-methylprop-2N-yl-1-sulfonate)amidoethylene))).
A strongly anionic, polyelectrolyte can be made from lignin by conducting a graft polymerization in the presence of 2-propenamide and 2,2-dimethyl-3-imino-4-oxohex-5- ene-1-sulfonic acid or its salts.

Poly(lignin-g-((1-amidoethylene)-co-(methylene 1N,1N-dimethyl-1-ammoniumcyclopenta-3,4-diyl chloride methylene))).
A Cationic Graft Copolymer of Lignin. The applications of the anionic graft copolymers described above are many but the negative charge on the polymer or the behavior of the anionic polymer under application conditions often limit the utility of these materials. One solution to this problem is to create new copolymers of lignin which have the appropriate functional groups to perform effectively in a given environment. One such class of new materials with the right parts to function where anionics fail is the class of cationic polymers. Data from such a copolymer will now be presented. This copolymer was made from 4N,4N-dimethyl-4-ammoniumhept-1,6-diene chloride. This cationic monomer was prepared according to the method of Negi, et. al. (70).

$$CH_2=CH \qquad CH=CH_2$$
$$\diagdown \qquad \diagup$$
$$CH_2 \quad CH_2 \qquad (13)$$
$$\diagdown \; + \; \diagup$$
$$N \qquad Cl^-$$
$$\diagup \; \diagdown$$
$$CH_3 \quad CH_3$$

Lignin has been reacted with 4N,4N - dimethyl-4-ammoniumhept-1,6 -diene chloride monomer to make a cationic graft copolymer. The structure of the copolymer is illustrated by the formula:

$$Lignin \; —(\; CH_2-CH{-}CH-CH_2 \,)_n{-}(\; CH_2-CH \;)_m{-}$$
$$CH_2 \quad CH_2 \qquad\qquad C=O$$
$$\diagdown + \diagup \qquad\qquad\qquad \diagup$$
$$N \qquad Cl^- \qquad\qquad NH_2$$
$$\diagup \; \diagdown$$
$$CH_3 \quad CH_3$$

The composition, reaction conditions, and yield of copolymer-1 (24-37 series) are shown in Table 12. The content of 4N,4N-dimethyl-4-ammoniumhept-1,6-diene chloride in the reaction ranges from 0 to 30 molar percent of total monomer content. The concentration of lignin , calcium chloride, and monomer in the

Table 12: Synthesis data of cationic lignin-graft copolymer

(lignin-g-2-propenamide-DMDAC)

| sample number | Reactant weight in gram | | | | volume in ml | | | Reaction parameter | | | | | |
	lignin	CaCl$_2$	A	B	DMSO	EDTA	E	Cl mmole	molar ratio	Cl/g	Cl/L	Cl/H	Yield %
24-37-1	0.66	0.33	2.35	1.03	29.28	1.30	0.65	12.35	83.8/16/2	0.35	18.7	2.26	68.43
24-37-2	0.65	0.33	2.25	1.25	28.23	1.50	0.65	13.59	80.4/19.6	0.39	20.91	2.49	68.43
24-37-3	0.66	0.33	2.07	1.55	28.82	1.90	0.65	15.69	75.4/24.6	0.43	23.77	2.88	51.37
24-37-4	0.66	0.33	1.93	1.88	33.25	2.30	0.65	17.61	70/30	0.43	26.68	3.23	49.61
24-37-5	0.68	0.34	2.69	-	25.26	-	0.65	6.13	100/0	0.21	9.01	1.12	95.37

Note: 1. A: 2-propenamide; B: DMDAC
        2. E: 30% hydrogen peroxide (equivalent weight: 8.383 meq/ml)
        3. The concentration of EDTA Na$_2$ is 200 ppm based on DMDAC
        4. Cl/g: Chloride content per unit weight of total reaction mass (mmole/g)
        5. Cl/L: Chloride content per unit weight of lignin (mmole/g)
        6. Cl/H: Molar ratio of chloride to hydrogen peroxide

reaction are around 1.8, 0.9, and 9.5 percent by weight of total
reaction mass respectively, as shown in Table 13. In the reactions
listed in Table 12, the ratio of lignin to calcium chloride is 2 to 1
by weight. The yield of reaction ranges from 49.61 to 95.37 weight
percent. By comparing yield to reaction chloride ion content, these
results indicate that yield of reaction decreases when content of
chloride ion (on a molar basis) increases. Simultaneously, however,
the yield increases when content of 2-propenamide increases.

**Table 13: The composition of Reaction Mixture Used to Make Lignin
Graft Copolymers**

			(basis: weight percent of total mass)			
Sample number	Total mass	Lignin Wt%	$CaCl_2$ Wt%	Monomer Wt%	Monomer mmole/g	Yield %
24-37-1	35.60	1.85	0.93	9.49	1.11	68.43
24-37-2	34.86	1.86	0.94	10.04	1.13	63.78
24-37-3	35.98	1.83	0.92	10.06	1.08	51.37
24-37-4	41.00	1.60	0.80	9.29	0.95	49.61
24-37-5	29.62	2.30	1.15	9.08	1.28	95.37

CONCLUSIONS.
     Poly(starch-g-(1-amidoethylene)) copolymers can be formed by
ceric-ion- initiated, free-radical polymerization of 2-propenamide on
starch. Poly(starch- g-[partially hydrolyzed 1-amidoethylene]) can
be formed by treatment of an aqueous solution of the copolymer with
0.5 M sodium hydroxide at $40°C$ under anaerobic conditions. Treatment
of the copolymer under these conditions for 10 minutes produces a
hydrolyzed copolymer with a degree of hydrolysis between 9.5 and 14.5
percent.
     Limiting viscosity numbers of the copolymers increase with
increasing molecular weight and upon hydrolysis. At fixed copolymer
molecular weight, limiting viscosity number increases with number of
grafts per starch molecule. Radii of gyration show the co- and
hydrolyzed polymers to be low mass, high- radius-of-gyration
viscosifiers. Co- and hydrolyzed polymer solutions are viscoelastic
with a relaxation time that increases with increasing polymer
concentration in solution or polymer molecular weight. Hydrolyzed
copolymers form more viscoelastic solutions than do copolymers.
     Viscosity of copolymer solutions decreases by, at most, 3 percent
when electrolyte concentration changes from 0 to 0.342 M sodium
chloride or 2.45×10^{-2} M calcium chloride. Viscosity of hydrolyzed
polymer solutions decreases exponentially with increasing electrolyte
concentration in water.
     Poly(lignin-g-(1-amidoethylene)), previously made by reacting
Kraft, pine lignin, and 2-propenamide in oxygen-bubbled, irradiated
dioxane, can be prepared by combining lignin, 2-propenamide, calcium
chloride, and a hydroperoxide in such solvents as
1-methyl-2-pyrrolidinone, dimethyl sulfoxide, dimethylacetamide,
dimethylformamide, and pyridine. The product, a highly water-
soluble, brown solid containing 4 to 7 wt % lignin and 70+ wt %
1-amidoethylene repeat units, is purified by precipitation in 2-
propanone and dialysis against water through a 1,000 mol. wt.
permeable membrane.

Poly(lignin-g-(1-amidoethylene)) and its hydrolyzed derivative,
poly(lignin-g-((1-amidoethylene)-co-(sodium 1-carboxylatoethylene))),
lower yield point, lower gel strength, and lower API filtrate volume
in a bentonite drilling mud. After hot rolling at 121^{0}C for 16 hr,
the graft copolymer outperforms an equal concentration of chrome
lignosulfonate when the compounds were tested as thinners or as
filtrate control agents.

Products synthesized in reaction mixtures containing 2 g of
2-propenamide/g of lignin acted as thinners in bentonite-based mud.

The performance requirements which must be met by polymers used in
resource recovery are job-specific and complex. These requirements
have been partially met by creating complex polymers with physical
properties which can provide the demanded performance.

While performance of resource recovery processes have been
significantly improved in the past two decades, further efforts must
be directed to 1. minimize shear degradation, 2. control adsorption,
3. control process rheology, and 4. provide appropriate chemical
stability. These are the properties which fundamental and applied
research must target during the next decade.

Acknowledgment

This work was partially supported by the National Science
Foundation under award number CPE-8260766 and under National Science
Foundation grant CBT-8417876. Support of the copolymer testing
program by A and R Pipeline Company is gratefully acknowledged.

Keven Anderle, George Merriman, James Z. Lai, Damodar R. Patil, Mu
Lan Sha, Nancy Chew, Chin Tia Li, Thomas Buchers, Cesar Augustin,
Harvey Channell, and others completed a sizable portion of this work
and their aid and effort is greatly appreciated and acknowledged.

LITERATURE CITED.

1. Kirk-Othmer Encyclopedia of Chemical Technology, Vol. 10,
Wiley-Interscience, New York, 2nd ED. 1966 pp.741-751.
2. E. Ott, H. M. Spurlin, M. W. Grafflin, High Polymers, Volume V,
Cellulose, Wiley-Interscience, New York 1955 Part III.
3. Robert L. Davidson Ed., Handbook of Water-Soluble Gums and Resins
Stach and Its Modifications, by M. W. Rutenberg, McGraw-Hill Book
Comp, New York, 1980 Chapter 22,ISBN 0-07-015471-6.
4. S. Erlander, D. French, J. Poly. Sci., 1956 20, 7-28.
5. D. French, Chemistry and Biochemistry of Starch, in W. J.
Whelan(ed.), Biochemistry of Carbohydrates, Biochemistry Series one,
Vol. 5, Universty Park Press, Baltimore, 1975 pp. 267-335,.
6. D. French, Denpun Kagaku, J. Jpn. Soc. Starch Sci., 19, 8-25
(1972).
7. D. French, J. Anim. Sci., 1973 27, 1048-1061.
8. Henry I. Bolker, Natural and Synthetic Polymers, An Introduction,
p. 580, Marcel Dekker,New York, 1974 ISBN 0-8247-1060-6.
9. Eero Sjostrom, Wood Chemistry, Fundamentals and Applications,
p.69, Academic Press, 1981 ISBN 0-12-647480-X.
10.K. V. Sarkanen, C. H. Ludwig, Lignins' Occurrence, Formation,
Structure, and Reactions, p.1, J. Wiley, 1971 ISBN 0-471-75422-6.
11. Joseph Haggen, Chem. Eng. News, (May 6,1985) 63 (#18), p.33-34.
12. T. Kent Kirk, T. Higuchi, H. Chang, Lignin Biodegradation:

Microbiology, Chemistry, and Potential Applications, CRC Press, 1980
Vol.1, p. 5,   ISBN 0-8493-5459-5.
13. A. Bjorkman, Svensk Papperstidn., 1956 59, 477.
14. J. C. Pew, Tappi, 1957 40, 553.
15. F. F. Nord, W. J. Schubert, Holz Froschung, 1951 5, 1.
16. F. F. Nord, W. J. Schubert, Tappi, 1957 50,285.
17. G. de Stevens, F. F. Nord, Fortschr. Chem. Forsch., 1954 3, 70.
18. G. de Stevens, F. F. Nord, J. Am. Chem. Soc., 1951 73, 4622.
19. S. F. Kudzin, F. F. Nord, J. Am. Chem. Soc., 1951 73, 690,
4619.
20. F. F. Nord, G. de Stevens, Naturwissenschaften, 1952 39, 479.
21. J. C. Pew, J. Am. Chem. Soc., 74, 2850, (1952).
22. E. Hagglund, Cellulosechemic, 1923 4, 84.
23. A. Sakakibara, N. Nakayama, J. Jpan, Wood Res. Soc. 1962 8, 153.
24. David N. S. Hon, Ed., Graft Copolymerization of Lignocellulosic
Fibers, ACS Symposium Series #187, Am. Chem. Soc., (1982) ISSN
0097-1656; 187.
25. Chem. and Eng. News, 1984 62 (#39), 19-20.
26. R. L. Whistler, Industrial Gums, Academic Press, New York (1973).
27. F. Smith, R. Montgomery, The Chemistry of Plant Gums and
Mucilages, Bk. Reichhold Publishing Corp., New York (1959).
28. C. J. Githens, J. W. Burnham, Soc. Pet. Eng. J., 1977 17, 5.
29. H.L. Chang, Polymer Flooding Technology - Yesterday, Today, and
Tomorrow, Proceedings of the Soc. Petrol. Eng. Symposium on Improved
Oil Recovery, 4/16-19/78, Tulsa, OK. SPE Paper #7043.
30. F. Bueche, Mechanical Degradation of High Polymers, J. Appl.
Polym. Sci., 1960 4, (#10), 101-106.
31. D.J. Pye, J. Petrol. Tech, 1964 16, (#8), 911.
32. B.B. Sandiford, J. Petrol. Tech., 1964 16, (#8), 917.
33. B.A. Toms, Proc. Internat. Rheological Congr., Holland,
North-Holland Pub. Co., Amsterdam, 1949 Vol. II, pp 135-141.
34. W. Barney Gogarty, Micellar/Polymer Flooding - An Overview,
Proceedings of The Soc. Petrol. Eng. Symposium on Improved Oil
Recovery, SPE Paper #7041, Tulsa, OK, 4/16-19/78.
35. E.H. Immergut, in Encyclopedia of Polymer Science and Technology,
H. F. Mark, N. G. Gaylord, N. M. Bikales, Eds., Interscience, New
York, 1965, Vol. 3, p. 242.
36. G. Mino, S. Kaizerman, E. Rasmussen, J. Am. Chem. Soc., 1959 81,
1494.
37. G. Mino, S. Kaizerman, E. Rasmussen, J. Polymer Sci., 1959 38,
393.
38. F. R. Duke, A. A. Forist, J. Am. Chem. Soc., 1949 71, 2790.
39. F. R. Duke, R. F. Bremer, J. Am. Chem. Soc., 1951 73, 5179.
40. A. A. Kaitai, U. Kulshrestha, R. H. Marchessault, in Fourth
Cellulose Conference (J. Polymer Sci. C, 2), R. H. Marchessault,
Ed., Interscience, New York, 1963, P. 403.
41. J. J. Meister, M. L. Sha, J. Appl. Polym. Sci., 1987 33,
1859-1871.
42. J. J. Meister, M. L. Sha, J. Appl. Polym. Sci., 1987 33, 1873.
43. J. J. Meister, D. R. Patil, M. C. Jewell, K. Krohn, J. Appl.
Polym. Sci., 1987 33, 1887.
44. H. Pledger Jr., John J. Meister, T. E. Hogen-Esch, G. B. Butler,
Proceedings of the 54th Annual Fall Tech. Confer., Soc. Petrol. Eng.,
SPE Paper # 8422, Las Vegas, NV, 9/23-26/79

45. G.B. Butler, T.E. Hogen-Esch, J.J. Meister, H. Pledger, Jr., U.S. 4,400,496, 8/23/1983;   B. L. Knight, J. Petrol. Tech., 1973 **25**, 618, May.
46. A. A. Berlin, E. A. Penskaya, Dokl. Akad. Nauk SSSR 1956 110, 585;   H. H. G. Jellinek, S. Y. Fok, Die Makromol. Chem. 1967 194, 18.
47. G. Muller, J. P. Lane, J. C. Fenyo, J. Poly. Sci., Poly. Chem. Ed., 1979 17, 659-672.
48. John J. Meister, A Rewiew of Synthesis, Characterization, and Properties of Complex Polymers for Use in the Recovery of Petroleum and Other Natural Resources, Chapter 2 in the Book: Water-Soluble Polymers for Petroleum Recovery, Dr.D.N. Schulz, Dr.G.A. Stahl, Editors, Plenum Publishing Corp. 1988 ISBN 0-306-42915-2.
49. J. Francois, D. Sarazin, T. Schwarts, G. Weill, Polymer 1973 20, (#8), 106-124.
50. H. Mark,Z. Elektrochem, 1934 40, 499;   R. Houwink, J. Prakt, Chem., 1940 157, 15.
51. D. C. MacWilliams, J. H. Rogers, T. J. West, in Water Soluble Polymers, N. M. Bikales, Editor, Plenum Publishing Company, New York, 1973 106-124.
52. J. Francois, D. Sarazin, T. Schwarts, G. Weill, Polymer 1979 20, No. 8, 969-975.
53. N. Narkis, M. Rubhun, Polymer, 1966 7, 507-512.
54. Edward T. Severs, Rheology of Polymers, Reinhold Publishing, 1962 92.
55. C. U. Ikoku, Transient Flow of Non-newtonian, Power Law Fluids, Ph.D. Thesis, Petroleum Engineering Department, Stanford University, (1978).
56. John J. Meister, Damodar R. Patil, Margaret C. Jewell, Kyle Krohn, Synthesis, Characterization, and Properties in Aqueous Solution of Poly(Starch-g- [(1-amidoethylene)-co-(sodium 1-(2-methylprop-2N-yl-1-sulfonate)amidoethylene)]), Proceedings of the American Chemical Society Division of Polymeric Materials, 1986 55, 380-384.
57. J.J. Meister, D.R. Patil, H. Channell, Proceed. Intern. Symp. on Oilfield and Goetherm. Chem., Soc. Petrol. Eng. Paper #13559, Phoenix, AZ, 4/9-11/85.
58. S. Y. Lin, 1983 In Progress in Biomass Conversion (K. V. Sarkanen, D. A. Tilman, E. C. Jahn, eds.), Vol. 3, pp. 31-78, Academic Press, New York, NY.
59. J.J. Meister, D.R. Patil, Macromolecules, 1985 18, 1559-1564.
60. J.J. Meister, Review of the Synthesis, Characterization, and Testing of Graft Copolymers of Lignin, p.305-322 of Renewable-Resource Materials: New Polymer Sources, C.E. Carraher, Jr., L.H. Sperling, Ed., Plenum Press, N.Y. (1985) ISBN 0-306-42271-9.
61. J.J. Meister, D.R. Patil, L.R. Field, J.C. Nicholson, J.Polym.Sci., Poly. Chem. Ed., 1984 22, 1963-1980.
62. J.J. Meister, D.R. Patil, H. Channell, J. Appl. Polym. Sci., 1984 29, 3457-3477.
63. Goheen, D. W.; Hoyt, C. H. Lignin in the "Kirk-Othmer Encyclopedia of Chemical Technology"; 3rd ed., Wiley-Interscience: New York, 1981; Vol. 14, pp 298-299.
64. J.J. Meister, D.R. Patil, Ind. Eng. Chem. Prod. Res. Devel., 1985 24, 306-313.

65. J.C. Nicholson, J.J. Meister, D.R. Patil, L.R. Field, Anal. Chem., 1984 56, 2447-2451.
66. J.J. Meister, D.R. Patil, Macromolecules, 1985 18, 1559-1564.
67. J.J. Meister, Review of the Synthesis, Characterization, and Testing of Graft Copolymers of Lignin, p.305-322 of Renewable-Resource Materials: New Polymer Sources, C.E. Carraher, Jr., L.H. Sperling, Ed., Plenum Press, N.Y. 1985 ISBN 0-306-42271-9.
68. J.J. Meister, J.C. Nicholson, D.R. Patil, L.R. Field, Macromolecules, 1986 19, 803-809.
69. J. Chatterji, J. K. Borchardt, Proceedings of the 55th Annual Fall Technical Conference, Soc Petrol. Eng. Paper 9288, Dallas, Tx, 9/21-24/80
70. Youji Negi, S. Harada, O. Ishizuka, Jo. Applied Poly. Sci., 1967 5, 1951-1985.

RECEIVED January 27, 1989

Chapter 10

Cationic Organic Polymer Formation Damage Control Chemicals

A Review of Basic Chemistry and Field Results

John K. Borchardt

Westhollow Research Center, Shell Development Company, Houston, TX 77251-1380

Polymers containing quaternary ammonium salt groups in some or all of the polymer repeat units have been evaluated as formation damage control agents in drilling, completion, acidizing, and hydraulic fracturing fluids as well as in enhanced oil recovery. Important chemical structure properties determining effectiveness of formation damage control polymers include polymer molecular weight and location of the quaternary ammonium group in the polymer repeat unit structure. Statistically significant sets of field results indicate two of these polymers are effective formation damage control agents in acidizing oil and gas wells.

Formation damage has become widely recognized as a major contributing factor to rapid productivity decline after well completion or workover. The reduction of formation damage during and after well completion and workover treatments will improve long-term well productivity. The use of certain quaternary ammonium salt polymers to substantially reduce clay swelling and fines migration has become common practice during drilling and well completion and stimulation operations (1).

Gabriel and Inamdar found that quaternary ammonium salt polymers which were highly effective stabilizers of water-swelling clays did not protect test cores from permeability damage caused by fines migration in the substantial absence of water-swelling clays (2). Earlier Reed and Coppel noted similar results when evaluating an effective water-swelling clay stabilizer, hydroxyaluminum, as a silica fines stabilizer (3). Silica fines were stabilized only when the unconsolidated test sand contained at least 2% smectite, a water-expandable clay.

More recently certain quaternary ammonium salt polymers have been claimed to be effective in substantially reducing fine particle migration even in the absence of water-swelling clays (1). However, there is little information readily available concerning

the effect of chemical structure on polymer performance. Therefore, it is worthwhile to review the technical and patent literature on these polymers to determine the effect of polymer molecular
weight and repeat unit structure on the performance of these
organic polymers as swelling clay and mineral fine particle stabilizers.

Formation Damage. While the causes of formation damage are numerous and may vary from one field to another, they may be grouped
into the following general categories:
        1. the migration of existing mobile fines within the
formation. This is caused by fines entrainment in rapidly flowing
fluids. Later fine particle deposition in capillary constrictions
causes reduced permeability.
        2. generation of mobile fines as a result of acidizing a
formation. Fines may also be created by crushing of proppant
grains during mixing and pumping of fracturing fluid gels or by the
effect of overburden pressure on proppant grains after fracture
generation.
        3. The creation of mobile fines resulting from low
salinity aqueous fluids contacting water-expandable clays. While
clay swelling will reduce capillary diameter thus decreasing rock
permeability, the major mode of formation damage is thought to be
the generation of mobile fine particles. This occurs when the
swelling clays act as the primary cementing medium of the formation. Clay expansion increases the mobility of pre-existing fine
particles formerly cemented in place. In addition, the expanded
clay itself is more likely to undergo disintegration and subsequent
migration in the presence of rapidly flowing fluids.
        4. fines contained in completion, workover, and stimula-
tion fluids may be introduced into the formation. These fines
usually plug flow channels before they penetrate deeply into the
formation, Filtration of treatment and injection fluids has become
a widespread practice to deal with this problem. Perforating also
causes near-wellbore formation damage due to rock crushing and fine
particle generation.
        Other minerals beside water-swelling clays have been found to
undergo fines migration. The permeability damage caused by essentially non-swelling clays such as kaolinite and chlorite is a
well-known phenomenon. Silica fines have been identified as a
potential source of permeability damage in various poorly consolidated U.S. Gulf Coast formations (1). Other minerals identified as
constituents of mobile fine particles include feldspar, calcite,
dolomite, and siderite (4,5).
        The migration of iron mineral fines, primarily hematite and
magnetite, is a common occurrence in portions of the Appalachian
Basin. The phenomenon often occurs after well stimulation and can
result in the continuing production of iron mineral fines which
pose a significant disposal problem. The migration of iron mineral
fines through propped fractures can substantially reduce the
fracture flow capacity. Many of these are mineral fines are native
to the formation and are not formed by precipitation of acid-soluble iron salts present in injection waters during or after acidi-

zing. (These iron salts can precipitate as the acid spends and the
pH increases.)

Organic Polymer Structures

Some of the structures of the organic polymers claimed in the
patent literature to be effective swelling clay or mineral fine
particle stabilizers are detailed in Table I. Chemical structure
considerations for field use include compatibility with fracturing
polymer (polysaccharide) crosslinkers and enhanced oil recovery
(EOR) chemicals such as anionic surfactants and partially hydrolyzed polyacrylamides and high pH solutions used in caustic flooding. Structures which cause the polymer to increase aqueous fluid
viscosity are undesirable. High temperature $(500^{\circ}-600^{\circ}F)$ can also
be an important consideration for high temperature formations or
use in thermal EOR processes such as steam injection.
    Organic polymers claimed to be effective swelling clay and
mineral fine particle stabilizers in the patent literature can be
divided into four classes. The polymers of class 1 have the
quaternary nitrogen atom as part of the polymer backbone (6-10).
Polymers in this class include poly(dimethylamine-co-epichlorohy-
drin, abbreviated poly(DMA-co-EPI), and poly(N,N,N',N'-tetramethyl-
1,4-1,4-diaminobutane-co-1,4-dichlorobutane), abbreviated poly
(TMDAB-co- DCB). These low molecular weights are not surprising
since these are condensation polymers. Molecular weights cited
range from 800 to 800,000 daltons.
    Polymers in which the quarternary nitrogen atom is part of a
five- or six-membered ring comprise the second class of polymers.
The ring forms part of the polymer backbone as indicated by the
second and third polymer repeat units given in Table I. The member
of this class cited in several patents is poly(diallyldimethyl-
ammonium chloride) abbreviated poly(DMDAAC).
    Both five- and six-membered ring structures (see Table I) have
been proposed for poly(DADMAC). The most recent work, a ^{13}C NMR
study, supports a five-membered ring structure for the polymer
repeat unit (10). High molecular weight products may be synthesized by free radical polymerization. DADMAC polymers having
molecular weights as high as 2.6×10^{6} daltons have been evaluated
as clay stabilization agents (12).
    Both poly(DMA-co-EPI) and poly(DADMAC) have been widely used
in the field in acidizing, hydraulic fracturing, sand control, and
other well treatments (12).
    The third class of polymers contains one or more nitrogen
atoms on a pendant sidechain in the polymer repeat unit (13,14).
The nitrogen may or may not be quaternary. In addition to being
swelling clay stabilizers, these polymers also stabilize non-
swelling mineral fine particles. Limited molecular weight data is
available but molecular weight values from 50,000 to 1×10^{6}
daltons have been cited for various polymers.
    The final class of polymers are copolymers containing one or
more of the repeat units of classes 2 and 3 (15-18). Copolymer
effectiveness would presumably be a function of the chemical
structures of each comonomer, comonomer sequence distribution, and
polymer molecular weight. The comonomer could be a relatively

Table I. Chemical Structures of Clay and Mineral Fines Stabilizers

Polymer Repeat Unit	*Reference*

$$\left[CH_2CHCH_2\overset{\overset{\displaystyle OH}{|}}{\underset{}{}}\;\; \overset{\overset{\displaystyle CH_3}{|}}{\underset{\underset{\displaystyle CH_3}{|}}{N^+}}\;\; Cl^- \right]_n$$

6-9, 12, 13

(cyclic structures)   or   (cyclic structure)

6-10, 12, 13

$—\{(CH_2)_4\, N^+ (CH_3)_2\}_n\quad nCl^-$

6-9

$—\{(CH_2)_4\, N^+ (CH_3)_2\{CH_2\}_2\, N^+ (CH_3)_2\}_n\quad 2nCl^-$

6-9

$—\{CH_2CH_2N^+ (CH_3)_2\}_n\quad nCl^-$

6-9

$—\{N^+ (CH_3)_2\{CH_2\}_6\, N^+ (CH_3)_2\{CH_2\}_3\}_n\quad 2nBr^-$

12

$$\left[CH_2\text{--}\overset{\overset{\displaystyle CH_3}{|}}{\underset{\underset{\displaystyle NH\{CH_2\}_3\, N^+ (CH_3)_2\text{--}CH_2CHCH_2\text{--}N^+ (CH_3)_2}{|}}{\underset{\underset{\displaystyle OH}{}}{C=O}}}\quad Cl \right]_n$$

12, 14

$$\left[CH_2\text{--}\overset{\overset{\displaystyle CH_3}{|}}{\underset{\underset{\displaystyle NH\{CH_2\}_3\, N^+ (CH_3)_2\{CH_2\}_2\, N^+ (CH_3)_2\text{--}CH_2CHCH_2\text{--}N^+ (CH_3)_2}{|}}{C=O}}\quad Cl \qquad OH \right]_n$$

15

$$\left(\overset{}{\underset{\underset{\displaystyle C=O}{\overset{\displaystyle |}{N}}}{CH\text{--}CH_2}} \right)_x \left(CH_2\text{--}\overset{}{\underset{\underset{\displaystyle CO_2 CH_2CH_2N(CH_3)_2}{|}}{C(CH_3)}} \right)_y \quad y\, Cl^-$$

17

$$\left(CH_2\text{--}\overset{\overset{\displaystyle CH_3}{|}}{\underset{\underset{\displaystyle CO_2H}{|}}{C}} \right)_x \left(CH_2\text{--}\overset{\overset{\displaystyle CH_3}{|}}{\underset{\underset{\displaystyle CO_2CH_2CH_2\, N(CH_3)_2}{|}}{C}} \right)_y$$

16

$$\left[CH_2\text{--}\overset{\overset{\displaystyle CH_3}{|}}{\underset{\underset{\displaystyle NH\{CH_2\}_3\, N^+ (CH_3)_2\text{--}CH_2CHCH_2N^+ (CH_3)_3}{|}}{\underset{\underset{\displaystyle OH}{}}{C=O}}} \right] \left[CH_2\text{--}\overset{\overset{\displaystyle CH_3}{|}}{\underset{\underset{\displaystyle O\{CH_2\}_2\, N(CH_3)_2}{|}}{C=O}} \right] \text{--} A$$

18

$$A = \left[CH_2\text{--}\overset{\overset{\displaystyle CH_3}{|}}{\underset{\underset{\displaystyle CO_2\, H\; or\; Na}{|}}{C}} \right]$$

inexpensive chemical as compared to the nitrogen-containing monomer. Only copolymers of class 3 repeat units appear to have been
studied.

Laboratory Test Procedures

Swelling Clay Stabilization. The laboratory evaluation test for
swelling clay stabilizers is described in detail in reference (18).
Briefly, a test column containing a blend of 85% (by weight) 70-170
U.S. mesh sand, 10% silica flour (<325 U.S. mesh), and 5% Wyoming
bentonite is prepared. A synthetic brine (see Table II) is injected to establish a standardized flow rate to which subsequent
flow rates are compared. The subsequent fluid injection sequence
is given in the various data tables. The post-treatment brine flow
rate indicates whether the polymer treatment has reduced the test
column permeability. Fresh water flow rate determines the effectiveness of the test polymer as a water-swelling clay stabilizer.
The fresh water flow rate after aqueous 15% hydrochloric acid
injection indicates the durability of the treatment to conditions
likely to be encountered in sandstone acidizing.

Mineral Fine Particle Stabilization. Experiments were performed
using test columns packed with a well blended mixture of 85% (by
weight) 70-170 U.S. mesh sand and 15% mineral fine particles, The
size distributions of the mineral fine particles are summarized
below:

Table II. Mineral Fine Particle Properties

Mineral	Median Particle Diameter (microns)	Surface Area (meter^2/gram)
silica	22.4	1.200
2:1 silica:kaolinite	0.88[a]	11.992
calcite	8.9	10.892
hematite	4.4	6.440

a. kaolinite only

A screen with 100 U.S. mesh (149 micrometer) openings was
placed at the bottom of the test column to prevent the production
of coarse sand particles from the test column. To avoid injection
fluid turbulence disturbing the test sand, a 7.5g layer of 20-40
U.S. mesh sand was placed on top of the test sand. All fluids
except polymer solutions were filtered prior to injection. Polymer
solutions were injected at 5 psia and immediately followed by
aqueous fluid at 40 psig. Effluent fluids were collected and
filtered through 0.45 micron paper to collect the produced fine
particles.
    Core floods were performed to determine if treatment polymers
would prevent permeability damage caused by fines migration within
consolidated rock and whether the adsorbed polymers would themselves reduce core permeability. The tests were performed using
Hassler sleeve chambers. With the exception of the polymer

treatment fluid, all injection fluids were passed through a two micron in-line filter prior to injection. There was no shut-in time following polymer treatment of the cores. Flow directions were adjusted to model treatment of both injection and production wells.

Fines production from untreated test sands and permeability damage observed in untreated cores indicated that the laboratory test flow rates were above the critical flow velocity required to initiate fines migration.

Laboratory Test Results - Swelling Clay Stabilizers

The performance properties of two Class 1 polymers, poly(DMA-co-EPI) and poly(TMDAB-co-DCB) as swelling clay stabilizers are summarized in Table III. Results using aqueous 5% hydrochloric acid as the polymer solvent and results noted when flowing aqueous 15% hydrochloric acid through polymer-treated test columns suggest that increasing the polymer molecular weight from 1750 to 7500 daltons may improve poly(DMA-co-EPI) performance in acidizing applications.

Similar molecular weight poly(DMA-co-EPI), 1750 daltons, ca. 13 repeat units, and poly(TMDAB-co-DCB), 1500 daltons, ca. 11 repeat units were compared. The two condensation polymers appeared to be about equally effective in preventing the swelling of Wyoming bento-nite. Any small differences are probably due to repeat unit chemical structure differences rather than the small variations in polymer molecular weight. The presence of the hydroxyl group and the smaller N^+ - N^+ distance in poly(DMA-co-EPI) could affect polymer conforma-tion in solution, geometry of the polymer - clay complex, and surface properties of the polymer - clay complex as compared to poly(TMDAB-co-DCB).

A series of DADMAC polymers was studied using the sample flow test design. The results summarized in Table III suggested that polymer-promoted clay stabilization in neutral media decreased with an increase in molecular weight from 37,000 to 75,000 daltons. However, performance in strongly acidic media such as encountered in acidizing appeared to improve with the same change in polymer molecu-lar weight.

DADMAC polymers were also evaluated as cement slurry addi-tives. The low viscosity polymers had little effect on cement fluid loss rates. The cement filtrate was collected and injected into the test column described above. In the absence of polymer, injection of cement filtrate resulted in rapid plugging of the test column. However, good permeability retention was observed during injection of cement filtrates containing poly(DADMAC). This polymer apparently flocculated fine particles in the cement slurry. This resulted in fewer formation damaging cement fine particles being present in the cement slurry. These particles caused rapid column plugging upon injection of the filtrate which did not contain poly(DADMAC). As before, polymer treated stabilized the test column to fresh water injection (see Table IV). In the absence of poly(DADMAC), fresh water injection would cause rapid column plugging as was observed in the first experiment summarized in Table III.

TABLE III. EFFECTIVENESS OF QUATERNARY AMMONIUM SALT POLYMERS AS SWELLING CLAY STABILIZERS[a]

	none	poly(DMA-co-EPI)	poly(DMA-co-EPI)	poly(DMA-co-EPI)	poly(DMA-co-EPI)	poly(TMDAB-co-DCB)	poly(DADMAC)	poly(DADMAC)	poly(DADMAC)
molecular weight	--	1750	7500	1750	7500	1500	37,000	50,000	75,000
polymer concentration % by weight	0	0.37	0.37	0.37	0.37	0.37	2.0	2.0	2.0
polymer solvent	SB	SB	SB	5% HCl	5% HCl	SB	SB	SB	SB
initial SB flow rate, cc/min	14.6	21.2	19.4	26.3	35.3	23.1	14.4	12.8	15.2
post-treatment flow rates, % of initial									
SB	100.0	121.2	134.5	85.2	186.2	116.9	102.8	100.0	92.1
fresh water	1.0	159.4	182.0	108.7	124.9	124.2	104.2	100.0	81.6
15% HCl	---	44.8	93.8	10.9	85.5	51.5	73.6	79.6	82.4
fresh water	---	184.0	156.2	117.5	113.1	160.2	91.7	85.9	68.4
diesel oil	---	---	---	139.2	72.2	---	---	---	---
fresh water	---	---	---	24.1	24.2	---	---	---	---
diesel oil	---	---	---	132.1	72.2	---	---	---	---

a. Data taken from references 6-9. The synthetic brine (SB) contained by weight 7.50% NaCl, 0.55% CaCl$_2$, and 0.20% MgCl$_2$. Injection volumes of polymer solution, standard brine, fresh water, aqueous 15% HCl, fresh water, and diesel oil were 300cc, 500cc, 500cc, 500cc, 400cc, 500cc, and 500cc respectively. T = 60°C (140°F).

Table IV.   Effectiveness of poly(DADMAC) as a Swelling Clay
            Stabilizer When Used in Cement Slurries

Fluid Injected	Flow Rate After poly(DADMAC) Treatment poly (DADMAC) Molecular Weight (% of initial brine before filtrate injection)	
	600,000	2,600,000
Standard Brine	47.1	25.6
Fresh Water	56.7	22.3
15% HCl	152.6	75.9
Fresh Water	182.5	78.9

a. The base fluid for the cement slurry was aqueous 5% KCl.
Polymer concentration was 6555-6880 ppm.

Results indicated that poly(DADMAC) will reduce damage caused
by contact of low salinity fluid lost from the cement slurry with
swelling clays present in the formation.   An increase in poly
(DADMAC) molecular weight from 600,000 to 2.6×10^6 daltons resulted in a decreased polymer effectiveness.   The test columns were of
relatively high permeability so the thickness of the adsorbed
polymer layer, predicted to be greater for the higher molecular
weight polymer, would have little effect on the observed flow
rates.

Laboratory Test Results - Mineral Fines Stabilizers

While clay swelling and concomitant fine particle migration is a
major cause of permeability damage, reduction of permeability
caused by fines migration in the absence of swelling clays can also
occur.   This fines migration is due to a combination of chemical
and mechanical forces and is greatest in the near-wellbore region
where fluid flow rates are highest.   Mineral fine particles which
can cause permeability damage include silica, relatively non-
swelling clays such as kaolinite, carbonates such as calcite and
dolomite, and iron minerals such as hematite, siderite, and magnetite.
      A series of U.S. patents (13-17) indicate that polymers
containing nitrogen atoms in relative long sidechains are effective in reducing the migration of the various mineral fine particles enumerated above.

Homopolymers.   Polymers such as poly(methacrylamido-4,4,8,8-tetra-
methyl-4,8-diaza-6-hydroxynonamethylene   dichloride),   abbreviated
poly(MDTHD), and a triaza analog, abbreviated poly(MTHHDT), have
been shown to be effective stabilizers of silica, calcite, and
hematite (14,15) as indicated by the data summarized in Table V.
      Results indicated that swelling clay stabilizers such as poly
(DMA-co-EPI) which do not possess a quaternary nitrogen atom in a
pendant chain may not be very effective at preventing permeability
damage due to fines migration in the absence of water-swelling
clays.

Table V. Mineral Fines Production From Unconsolidated Test Columns[a]

Polymer (N atoms in sidechain)	Mineral Fines Production (% of untreated column), mineral fines used:		
	silica	calcite	hematite
poly(MDTHD) (2)	5.8	28.0	39.5
poly(MTHHDT) (3)	14.3	83.3	66.7
poly(APTMAC) (2)	65.0	49.6	139.5
poly(DMA-co-EPI) (0)	65.0	165.4	139.5

a. The polymer solvent was aqueous 2% NH_4Cl. Polymer concentration was 2% by weight active material. Test temperature was 62.8°C (145°F). Mineral fines production was measured during injection of 27 pore volumes of fresh water.

The methacrylamide backbone of the most effective polymers suggests they are produced by free radical copolymerization. Relatively little molecular weight information on these polymers is available although the molecular weight of a triaza analog of MDTHD has been given as 135,000 daltons (15). Similar polymers having acrylamide backbones such as poly(acrylamido-3-propyltrimethylammonium chloride), abbreviated poly(APTMAC) appear less effective in this application (14) probably due to their greater tendency to increase aqueous fluid viscosity and thus suspend solids (Table V).

Copolymers. Copolymers have also been studied (16-18). While one comonomer contains 1-2 quaternary nitrogen in a flexible pendant chain, the other comonomer was nonionic. Copolymers of the methyl chloride salt of dimethylaminoethyl methacrylate (one quaternary nitrogen atom) and dimethylaminoethyl methacrylate (DMAEMA) and of MDTHD (2 quaternary nitrogen atoms) and DMAEMA, N,N-dimethylacrylamide (NNDMAm) or dimethylaminopropyl methacrylate (DMAPMA) have been studied and the results summarized in Table VI.

Copolymers of MDTHD and DMAPMA appeared to be the most effective silica, calcite, and hematite mineral fines stabilizers. Increasing the copolymer MDTHD content had little effect on polymer performance. Similar results were observed for a series of MDTHD - DMAEMA copolymers and a series of DMAEMA·CH_3Cl salt - DMAEMA copolymers (Table VI). In contrast, increasing the MDTHD content of MDTHD - NNDMAm copolymers from 67% to 90% improved copolymer performance as a silica fines and hematite fines stabilizer.

Terpolymers of DMAEMA, the methyl chloride salt of DMAEMA, and a termonomer: the sodium salt of 2-acrylamido-2-methylpropane sulfonic acid (AMPS), methacrylamidopropyltrimethylammonium chloride (MAPTAC), or NNDMAm, were studied (Table VI). For 9% termonomer, AMPS-containing polymers provided the most effective silica fines stabilization. This was somewhat surprising since MAPTAC contains a quaternary ammonium group on a flexible sidechain. Consistent with the relatively poor performance of the 9% MAPTAC terpolymer was the observation that, when the terpolymer MAPTAC content was increased from 9% to 33%, silica fines stabilization was not substantially increased.

TABLE VI. EFFECTIVENESS OF COPOLYMERS AS MINERAL FINES STABILIZERS[a]

Copolymer	Treatment Fluid Viscosity (cps)	Silica before 15% HCl injection	Silica after 15% HCl injection	Mineral Fines Production (% of untreated test column)	
				Calcite	Hematite
1:2 MDTHD:NNDMAm	1.6	38.1	32.4	77.8	88.9
2:1 MDTHD:NNDMAm	1.9	42.3	35.3	---	72.2
2:1 MDTHD:NNDMAm	2.2	19.1	20.6	---	---
9:1 MDTHD:NNDMAm	---	14.3	26.4	---	---
1:2 MDTHD:DMAEMA	1.7	14.3	23.5	---	94.4
1:1 MDTHD:DMAEMA	2.6	14.3	23.5	---	---
2:1 MDTHD:DMAEMA	---	23.8	23.5	---	100.0
7:3 MDTHD:DMAPMA	---	9.5	---	37.2	57.1
8.75:1.25 MDTHD:DMAPMA	---	19.1	---	---	---
9.0:1.0 MDTHD:DMAPMA	---	9.5	---	---	---
9.5:0.5 MDTHD:DMAPMA	---	9.5	---	---	---
1:9 DMAEMA CH_3Cl:DMAEMA	2.7	23.8	29.4	---	---
1:3 DMAEMA CH_3Cl:DMAEMA	2.6	23.8	26.5	81.1	77.8
1:1 DMAEMA CH_3Cl:DMAEMA	2.6	38.1	29.4	---	---
3:1 DMAEMA CH_3Cl:DMAEMA	2.6	28.6	32.4	114.6	100.0
1:1 DMAEMA $CHCl_3$:DMAEMA	5.1	23.8	35.3	---	---
1:2 MDTHD:DMAEMA[b]	---	27.6	---	---	---
1:1 MDTHD:DMAEMA[b]	---	20.7	---	---	---

Continued on page 214

TABLE VI (continued). EFFECTIVENESS OF COPOLYMERS AS MINERAL FINES STABILIZERS[a]

Copolymer	Treatment Fluid Viscosity (cps)	Mineral Fines Production (% of untreated test column)			
		Silica before 15% HCl injection	Silica after 15% HCl injection	Calcite	Hematite
DMAEMA (CH$_3$)$_2$SO$_4$:DMAEMA:AMPS 45.5:45.5:9	1.9	42.9	38.2	----	----
DMAEMA (CH$_3$)$_2$SO$_4$:DMAEMA:NNDMAm 45.5:45.5:9	2.0	33.3	29.4	----	----
DMAEMA CH$_3$Cl:DMAEMA:MAPTAC 45.5:45.5:9	2.0	23.8	29.4	----	----
DMAEMA CH$_3$Cl:DMAEMA:MAPTAC 33.3:33.3:33.3	1.8	33.3	29.5	----	----

a. See footnote a, Table V for experimental details. Data taken from reference 18.
b. The polymer solvent was aqueous 15% HCl.

The observation that the quaternary ammonium monomer content of MDTHD:DMAEMA and DMAEMA CH_3Cl:DMAEMA copolymers had little effect on their silica fines stabilization properties of prompted an investigation of nonionic polymers as mineral fines stabilizers (17,18). A series of N-vinylpyrrolidinone (NVP) copolymers with DMAEMA have been studied. Results are summarized in Table VII.

Table VII. Reduction of Mineral Fines Production Using NVP Copolymers[a]

	Mineral Fines Production (% of untreated test column)			
NVP Comonomer % by weight	DMAEMA	DMAEMA	DMAEMA	DMAEMA·DMAEMA$(CH_3)_2SO_4$
comonomer	20	20	8	12	8
Molecular Weight	$1x10^5$	$1x10^6$	$1x10^6$	$1x10^6$
Mineral Fines				
Silica				
before 15% HCl	9.5	17.6	9.5	14.3
after 15% HCl	9.5	----	20.6	14.4
Silica/Kaolinite				
before 15% HCl	---	75.6	----	4.4
after 15% HCl	---	52.9	----	5.3
Calcite	---	69.8	----	51.2
Hematite	---	32.4	----	47.6

a. See footnote a, Table V and Laboratory Test Procedures section for experimental details.

Limited silica fines stabilization data indicated that increasing copolymer molecular weight from 100,000 to 1,000,000 daltons had, if anything, a negative effect on silica fines stabilization. At a molecular weight of 1,000,000 daltons, this copolymer appeared to be more effective in stabilizing silica fines than silica/kaolinite, calcite, or hematite fines. However, the results may be due in part to the larger particle size and lower surface area of the silica fines (see Table II).

When the DMAEMA content of NVP - DMAEMA copolymers was reduced from 20% to 8%, the silica fines stabilization effectiveness appeared to improve slightly. When the 80/20 NVP - DMAEMA copolymer was converted to a terpolymer containing 8% DMAEMA $(CH_3)_2SO_4$, silica fines stabilization was substantially unaffected. However, stabilization of silica/kaolinite fines was greatly improved. This suggested that the interaction of polymer quaternary nitrogen atoms with anionic sites on mineral surfaces was important for the stabilization of migrating clays but a different interaction was important for the stabilization of silica fines. Calcite fines stabilization improved while hematite fines stabilization effectiveness decreased. This also indicated the nature of the adsorbed polymer - fine particle complex varied for different minerals.

Berea core flood test results (Table VIII) suggested that the presence of DMAEMA $(CH_3)_2SO_4$ improved the permeability damage characteristics of 80% NVP copolymers. The kerosene flow rate

TABLE VIII. PERMEABILITY DAMAGE CHARACTERISTICS OF NVP COPOLYMER MINERAL FINES STABILIZERS - BEREA CORE TESTS[a]

Injected Fluid	80:20 poly(NVP-co-DMAEMA)		80:12:8 poly(NVP-co-DMAEMA-co-DMAEMA $(CH_3)_2SO_4$	
	Cumulative Volume Injected (cc)	Stabilized Permeability (md)	Cumulative Volume Injected (cc)	Stabilized Permeability (md)
Synthetic Brine	400	104	800	552
Kerosene	800	246	1200	243
Synthetic Brine	1350	27	1600	34
0.25%w polymer[b]	1450	- - -	1700	- - -
Synthetic Brine	2560	22	2115	45
Kerosene	2960	185	2515	239

a. See reference 17 for experiment details. T = 60°C (140°F). Polymer molecular weight was 1,000,000 daltons. The Berea sandstone test cores contained 5-10% kaolinite, 2-5% illite, 0-2% chlorite, amd 0-5% mixed layer clays.

b. The polymer solvent was aqueous 2% by weight ammonium chloride solution.

after polymer treatment of the core was 98% of the pretreatment flow rate when the polymer contained 8% DMAEMA $(CH_3)_2SO_4$ compared to 75% when no DMAEMA $(CH_3)_2SO_4$ was present in the polymer. The results summarized in Table IX indicated that another copolymer which does not contain quaternary nitrogen atoms. poly (DMAEMA - co - methyl acrylate) was also an effective silica fines stabilizer.

Increasing the molecular weight of a copolymer containing 5% methyl acrylate (MA) from 100,000 to 1,000,000 daltons had little effect on silica stabilization effectiveness (see Table IX). Increasing the methyl acrylate content from 5% to 30% had also little effect on silica fines stabilization effectiveness. Acidizing substantially reduced the effectiveness of this class of copolymer. Results for the injection of 10,000 pore volumes of water indicated that silica fines elution from the test column was substantially reduced on a long-term basis.

The effectiveness of nonionic polymers as migrating clay stabilizers and the geometry of the adsorbed polymer - mineral complex may be substantially different for the nonionic polymers and the quaternary ammonium salt polymers. The observation that some quaternary ammonium salt polymers, while effective swelling clay stabilizers, are ineffective mineral fines stabilizers is consistent with a different adsorbed polymer - particle complex geometry on different mineral surfaces.

Monomer reactivity ratios and thus comonomer sequence distributions in copolymers can vary with copolymerization reaction conditions. The comonomer distribution could affect the geometry of the adsorbed polymer - mineral complex and the fines stabilization properties.

Field Test Results

Experiment 1. While there have been a number of reports concerning the effectiveness of quaternary ammonium salt polymers as swelling clay stabilizers in controlling formation damage, the number of wells involved was usually too small for the comparison of results (between well treatments which utilize quaternary ammonium salt polymers and those which do not) to be statistically significant. However, two sets of statistically significant field results for the stimulation of a large number of wells in the Fordache Field (Pointe Coupee Parish, Louisiana) are available (19). The wells were completed in the Wilcox W-8 and Sparta A formations. Average formation permeability was 8.6 and 180 millidarcies respectively. Formation temperatures were $109^{\circ}C$ ($228^{\circ}F$) and $132^{\circ}C$ ($270^{\circ}F$) respectively. Swelling and migration of silicate mineral fines were cited as the cause of rapid production declines in these wells.

The stimulation treatments were performed using retarded hydrofluoric acid. A typical retarded hydrofluoric acid treatment consisted of:

    1. 100 gal/ft of perforated interval of aqueous 5% HCl
    2. 50 gal/ft of perforated interval of aqueous 3%HF/12% HCl
    3. 25 gal/ft of perforated interval of 2.8% NH_4F (pH 7-8)
    4. 25 gal/ft of perforated interval of aqueous 5% HCl
    5. Repetition of steps 3 and 4 five times

TABLE IX. EFFECTIVENESS OF DIMETHYLAMINOETHYLACRYLATE METHYL ACRYLATE COPOLYMERS AS SILICA FINES STABILIZERS[a]

weight % DMAEMA	Polymer Molecular Weight	Polymer Concentration (% by weight)	Silica Fines Production (% of untreated column)	
			before 15% HCl	after 15% HCl
95	100,000	0.20	23.8	44.1
95	300,000	0.19	9.5	50.0
95	500,000	0.19	14.3	44.1
95	1,000,000	0.19	9.5	47.1
70	200,000	0.40	38.1	41.2
70	>1,000,000	0.45	14.3	50.0
95	1,000,000	0.19	59.5[b]	----
70	1,000,000	0.20	55.0[c]	----

a. See reference 16 for experimental details. T = 62.8°C (145°F).
b. Total fines production after injection of 10,015 pore volumes fresh water.
c. Total fines production after injection of 10,502 pore volumes fresh water.

6. 100 gal/ft perforated interval of diesel oil or aqueous NH_4Cl

7. the required volume of diesel oil, aqueous NH_4Cl, or nitrogen to displace fluids from the tubing.

Average treatment volume was 600 gallons. All fluids contained 1% (by volume) of water wetting non-emulsifier. The treatments utilizing a cationic organic polymer included the polymer in all aqueous based fluids. The reported polymer concentration of one percent by volume of the aqueous polymer solution as supplied. Active polymer concentration is actually less than this. When the clay stabilization polymer was part of the well treatment, a non-ionic water wetting nonemulsifier was used.

The first set of data is for oil production from 22 wells. A quaternary ammonium salt polymer clay stabilizer was utilized in five of the well treatments. Otherwise the 22 well treatment designs were identical. Use of the clay stabilizer in 5 well treatments resulted in a 131% production increase compared to a 156% increase after stimulation of 17 wells without clay stabilizer. Although the initial overall production response of the five clay stabilizer treated wells was less, the overall production decline rate was 4% per year compared to 16%/yr for the treatments which did not include the clay stabilizing polymer. This decline rate was determined for the period 4 to 24 months after well treatment. It is tempting to speculate that the lower initial production response of the five polymer treated wells was due to the formation of an adsorbed polymer layer which reduced formation permeability (particularly of the Wilcox Formation) significantly.

Gas production from sixteen wells was also analyzed. Twelve retarded hydrofluoric acid treatments did not include the clay stabilization polymer. The overall gas production increase was 116% compared to an overall increase of 200% obtained from four wells for which the clay stabilization polymer was included in the well treatment. With the exception of the use of the clay stabilizer, the sixteen well treatment designs were identical.

Experiment 2. A cationic organic polymer has also been evaluated as a mineral fines stabilizer in a statistically significant number of acidizing treatments utilizing hydrofluoric acid (20). This polymer was reported to remain cationic in both acidic and basic media. Thus the cationic sites appear to be quaternary nitrogen atoms. This study involved twenty offshore Louisiana wells completed in a Miocene sand having a formation temperature of $79°$ - $82°C$ ($175°$ - $180°F$). Wells in this area had a history of production declines apparently caused by mineral fines migration. A large treatment volume, 18,000 - 28,000 gallons, was used to provide substantial radial penetration of the formation. The treatment design was:

1. 50 gal/ft of formation 15% aqueous HCl containing 5% mutual solvent and 1% water wetting non-emulsifier and 50 lb/1000 gallons citric acid

2. 50 gal/ft of formation aqueous 3% HF/12% HCl

3. 300 gal/ft of formation aqueous retarded HF acid

4. the required volume of fluid required to displace all the acid from the tubing into the formation

A corrosion inhibitor was present in all acid fluids. Eight of the
well treatments incorporated a cationic organic polymer mineral
fines stabilizer in the first three treatment stages. The active
polymer concentration was less than the reported aqueous polymer
concentration of one percent (by volume). Again, this was because
the polymer was not supplied as a 100% active product.

Production was monitored for 4-6 months after the stimulation
treatments. Total oil production from the twelve wells treated
without the polymeric mineral fines stabilizer (1100 bbl/day) was
decreasing at a rate of 0.13%/day while the produced water:oil
ratio remained fairly constant. The eight wells treated using the
cationic organic polymer mineral fines stabilizer exhibited a total
oil production of 7700 bbl/day which was increasing at a rate of
0.32% per day. The water:oil ratio remained constant.

The twelve wells for which no cationic organic polymer fines
stabilizer was used were exhibiting increasing gas production
(0.15% per day) four to six months after the well treatments. This
increase was due to the performance of one well from which gas
production had more than doubled (from 1.46 MM scf/day to 2.14
scf/day). If this well is omitted from consideration, total gas
production from the remaining wells was 4.92 MM scf/day and was
decreasing at a rate of 0.06%/day. In contrast, total gas production from the eight wells treated using the cationic organic
polymer mineral fines stabilizer, 3.39 MM scf/day, was increasing
at a rate of 0.49%/day.

Conclusions

Results indicate that the effectiveness of quaternary ammonium salt
polymers in stabilizing swelling clays and mineral fine particles
is dependent on monomer chemical structure and polymer molecular
weight. Long flexible pendant sidechains containing quaternary
nitrogen atoms appear to be required for these polymers to function
as mineral fine particle stabilizers.

Nonionic copolymers of N-vinylpyrrolidinone also functioned as
mineral fine particle stabilizers.

The results of two field experiments involving a statistically
significant number of wells indicated that quaternary ammonium salt
polymers can function well as swelling clay and mineral fine
particle stabilizers under actual field conditions.

Literature Cited

1. Borchardt, J.K.; Roll, D.L.; and Rayne L.M. Proceedings of the
    55th Annual California Regional Meeting of the Society of
    Petroleum Engineers, 1984, pp. 297-310, Paper No. SPE 12757 and
    references therein.
2. Gabriel, G.A.; Inamdar, G.R. Proceedings of the 56th Annual
    Fall Technical Conference and Exhibition of the Society of
    Petroleum Engineers, 1983, Paper No. SPE 12168.
3. Reed, M.G.; Coppel, C.P. Proceedings of the 43rd Annual Cali
    fornia Regional Meeting of the Society of Petroleum Engineers,
    1972, Paper No. SPE 4186.
4. Muecke, T.W. J. Pet. Technol. 1979, 31, 144.

5. Khilar, K.C.; Fogler, H.S.   Soc. Pet. Eng. J., 1983, 23, 55.
6. McLaughlin, H.C.; Weaver, J.D.   U.S. Patent 4 366 071, 1982.
7. McLaughlin, H.C.; Weaver, J.D.   U.S. Patent 4 366 072, 1982.
8. McLaughlin, H.C.; Weaver, J.D.   U.S. Patent 4 366 073, 1982.
9. McLaughlin, H.C.; Weaver, J.D.   U.S. Patent 4 366 074, 1982.
10. Anderson, R.W.; Kannenberg, B.G.   U.S. Patent 4 158 521 (1979).
11. Lancaster, J.E.; Baccei, L.; Panzer, H.P.   J. Polym. Sci.
    Polym. Lett. Ed., 1976, 14, 549.
12. Smith, C.W.; Borchardt, J.K.   U.S. Patent 4 393 939, 1983.
13. Hall, B.E.   World Oil December, 1986, 49.
14. Borchardt, J.K.; Young, B.M.   U.S. Patent 4 497 596, 1985.
15. Borchardt, J.K.; Young, B.M.   U.S. Patent 4 536 305, 1985.
16. Borchardt, J.K.; Young, B.M.   U.S. Patent 4 558 741, 1985.
17. Borchardt, J.K.   U.S. Patent 4 536 303, 1985.
18. Borchardt, J.K.   U.S. Patent 4 563 292, 1986.
19. Holden,III, W.W.; Prihoda, C.H.; Hall, B.E.   J. Pet. Technol.,
    1981, 33, 1485.
20. Presented at the 55th Annual California Regional Meeting of the
    Society of Petroleum Engineers and available as an addendum to
    reference 1.

RECEIVED November 28, 1988

ADSORPTION OF POLYMERS AND SURFACTANTS

Chapter 11

Influence of Calcium on Adsorption Properties of Enhanced Oil Recovery Polymers

L. T. Lee, J. Lecourtier, and G. Chauveteau

Institut Français du Pétrole, B.P. 311, 92506 Rueil-Malmaison, Cedex, France

The influence of calcium on the adsorption of high molecular weight EOR polymers such as flexible polyacrylamides and semi-rigid xanthans on siliceous minerals and kaolinite has been studied in the presence of different sodium concentrations. Three mechanisms explain the increase in polyacrylamide adsorption upon addition of calcium: (i) reduction in electrostatic repulsion by charge screening, (ii) specific interaction of calcium with polymer in solution, decreasig its charge and affinity for solvent, and (iii) fixation of calcium on the mineral surface, reducing surface charge and creating new adsorption sites for the polymer. The intrinsic viscosities of polyacrylamide solutions are significantly lowered in the presence of calcium, and the increase in Huggins constant at high calcium concentrations suggests attractive polymer-polymer interactions. The effects of calcium on polymer-solvent and polymer-surface interactions are dependent on polymer ionicity; a maximum intrinsic viscosity and a minimum adsorption density as a function of polymer ionicity are obtained. For xanthan, on the other hand, no influence of specific polymer-calcium interaction is detected either on solution or on adsorption properties, and the increase in adsorption due to calcium addition is mainly due to reduction in electrostatic repulsion. The maximum adsorption density of xanthan is also found to be independent of the nature of the adsorbent surface, and the value is close to that calculated for a closely-packed monolayer of aligned molecules.

A controlling factor in the success of polymer flooding in enhanced oil recovery (EOR) is the level of polymer adsorption on reservoir rocks. Adsorption depletes polymer from the mobility control slug leading to delayed oil recovery and too high a level of adsorption renders the EOR process uneconomical. Although there has been extensive research in the field of polymer adsorption, a comprehensive study of adsorption of high molecular weight EOR polymers under imposed field conditions has been few (1-10). One of the commonly encountered conditions is the presence of high levels of monovalent and even multivalent ions which can interact with both polymers and solid surfaces, hence complicating further the understanding of the adsorption mechanism. In our previous studies on the

0097–6156/89/0396–0224$06.00/0

influence of pH and monovalent ions on adsorption of polyacrylamides and xanthan
(9, 10), it has been shown that adsorption of these polymers is mainly governed by
a competition of attractive H-bonding and repulsive electrostatic interactions, and
that monovalent ions increase adsorption by screening polymer and surface charges
and thus reducing electrostatic repulsion.

This study aims at determining the effects of calcium on the adsorption of
polyacrylamides and xanthans on siliceous minerals and kaolinite.

Polymers

The polyacrylamides are homopolymer (PAM) and copolymers (HPAM) of
acrylamide and acrylate varying from 0 to 50% acrylate content as determined by
potentiometric titrations.   The average molecular weight for all the samples
measured by low angle light scattering is about 8×10^6 daltons. Solutions are
prepared by gentle stirring using a magnetic stirrer in de-ionized water containing
the required salts and 400 ppm of NaN_3 as stabilizer.

The xanthans (XCPS I, XCPS II) are fully pyruvated samples with mean
molecular weights of 1.8×10^6 and 4.2×10^6 daltons respectively (11, 12). Solutions
are prepared by diluting a fermentation broth. Any possible existing microgels are
removed by a filtration method described elsewhere (13), and low molecular weight
impurities are eliminated by ultrafiltration.

Minerals

Siliceous minerals carrying surface silanols (sand and silicon carbide (SiC)) and
kaolinite carrying silanols and aluminols are used in this study. The sand is from
Entraigues, France. The particle size ranges from 80 to 120 μm and the average
specific surface area is ~ 0.1 m^2/g. It is washed in 1 M HCl before use. The SiC has
a particle size of 18 μm and a specific surface area of 0.4 m^2/g. The surface of SiC
after heat treatment at 300°C and acid washing is oxidized and resembles that of
silica (14). The kaolinite is from Charentes (France) and has undergone ion
exchange with Na to obtain a homoionic Na-kaolinite. The particle size ranges from
0.2 to 0.8 μm and the total specific surface area is 20 m^2/g (15 and 5 m^2/g for basal
and lateral surface respectively (15)).

Adsorption Measurements

Adsorption is determined by the depletion method using a Dohrmann DC 80 carbon
analyzer. The mineral is contacted with the polymer solution and agitated with a
mechanical tumbler for 24 hours, a time which has been verified to be sufficient for
adsorption to be complete (9).   A more detailed description of experimental
procedures is given elsewhere (10). All the data reported in this study are taken in
the plateau region of the adsorption isotherm.

Polymer Solution Properties

The properties of polymers in solution are investigated in conjunction with
adsorption since solution properties are dependent on polymer-solvent interactions

which affect the polymer behavior at the interface. The reduced specific viscosities of HPAM (ionicity = 30%) in 2 g/l NaCl and various concentrations of $CaCl_2$ are plotted as a function of polymer concentration in Figure 1. For a flexible polyelectrolyte, an increase in salt concentration significantly reduces the intrinsic viscosity, [η], due to screening of charged groups on the polymer and thus decreasing the electrostatic persistence length. This is observed upon increasing NaCl from 2g/l (solid line) to 20g/l (dotted line) in the absence of $CaCl_2$ where [η] decreases from 12700 to 4200 cm^3/g respectively. For HPAM in the presence of monovalent ions, the [η] has been found to be proportional to $c_s^{-1/2}$, where c_s is the monovalent salt concentration (9, 10). In the presence of $CaCl_2$ (2, 5 and 10 g/l) at 2g/l NaCl, the [η] of HPAM decreases even more significantly than expected from ionic strength effect due to specific interactions of the divalent ions with the polymer. Such interactions of HPAM with divalent ions which have also been studied elsewhere (16) using other techniques such as conductivity, densimetry and light scattering alter not only the charge and dimension of the macromolecule but also inter-molecular interactions. The latter is characterized by the Huggins constant, k'(17), which is deduced from the slope of the reduced specific viscosity versus concentration curve. In Figure 2 are [η] and k' values (in parentheses) of HPAM (ionicity = 30%) measured in 20g/l NaCl as a function of $CaCl_2$ concentration. In the absence of Ca^{2+}, k' is slightly less than 0.4, a value which corresponds to the theoretical value for free-draining, strictly repulsive non-deformable molecules (17). Upon addition of Ca^{2+}, this value increases and reaches 1 at 20g/l $CaCl_2$, and remains at around the same value at 40g/l $CaCl_2$. Since the intrinsic viscosities do not decrease significantly in this calcium concentration range and remain at values relatively high compared to θ conditions ([η]$_θ$ ~ 500 cm^3/g), this high value of k' indicates a tendency for inter-molecular attraction of the macromolecules.

For xanthan (XCPS I), even though its degree of ionicity is higher than that of HPAM, the [η] does not appear to be as sensitive to Ca^{2+} (Figure 3). The presence of 2 and 20g/l $CaCl_2$ produces the same slight effect of reducing the [η] from 4300 to 3400 cm^3/g, the limiting value at high salinity (11). This is due to the high rigidity of xanthan and thus its large structural persistence length (11, 12) which limits the change in molecular conformation and which renders relatively weak the effects of electrostatic repulsions between charged groups on the polymer chain. The k' in this case remains at 0.4 even up to 20g/l $CaCl_2$, showing the absence of attractive polymer-polymer interactions.

Polymer Adsorption Properties

Polyacrylamides

Adsorption on Siliceous Minerals. The adsorption of polyacrylamides on siliceous minerals in the presence of monovalent ions has been discussed previously (9, 10). While PAM adsorption is unaffected by monovalent ions since it is not governed by electrostatic factors, HPAM adsorption is increased due to reduction in electrostatic repulsion by charge screening.

In the presence of divalent ions, apart from charge screening, adsorption can be modified due to additional factors arising from specific interactions of the divalent ions with the polymer and the surface. In this case, adsorption of calcium on the SiO_2 surface (15, 18) not only reduces the surface charge but can also

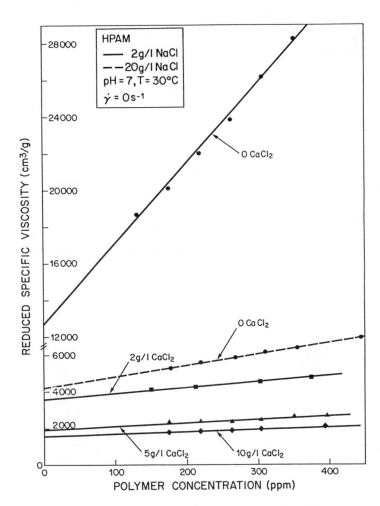

Figure 1. Effect of calcium on zero shear rate reduced specific viscosity of HPAM.

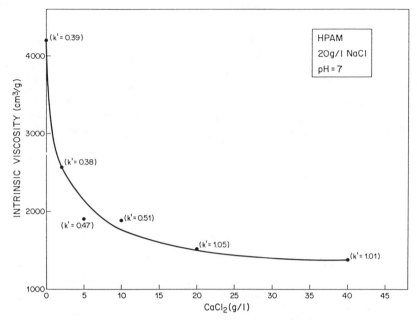

Figure2. Effect of calcium on intrinsic viscosity and Huggins constant of HPAM.

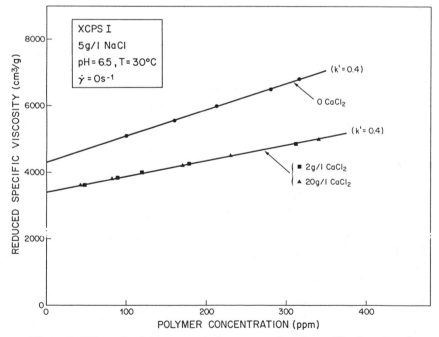

Figure 3. Effect of calcium on zero shear rate reduced specific viscosity of XCPS.

activate polymer adsorption on otherwise non-adsorbent sites (dissociated silanols). The effects of calcium on PAM and HPAM (ionicity = 30%) adsorption at pH 7 in 20g/l NaCl on sand and SiC are shown in Figures 4 and 5 respectively. Although the adsorption of PAM is not governed by electrostatic factors and the [η] remains unchanged with Ca^{2+} (see Figure 8) the presence of Ca^{2+} nevertheless provokes a slight but definite increase in its adsorption. This is attributed to the specific interaction of the nonionic polymer with the adsorbed Ca^{2+}. Indeed, weak interaction of Ca^{2+} with PAM has been detected by UV spectroscopy (19). The enhancement in adsorption of PAM by Ca^{2+} is also observed for SiC (Figure 5). Since oxidized SiC surface is similar to SiO_2 surface with a different charge density, Ca^{2+} can be expected to adsorb also on the SiC surface.

The adsorption of HPAM on sand (Figure 4) is not detected below a threshold value of Ca^{2+} due to strong electrostatic repulsion between the polyelectrolyte and the highly charged negative surface. This threshold value, which was also observed in the case of monovalent ions (9), represents the point where the critical adsorption energy is overcome, and once this value is surpassed, adsorption increases sharply. This form of adsorption behavior is in line with predictions of theories on polyelectrolyte adsorption (20).

Mechanistically, calcium increases HPAM adsorption by (i) screening of polymer and surface charges thus reducing electrostatic repulsion, (ii) specific interaction with the polymer in solution decreasing the polymer charge and intrinsic viscosity as shown from the results above, and changing also the polymer affinity for solvent, and, (iii) fixation on the mineral surface serving as a bridge between the negative surface site and polymer. From 1 to 8g/l $CaCl_2$, adsorption increases due to the reasons stated above. Between 8 and 15g/l $CaCl_2$, adsorption reaches a first plateau where the adsorption density is only about half the value obtained for nonionic PAM; this shows that there is a residual repulsion between the free HPAM and the partially polymer covered sand surface. With further increase in $CaCl_2$ beyond 15g/l however, adsorption increases until it reaches a second plateau which coincides with the maximum adsorption level for PAM. Interestingly, the region beyond 15g/l $CaCl_2$ corresponds to the region where the k' increases significantly (see Figure 2). Hence, the additional force of adsorption beyond the first plateau can be related to the occurrence of slight attractive polymer-polymer interactions. An alternative interpretation of the two plateaus observed in the adsorption curve is the existence of two different types of adsorption sites with different reactivities towards the polymer.

Unlike the sand used above, the adsorption of HPAM on SiC at 20g/l NaCl is significant even in the absence of Ca^{2+} (Figure 5). This is mainly due to the lower charge density of SiC, hence the weaker electrostatic repulsion. The higher affinity of HPAM for SiC may also explain the attainment of maximum adsorption at lower Ca^{2+} level, and may also be the reason that the higher interaction of HPAM with Ca^{2+} can induce an adsorption level higher than that of PAM.

Adsorption on Kaolinite. For kaolinite, the polymer adsorption density is strongly dependent on the solid/liquid ratio, S/L, of the clay suspension. As S/L increases, adsorption decreases. This S/L dependence cannot be due totally to auto-coagulation of the clay particles since this dependence is observed even in the absence of Ca^{2+} at pH 7 and at low ionic strength where auto-coagulation as measured by the Bingham yield stress is relatively weak (21). Furthermore, complete dispersion of the particles in solvent by ultra-sonication before addition of

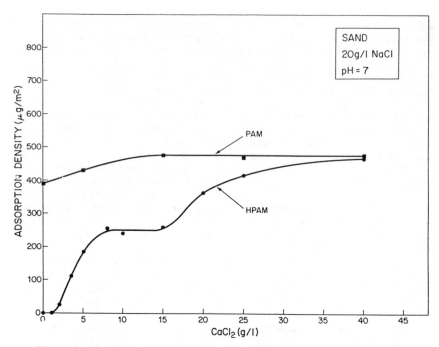

Figure 4. Influence of calcium on adsorption of PAM and HPAM on sand.

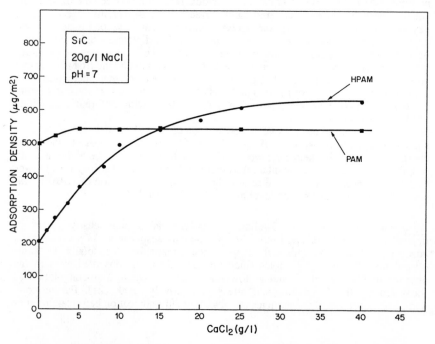

Figure 5. Influence of calcium on adsorption of PAM and HPAM on SiC.

polymer does not eliminate such dependence. The dependence of polymer adsorption on S/L is therefore attributed mainly to the flocculation of clay particles by the adsorbed polymers, a process which is favored at high S/L due to reduced inter-particle distance, and which renders some of the adsorption surface inaccessible. In the presence of calcium however, the effect of coagulation by Ca^{2+} is expected to play an important role. Therefore, in order to obtain a representative level of adsorption density on clay particles, the adsorption of polymers on kaolinite throughout this study is conducted as a function of S/L and the results extrapolated to S/L=0.

In the absence of Ca^{2+}, it has been claimed that adsorption of PAM on kaolinite takes place only on the lateral surface while the basal surface is non-adsorbent (22). However, even though adsorption may take place predominantly on the edge surface due to the presence of the more reactive aluminols, this does not exclude some adsorption on the basal surface. In fact, some recent studies which are in progress have indicated the adsorption of PAM on the kaolinite basal surface (23). For HPAM at low salinities, however, adsorption takes place only on the more reactive aluminols of the edge surface (23).

The adsorption of PAM and HPAM on kaolinite (Figure 6) is increased in the presence of Ca^{2+} due to the reasons stated above for sand and SiC. The fixation of Ca^{2+} on kaolinite is verified experimentally and the results plotted in the same figure.

Effect of Monovalent to Divalent Ratio. The above mentioned interactions of Ca^{2+} with the polyelectrolyte and with the negatively charged surface are electrostatic in nature, which means that an increase in ionic strength by monovalent ions should decrease such interactions. This is shown for the case of SiC at 2 and 20g/l NaCl at pH 7 (Figure 7). In the absence of Ca^{2+}, adsorption of HPAM is lower at 2g/l than at 20 g/l NaCl, but in the presence of Ca^{2+}, the higher interactions of the divalent ion with the negative surface and polyelectrolyte at lower NaCl concentration result in a higher adsorption. Moreover, the adsorption level at high calcium content (750 $\mu g/m^2$) is higher than the maximum level obtained in the presence of NaCl alone (500 $\mu g/m^2$) (10). This clearly demonstrates that specific interactions of calcium with polymer and surface are significant factors in enhancing HPAM adsorption.

Effect of Polymer Ionicity. The influence of polymer ionicity on solution and adsorption properties is investigated for polyacrylamides varying from 0-50% ionicity in the absence and presence of calcium.

The intrinsic viscosities of the polyacrylamides as a function of polymer ionicity in 2 g/l NaCl and at various levels of Ca^{2+} are shown in Figure 8. In NaCl solution, [η] increases continuously with polymer charge due to the increase in electrostatic persistence length of the polymer. In the presence of Ca^{2+}, the [η] is reduced for every polymer but at low ionicity and low Ca^{2+} content, the increase in [η] with charge is maintained. At higher linear charge density however, a stronger interaction of the polymer with Ca^{2+} significantly reduces the polymer charge and its solubility, resulting in a larger decrease in [η] and even precipitation for hydrolysis degrees greater than 30% when more than 2g/l $CaCl_2$ is added. The [η] thus exhibits a maximum as a function of ionicity.

The influence of Ca^{2+} on the variations of adsorption with the degree of ionicity of HPAM on sand at pH 7 in 2 g/l NaCl is given in Figure 9. Without calcium, adsorption decreases with polymer charge because of the increase in

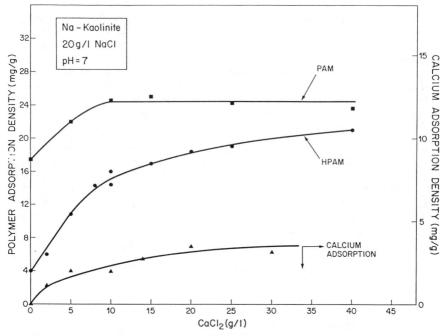

Figure 6. Adsorption of calcium and its influence on adsorption of PAM and HPAM on Na-kaolinite.

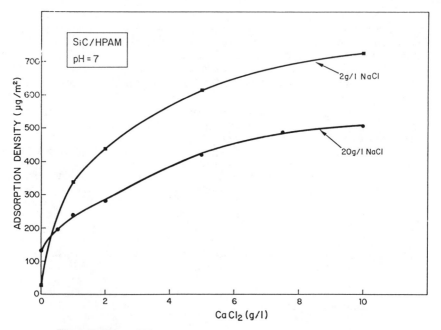

Figure 7. Effect of NaCl on adsorption of HPAM on SiC in the presence of calcium. (Reproduced with permission from ref. 26. Copyright 1988 Institut Français du Petrole.)

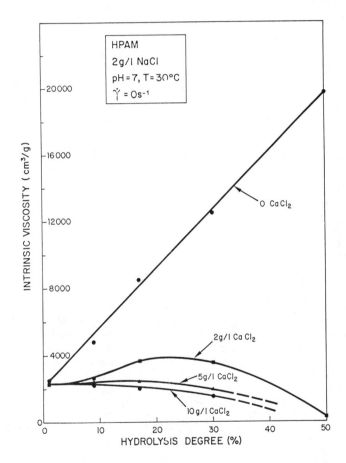

Figure 8. Intrinsic viscosity of HPAM versus ionicity in the presence of calcium. (Reproduced with permission from ref. 26. Copyright 1988 Institut Français du Petrole.)

electrostatic repulsion between the polymer and the surface. The addition of Ca^{2+} increases the adsorption of each polymer, but the decrease in adsorption with polymer charge is preserved until the degree of ionicity exceeds 30%, at which point the high interaction of polyelectrolyte with divalent ion reduces the solubility of the polymer, and increases its affinity for the surface. Hence, the adsorption of HPAM with ionicity exhibits a minimum. For SiC, on which adsorption is higher, this phenomenon is more pronounced (Figure 10) and a slight adsorption increase is detected even in the absence of Ca^{2+}. This is attributed to a decrease in solubility of the HPAM at high acrylate content (24). Such a minimum is enhanced in the presence of calcium due to the lower solubility of calcium acrylate.   As a consequence, the minimum is shifted to a lower value of ionicity.

In the case of kaolinite (Figure 11) where polymer adsorbs strongly and predominantly on the edge surface (see above discussion), the reduction in adsorption with ionicity shows that the overall adsorption is nevertheless governed by the net charge of the clay. Interestingly, an increased adsorption at high ionicity is not observed. A possible explanation is that the nature of interaction of polymer and clay surface is of higher affinity than that of polymer-siliceous surface due to the presence of more reactive aluminols (22). As such, the effect of change in polymer activity due to change in ionicity is suppressed by the more prominent polymer surface interaction.

Xanthans

Adsorption on Siliceous Minerals. All adsorption studies of xanthan (XCPS) in the presence of calcium are conducted at pH 6.5 to avoid precipitation which has been reported at pH$\geq$7 for xanthan solutions containing calcium (25).

The adsorption results of both xanthan samples on sand at pH 6.5 in 20g/l NaCl are shown in Figure 12. The presence of calcium is seen to increase the adsorption of both xanthan samples and a maximum in adsorption is reached at high calcium concentrations. These variations are very similar to those observed for XCPS adsorption in the presence of NaCl (26).

Xanthan adsorption is governed by a competition of electrostatic repulsions and attractive H-bonding forces between polymer and surface. For XCPS II, at very low ionic strength, electrostatic repulsion dominates and thus adsorption is detected only after a threshold value of calcium. The maximum adsorption level is attained at a higher Ca^{2+} concentration, and the maximum adsorption of XCPS II exceeds that of XCPS I. This may be an effect of differences in polymer molecular weight and/or structure which can be detected on sand surface due to its heterogeneity in adsorption site density which permits only partial surface coverage by XCPS I. For SiC, which has a more homogeneous adsorption site density, such a difference in maximum adsorption densities for the two samples is insignificant (Figure 13).

In comparison to HPAM where Ca^{2+} increases adsorption by both charge screening and specific interaction with polymer and surface, XCPS adsorption seems to be increased by Ca^{2+} by charge screening only. This is deduced from solution and adsorption properties of the xanthan. Firstly, in solution, even though not much information on the degree of calcium-xanthan interaction can be derived from intrinsic viscosity data due to the structural rigidity of XCPS, studies on XCPS solution stability in the presence of calcium have shown that at pH<7, no precipitation occurs even at very high calcium concentration (25). Secondly, the maximum adsorption density of XCPS in the presence of calcium does not exceed

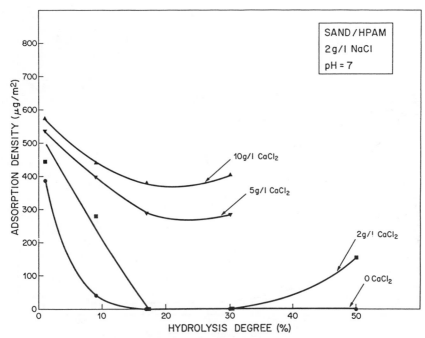

Figure 9. Adsorption of HPAM on sand versus ionicity in the presence of calcium.

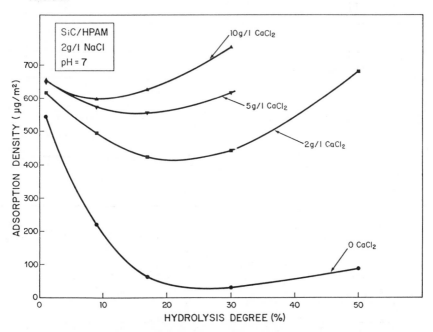

Figure 10. Adsorption of HPAM on SiC versus ionicity in the presence of calcium. (Reproduced with permission from ref. 26. Copyright 1988 Institut Français du Petrole.)

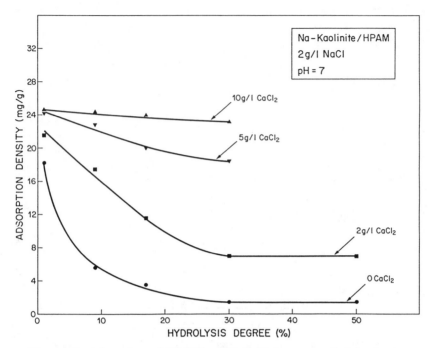

Figure 11. Adsorption of HPAM on Na-kaolinite versus ionicity in the presence of calcium.

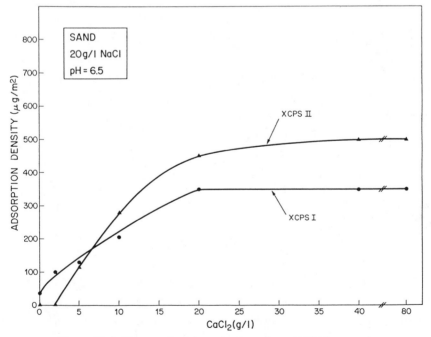

Figure 12. Influence of calcium on adsorption of XCPS on sand.

that obtained at high NaCl concentrations where adsorption is enhanced only by charge screening (26). This is in contrast to the case for HPAM where Ca^{2+} induces an adsorption level which exceeds the maximum adsorption density attained at high monovalent ion concentrations (10).

The charge screening effect of Ca^{2+} in increasing XCPS adsorption is more clearly seen in Figure 13 for the case of SiC where the NaCl content is lower (1g/l). Due to the low initial ionic strength, no adsorption is detected for both XCPS samples at low calcium concentration. The maximum adsorption level in this case is the same as that obtained in the presence of monovalent ions only (26).

Adsorption on Kaolinite. As for polyacrylamides, adsorption of XCPS on kaolinite is conducted as a function of S/L and the results extrapolated to S/L=0. However, the S/L dependence of XCPS adsorption on kaolinite is considerably less than that for HPAM. This is due to the flat conformation of the adsorbed molecules of semi-rigid xanthan (25) compared to the more extended conformation of flexible HPAM (27). The absence of loops and tails in the adsorbed XCPS layer thus diminishes the probability of flocculation of particles by polymer bridging. The slight dependence in adsorption on S/L may therefore be attributed to coagulation of particles induced by Ca^{2+}.

Figure 14 shows the adsorption results of XCPS I on kaolinite at 2 and 20g/l NaCl in the presence of calcium. In this case, the maximum adsorption density (3 mg/g) is significantly less than that of HPAM (~ 21 mg/g). This may be attributed to the rigidity of the xanthan and the relative insensitivity of its adsorption to calcium in comparison to HPAM. The enhancement of XCPS adsorption by the simple effect of charge screening by Ca^{2+} is again evident here from the higher rate of increase in adsorption at higher total ionic strength (in contrast to HPAM - see Figure 7) and the same final adsorption level. Interestingly, applying the fact that adsorption of XCPS takes place on the lateral surface only (23), the total polymer adsorption divided by the lateral surface (5 m^2/g) yields an adsorption density of 600 $\mu g/m^2$. This maximum value of adsorption induced by calcium, as in the case for SiC, corresponds to that obtained in the presence of NaCl only (26), and also to the results of XCPS I and XCPS II on SiC where the maximum adsorption is 550-600 $\mu g/m^2$. In addition, it is close to the value calculated (720 $\mu g/m^2$) for a close-packing monolayer of aligned XCPS molecules. From the above results, it can be concluded that adsorption of semi-rigid xanthan is not very sensitive to the type of adsorbent surface, and is insensitive to calcium provided that the surface has a homogeneous adsorption site density.

Conclusions

1. The adsorption of anionic polyacrylamides on sand, SiC and kaolinite is significantly increased in the presence of calcium by: (i) reduction of electrostatic repulsion between ionic polymers and surface, (ii) specific interaction of Ca^{2+} with polymers in solution as shown by the significant decrease in intrinsic viscosities and a decrease in affinity of the polymers for solvent, and, (iii) fixation of Ca^{2+} on the negative surface sites acting as a bridge and thus activating polymer adsorption on otherwise non-adsorbent sites. The relative importance of these mechanisms is dependent on polymer ionicity and overall salinity.

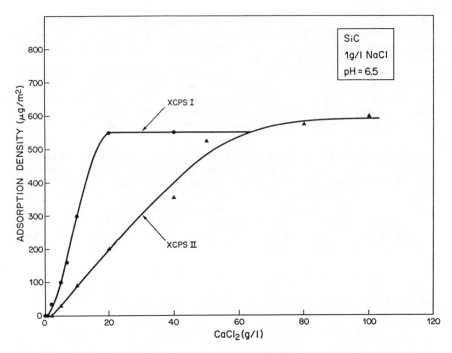

Figure 13. Influence of calcium on adsorption of XCPS on SiC.

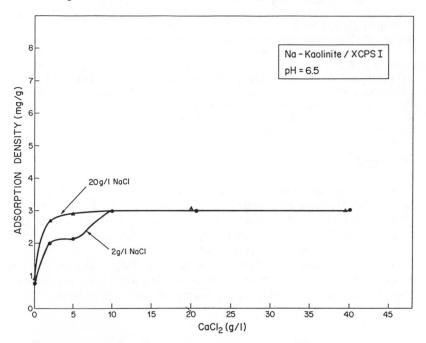

Figure 14. Influence of calcium on adsorption of XCPS on Na-kaolinite.

2. The adsorption of nonionic polyacrylamide is also increased in the presence of calcium, indicating specific interaction of calcium and the polymer.

3. The adsorption of xanthan is increased by calcium but to a less extent than that for HPAM. The increased adsorption seems to be due mainly to the effects of screening of polymer and surface charges by calcium, and the maximum adsorption density is equivalent to that induced by monovalent ions.

4. The maximum adsorption density of semi-rigid xanthan is not very sensitive to the nature of the adsorbent surface provided that the surface has a homogeneous adsorption site density. This maximum level is close to the value calculated for a closely-packed monolayer of xanthan molecules.

5. For polyacrylamides, as a function of polymer ionicity, the presence of calcium induces a maximum in intrinsic viscosity and a minimum in adsorption density on siliceous minerals. This holds important practical implications in EOR since an optimal polymer ionicity can be selected according to field conditions.

Acknowledgments

This work is conducted under the financial support of ARTEP (Association de Recherches sur les Techniques d'Exploitation du Pétrole). The authors would like to thank Ph. Delaplace and M. Nitabah for performing some of the experiments, and Rhone-Poulenc for supplying the polymers.

Literature Cited

1. Willhite, G. P.; Dominguez, J. G. Improved Oil Recovery by Surfactant and Polymer Flooding; Acad. Press Inc.; New York, 1977.

2. Klein, J.; Westerkamp, A. Die Ang. Makr. Chem. 1980, 92, 15.

3. Martin, F. D.; Sherwood, N. S. SPE Paper No. 5339; Rocky Mtn. Mtg., Denver, 1975.

4. Szabo, M. T. SPEJ; 1975, 15, 323.

5. Hollander, A. F.; Somasundaran, P.; Gryte, C. C.; J. Appl. Polym. Sci. 1981, 26, 2123.

6. Shah, S.; Heile, S. A.; Glass, J. E. SPE Paper No. 19561, Phoenix, 1985.

7. Volz, A. V.; ACS Meeting, Annheim, 1986.

8. Chauveteau, G.; Lecourtier, J. The Role of Polymers in Enhanced Oil Recovery; Schulz, D. N. ; Stahl, G. A., Eds., Plenum, 1987.

9. Lecourtier, J.; Chauveteau, G. 3rd. European Symposium on EOR, Rome, April, 1985.

10. Lecourtier, J.; Lee, L. T.; Chauveteau, G. AIChE Mtg., Houston, April, 1987.

11. Muller, G.; Anrhourrache, M.; Lecourtier, J.; Chauveteau, G. Int. J. Bio. Macromol. 1986, 8, 167.

12. Lecourtier, J.; Chauveteau, G.; Muller, G. Int. J. Bio. Macromol. 1987, 8, 306.

13. Chauveteau, G.; Kohler, N. SPEJ 1984, 24, 361.

14. Whitman, P. K.; Feke, D. L. Adv. Ceramic Materials 1986, 1 (4), 366.

15. Poirier, J. E.; Thesis, Nancy, France, 1984.

16. Schwartz, T.; François, J. Makromol. Chem. 1981, 182, 2775.

17. Russel, W. B. J. Fluid Mech. 1979, 92, (3), 401.

18. James, R.O.; Healy, T. W. J. Coll. Int. Sci. 1972, 40, (1), 42.

19. Truong, D. N.; Thesis, Strasbourg, France, 1984.

20. Van der Schee, H. A.; Lyklema, J. J. Phys. Chem. 1984, 88, 6661.

21. Rand, B.; Melton, I. E. J. Coll. Int. Sci. 1977, 60, (2), 308.

22. Nabzar, L.; Carroy, A.; Pefferkorn, E. Soil Sci. 1986, 141, (2), 113.

23. Lee, L. T.; Lecourtier, J.; Chauveteau, G.; Unpublished Results, Institut Français du Pétrole.

24. Kulicke, W. M.; Horl, H. H. Coll. and Polym. Sci. 1985, 263, 530.

25. Lecourtier, J.; Noik, C.; Barbey, P.; Chauveteau, G. 4th European Symposium on EOR, Hamburg, Oct. 1987.

26. Chauveteau, G.; Lecourtier, J.; Lee, L. T. Revue de l'Institut Francais du Petrole 1988, 43, (4).

27. Chauveteau, G.; Tirrell, M.; Omari, A. J. Coll. Int. Sci. 1984, 100, (1), 41.

RECEIVED November 28, 1988

Chapter 12

Retention Behavior of Dilute Polymers in Oil Sands

Jitendra Kikani[1] and W. H. Somerton[2]

[1]Department of Petroleum Engineering, Stanford University, Stanford, CA 94305
[2]Department of Mechanical Engineering, University of California, Berkeley, CA 94720

This study investigates the retention behavior of dilute polymer solutions in oil sands. Results indicate that the presence of a large amount of fines and/or a variety of minerals in the sand may result in high adsorption and retention causing excessive loss of polymer and high injection pressures. Injection of a surfactant with the polymer leads to increased oil recoveries because the dilute polymer may selectively adsorb on mineral grain surfaces leaving the surfactant to act at liquid/liquid contacts.

For this study flow (dynamic) and static (batch) tests were carried out on Wilmington oil field unconsolidated sands at reservoir temperatures and flow rates with polyacrylamide (Dow Pusher-500) polymers. Effluent concentration, viscosity, and pH were monitored as a function of time. Extensive characterization studies for the sand were also carried out.

Adequate mobility control between fluid banks is a pertinent factor in the successful application of secondary and tertiary oil recovery processes. During a waterflood, oil viscosities are normally higher than that of the driving aqueous phase and thus there exists an adverse mobility ratio. This causes fingering and/or channeling of the displacing fluid which reduces pattern conformance and results in low sweep efficiencies. In order to improve this situation the mobility of the displacing fluid needs to be reduced. Reduction in the mobility of the displacing fluid can be obtained by increasing the viscosity of the displacing phase or reducing the permeability to the displacing phase(1,2). Polyacrylamide and bio-polymers have proved to be useful for these purposes. These polymers increase the water viscosity substantially at low polymer concentrations. The resulting reduced mobility of the displacing phase suppresses the fingering phenomenon and improves piston-like displacement. In addition to the mobility control provided by viscosity enhancement of water, there is a selective reduction in the permeability to the aqueous phase due to the selective blocking of pores by the polymers(3). Both, viscosity enhancement and permeability reduction to the aqueous phase, act in the direction of decreasing mobility and even for fairly dilute polymer solutions, hold promise that adequate mobility control of fluid banks could be achieved.

As the displacing fluid front advances, the structural complexity of these

0097–6156/89/0396–0241$06.00/0
© 1989 American Chemical Society

polymers coupled with the complexity of the flow channels in the porous medium cause part of these polymers to be retained in the reservoir. This causes a reduction in the concentration of the polymer solution at the front and consequently a loss of mobility control. In addition to the mechanical filtering of the polymer molecules, adsorption on the grain surfaces reduces the polymer concentration in the displacing fluid. Various retention mechanisms of polymer in porous media are discussed in detail by Willhite and Dominguez(4).

The nature of the polymer controls, to a large extent, the retention behavior in porous media. Partially hydrolyzed polyacrylamide polymer molecules have the capacity of assuming a variety of three dimensional configurations. Also because of competing mechanisms in the polymer formation, there is a wide range of chain lengths. The weight average molecular weight of the polymer used for the present work was 5.89 million (Dow Chemical Co., Personal Communication, 1985). Polyacrylamide polymers used in this study have been studied extensively by a number of researchers(5,6). Susceptibility of these polymers to salinity, pH, shear, temperature, etc. is well documented(7,8). Mechanical entrapment, retention, degradation and adsorption behavior on a number of porous media including fired Berea sandstone(2), bead packs(9) and Ottawa sands(10) have been reported. It is interesting to note here that although numerous studies have been carried out with polyacrylamide polymers, not many studies have been performed on partially oil-saturated reservoir sands. Also, noticeably conflicting results, for polymer retention and adsorption, have been reported in the literature.

The present study investigates the adsorption and trapping of polymer molecules in flow experiments through unconsolidated oil field sands. Static tests on both oil sand and Ottawa sand indicates that mineralogy plays a major role in the observed behavior. Effect of a surfactant slug on polymer-rock interaction is also reported. Corroborative studies have also been conducted to study the anomalous pressure behavior and high tertiary oil recovery in surfactant dilute-polymer systems(11,12).

Experimental

Static(batch) and dynamic(flow) tests were carried out on toluene - extracted and peroxide - treated Wilmington oil field unconsolidated sands with dilute solutions of polyacrylamide (Dow Pusher-500) polymer in 1 wt% NaCl at 50^0 C and 1.5 ft./day, simulating reservoir temperature and flow rates. In the static tests, Ottawa sand, with particle size distributions similar to the Wilmington sand, were also used for comparison purposes.

The core - flood apparatus is illustrated in Figure 1. The system consists of two positive displacement pumps with their respective metering controls which are connected through 1/8 inch stainless steel tubing to a cross joint and subsequently to the inlet end of a coreholder 35 cm. long and 4 cm. in diameter. On-line filters of 7 μm size were used to filter the polymer and brine solutions. A bypass line was used to inject a slug of surfactant solution. Two Validyne pressure transducers with appropriate capacity diaphragms are connected to the system. One of these measured differential pressure between the two pressure taps located about one centimeter from either end of the coreholder, and the other recorded the total pressure drop across the core and was directly connected to the inlet line. A two - channel linear strip chart recorder provided a continuous trace of the pressures. An automatic fraction collector was used to collect the effluent fluids.

Sand and polymer characterization .   The oil sands were extracted in a Soxhlet extraction apparatus with toluene as the refluxing liquid. After the extr-

Table I. Mineralogy of Wilmington Oil Sand

Property	Count
Grain Size	
> 210 µm	39.4 %
210 - 74 µm	35.8 %
74 - 43 µm	13.3 %
43 - 2 µm	11.2 %
< 2 µm	0.3 %
Mineral Content	
Quartz	43.0 %
K-Feldspar	21.0 %
Plagioclase	15.0 %
Biotite Mica	10.0 %
Others	11.0 %
Clay Minerals	
Montmorillonite	major
Kaolinite	minor
Illite	trace

-action was complete, the thimbles were dried and the sand poured into a beaker to which distilled water was added just to cover the surface of the sand. The water was brought to a slow boil and a small amount of hydrogen peroxide was added. This was continued at intervals until no more effervescence was observed. This was done because peroxide treatment oxidizes the remaining organics coating the surface of the sand.

Mineralogy of the unconsolidated Wilmington oil sand was obtained by grain size analysis, thin-section study of an impregnated color-stained sample, scanning electron-microscope(SEM), and X-ray diffraction studies. A summary of the results of the study is given in Table I. Surface area of the whole sand sample, the < 2 micron fraction, and the Ottawa sand (quartz) was determined by nitrogen adsorption using helium as the carrier gas in a BET apparatus. The samples were prepared by the freeze - drying technique in a vacuum freeze dryer. In this technique, a suspension of sand sample in deionized water is instantaneously frozen in a liquid nitrogen atmosphere. The sample holder is then attached to the vacuum freeze dryer kept at a temperature of -50^0 C. This procedure is used to minimize agglomeration of the mineral grains and expose the greatest surface area. The surface areas are reported in Table II. This table also shows the particle size distribution of Ottawa sand used in the static tests.

Table II. Surface Area Measurements

Wilmington Oil Sand, whole rock	0.95 m^2/g
Wilmington Oil Sand, < 2 μm size	14.9 m^2/g
Ottawa Sand	0.12 m^2/g
composition : > 48 - 65 mesh	40 %
100 - 150 mesh	15 %
150 - 200 mesh	20 %
200 - 325 mesh	15 %
325 - 400 mesh	10 %

Polymer solutions were prepared by standard techniques(13). Viscosities were measured by a Brookfield LVT viscometer with UV adapter. Shear rates of 30 rpm were found to be the most suitable for viscosity measurements up to 400 ppm above which the pseudo-plastic behavior caused significant shear rate dependency on the apparent viscosity. pH of the effluent solutions were measured by a digital pH/ionmeter with combination electrodes. Polymer concentration measurements were performed on a double-beam ratio recording UV-visible spectrophotometer with micro-computer electronics by the use of turbidimetric method presented by Foshee et. al.(13). Wavelength, slit sizes, pH and mixing time were all properly calibrated before use. The calibration curve is shown in figure 2. A detailed description of the procedures and results can be found in reference 14. Figure 2 shows the effect of mixing time on the calibration curve.

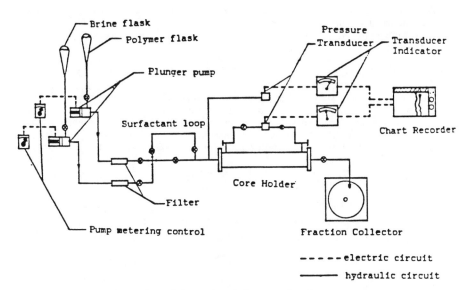

Figure 1. : Schematic of the core flood apparatus

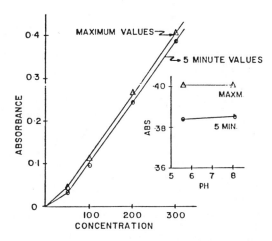

Figure 2. : Effect of pH and time on turbidity measurements for polymer concentration

We note that there is a constant shift in the curve for different mixing times. Experiments were run for various mixing times, and the maximum concentration was observed after about 25 minutes after which concentration started decreasing owing to settling of particles. If we are consistent in the use of either one of the curves the final results are not affected. The inset figure shows the effect of pH on the calibration curve. It is worthwhile noting that, for the range of hydration pH we are interested in, the concentration measurements are independent of the hydration pH.

Flow Tests . One foot long sand packs using Wilmington oil field unconsolidated sand were prepared for each of the flow tests. Porosity and permeability of all the sand packs were within 30-35% and 100-300 md, respectively. All core packs were evacuated to about 1 mm of mercury (Hg) before saturating them under gravity to assure complete water saturation. Table III gives the core and fluid properties for the flow tests. The properties of the cores were chosen so that they are close to the field conditions reported by Krebs(15).

Table III. Core and Fluid Properties

Run				Polymer	
No.	Pore Volume	Porosity(ϕ)	Brine Perm.(k)	Conc.	Visc
	cc.	%	md	ppm	cp
1	129.8	30.6	155.0	300.0	1.13
2	132.5	30.8	120.0	300.0	1.12
3	133.9	31.7	112.0	300.0*	1.20
4	142.5	33.1	323.5	300.0	1.18

* 2.0 % PV slug of surfactant

Brine was flushed through the core at a rate of 1.5 ft/day in a thermostatically controlled temperature bath (at 50^0 C) until steady state conditions on pressure drop, effluent viscosity and pH were obtained. In Figs. 3 through 6, this is indicated by the steady value of these variables prior to injection of the polymer. This was done for comparison purposes. Prefiltered polymer solution was then injected into the system and the effluent was monitored for concentration of polymer, pH, and viscosity. Pressure drop data were recorded on a strip chart recorder. Positive displacement pumps lined with nitronic alloy were used for displacement experiments. This was done to ensure that ferrous related degradation of polyacrylamides were minimized. In runs carried out with 8.6 cp Ranger zone crude oil (diluted by Chevron 410-H solvent) the core was waterflooded to residual oil saturation before starting polymer injection . A 2.0 percent pore volume slug of 40%(v/w) detergent alkylate sulfonate surfactant was used in one of the runs to evaluate its effect on adsorption and to corroborate earlier observations on residual oil recovery(11).

Static Tests . Static batch tests were carried out in amber colored pyrex bottles cleaned with doubly distilled and deionized water and dried in the oven. Six different samples were prepared for these tests. The significance of each sample is explained later in this paper. About 200 cm^3 of 300 ppm polymer solution was poured onto 50 grams of sand and kept in the oven for about 24 days. The polymer solutions were monitored for viscosity loss, pH, and concentration periodically to study the adsorption behavior. As the static tests are more diagnostic of adsorption on all the grain surfaces of the sands, results could be compared with the dynamic test results, where polymer solution is not accessible to small pore throats in which the polymer is mechanically obstructed from entry.

Results

Flow Tests .   Results of the flow tests are shown in Figures 3 through 6. Figure 3 shows the results of a typical run with a brine saturated sand pack wherein a 300 ppm polymer solution in 1 wt% NaCl was injected at a pH of 8.26. Before this, steady state conditions were established in the core by injecting 1 wt% NaCl. The pH values were stabilized at 8.0 and viscosity at around 1.1 cp. The pressure drop across the core stayed constant up to about 8 PV of polymer injection, the pH stayed in the acidic range, and effluent viscosity was consistently lower than the influent value. At about 8 PV the pressure drop started to build and within 2 PV, increased up to about 100 psi essentially plugging the core. No polymer was eluted until the end of the run.

Since no polymer was detected in the effluent, it is clear that polymer was adsorbed and, when the active adsorption sites were filled, pore throats became smaller and as a result filtering of large molecules ocurred causing the large increase in presure drop due to plugging of the sand pack. It is improbable that low effluent viscosity is caused by scission of polymer chains because of the low shear rates used, however, this could be due to filtering of high molecular weight fractions of the polymer which contribute most to the solution viscosity. Also, random chain-scission resulting from rock-polymer interactions could partly account for low effluent solution viscosities. The lowering of pH while passing through the core may be partly responsible for low solution viscosities(8). Also, plugging of the core cannot be explained by the reformation of microgels due to rehydration of the polymer at a lower pH (16) within the core. This is due to the fact that the actual preparation of the polymer solution was done at a pH ≥ 9.0 and the polymer was stored for a few days before being injected into the core.

In the next run, a core pack was saturated with 8.6 cp (at 50^0 C) Rangerzone crude oil and water flooded to residual oil saturation. Polymer flood was then initiated and about 1.2% of the original oil in place (OOIP) was recovered. The results are shown in Figure 4. The pressure profiles show behavior essentially similar to the previous run except that the pressure drop across the core increased to 100 psi within 4 PV of injection of polymer. The steady state values of pH and viscosity were 7.0 and 0.7 cp. respectively. The oil ganglia retained in larger pores resisting displacement probably reduced the amount of polymer adsorbed and reduced the number of pores that the polymer molecules needed to seal off in order to block the core. This could explain the more rapid plugging of the core. Effluent pH and viscosities remained much lower than influent values.

Figure 5 shows the results of the run performed using a surfactant slug. For this purpose, a 0.02 PV slug of 40%(v/w) detergent alkylate sulfonate surfactant(DAS), synthesized by the Morgantown Energy Technology Center of the U.S. DOE and from the waste products of the detergent industry, was injected into a water-flooded core stabilized at an effluent pH of 7.0 and viscosity of 0.7 cp, and the slug was displaced by a 300 ppm polymer solution. Tertiary oil recovery of 4.6% OOIP was obtained which is much higher than the run without the surfactant. This confirms the observations of Williams(11) who obtained higher oil recoveries with dilute polymer solutions in the presence of surfactant but with anomalous pressure behavior. Pressure drop across the core

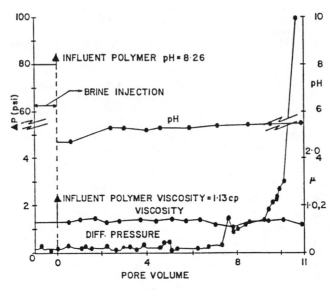

Figure 3. : Effluent Profiles for brine saturated Wilmington sand

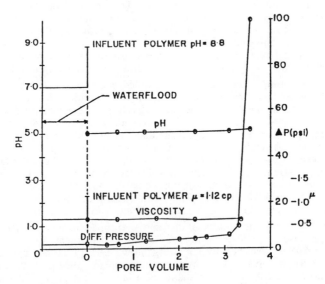

Figure 4. : Effluent Profiles in residual oil-brine saturated Wilmington sand

increased to 148 psi before 2 PV were injected indicating plugging. The pressure behavior in this run is slightly counter-intuitive because one would expect that the breaking up of the oil ganglia by the surfactant should have delayed the pressure increase until plugging by the polymer occurred. This could be explained by rock-polymer-surfactant interactions. The polymer probably acts as a sacrificial agent adsorbing on the grain surfaces and thus, reducing the loss of surfactant. Somasundaran(17,18) has observed such behavior for studies with kaolinites where the surfactant adsorption is considerably reduced in the presence of polymer solution. This would cause the low tension flood to be more effective, yielding higher recoveries. In fact, it is the Capillary number which controls the residual oil saturation. An increase in capillary number reduces the residual oil saturation. The capillary number is directly proportional to the flow velocity and is inversely related to the interfacial tension. Since preferential adsorption of polymer is suspected, the surfactant acts more effectively to reduce the interfacial tension which increases the capillary number. Another way of getting higher capillary numbers is by increasing the flow velocity. Note that, although the pressure drop in the above run was high the velocity was fairly low and was kept constant thoroughout the run. Thus, the velocity did not contribute to the increase in the capillary number.

In-situ emulsion formation, as proposed by Kamath et al(19), with DAS surfactants may cause higher pressure drops across the core. This is because of the blocking tendency of the emulsion which has lower mobility. This could explain the earlier plugging of the core compared to other runs. Effluent pH and viscosity showed behavior similar to the previous runs. It is worthwhile noting here that such pressure drops were not manifested by face plugging of the core near the entrance. This was confirmed by simultaneously monitoring the pressure at the inlet end of the core as well as the differential pressure across the two pressure taps located about 1 cm. from each end of the core. The inlet end pressure transducer showed reasonably low pressures throughout the run for each experiment.

Low effluent pH, seen consistently in all the flow tests, may be due to Na^+/H^+ exchange which could release hydrogen ions to the flowing fluid and subsequently reduce its pH. This effect has been observed earlier by Somerton, et al(20) in their study of San Joaquin Valley(Kern Front and Midway-Sunset) cores. A reduction in pH may have increased polymer adsorption on clay minerals and may partly account for the high adsorption values observed. This is consistent with the observations of Michaels and Morelos(6) who postulated that the adsorption of polyacrylamide on kaolinite occurs via hydrogen bonding between the un-ionized carboxyl or amide groups on the polymer chains and oxygen atoms on the solid surface. Adsorption is hindered by electrostatic repulsion between the negatively charged clay surfaces and the ionized carboxyl group on the polymer. The adsorption of polymer is thus favored by the reduction in the degree of carboxylate ionization which occurs on reduction of pH.

Expecting a major role particle size distribution and of clay minerals in the anomalous pressure behavior and high adsorption of polymer in the earlier tests, another run (Fig. 6) was carried out without the clay size sand particles (<43 microns) in the core. A 300 ppm polymer solution was injected into the brine saturated core. Polymer broke through after 2.5 PV of polymer was injected, although it never quite achieved the influent concentration. No anomalous pressure behavior was observed; the pressure drop increased from about 0.3 to 1.7 psi across the core which could partly be explained by the greater viscosity of the polymer and perhaps some permeability reduction by selective blocking of the pores. This suggests that the clay-size sand fraction was instrumental in causing plugging in the previous runs. Surface area measurements indicated that

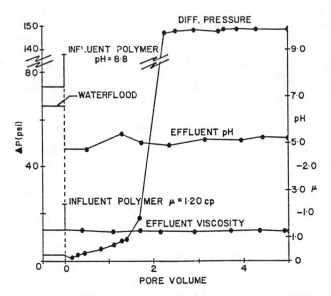

Figure 5. :   Effluent profiles for oil-water-surfactant saturated Wilmington sand

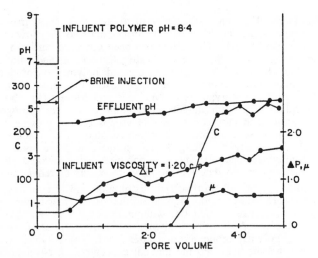

Figure 6. :   Effluent profiles in coarse grained fraction of Wilmington sand

although only 11.5% of the sample was clay-size particles, their contribution to the surface area of the whole system was about 70-80%.

Static Tests .   Several static tests were carried out to delineate the factors that might have caused the behavior observed in the dynamic tests. Care was taken in these tests to alleviate some of the degradation problems associated with polyacrylamide polymers. Six test samples were prepared. One was a control sample which consisted of polymer solution only, in a pyrex bottle. Measurements made on this control sample should determine whether the environment, i.e., the pyrex bottle, the oxygen the system was being exposed to, or the oven temperature(50^0 C), had any effect on the polymer. The second sample consisted of Ottawa sand and polymer solution. This could act as a base case for the study of the effect of mineral content and grain size on adsorption. The third and fourth samples consisted of extracted Wilmington oil sands with polymer solutions hydrated at a pH of 9.4 and 5.05, respectively. The other two samples were prepared from extracted oil sand saturated with Ranger-zone crude oil used in the flow tests, and hydrated at pH values of 9.4 and 5.05. The results of the tests at a pH of 9.4 are shown in Figure 7. The control sample stayed essentially at the original concentration indicating that the environment did not contribute to any polymer degradation or loss. The Ottawa sand showed some adsorption at early times but stabilized thereafter. Extracted sand showed a very sharp drop in concentration at lengths of time, representative of residence times of the polymer in the flow tests. The partially oil-saturated sands showed slightly less stabilized adsorption. Figure 8 depicts the adsorption behavior for the sands hydrated at a pH of 5.05. Comparing Figures 7 and 8 we observe that hydration at a lower pH value showed a greater polymer adsorption. Figure 9 shows the adsorption loss for sand with grain sizes greater than 43 microns. A significantly lower adsorption was observed. The surface area of the fines was 14.9 m^2/g compared to the whole sand surface area of 0.95 m^2/g. This confirms the observations in the flow tests that the fines had a significant effect on polymer losses.

The amount of polymer retained in the static tests was calculated by a simple material balance. The results of the calculation are shown in Table IV. This table also shows the polymer loss in dynamic tests.

In the static tests the results differ slightly depending on the hydration pH of the polyacrylamide solution(6). The adsorption losses in run 3-S for the Ottawa sand is fairly low compared to the losses with the extracted oil sand. This result was observed in runs 4-S and 5-S too. In run 5-S the < 43 μm grains were removed. This results in lower adsorption but not as low as one would expect, taking into account the reduction in surface area. Willhite and Dominguez(4) documented results from polymer retention tests done by a number of researchers for unconsolidated sandstones and Berea cores. Our static tests showed substantially higher values of adsorption compared to theirs, though they were of comparable values for Ottawa sands. Comparing the adsorption in static tests with the retention in dynamic tests up to the time when the cores were blocked, we note that in the dynamic tests the retention was considerably lower than those in the static tests.

On equating the adsorption with the surface area of the sand and observing the large time constant of decay of adsorption evident from Figs. (7) through (9), it appears that there is a multilayer adsorption process or an inter-layer adsorption within the clay particles which is causing very high polymer losses. Low adsorption on Ottawa sand suggests that there is an effect of the types of minerals present in the sand.

The retention was much higher in the static tests compared to the flow tests. This is partly due to the inaccessibility of the smaller pores by the polymer

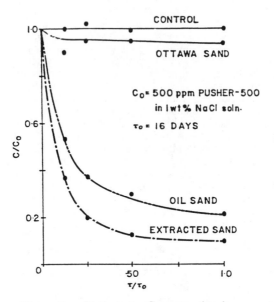

Figure 7. : Static test - Concentration loss

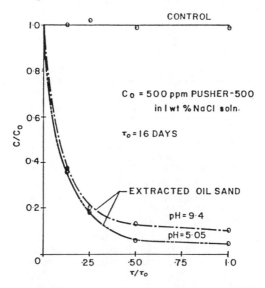

Figure 8. : Effect of rehydration pH on concentration loss in static tests

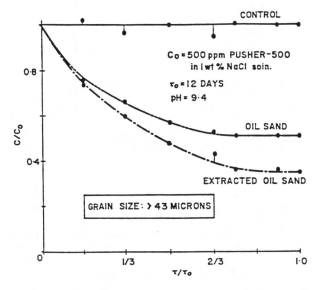

Figure 9. :   Static test - Concentration loss (coarse sand)

Table IV. Polymer Loss in Static and Dynamic Tests

Run (Dynamic)	Sand Type	Polymer Loss μg/g	Loss Until Polymer Breakthrough μg/g
2-D	Extracted Oil Sand	515.3	-
3-D	Oil Sand	180.7	-
4-D	Oil Sand	288.0	-
5-D	Extracted Oil Sand (>43 μm)	194.1	141.3

Test (Static)	Sand Type	pH	Polymer Loss μg/g
3-S	Ottawa	9.4	45.0
	Ottawa	4.8	45.0
	Extracted Oil Sand	9.4	1170.0
	Extracted Oil Sand	4.8	1309.5
4-S	Ottawa	9.4	160.0
	Extracted Oil Sand	9.4	1800.0
	Extracted Oil Sand	5.05	1920.0
	Oil Sand	9.4	1580.0
	Oil Sand	5.05	1720.0
5-S	Extracted Oil Sand	9.4	1300.0
	Oil Sand	9.4	980.0

molecules in the flow tests. This is partly offset by trapping of molecules in the pore throats in the case of the flow tests run in low permeability cores having small pore sizes. The second cause for lower retention in the flow tests is the early plugging of the upstream part of the core, preventing polymer from contacting the entire core. As explained earlier, the experiments did not indicate that the retention was confined to an entrance face - plugging effect.

Conclusions

Filtering as well as high adsorption losses of polymers in formations containing substantial amounts of fines may cause a significant loss of mobility control in polymer floods. The mobility control provided by the polymer solutions is also reduced by the decrease in viscosity due to the filtering of the high molecular weight fractions of the polymers. Effect of a variety of minerals present in the oil sands has been observed. High injection pressures could be encountered during polymer injection in oil fields if significant amounts of fines are present. This is caused by pore blockage leading to rapid pressure buildups. Shuler et al(21) report on such problems at West Coyote field single well evaluation program. They attribute the low injectivity and high pressure drops to formation damage caused by the high resistance factor of the polymer used.

High adsorption loss observed in the present work in both dynamic and static tests indicates a possibility of multilayer adsorption. However, the long times required to achieve adsorption equilibrium may indicate interlayer adsorption in the clay minerals.

In the field, such effects as are observed in the laboratory may be mitigated due to the existence of very high shear rates at and near the well bore. Far away from the well bore, the polymer sees a larger flow area and this may preclude agglomeration and blocking of pore channels. Thus, extrapolation to the field has to be done very carefully.

Presence of a DAS surfactant improves tertiary oil recovery although this may cause additional pressure drops across the system.

Acknowledgments

The fellowship support provided by the Jane Lewis Fellowship Committee is appreciated. This research was funded in part by a U.S. Department of Energy grant through Lawrence Berkeley Laboratory.

Literature Cited

1.   Pye, David J. J Pet. Tech. Aug. 1964, 911-16.

2.   Mungan, N.; Smith, F. W. J Pet. Tech. Sept. 1966, 1143-50.

3.   Sandiford, B. B. J Pet. Tech. Aug. 1964, 917-24.

4.   Willhite, G.P.; Dominguez, J.G. In Improved Oil Recovery by Surfactant and Polymer Flooding, Ed.; Academic Press: New York, 1977, 511-55.

5.   Chauveteau, G. Proc. 56th Ann. Fall Tech. Conf. of SPE, 1981.

6.   Michael, A.S.; Morelos, O. Ind. and Eng. Chem., Sept. 1955, 1801-08.

7.   Shupe, R. D. J Pet.Tech. Aug. 1981, 1513-28.

8.   Mungan, N. J Can. Pet. Tech. April-June 1969, 45-50.

9.   Jennings, R.R.; Rogers, J.H.; West, T.J. J Pet. Tech., March 1971, 391-401.

10.   Maerkar, J. M. Soc. Pet. Eng. J Aug. 1976, 172-74.

11.   Williams, D. B. M.S. Report, Univ. of California, Berkeley, 1984.

12.   Ganapathy, S. M.S. Report, Univ. of California, Berkeley, 1982.

13. Foshee, W.C.; Jennings, R.R.; West, T.J. Proc. 51st Ann. Fall Tech. Conf. of SPE, 1976.
14. Kikani, J. M.S. Report, Univ. of California, Berkeley, 1985.
15. Krebs, H.J. J Pet. Tech., Dec. 1976, 1473-80.
16. Burcik, E.J.; Thakur, G.C. J Pet. Tech., May 1974, 545-48.
17. Somasundaran, P. Annual Report Submitted to NSF and a Consortium of Supporting Industrial Organizations, Columbia University, July 1978.
18. Somasundaran, P. US Dept. of Energy Report # DOE/BC/10082-2.
19. Kamath, K.I. et al, Proc. 51st SPE Calif. Reg. Mtg. Bakersfield, 1981.
20. Somerton, W.H.; Radke, C.J. J Pet. Tech., March 1983, 643-53.
21. Shuler, P.J.; Kuehne, D.L.; Uhl, J.T.; Walkup, G.W. Soc. Pet. Eng. J - Res. Eng., Aug. 1987, 271-80.

RECEIVED November 28, 1988

Chapter 13

Physicochemical Basis of Quantitative Determination of Anionic Surfactant Concentrations by Using an Autotitrator

Robert W. S. Foulser, Stephen G. Goodyear, and Russell J. Sims

Winfrith AEE, Dorchester, Dorset DT2 8DH, England

The application of an autotitrator for the determination of surfactant concentrations has been investigated based upon turbidimetric and photometric techniques. The measured titration curves for the two techniques were found to be very similar indicating a common underlying mechanism. Experiments to investigate this mechanism have shown that the light signal measured by the instrument depends on the texture of an emulsion formed by stirring the aqueous and chloroform phases in the titrator cup. Considering the physio-chemical basis of the measurement a rapid, automatic method has been developed and successfully applied to SDBS.

The quantitative determination of surfactant concentration in solution is an essential part of any experimental work on surfactant adsorption or phase behaviour. In the field of experimental enhanced oil recovery the technique employed should be capable of determining surfactant concentrations in sea water, and in the presence of oil and alcohols, the latter being frequently added as a co-surfactant.

As part of the studies undertaken in our laboratory it was necessary to be able to determine quantitatively the surfactant present in large numbers of samples ($>$ 100 per week) arising, for example, from core flooding experiments. The chosen method needed to be rapid to reduce analysis time, and to require little manipulation of the sample to reduce errors. In this paper we report the development of a method for the determination of anionic surfactants based upon autotitration and comment on the physico-chemical basis of the technique.

Numerous methods have been developed for the determination of anionic surfactants and these have been reviewed by Longman (1). The measurement of absorbance of light by a dyestuff-anionic surfactant complex, which has been extracted into an organic solvent is a key feature of many methods, and Sodergren has successfully used segmented flow colorimetry for an automated version of this procedure (2). An alternative is the two phase titration technique, pioneered by Herring (3) which uses dimidium

0097–6156/89/0396–0257$06.00/0

bromide (1) and disulphine blue (2) as the indicators. This
technique, originally introduced by Holness and Stone (4) for
qualitative purposes, has been thoroughly investigated by Reid and
co-workers (5, 6) and adopted as a standard method (7). More
recently potentiometric methods have been employed but these were
considered unsuitable because the membrane electrode may fail on
prolonged contact with organics as required in this application (8).

(1)

(2)

EXPERIMENTAL

APPARATUS AND MATERIALS. A Mettler DL40 memotitrator, DK19 Filter
Titrator with filters and a DK181 Phototrode were purchased from
M.S.E. Scientific Instruments, Crawley. Visible spectra were
recorded on a Perkin Elmer Lambda 3 spectrophotometer.
      Hyamine 1622 (4.0 mmol dm^{-3} solution), disulphine blue,
dimidium bromide, sodium chloride (AnalaR grade), sodium hydroxide
(AnalaR grade), chloroform (reagent grade) and decane (GPR grade)
were purchased from B.D.H. Ltd., Poole. Dodecylbenzenesulphonic
acid (98%, remainder sulphuric acid and unsulphonated oil) was
purchased from Alpha Chemicals, Coventry. Water (conductivity 18
MΩ^{-1}) was obtained from a reverse osmosis plant equipped with a
MILLI-Q polishing unit. The MILLI-Q polishing unit contained ion
exchange and charcoal cartridges, the latter to remove trace
organics.
      Aqueous solutions of sodium dodecylbenzenesulphonate (SDBS)
were prepared by neutralizing dodecylbenzenesulphonic acid solution
with sodium hydroxide solution.

PRINCIPLE OF OPERATION OF THE AUTOTITRATOR. The autotitrator
operates by detecting the change in light transmission through a
stirred solution to which titrant is added, the light intensity is
subsequently recorded as a voltage generated by a photocell. The
light source (Figure 1) generates high frequency modulated light
which passes through a filter, to select the required wavelength,
then down the fibre optic cable. The light is split at the probe;
one route passing through the solution in the titration vessel, the
other through a reference cell. It should be noted that the light
collected at the detector cell is the sum of the light transmitted
through the solution (reflected back by the concave mirror) and any
component which arises by back scattering of light directly from the
solution. In the detector the returned light is converted to a
signal which represents the change in transmittance of the sample in
the titration vessel as the titrant is added.

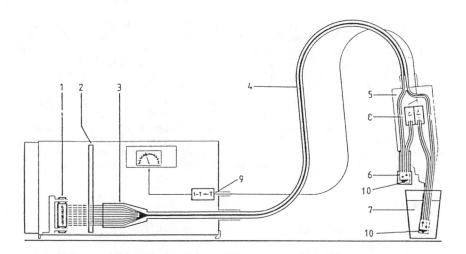

FIGURE 1   Schematic Diagram of Light Source and Probe

1	Light Source	6	Reference
2	Fixed Wavelength Filter	7	Titration vessel
3	Fibre Optic Converter	8	Detectors
4	Light conducting cable	9	Demodulator
5	Probe	10	Concave mirror

In a typical titration with the autotitrator a significant
fraction of the equivalence volume is added in a single aliquot and
allowed to equilibrate before further titrant is added. The end
point is detected by comparing changes in the millivolt output of
the light source (Figure 1) (9).

After the addition of the initial single aliquot the injection
of titrant can proceed in two different modes:
  (i) a drop of titrant (of fixed size) is added if the EMF changes
      by less than DE (mV) in a specified time interval DT (s).
      Unless otherwise stated this mode was used for the titrations
      reported with DE=1 and DT=1.
 (ii) a drop of titrant is added after a preset time has elapsed
      from the last drop. This mode of operation has been used in
      some carefully controlled comparative studies described later.

BASIS OF MANUAL PHOTOMETRIC TITRATION. The determination of anionic
surfactants by a photometric titration employs a cationic indicator
to form a coloured complex with the surfactant which is insoluble in
water but readily soluble in chlorinated solvents (1). The end
point of the titration occurs when there is a loss of colour from
the organic phase. A considerable improvement in this technique is
achieved by the use of a mixture of anionic and cationic dyes (4),
for example disulphine blue and dimidium bromide (Herring's
indicator (3)). The sequence of colour changes which occurs during
the two phase titration of an anionic surfactant (AS) with a
cationic titrant (CT) using a mixed indicator consisting of an
anionic indicator (AD) and cationic indicator (CD) is summarised in
Scheme 1.

Scheme 1

CD/AS   +   AD + CT   →   AD/CT   +   CT/AS   +   CD

Soluble in   Soluble in   Soluble in   Soluble in   Soluble in
  $CHCl_3$ H_2O $CHCl_3$ $CHCl_3$ H_2O
pink colour   green colour   blue colour   colourless   colourless

At the end the cationic indicator (CD) passes into the aqueous phase
and a small quantity of the anionic indicator/cationic titrant
complex (AD/CT) passes into the organic phase to give a grey/blue
tint.

BASIS OF MANUAL TURBIDIMETRIC TITRATION. When a cationic titrant is
added to an aqueous anionic surfactant solution a sparingly soluble
complex is produced, Scheme 2.

Scheme 2

$$RSO_3^- + R_4N^+ → RSO_3NR_4$$

The quantity of precipitate formed can be controlled by the
addition of an organic solvent such as chloroform to the system.
The chloroform forms a separate phase in which the complex is
soluble and a significant quantity of the complex can be produced

before the solubility limit in the aqueous phase is exceeded. A plot of turbidity versus volume of titrant added would be expected to have a steep fall to the end point followed by a shallow rise as the turbidity of the solution is diluted by excess titrant, ie a "tick" shape.

CHOICE OF FILTER FOR AUTOMATED PHOTOMETRIC TITRATION. At the end of a photometric titration using the above two indicators the colour of the chloroform phase changes from pink to blue. To choose a filter to detect this end point the visible spectra of the separated chloroform layers of surfactant titrations were recorded before, at and beyond the end point, see Figure 2. At 580 nm there was a greater change in absorbance than at 440 nm, thus the 580 nm filter was preferred.

The plot of light transmission versus volume of titrant added would be expected to be a step change, where the equivalence point might reasonably be taken as the position of greatest slope in the titration curve.

EXPERIMENTAL PROCEDURES: AUTOMATED PHOTOMETRIC TITRATION. Indicator solution (5 cm^3) (5) and chloroform (10 cm^3) were placed in the titration beaker together with the aqueous surfactant sample and water (30 cm^3 less the volume of the surfactant sample). The titration was then carried out with hyamine solution (4.0 mmol dm^{-3}) added in 0.05 cm^3 increments after the addition of an initial single aliquot.

EXPERIMENTAL PROCEDURES: AUTOMATED TURBIDIMETRIC TITRATION. A method for the automated aqueous turbidimetric titration of surfactants has been published (10) in which anionic surfactants are titrated against N-cetylpyridinium chloride to form a colloidal precipitate near the equivalence point. N-cetylpyridinium halides have a disadvantage in that they have the tendency to crystallise out of solution (5), consequently the strength of the solution may alter slightly without the knowledge of the operator, also the crystals suspended in solution may cause damage to the autotitrator. In view of these drawbacks hyamine was preferred as the titrant.

An aqueous surfactant sample was placed in the titration beaker together with chloroform (10 cm^3) and water (30 cm^3 less the volume of the surfactant sample). The titration was then carried out using hyamine solution as in the photometric case.

TITRATIONS FOR COMPARISON OF METHODS. The automated photometric and turbidimetric methods were compared using 30 cm^3 samples of surfactant solution containing a nominal 20 μmol SDBS to give an equivalence volume of 5 cm^3. The effect of salinity on the titrations was studied using samples prepared containing sodium chloride concentrations of 0.0, 0.14, 0.70 and 1.46 wt%. The influence of the choice of filter (580 or 620 nm) was also investigated.

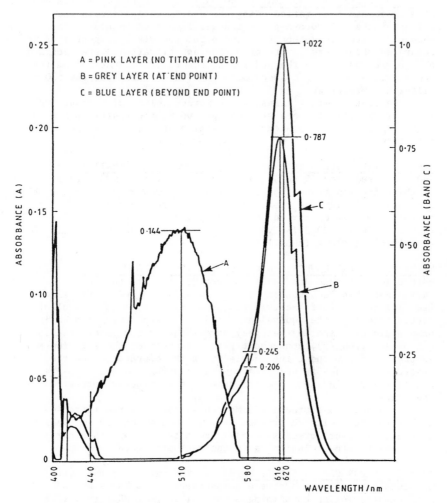

FIGURE 2   Visible spectra of chloroform layers, before, at and
beyond the end point of photometric titration

RESULTS AND DISCUSSION

COMPARISON OF METHODS. Both methods were fairly rapid with a
typical analysis time of 5 minutes per sample as compared with 30
minutes per sample for the manual method.
     A typical photometric titration of SDBS against hyamine at
low salinity is shown in Figure 3. It was noted, however, that the
titration curve was "V" shaped and not the anticipated step curve.
The turbidimetric titration of SDBS against hyamine afforded a
curve, Figure 4, very similar to that for the photometric
titration.
     Photometric and turbidimetric titrations at the same salt
level were always similar suggesting that only a turbidimetric
signal is seen. Also, there is little difference between the
results obtained at 580 and 620 nm wavelength, which further
demonstrates the absence of a photometric signal because there is a
significant difference in the absorption of the indicator dyes
between these wavelengths. Salt concentration, however, has a
marked effect on the shape of the titration curve. At low salt
concentrations all the curves have a sharp minimum, as shown in
Figure 5. As the salt concentration increases this sharp minimum
begins to broaden out, until the highest salt level studied (1.46
wt%), causes the curve beyond the titration point to show very
little increase, as shown in Figure 6.

EMULSION MECHANISM. The above results show that the indicator was
unnecessary because the method is essentially a turbidimetric one.
However, the occurence of a sharp minimum in the titration curve at
low salt concentrations, suggests that the formation of a colloidal
precipitate is not the underlying mechanism for the turbidity.
Consequently, efforts were made to clarify the underlying
physico-chemical mechanism of the titration.
     During surfactant titrations two observations were made:
  (i) The contents of the reaction vessel were turbid being milky
       white in appearance, although this was less intense near the
       endpoint of the titration.
 (ii) If the addition of hyamine and the stirring were stopped two
       clear phases separated. This occurred significantly faster
       when the titration was in the region of the endpoint.
It was apparent that when the two immiscible fluids were stirred
droplets of chloroform formed in the aqueous phase. It was
hypothesised that the response of the phototrode was dominated by
light scattered back from the droplets without reaching the mirror
and that, as the droplet size decreases, the intensity of the back
scattered light increases. This was confirmed in tests by
increasing the rate of stirring and so decreasing the droplet
size.
     This concept allows the shape of the titration curves to be
explained by postulating that the chloroform droplet size decreases
as the interfacial tension (ift) between the aqueous and chloroform
phases is decreased by the presence of active surfactant. As the
endpoint in a titration is approached the amount of active SDBS
decreases as it complexes with the injected hyamine. The reduction
in the amount of active surfactant material results in an increase

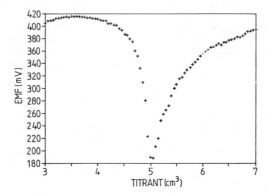

FIGURE 3   Typical photometric titration of SDBS against Hyamine
at low salinity

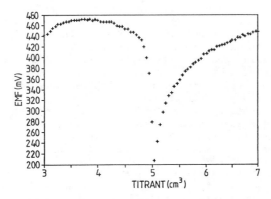

FIGURE 4   Typical turbidimetric titration of SDBS against Hyamine
at low salinity

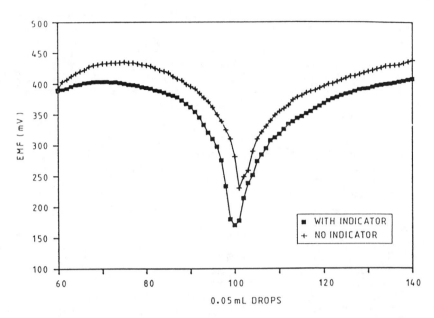

FIGURE 5   Titration curves at 0.0 wt% NaCl using 580 nm light

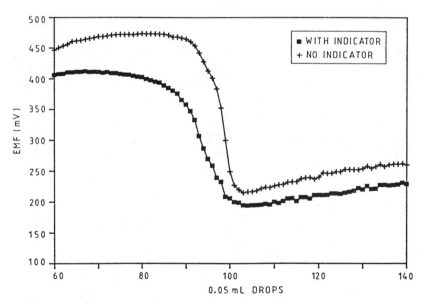

FIGURE 6   Titration curves at 1.46 wt% NaCl using 580 nm light

in droplet size and a corresponding lowering of the transmittance.
This process continues to the endpoint where there is no more
active surfactant present. At this stage the average droplet size
is maximised and, consequently, the transmittance is at a minimum.
Further addition of hyamine, which is a cationic surfactant, makes
possible the increase in transmittance observed in the titrations
(up to 0.70 wt% salt). It is assumed that the presence of salt
reduces the ability of the hyamine to lower the ift of chloroform
and water (much more so than any reduction for SDBS) and offers an
explanation for the variation in the shape of the titration curves
with salt level.

TEST OF THE EMULSION MECHANISM. The above hypothesis suggests that
at the equivalence point all the SDBS is complexed with hyamine
and, effectively, only a brine and chloroform system is present.
It also suggests that an excess of either anionic or cationic
surfactant causes a change in droplet size and an increase in light
scattering. Therefore, it should be possible to mimic the two
branches of the titration curve emanating from the equivalence
point by starting with pure brine (35 cm^3) and chloroform (10 cm^3)
and using either hyamine or SDBS as titrants. Experiments
undertaken to examine this hypothesis are described below.
     In order to achieve an initial droplet size distribution in
the titration cup the contents were stirred for 230 seconds before
injecting any titrant. The mode of operation of the autotitrator
was changed to inject 0.05 cm^3 of hyamine every 10 seconds to make
conditions for each titration exactly the same. The results when
hyamine was used as the titrant are shown in Figure 7. This figure
shows the expected increase in signal with increasing hyamine
concentration and that increasing the salt concentration decreases
the slope of the titration curves. The procedure was then repeated
using SDBS as the titrant to produce the results plotted in
Figure 8. Again the curves show the expected increase in signal
with increasing surfactant concentration and the decrease in slope
of the titration curve at higher salt levels is consistent with the
trend observed in the original tests, although it is less marked
than when hyamine is used as the titrant.
     These tests established that detection of the chloroform
emulsification was the principle underlying action of the
autotitrator. However, while there is agreement in the qualitative
dependence on the salt level there are differences in the apparent
rates of change in signal with aliquot addition. These can be
attributed mainly to non-equilibrium effects.
     The initial titration aliquots were added automatically on the
basis of the rate of change of EMF, mode (i), and the resulting
time between aliquot additions was usually shorter than the 10
second equilibration time allowed in the mode (ii) titrations
described above. The time differences were especially significant
when the transmittance change per 0.05 cm^3 aliquot was small, for
example when the hyamine was present in excess at high salt
concentrations. This means that the mode (i) titrations are more
influenced by kinetic effects and so the measured curves are less
distinctive as may be seen by comparing the results at 1.46% salt
in Figures 6 and 7.

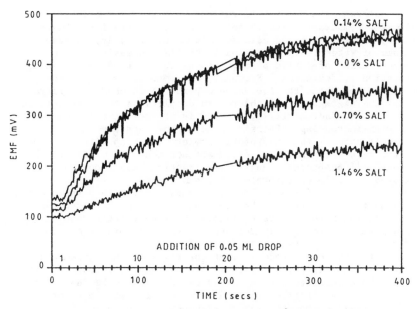

FIGURE 7   Titration of 35 cm³ brine and 10 cm³ chloroform against
hyamine.   230 seconds initial equilibration time and 10
seconds between successive titrant additions.

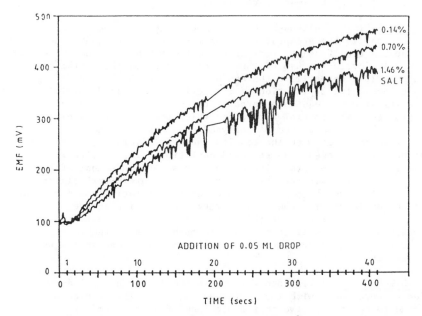

FIGURE 8   Titration of 35 cm³ brine and 10 cm³ chloroform against
SDBS.   230 seconds initial equilibration time and 10
seconds between successive titrant additions.

Although the hyamine titrations in Figure 7 are in principle directly comparable with those obtained for titrations (Figures 5 and 6), this is not the case for the SDBS. Firstly, because the volume of water in the reaction vessel is increasing as the concentration of active SDBS increases in Figure 8, while in a real titration it decreases. Secondly, and more importantly, the mechanism by which the equivalence point is reached in a titration is by the aggregation of chloroform droplets as the concentration of active surfactant falls, whereas in Figure 8 the increase in the EMF with increasing SDBS is related to the increase in transmittance as the chloroform droplets break up with stirring. If the rates for these processes are not equal then different curves will be produced unless the reaction vessel contents are allowed to fully equilibrate between aliquots.

An example of these effects is shown in Figure 9 where a mode (ii) surfactant titration has been performed in the absence of salt and allowing a 10 seconds equilibration time between each aliquot. This gives a more clearly defined equivalence point when compared to the mode (i) titration in Figure 5.

A further difference is that the value of the transmittance minima of the titration curves is higher than that found for the water and chloroform system. This is believed to be due to the fact that in a titration the equivalence point will not correspond to an integer number of hyamine aliquots and so, even if the contents of the reaction vessel were allowed to equilibrate fully, there would still be a small amount of uncomplexed surfactant present which will be sufficient to decrease the droplet size and so increase the back-scattered light.

ERROR ANALYSIS. Although the endpoint might be expected to correspond to the steepest point of the curve, the equivalence point has been found to correspond to the minimum of the curve. At low salt levels the use of the steepest point on the titration curve will underestimate the equivalence point by 1% but at higher salt levels the application of this criteria may lead to misleading results with the equivalence point being considerably under-estimated. The initial titrations which possess clear minima (i.e. 0 and 0.14 wt% salt) can be analysed using the minimum as the equivalence point with the results shown in Figure 10, where the error bars represent one standard deviation. The results were compared using t-tests. The use of the superfluous indicator gives significantly lower results compared to the purely turbidimetric method, because the mixed indicator has a net cationic dye content which complexes with the SDBS thereby reducing its active concentration. For the turbidimetric method consistent results are obtained at both salinities and with both filters, though the results show less scatter with the 580 nm filter and so its use is preferred.

EFFECT OF OIL. Titrations were performed in which small amounts of decane were added with the surfactant sample. The results were found to be insensitive to the presence of up to 2 cm^3 of the decane. This allows the application of the method to both simple aqueous solutions and microemulsions containing significant quantities of decane.

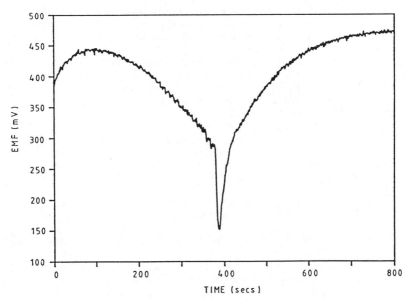

FIGURE 9 Turbidimetric titration SDBS against hyamine.
10 seconds between successive titrant additions.

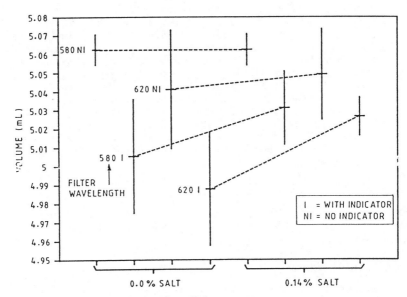

FIGURE 10 Analysis of equivalence volumes from titration curve
minima.

CONCLUSIONS

A detailed study of the determination of the anionic surfactant
SDBS by automated titration against hyamine using an autotitrator
has been undertaken. The following conclusions and observations
can be made.
   (i) An automated turbidimetric method using a chloroform phase
       and hyamine titrant without indicator dyes has been developed
       for the determination of SDBS concentrations.
  (ii) The mechanism producing the turbidity in the titration cup is
       postulated to be the formation of chloroform droplets in the
       aqueous phase under the action of vigorous stirring.
       Chloroform droplets back-scatter light at an intensity which
       decreases with drop size and the drop size is determined by
       the ift between the aqueous and chloroform phases. The ift,
       and hence the drop size, is decreased by the presence of
       active uncomplexed surfactant or hyamine.
 (iii) The contents of the titration cup should be stirred at the
       maximum rate available to enhance the rate of formation of
       the chloroform droplets.
  (iv) A light source of 580 nm wavelength gave a smaller error than
       a wavelength of 620 nm.
   (v) The equivalence point should be determined from the minimum
       in the titration curve which is well defined for salt levels
       below 0.14% wt% for SDBS. Thus distilled water should be
       used to dilute surfactant solutons.
  (vi) Sharper minima can be obtained by allowing a longer time
       interval between titrant additions.
 (vii) This study has been specific to the titration of SDBS against
       hyamine. However, it will probably apply to other anionic
       surfactants since the analytical method works by indirectly
       measuring interfacial tension, and it is a generic property
       of surfactants to alter ift. Whether or not the method is
       suitable for a particular surfactant will depend on the
       detail of the titration curve specific to the surfactant
       system and the accuracy to which the determination is
       required. Quantitative results will only be possible when
       the method possesses a well defined equivalence point
       determined by the minima of the titration curves. If no such
       clear definition exists then the method cannot be relied on
       to give accurate results.

ACKNOWLEDGMENT

This work was carried out for the UK Department of Energy under
contract number R5221 as part of its Enhanced Oil Recovery
Research.

LITERATURE CITED

1.   Longman, G. F.; The Analysis of Detergents and Detergent
     Products; J. Wiley & Sons, 1978.
2.   Södergren, A.; Analyst 1966, 91, 113.
3.   Herring, D. E.; Lab Pract. 1962, 11, 113.
4.   Holness, H.; Stone, W. R.; Analyst 1957, 82, 166.

5. Reid, V. W.; Longman, G. F.; Heinerth, E.; Tenside 1967, 4, 292.
6. Reid, V. W.; Longman, G. F.; Heinerth, E.; Tenside 1968, 5, 90.
7. Determination of Anionic - Active Matter - Direct Two-phase Titration Procedure, ISO 2771, 1972.
8. Synthetic Anionic Active Matter in Detergents by Potentiometric Titration. ASTM D4251-83.
9. Operating Instructions for the DL40RC; Mettler Instruments AG, CH-8606 Griefensee, Switzerland.
10. Turbidimetric Titration of Anionic Tensides, Mettler Application No 135; Mettler Instruments AG, CH-8606 Griefensee, Switzerland.

RECEIVED September 23, 1988

Chapter 14

Mixed Micellization and Desorption Effects on Propagation of Surfactants in Porous Media

L. Minssieux

Institut Français du Pétrole, B.P. 311, 92506 Rueil-Malmaison, Cedex, France

The micellization and adsorption properties of industrial sulfonate/ethoxylated nonionic mixtures have been assessed in solution in contact with kaolinite. The related competitive equilibria were computed with a simple model based on the regular solution theory (RST). Starting from this analysis, the advantage of adding a hydrophilic additive or desorbing agent to reduce the overall adsorption is emphasized. As a test, surfactant slug flow experiments were performed in clayey sandpacks with and without the injection of a desorbent behind the micellar slug. Results show that a substantial decrease in surfactant retention is obtained in calcic environment by such an additive. Likewise, the ethoxylated cosurfactant in the micellar slug can be remobilized simultaneously with sulfonatewithout any change in its ethylene oxide distribution. The application of the RST to sulfonate/ethoxylated alkylphenol mixtures explains semi-quantitatively the relationship between their properties and composition.

Surfactants are often used for oil production, either for well treatments or in EOR recovery processes such as micellar flooding, controlling the mobility of injected gas by the forming of foam, or huff-and-puff steam injection. Reducing the costs of such operations goes via improving the efficiency of the products injected into the reservoir rock. To achieve this, a good understanding of the properties of surfactants and of their mixtures is required. Considerable fundamental research efforts have been made in this area in recent years, particularly concerning binary mixtures of isomerically pure or purified surfactants (1 - 3) in solution in deionized water or water containing NaCl.

In this paper, we focused on behavior of industrial surfactant mixtures that can be used in calcic saline media, i.e. under conditions often encountered in onshore or offshore fields. The results obtained have been used for the laboratory testing and for interpreting the working mechanisms of a method of reducing losses of products in reservoir rocks and of propagating them more effectively in the formation to be treated.

0097–6156/89/0396–0272$06.00/0

Surfactants Investigated

Ethoxylated fatty alcohols and alkylphenols were used. The products available on the market make up homologous series containing an average of between 3 and 100 ethylene-oxide groups. They thus have a wide HLB (hydrophilic/lipophilic balance) range. Besides, they are among the least expensive surfactants on the market.

As an anionic surfactant, a synthetic alkylate-base sulfonate containing about 60 % active material (Synacto 476) was used. To make it compatible with the injection water considered (composition in Table I) containing 1500 ppm Ca^{++} and Mg^{++} ions, a nonionic cosurfactant was combined with it, i.e. an unsaturated ethoxylated fatty alcohol with 8 ethylene oxide groups (Genapol). Their main characteristics and properties are listed in Table II.

Micellar Properties of Aqueous Solution of Surfactants

Monomer/Micelle Equilibrium: Mixtures of surfactants, like any surfactant species in an aqueous solution, give rise to monomer or micelle aggregates provided that the concentration reaches a minimum value, called the critical micellar concentration (CMC). The micelles thus formed are mixed, i.e. made up of the different surfactant species in solution.

The corresponding monomer/micelle equilibria can be dealt with by the regular solution theory (RST), as shown in particular by Rubingh in 1979 (1). The application of this theory to numerous binary surfactant systems (2 - 4) has followed and led to a set of coherent results (5).

Without going into this theory in detail, let us reproduce here the equation proposed by Rubingh for the activity factor of surfactant species making up mixed micelles in a binary system :

$$f_1 = e^{\beta_{12}(1 - x_1)^2}$$
$$f_2 = e^{\beta_{12} x_1^2}$$

in which x_1 is the molar fraction of species 1 in the mixed micelles, and β_{12} is the molecular interaction parameter. This parameter takes into consideration the full interaction forces existing between the surfactant moles of a mixed micelle. It is thus entirely characteristic of the surfactant pair considered and of their ionic environment, as shown in the list of values compiled by Nagayaran (5).

Calculating the Characteristic Interaction Parameter of the Micellar Systems Used: To perform the calculation of β_{12} for the systems examined, i.e. Sulfonate/Genapol/ethoxylated nonylphenol mixtures, the following assumptions were made:

. Industrial surfactants can be assimilated with a single entity.

TABLE I :   Composition of Synthetic Brine

Cations	Conc. (ppm)	Salt concentration (g/l)	
Na^+	8,600	NaCl	17.00
K^+	65	KCL	0.13
Mg^{++}	290	$MgCl_2$, 6 H_2O	2.436
Ca^{++}	1,300	$CaCl_2$, 2 H_2O	4.777
Na^+	1,918	Na_2 SO_4	5.92
		TDS =	30.26

TABLE II : Characteristics of Surfactants Used

Additive	CMC		Ads Plateau		Molecular weight	% active material (A.M.)
	mg/l	M/l	mg/g	μM/g		
Micellar system:						
(Sulf/Gen.)	30		5.20			
as sulf. alone	9	$2.33.10^{-5}$	1.87	4.86	385	60
Genapol 8 E.O.	30	5.10^{-5}	5.2	8.5	608	>98
Desorbents						
Nonylphenols						
with: 14 E.O.	40	$4.78.10^{-5}$	0.37	0.44	836	>98
30 E.O.	160	10.4 "	0.30	0.20	1,540	>98
50 E.O.	500	20.7 "	0,44	0.18	2,420	>98
100 E.O.	1,300	28.1 "	0.77	0.17	4,620	97

. There is no molecular interaction between nonionic surfactants with an ethylene-oxide chain, i.e. Genapol and ethoxylated nonylphenols. Indeed, research by Nishikido (6) on polyoxyethylene laurylethers (5 < E.O. number < 49) has shown the ideal behavior ($\beta_{12} = 0$) of their mixtures. Likewise, Xia (7) has found very low β_{12} values for mixtures of ethoxylated fatty alcohols.

. Taking in account the results obtained by Graciaa (8) and Osborne-Lee (9) for alkylbenzene sulfonate and alkylphenols with an increasing degree of ethoxylation, we have considered that the interaction between sulfonate and associated highly ethoxylated nonylphenol (with 30, 50 or 100 E.O.) was predominant in mixed micelles of the mixtures investigated.

. The sulfonate/Genapol pair was assimilated with a pseudo-component, with the cosurfactant acting only as a solubilizer in the brine used.

The mixture of the three additives was then dealt with as a pseudo-binary system to which the RST theory was applied.

Figure 1 gives the measurements of surface tension used for determining the CMCs of sulfonate/Genapol and nonylphenol 30 E.O. mixtures, with the last surfactant being called a desorbent (this term will be justified below). Minimum in surface tension was seen only for a few nonionic solutions (e.g. NP 50 E.O.). In this case, we used dyes that, once solubilized in the micelles, cause the solution to change color, which is another way of measuring the CMC.

The value of the characteristic interaction parameter of these systems (30° C), adjusted from the CMC measurements in Figure 1, was calculated by means of RST and taken equal to -2.5. This value is effectively in the range of the ones found by Graciaa for similar anionic/nonionic mixtures (8).

Limit Concentration of Monomers in Solution: In the calcic environment considered, the CMC values of surfactants are low. For example, sulfonate and Genapol solutions reach their CMC at 30 ppm (Table II). The surfactant solutions injected in practice at concentration of about one or several percent are thus generally used well above their CMC. Under such conditions, the predominant fraction of each surfactant is the micellar form whose composition (x_i) is practically equal to the initial proportion of products (i.e. alpha 1 for sulfonate). At this concentration level of products, very small proportions of monomer species coexist, the limit concentrations of which are respectively :

$$\text{mono 1 conc} = x_1 \, f_1 \, CMC_1 = \alpha_1 \, CMC_1 . \, e^{\,\beta_{12}\,(1-\alpha_1)^2}$$

$$\text{mono 2 conc} = (1 - x_1) \, f_2 \, CMC_2 = (1 - \alpha_1) . \, CMC_2 . \, e^{\,\beta_{12}\,\alpha_1^{\,2}}$$

For the pseudo-binary mixture ($\alpha = 0.5$) of sulfonate and nonylphenol with 30 E.O., figure 2 shows how the concentration of each of their monomer calculated by the RST theory (1), varies as a function of the overall surfactant concentration. It can be expected that the asymptotic regime in which monomer concentrations are stabilized will correspond to a plateau of the adsorption isotherm for the surfactant mixtures considered.

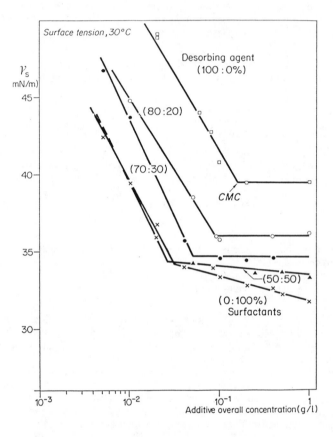

Figure 1. CMC determination of surfactant mixtures. (Reproduced with permission from ref. 16. Copyright 1987 Deutsche Wissenschaftliche Gesellschaft.)

Properties of Surfactant Solutions in Equilibrium with an Adsorbent Solid

Taking Simultaneous Micellization and Adsorption Phenomena into Consideration: In the presence of an adsorbent in contact with the surfactant solution, monomers of each species will be adsorbed at the solid/ liquid interface until the dual monomer/micelle, monomer/adsorbed-phase equilibrium is reached. A simplified model for calculating these equilibria has been built for the pseudo-binary systems investigated, based on the RST theory and the following assumptions :

. Each species is adsorbed individually according to a linear adsorption isotherm until there is an equilibrium concentration in solution equal to the CMC_i of the product considered.

. There are no interactions between species 1 and 2 adsorbed on the solid surface.

. Admicelle formation and associated CAC (Critical admicelle concentration) as proposed by Scamehorn (10) and Harwell (11) were not introduced here for a practical reason : a feasible and fast method of CAC measurement does not seem to exist at the moment. The difficulties related to such delicate determinations appear well from observation of the detailed adsorption isotherms of pure sulfates mixtures published by Roberts et alii (10).

The problem could be even more difficult in the case of industrial anionic/nonionic surfactants, due to their polydispersity and very low CMC in the salty environment considered here. So the corresponding needed CAC data was not available.

. The equilibria are instantaneous.

These assumptions are akin to those taken in account in the mixed adsorption model of Trogus (12). The difference between the two models lies in the relationship linking CMCs of single and mixed surfactants and monomer molar fractions : Trogus used the empirical equation proposed by Mysels and Otter (13); in our model, the application of RST leads to an equation of the same type.

Calculation examples of mixed surfactant adsorption: The solid chosen as the model adsorbent was made up of a natural sand (specific area = 380 cm^2/g) mixed with 5% clay (Charentes kaolinite with specific area = 26.8 m^2/g). This material was taken as a model of clayey sandstone reservoirs.

The adsorption plateaus on this solid, determined with each of the surfactants (Table II) and the individual CMC values, were used to calculate the adsorption constants input in the model. Figure 3 compares the total adsorption (sulfonate + NP 30 EO) of the pseudo-binary system investigated as a function of the initial sulfonate fraction of the mixtures under two types of conditions : (1) on the powder solid, batch testing with a solid/liquid ratio, S/L = 0.25 g/cc (2) in the porous medium made from the same solid, for which this solid ratio is much higher (S/L = 4.0 g/cc).

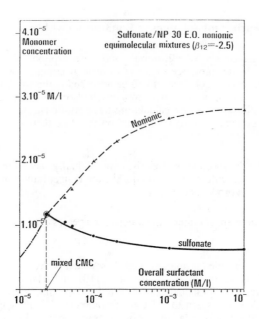

Figure 2. Monomer versus overall surfactant concentrations in micellar solutions.

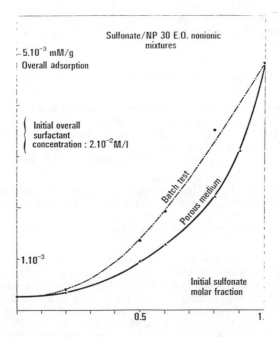

Figure 3. Mixed overall adsorption of pseudobinary systems.

The differences between the two curves can be explained by the sulfonate (the most adsorbed surfactant) monomer concentrations at equilibrium, which were reached in both cases, considering the amounts of surfactants, liquid and solid present. Figure 4 shows a distinct evolution of monomer concentrations for the two solid/liquid ratios considered.

In both cases, overall adsorption and especially that of sulfonate (or "primary" surfactant in the composition of most micellar systems used for EOR) are considerably reduced by simply adding a second product having low adsorption characteristics (NP 30 EO in the above example). This is why we have called this strongly hydrophilic surfactant a desorbent.

Such an idea was patented in 1981 (14). Besides research by Scamehorn and Schechter (15) provided an experimental illustration of this by batch adsorption tests of kaolinite with some purified anionic/nonionic products. Our objective was to enlarge and test this technique under the dynamic flow conditions of industrial surfactant injection in an adsorbent porous medium.

Surfactant Transport in Porous Media: Dynamic Adsorption/Desorption Equilibria

Circulation Test Conditions for Additive Solutions in Porous Media: The sand/kaolinite mixture described above was used to form sandpacks in a Rilsan cell (13 or 30 cm long, 2.5 cm in diameter, 36 % porosity). The corresponding solid/liquid ratio was then 4.72 g/cc.

The surfactant retention tests were performed in the porous medium at 43° C in sandpack (S_{or} = 0) saturated with brine (See composition in Table I). The injection flow rate used in these tests (2 cm³/h) corresponds to a front velocity of 30 to 40 cm/day.

A typical sequence followed in this test series consists in injecting : (1) a micellar slug of one pore volume of aqueous solution of 4% of the preceding pseudo-binary system (2% sulfonate/2% Genapol) ; (2) a slug of desorbent solution corresponding to a fixed amount of additive (e.g. equal to 1 PV at a concentration of 0.5 %) ; (3) at least 1.5 PV of brine with no additive.

Reference tests were also performed in the absence of any desorbent (Tests 1 and 2 in Table III). Likewise, the propagation of each desorbent was examined separately, without any prior micellar slug injection. The effluents were sampled for analysis by a fraction collector.

The sulfonate content was determined either by the well-known technique of two-phase titration with hyamine or by liquid chromatography (HPCL). Nonionic surfactants were analyzed by HPLC (16) in the reverse or normal phase mode depending on whether the aim was to determine their content in effluents or to compare their ethylene oxide distribution.

Such products were detected in the UV, using a Waters multiwavelength detector (Model 490 E). The wavelenght chosen for the ethylenic nonionic (Genapol), rich

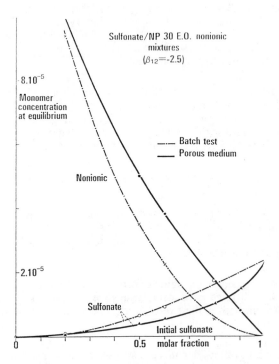

Figure 4. Monomer concentration versus initial sulfonate fraction for two typical solid–liquid ratios.

TABLE III : Adsorption/Desorption Data From Dynamic Flow Tests

Run n°	Desorbent concentration used		pH effluents	Surfactant retention (mg/g)					Associated concentration in effluents (g/l)	
	g/l	mM/l		sulfonate	Cosurf.	Desorbent mg/g	μM/g	sulfonate cosurf. ratio	Sulfonate	Desorbent
1	0	0	5.0 - 6.0	2.81	4.70			1.67		
2	0	0	5.8	1.85	3.42			1.85		
NP 30 EO										
3	2.7	1.75	6.7 - 7.0	1.22	1.78	0.50	0.32	1.46	1.3 - 1.8	2.3 - 2.4
4	5.0	3.25	5.0 - 5.6	1.55	2.60	0.17	0.11	1.68	4.2 - 5.2	3.2 - 4.2
5	5.0	3.25	5.8	1.17	2.30	0.46	0.30	1.96	4 - 5	2.5 - 3.7
6	5.0	3.25	7.0	1.08	1.92	0.16	0.10	1.78	3.6 - 4.5	3.5 - 4.8
NP 50 E.O.										
7	5.0	2.07	6.2	1.08	1.07	0.48	0.20	1	3.5 - 4.5	3 - 4.0
NP 100 E.O.										
8	5.0	1.08	6.9	1.30	1.90	0.96	0.21	1.46	1.7 - 2.4	0.7 - 1.3

in double bonds, was 229 nm. 278 nm, wavelenth characteristic of aromatic ring absorption, was taken for the detection of ethoxylated alkylphenols or of the synthetic aromatic sulfonate used.

To mark the displacement front, 150 ppm of sodium iodide was incorporated in the surfactant micellar slug. This tracer can easily be detected in effluents with a UV detector at 229 nm.

Elution of a Surfactant Slug in the Presence of a Desorbent: A mass balance for each of the three additives was used to and results obtained during each test are given in Table III.

The first two tests were performed without any desorbent and used as references, at two distinct levels of equilibrium pH. The performances of three desorbents (NP 30, 50 and 100 E.O.) having an increasing ethoxylation degree were compared at the same mass concentration (0.5%).

Figures 5 and 6 illustrate the surfactant elutions obtained in Tests 4 and 7.

Surfactant Remobilization by Means of Desorbent: It appears from Table III, that, in the presence of a desorbent, 30 to 45% of the sulfonate and 33 to 69% of the Genapol can be remobilized. This assessment is made by comparison with the reference tests performed at the closest pH values.

Figures 5 and 6 show that the concentration of the two surfactants in the effluents increases simultaneously with the production of the desorbent, which confirms the mixed micellization mechanism described above. Figure 5, where the three additives are produced lately, illustrates the phenomenon particularly well. At the lower pH corresponding to strong adsorption conditions for sulfonate (test 4), the one pore-volume micellar slug would have been entirely consumed by the medium in the absence of any desorbent.

Surfactant Transport in an Adsorbent Porous Medium. Chromatographic Aspects: A first observation was made in all the tests in Table III. The breakthrough of both surfactants from the micellar slug always occurs simultaneously without any chromatographic effect (Figures 5 and 6). This stems both from the chemical nature of the two products selected and also from the fact that the injected concentration is much greater than the CMC of their mixtures.

Likewise, in order to evaluate nonionics transport, ethylene-oxide distribution in the cosurfactant (Genapol) was determined by HPLC at two stages of production in test 7 : (1) before breakthrough of the desorbent, i.e. in the presence of sulfonate in the effluent; and (2) after its breakthrough when the three additives coexist in solution in the form of mixed micelles.

Figure 7 shows the quasi identity of the ethylene-oxide distributions of the Genapol samples, analyzed at the outlet of the porous medium. For nonylphenols with 14 and 30 EO, we also checked that the distribution of these nonionic agents (injected in a concentration of 5 g/l) was not appreciably changed after transit via the adsorbent porous medium. Under these conditions, the mixed micelles formed

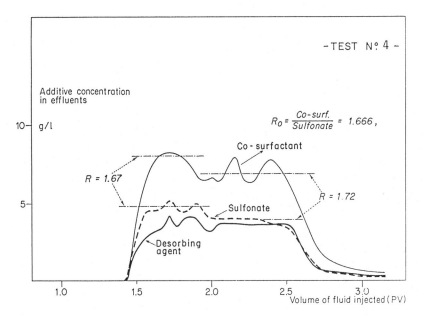

Figure 5. Elution of surfactants by means of desorbent NP 30 E.O. (Reproduced with permission from ref. 16. Copyright 1987 Deutsche Wissenschaftliche Gesellschaft.)

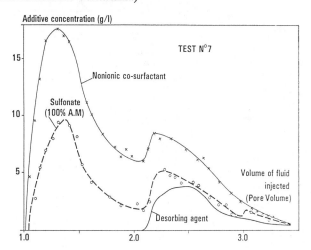

Figure 6. Elution of surfactants by means of desorbent NP 50 E.O.

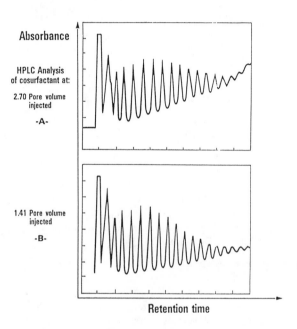

Figure 7. Ethylene oxide distribution of cosurfactant produced before (A) and after (B) desorbent breakthrough. (Reproduced with permission from ref. 16. Copyright 1987 Deutsche Wissenschaftliche Gesellschaft.)

contain all the monomer species of the different constituents of industrial products moving at the same speed. The same analysis remains to be done with the industrial sulfonate used.

It can be mentioned here that we found in another study (17), that these surfactant remobilization mechanisms by mixed micellization also operated in the presence of crude oil in the medium and thus help increase oil recovery.

Comparison of Retention Properties of Three Desorbents with an Increasing Degree of Ethoxylation: The individual behavior of three nonionic desorbents (NP 14, 30 and 100 E.O) is compared in Figure 8. Slug size was 1.16 PV in those tests. The outflow of the tracer indicates the slug front of the additive injected. The concentration used was 5 g/l in all tests. On a weighted basis, it was the NP 30 E.O., that led to the lowest final retention, i.e. 0.30 mg/g of rock (Table II).

Interpretation of Porous Medium Results

As suggested above, the main recovery mechanism of surfactants retained in the rock can be interpreted as a micellization phenomenon inside the pores. Upon contact with micelles from the desorbent agent, the adsorbed surfactants are solubilized in the form of mixed micelles. This also explains the effectiveness of the desorbent still observed at low concentration (0.27% in Test 3 in Table III, concentration much higher than the CMC of NP 30 EO equal to 0.016%).

Comparison: Theoretical Equilibrium Calculations and Results of Circulation Tests in Porous-Media: To make this interpretation more quantitative, the regular solution theory (RST) was applied to sulfonate/desorbent dynamic equilibria reached inside porous media by using the approach described above. In so doing, we assumed that the slugs injected were sufficiently large and that a new equilibrium was reached at the rear of micellar slug in the presence of desorbent.

Calculations were made at the desorbent concentrations used in Tests 3, 6,7 and 8 in Table III. Table IV below gives the respective adsorptions of sulfonate and desorbent as well as their equilibrium concentration. A comparison with the corresponding experimental values in Table III shows good agreement with regard to sulfonate from the micellar slug. On the other hand, losses of desorbent are systematically underestimated.   This shows that the assumption of the independent adsorption of both surfactants on the solid is incorrect and that presumably cooperative adsorption of desorbent and sulfonate takes place. Accordingly the model used needs to be improved.

Conclusions

The present study suggests the potential application of a method for reducing surfactant losses in reservoirs, thus, ipso facto increasing their effectiveness. This method consists in incorporating a suitable desorbent in the water used to drive the surfactant slug injected into the formation to be treated.

Such a desorbent may be, for example, a hydrophilic nonionic surfactant, which is among the least expensive on the market and is suitable in calcic environment.

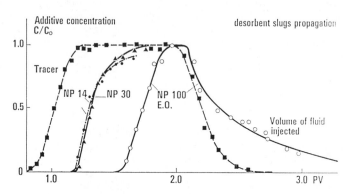

Figure 8. Profiles of additive produced in slug injection runs of three distinct desorbents.

TABLE IV: Computation of Solid/Micellar Solutions Equilibria

Run n°	Desorbent used	Coefft. β_{12}	Sulfonate		Desorbent	
			Loss (mg/g)	Equil. conc. (g/1)	Loss (mg/g)	Equil. conc. (g/1)
3	NP 30 E.O.	- 2.5	1.30	2.40	0.024	2.38
6	NP 30 E.O.	- 2.5	1.14	3.43	0.042	4.80
7	NP 50 E.O.	- 4.45	1.21	3.14	0.006	4.96
8	NP 100 E.O.	- 5.73	1.39	2.26	0.002	4.99

The choice should be optimized as a function of the type of surfactants to be desorbed and the temperature and characterictics of reservoirs.

The critical micellar concentrations of anionic/nonionic surfactant mixtures examined are low in a saline medium, so that, at the concentrations injected in practice, the chromatographic effects resulting from the respective adsorption of monomers are masked. Such surfactants propagate simultaneously in the medium in the form of mixed micelles.

In the same way, after transit through a porous medium, non appreciable change was found in the ethylene oxide distribution of nonionic surfactants used as a cosurfactant or desorbent.

The theory of regular solutions applied to mixtures of aromatic sulfonate and polydispersed ethoxylated alkylphenols provides an understanding of how the adsorption and micellization properties of such systems in equilibrium in a porous medium, evolve as a function of their composition. Improvement of the adjustment with the experimental results presented would make necessary to take also in account the molar interactions of surfactants adsorbed simultaneously onto the solid surface.

Acknowledgments :

This study was partly supported by the European Economic Community.

We wish to express here our appreciation to N. MONIN and G. SYLVESTRE who performed the laboratory tests and analyses.

LITERATURE CITED:

(1) Rubingh, D. N., Solution Chemistry of Surfactants, K. L. Mittal, ed., vol. 1, 337-354, New York, 1979.

(2) Holland, P. M. and Rubingh, D. N., "Nonideal Multicomponent Mixed Micelle Model, J. Phys. Chem., vol. 87, 1984-1990, 1983.

(3) Scamehorn, J.F., Schechter, R.S. and Wade, W. H., J. Disp. Sc. Tech., 3 (3), 261-178, 1982.

(4) Zhu, B. Y. and Rosen, M. J., J. Coll. Sc., vol. 99 No. 2, June 1984.

(5) Nagajaran, Microemulsions, Adv. Coll. Int. Sc. November 1986.

(6) Nishikido, N. et al., Bul. Chem. Soc. Japan, vol 48 (5), 1387-1390, 1975.

(7) Xia J. et alii, "Effects of Different Distributions of Lyophobic Chain Length on the Interfacial Properties of Nonaethoxylated Fatty Alcohol" in "Phenomena in Mixed Surfactant Systems", J.F. Scamehorn, Ed. 1986, ACS Stmposium Series 311, Wash.

(8) Graciaa A. et alii, "The Partitioning of Nonionic and Anionic Surfactant Mixtures Between Oil/Microemulsion/Water Phases", n° SPE 13 030, Houston, 1984.

(9) Osborne-Lee I.W. et alii, J. Coll. Int. Sc., Vol. 1O8, n° 1, Nov. 1985.

(10) Scamehorn, J.F. Phenomena in Mixed Surfactant Systems; American Chemical Society, Symposium Series : Washington, DC 1986.

(11) Harwell, J.H. et alii, Aiche J. 1985, 31, 415.

(12) Trogus, F.J.; Schechter, R.S. ; Pope, G.A.; Wade W.H., J. Petr. Tech. 1979, June, 769.

(13) Mysels, K.J.; Otter, R.J., J. Coll. Sc. 1961, 16, 474.

(14) Kudchadker, US Patent 4, 276, 933, July, 1981.

(15) Scamehorn J.F., Schechter R.S. and Wade WH., "Adsorption of Surfactants on Mineral Oxide Surfaces from Aqueous Solution", J. Coll. Int. Sc., Vol. 85, n° 2, Feb. 1982.

(16) Minssieux L., "Method for Adsorption Reduction of Mixed Surfactant Systems", Proc. 4th. Eur. EOR Symp., 1987, p. 293.

(17) Minssieux L., "Surfactant Flood with Hard Water : A Case Study Solved by HLB Gradient", SPE Res. Eng., Vol. 2, N° 4, 605-612, Nov. 1987.

RECEIVED November 28, 1988

SURFACTANT   PHASE   BEHAVIOR

Chapter 15

Calorimetric Phase Studies of Model Amphiphilic Petroleum Recovery Systems

Duane H. Smith, G. L. Covatch, and R. O. Dunn[1]

Enhanced Oil Recovery Group, U.S. Department of Energy, Morgantown Energy Technology Center, Morgantown, WV 26507-0880

Ways are discussed of measuring both compositions and heats of formation (i.e., excess enthalpies) of two conjugate phases in model amphiphile/water systems by isoperibol titration calorimetry. Calorimetric and phase-volume data are presented for n-C_4H_9OH/water at 30 and 55 °C and for n-$C_4H_9OC_2H_4OH$/water at 55 and 65 °C, and compared to results in the literature. Some considerable practical advantages of calorimetry for the development of oil recovery technologies are pointed out.

Petroleum recovery typically deals with conjugate fluid phases, that is, with two or more fluids that are in thermodynamic equilibrium. Conjugate phases are also encountered when amphiphiles (e.g., surfactants or alcohols) are used in enhanced oil recovery, whether the amphiphiles are added to lower interfacial tensions, or to create dispersions to improve mobility control in miscible flooding (1,2).

Experimental and theoretical studies, as well as computer simulators, all require knowledge of the number and compositions of the conjugate phases, and how these change with temperature, pressure, and/or overall (i.e., system) composition. In short, all forms of enhanced oil recovery that use amphiphiles require a detailed knowledge of phase behavior and phase diagrams.

However, the measurement of phase diagrams is often tedious and time consuming; and the number of variables and combinations of

[1]Current address: Northern Regional Research Center, U.S. Department of Energy, Peoria, IL 61604

compounds is very large. Hence, equations-of-state and other thermodynamic models are used to correlate and predict phase behavior. Extensive experience with CO_2/hydrocarbon systems has shown that the correlations are substantially improved when they are based not only on compositions, but on data for other thermodynamic variables as well.

Furthermore, the experimental difficulties of studying phase behavior can be greatly compounded when surfactants are added to a system of oil and/or CO_2 and water, because the surfactant system often forms stable dispersions (emulsions and/or foams). These difficulties may prove to be especially severe in the development of surfactant-based mobility control for miscible flooding, in which the EOR process <u>requires</u> the formation of dispersions, and in which eventually many measurements will have to be made at miscible-flood pressures (P > 7 MPa).

For these various reasons one may ask if it is possible to develop automated or semi-automated methods that will simultaneously measure compositions and other thermodynamic parameters at reservoir pressures and temperatures, without waiting for bulk separation of dispersed phases in the apparatus.

In an earlier study calorimetry achieved this objective for the compositional boundaries between two and three phases (3). Such boundaries are encountered both in "middle-phase microemulsion systems" of low tension flooding, and as the "gas, oil, and water" of multi-contact miscible EOR systems (1,2). The three-phase problem presents by far the most severe experimental and interpretational difficulties. Hence, the earlier results have encouraged us to continue the development of calorimetry for the measurement of phase compositions and excess enthalpies of conjugate phases in amphiphilic EOR systems.

This paper considers systems of lesser dimensionality than the previous study, namely, systems of two compounds, which (ignoring the vapor) can form only one or two phases. Specifically, excess enthalpies and phase compositions have been measured (at ambient pressure) by isoperibol calorimetry for n-butanol/water at 30.0 and 55.0 $^\circ$C and for n-butoxyethanol/water at 55.0 and 65.0 $^\circ$C. (Butanol, or C4E0, is C_4H_9OH; butoxyethanol, or C4E1, is $C_4H_9OC_2H_4OH$.) The miscibility gap of each of these systems is shown in Figure 1. For each system, titrations have been performed from the neat compound (water or amphiphile) across the phase boundary (aqueous or amphiphilic conjugate

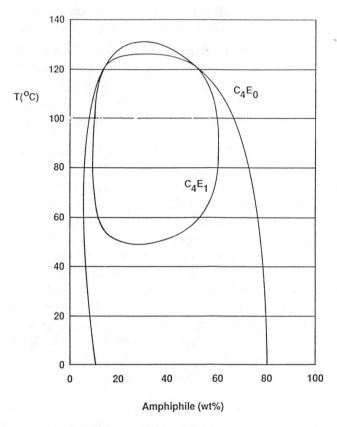

Figure 1. Temperature-composition phase diagrams of n-butanol (C4E0) and n-butoxyethanol (C4E1) , respectively, with water. (redrawn from data in refs. 4 and 5.)

phase) and into the two-phase region. The results are compared (where possible) with previously reported excess enthalpies, or with phase compositions obtained by more conventional phase-diagram methods.

Calorimetry has been successfully used to measure enthalpies and phase boundaries in several CO_2/hydrocarbon systems at reservoir pressures (6). Together, the present and previous studies encompass all of the classes of compounds of the amphiphile/CO_2/hydrocarbon/water systems that are encountered in dispersion-based mobility control for miscible-flood EOR (2).

Experimental

Calorimetric measurements were performed with two Tronac 450 isoperibol calorimeters, and calibrations were made before and after each titration. To calculate heats from changes in thermistor voltage, corrections were made for stirring heat, heat leakage, and change in the energy equivalence ("heat capacity") of the calorimeter and contents due to addition of titrant (7). Titrations of 1.999 cm^3/300 s (calibrated at 55 °C) and 1.998 cm^3/1200 s (calibrated at 30 °C) were used in the respective calorimeters. Thermistor voltages were recorded at 1 s intervals for the faster titration rate and at 5 s intervals for the slower rate. In a typical measurement sequence two successive 2 cm^3 additions of titrant were made to an accurately weighed titer of about 20 cm^3. This sequence was then repeated with a different titer composition as many times as necessary to cover the required range of amphiphile/water compositions. An MS-DOS desk-top computer with 640K of random-access memory (RAM) was used to control the calorimeter and to collect thermistor voltages at the fixed time intervals. Spread-sheet macros running on this same computer were used to make various plots, including thermistor voltage vs. time and corrected heat vs. time.

Phase-volume samples were prepared gravimetrically, thermostatted in constant-temperature water baths ($\pm$ 0.02 °C), and examined visually for the number of phases.

The n-butanol (99.7 mole% stated purity) was from Malinckrodt, and n-butoxyethanol (99+ % stated purity) was from Aldrich. Each was analyzed by gas chromatography for impurities and by Karl Fischer titration for water, and used without further purification. The water was distilled.

Data Treatment and Results

Figure 2 shows measured excess enthalpies for aqueous phases and two-phase mixtures of n-butoxyethanol/water at 55.0 $^\circ$C. Data from four replicate additions of 2 cm^3 of butoxyethanol to water are shown in Figure 2a; Figure 2b shows data from the subsequent addition of an additional 2 cm^3 of butoxyethanol to each of the solutions of Figure 2b. (To reduce overlap between the symbols, only every tenth experimental point of each titration has been plotted.) The large discontinuity between successive titrations is caused by the fact that the instrument necessarily takes the enthalpy of the initial composition of each titration to be zero. (The treatment of this discontinuity is discussed below.)

From the phase rule and the compositional path taken in the measurements, one can show that theoretically the system enthalpy (or volume) must be a linear function of composition in the two-phase region (3.8). Hence, one can find the phase boundary (in this case, the composition of the aqueous phase) by fitting a linear regression to experimental points of the two-phase region, and estimating the composition at which the difference between the measurement and the regression exceeds the instrumental noise. We used an iterative spread-sheet computer program that first provided five-point smoothing of the data. Then the program (1) regressed on five points known to be in the two-phase region; (2) checked for collinearity of the next datum; (3) substituted the latter datum for one of the five regression points (1), if collinearity were found; and (4) returned to step (1), continuing until the deviation from linearity was found. As a refinement of this procedure, one may fit a smoothing function to the single-phase enthalpies, and take the point of intersection of the linear and nonlinear smoothing functions as the phase boundary. Unlike the rigor of the linear function of the two-phase region, the choice of smoothing function for the single-phase region is necessarily model-dependent.

The single-phase enthalpies of Figures 2a and 2b are segments of one theoretical curve. Hence, to find the phase boundary with the aid of a smoothing-function fit to single-phase data, one may choose either to fit only the single-phase data of Figure 2b, or to include the data from Figure 2a, as well. Regardless of the choice made for the phase boundary computations, calculation of the enthalpy of formation of the aqueous phase with water as the reference state requires determination of the unknown constant (i.e., the size of the discontinuity) between the data of Figure 2a and the data of Figure 2b.

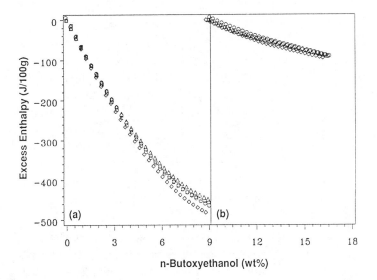

Figure 2. (a) Excess enthalphies of n-butoxyethanol/water at 55.0 °C from four replicate titrations, starting at neat water; (b) continuations of the titrations of (a).

One way to estimate the discontinuity is to simply average the final enthalpies of all of the runs of Figure 2a. However, such a procedure gives undue weight to just a few of the total number of data.

A second procedue is to fit a smoothing function to each group of replicate titrations of a given concentration range; the unknown constant then can be calculated from the different values of the two functions at the concentration that marks the end of the first group of titrations and the beginnning of the second group of replicates.

Cubic equations have been used for several alcohol/water systems, and also for the 2-propanol/CO_2/water system (as well as for numerous CO_2/hydrocarbon systems). (6,9,10) Furthermore, although a critical micelle concentration (CMC) has been reported for butoxyethanol and water, and we have calorimetrically measured the CMC of decane-saturated butoxyethanol micelles at 20 $^{\circ}$C, there was no evidence of micelle formation at the higher temperatures of the present study (11,12; Smith, D. H., et al., presented at the symposium "Use of Surfactants for Mobility Control in CO_2 and N_2 EOR, Ann Arbor, June 23-24, 1987). These facts suggest fitting simple cubic polynomials

$$\Delta H^E = \sum_{i=0,3} a_i X^i \tag{1}$$

to the measured excess enthalpies, ΔH^E, as a function of the wt% amphiphile, X, to calculate the values of the unknown "discontinuity" parameters a_0. For heuristic reasons, both of these fitting procedures were tried.

Figure 3 shows the excess enthalpy vs. composition measured for aqueous solutions of n-butanol and water at 30.0 $^{\circ}$C; Figure 4 shows corresponding results for the amphiphilic side of the miscibility gap at 55.0 $^{\circ}$C. Each Figure shows titration data for compositions inside of the miscibility gap (where the "curve" is linear), as well as enthalpies in a single-phase region. Data from the literature are also shown for comparison with the present results (13-16). Table I shows values of the compositions of the aqueous and amphiphilic phases for n-butanol/water at 30.0 and 55.0 $^{\circ}$C and for n-butoxyethanol/water at 55.0 and 65.0 $^{\circ}$C. Shown for each composition are the value from calorimetry, as obtained in the present study; the value from phase volume measurements, also obtained in the present study; and a value

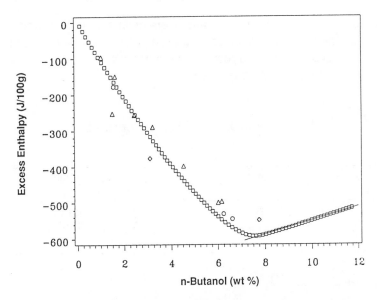

Figure 3. Excess enthalphies of aqueous solutions of n-butanol/ water at 30.0 °C, compared to results from the literature: squares, present results; triangles, ref. 13; circles, ref.14; diamonds, interpolated from ref. 16.

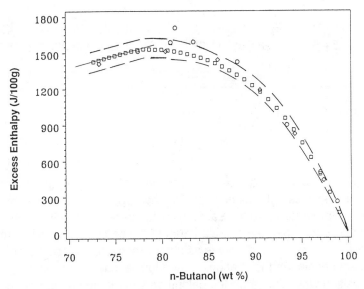

Figure 4. Measured excess enthalpies for the amphiphilic side of the n-butanol/water diagram at 55.0 °C compared to results from the literature.(14,16) (Symbols same as for Fig. 3.)

obtained from the literature. The latter were estimated from a graph of phase-volume measurements for n-butoxyethanol/water, and taken from a table of critically evaluated results for n-butanol/water (4,5). Confidence intervals are also shown.

Table I. Two-Phase Compositions

	Aqueous			Amphiphilic		
T (^{o}C)	Calorimeter	Phase Volume	Liter.	Calorimeter	Phase Volume	Liter.
n-Butanol (wt %)						
30	8.35±.29	6.79±.21	7.05[a]	78.07±.25	78.88±.13	79.37[a]
55	5.98±.80	5.75±.25	6.39[a]	75.79±.79	79.75±.25	76.8[a]
n-Butoxyethanol (wt %)						
55	13.0±1.3	12.58±.02	13.45[b]	49.35±.45	50.79±.02	49.4[b]
65	10.5±.5	10.28±.02	11.08[b]	56.68±2.0	58.28±.02	57.37[b]

a - Ref. 4.
b - Ref. 5.

Table II shows the "heats of formation" of the conjugate phases, that is, the excess enthalpies for mixing the appropriate amounts of water and amphiphile (at the same initial temperature and pressure as the final system) to make a unit amount of the conjugate phase. Values labeled "calorimeter" and "phase volume," respectively, are based on the same set of calorimetric titrations. In the former case the phase composition was taken from the calorimetric measurements, and in the latter case the composition was taken from our phase-volume compositions. Literature values for the heats of formation are based on data from references 13-16.

Discussion

Examination of Table I shows that the compositions that we obtained by calorimetry were an average of 0.4 wt% amphiphile lower than the values

Table II. Excess Enthalpies

T		Aqueous	Amphiphilic
(°C)		(J/100g)	(J/100g)
	n-Butanol:		
30	Calorimeter	-573.±13.1	921.6±10.6
	Phase Volume	-582.5 ±11.0	937.3±4.5
	Literature	-546.7±6.9[a]	736.7±42.4
55	Calorimeter	9.0±.95	1492.5±32.7
	Phase Volume	4.8±.64	1523.0±3.0
	Literature	43.0±67.5[a]	1524.9±36.4[a]
	n-Butoxyethanol:		
55	Calorimeter	-518.8±36.8	-705.2±26.4
	Phase Volume	-514.6±36.5	-679.7±26.3
65	Calorimeter	-224.8±24.5	-525.6±29.3
	Phase Volume	-222.3±24.9	-501.2±29.8

a - Ref. 14.

in the literature, whereas the compositions that we obtained from phase volumes were an average of 0.1 wt% amphiphile lower than the literature values. The standard deviations (for the differences between the measurements of this paper and those in the literature) were essentially identical for the calorimetric and phase-volume techniques (0.35 wt%).

Figure 3 shows thirteen previously reported points for the heats of mixing of aqueous solutions of n-butanol and water at 30.0 °C (13,14,16). (Some of these data are obscured by overlap with our results.) The agreement with the previous data appears excellent, although around 7 wt% butanol our enthalpies are somewhat more negative than previously reported values.

In Figure 4 the results from the three different groups are in excellent agreement for butanol concentrations of 90 wt% and greater, although the data from the Russian group scatter somewhat more around our results than do the values interpolated from Westmeier's data.(14,16). At lower amphiphile concentrations the isoperibolic calorimeter measurements are in noticeably better agreement with the data of ref. 16 than with the Russian work (14-16). However, almost all results fall within the 95% confidence interval (dashed lines) for our results.

Although heat capacities have been reported for the butoxyethanol/water system, excess enthalpies that could be compared directly with our results apparently have not been measured (12).

Isoperibol titration calorimetry provides a convenient and rapid way to measure conjugate phase compositions for amphiphile/water systems that have miscibility gaps. When a simple desktop computer is used to control the calorimeter and to record the data, a typical titration takes about one hour to complete, covers a composition range of about 10 wt%, and provides a compositional resolution of 0.03 wt%. (Resolution was limited by the 640K contiguous RAM of our MS-DOS computer; higher resolution would be possible with use of a more sophisticated computer system.) Moreover, the treatment of the data, including determination of the phase compositions, is reduced to simple, documented algorithms, instead of to undocumentable judgements that may not be reproducible from one operator to another.

Often the experimentalist will have an approximate idea of the location of a phase boundary from data taken at other temperatures or from measurements on closely related compounds. In such cases, a reasonably accurate guess about the location of the phase boundary may allow one to choose a starting composition that crosses the boundary with only a single titration. This can be considerably more efficient than starting with a neat compound and performing titrations over successive compositional ranges until the phase boundary is found.

However, the titrations must begin with the neat compound, if one desires to measures excess enthalpies referred to a single-compound standard state. In this case the number of experiments and the data reduction effort are necessarily greater than when it is sufficient to measure only the phase boundary. Much of the extra effort is inherent in titration methods:

The initial and final compositions of a titration experiment are given by

$$X_1^s = W_1^s / (W_1^s + w_2^s) \qquad (2a)$$

$$X_1^f = (W_1^s + \delta W_1) / (W_1^s + w_2^s + \delta W_1 + \delta W_2) \qquad (2b)$$

where the subscripts refer to the compounds, superscripts s and f denote starting and final weight fractions (X), and δW_i is the weight of component i added in the titration. As shown by Equations 2, a titration can cover only a limited range of composition, and the range becomes more limited as X_1^s approaches the value 0.5. Thus the smaller the mutual solubilities of the two components, the greater is the convenience of titration calorimetry for measuring excess enthalpies referenced to neat-compound reference states. One alternative to multi-step titrations is to use flow calorimetry instead. (6).

Sometimes another possibility, which has several advantages of its own, is to use a critical point as the standard state. For example, the lower consolute point is a convenient standard state for the butoxyethanol/ water system, whose lower consolute solution temperature is at about 49 °C. (See Figure 1.) The compositions of the conjugate phase pairs are found at several temperatures from titrations that need only cross the phase boundary. Equations from critical scaling theory, which contain the critical-point composition and/or temperature, are then fit to the compositions. These fits give the critical point, which can also serve as the standard state for the enthalpy data. (Smith, Duane H.; Dunn, R. O. "Excess Enthalpies of Formation of Triconjugate Phases of a Microemulsion System," 43rd Annual Calorimetry Conference, Bartlesville, OK, August 15-19, 1988)

Nonionic and ionic surfactants that contain "ethoxy" ($-OC_2H_4-$) and "propoxy" ($-OC_3H_6-$) groups are excellent (i.e., apparently the best available) materials for miscible-flood mobility control and for ultra-low tension flooding. (1, 18-20). However, both ionic and nonionic alkoxylated surfactants commonly exhibit miscibility gaps in water and oilfield brines (18,20-22). (A temperature on the lower part of a miscibility gap defined by a chosen composition of surfactant, water, and electrolyte is commonly referred to as "the cloud point" of that composition.) To avoid problems of surface handling, injection, and undesired mixing and compositional paths in the formation, it is usually deemed desirable that the miscibility gap of an EOR surfactant/water system be at a sufficiently high temperature that the gap is not encountered in oilfield use. Moreover, important physical properties, such as surfactant adsorption, have been shown to correlate with the distance of the system from the miscibility gap, even when the system remains in the single-phase region (18). Hence amphiphile/water miscibility gaps of the type studied in this paper are of considerable importance in all surfactant-based EOR processes, and the results of this paper show that calorimetry is a useful technique for studying these miscibility gaps. Two further advantages of calorimetric methods are that they require neither that samples transmit light, nor that dispersions cream and break. Since crude oils commonly form stable emulsions and opaque fluids, both of these advantages are of considerable potential value in requisite thermodynamic studies of enhanced oil recovery systems.

Conclusion

Isoperibolic calorimetry measurements on the n-butanol/water and n-butoxyethanol/water systems have demonstrated the accuracy and convenience of this technique for measuring consolute phase compositions in amphiphile/water systems. Additional advantages of calorimetry over conventional phase diagram methods are that (1) calorimetry yields other useful thermodynamic parameters, such as excess enthalpies; (2) calorimetry can be used for dark and opaque samples; and (3) calorimetry does not depend on the bulk separation of conjugate fluids. Together, the present study and studies in the literature encompass all of the classes of compounds of the amphiphile/CO_2/hydrocarbon/water systems that are encountered in dispersion-based mobility control for miscible-CO_2 enhanced oil recovery and in ultra-low tension flooding (1,2,6).

Literature Cited

1. Smith, Duane. H. In Proc. Fifth SPE/DOE Symp. Enhanced Oil Recovery, SPE/DOE 14914, 1986; vol. I, pp. 441-56.

2. Surfactant-Based Mobility Control: Progress in Miscible-Flood Enhanced Oil Recovery; Smith, Duane H., Ed.; ACS Symposium Series No. 373; American Chemical Society: Washington, D.C., 1988.

3. Smith, Duane. H.; Allred, G. C. J. Colloid Interface Science 1988, **124**, 199-208.

4. Alcohols with Water; Barton, A. F. M., Ed.; Pergamon Press: New York, 1984; pp. 32-91.

5. Kahlweit, M.; Strey, R.; Firman, P.; Haase, D. Langmuir 1985, **1**, 281-8.

6. Cordray, D. R.; Christensen, J. J.; Izatt, R. M. Separation Science Technol. 1987, **22**, 1169-81.

7. (a) Eatough, D. J.; Christensen, J. J.; Izatt, R. M. Thermochim. Acta 1972, **3**, 219-232. (b) Christensen, J. J.; Izatt, R. M.; Hansen, L. D.; Partridge, J. A. J. Phys. Chem. 1966, **70**, 2003-2010.

8. Smith, Duane H.; Lane, B. J. J. Dispersion Sci. Technol. 1987, **8**, 217-47.

9. Davis, M. I. Thermochim. Acta 1985, **90**, 313-29.

10. DiAndreth, J. R.; Paulaitis, M. E. In Surfactant-Based Mobility Control: Progress in Miscible-Flood Enhanced Oil Recovery; Smith, Duane H., Ed.; ACS Symposium Series No. 373; American Chemical Society: Washington, D.C., 1988; ch. 4.

11. Mukerjee, P.; Mysels, K. J. Critical Micelle Concentrations of Aqueous Surfactant Systems, U. S. Department of Commerce, NSRDS-NBS **36**,1971.

12. Roux, G.; Perron, G.; Desnoyers, J.E. J. Solution Chem. 1978, **7**, 639-54.

13. Goodwin, S. R.; Newsham, D. M. T. J. Chem. Thermodyn. 1971, **3**, 325-34; tabulated in ref. 17.

14. Belousov, V. P.; Ponner, V. Vestn. Leningr. Univ., Fiz., Khim. 1970, **10**, 111-15; tabulated in ref. 17.

15. Belousov, V. P.; Panov, M. Yu. Vestn. Leningr. Univ., Fiz., Khim. 1976, **2**, 149-50; tabulated in ref. 17.

16. Westmeier, S. Chemische Techn. 1978, **30**, 354-7.

17. Handbook of Heats of Mixing, Christensen, J. J.; Hanks, R.W.; Izatt, R.M., Eds.; John Wiley and Sons: New York, 1982.

18. Lewis, S. J.; Verkruyse, L. A.; Salter, S. J. In Proc. Fifth SPE/DOE Symp. Enhanced Oil Recovery, SPE/DOE 14910, 1986; vol. I, pp. 389-98.

19. Borchardt, J. K.; Bright, D. B.; Dickson, M. K.; Wellington, S. L. In
     Surfactant-Based Mobility Control: Progress in Miscible-Flood
     Enhanced Oil Recovery; Smith, Duane H., Ed.; ACS Symposium
     Series No. 373; American Chemical Society: Washington, D.C.,
     1988; ch. 8.
20. Borchardt, J. K. In Surfactant-Based Mobility Control: Progress in
     Miscible-Flood Enhanced Oil Recovery; Smith, Duane H., Ed.; ACS
     Symposium Series No. 373; American Chemical Society:
     Washington, D.C., 1988; ch. 9.
21. Smith, Duane H. J. Colloid Interface Sci. 1985, **108**, 471-83.
22. Smith, Duane H.; Fleming, P. D. III J. Colloid Interface Sci. 1985,
     105, 80-93.

RECEIVED November 28, 1988

Chapter 16

Structure–Performance Characteristics of Surfactants in Contact with Alkanes, Alkyl Benzenes, and Stock Tank Oils

Thomas A. Lawless and John R. Lee-Snape

Winfrith Petroleum Technology Centre, Winfrith AEE, Dorchester, Dorset, England

The assessment of surfactant structures and optimal mixtures for potential use in tertiary flooding strategies in North Sea fields has been examined from fundamental investigations using pure oils. The present study furthermore addresses the physico-chemical problems associated with reservoir oils and how the phase performance of these systems may be correlated with model oils, including the use of toluene and cyclohexane in stock tank oils to produce synthetic live reservoir crudes. Any dependence of surfactant molecular structure on the observed phase properties of proposed oils of equivalent alkane carbon number (EACN) would render simulated live oils as unrepresentative. Both commercial grade and pure nonionic and anionic surfactants have been evaluated by phase inversion and optimal salinity screening procedures to establish relationships to their molecular structures.

The optimal structure of surfactants for practical and efficient EOR flooding strategies in North Sea oil reservoirs remains largely unresolved. Previous research studies (1–4) have attempted to assess surfactant performance potential using pure synthesised materials. These have been successful in focussing on molecular structural benefits and indentifying some shortcomings associated with differing functional moieties. The present study attempts to probe the relationship between structure and performance of 8 surfactants; 7 of which are commercial in origin.

Cosurfactants have not been employed in the present study. However, surfactants from commercial sources will contain isomers and manufacturing impurities. Nevertheless, a major aim of this study has been to address the performance characteristics of commercial formulations. Wherever appropriate, hydrophobic

0097–6156/89/0396–0305$06.00/0
Published 1989 American Chemical Society

structural assemblies were selected that were available in both anionic and nonionic form. Variations in the ethoxylate chain length, the degree of anionic substitution, inorganic salt content and unreacted products will all affect performance behaviour, and therefore demand careful attention. A methodical assessment of all surfactant formulations has been undertaken using the technique of conductivity to determine the temperature or salinity required for phase inversion to occur. Of direct interest is the EACN concept (5) and how pure oils may be related to reservoir crudes. Furthermore the ability of certain aromatics and cyclics to act as separator gas equivalents is also addressed. The influence of ethoxylate inclusion to the surfactant hydrophile and the observed concomitant equivalences, for toluene and cyclohexane have been investigated to follow the applicability of such concepts. These optimal correlations and their inherent sensitivities aid the interpretation of formulation potential for field injection.

Experimental : Materials and Methods

Oils The n-alkane series C_6-C_{14}, toluene and cyclohexane were purchased from BDH, Poole, UK, each with a stated purity of 99%; reagents were used as received. Crude oil samples were obtained from two North Sea fields; one located in the Norwegian sector and the other from the UK sector. Stock tank oil from the Gullfaks field was supplied by Statoil, Norway and the other stock tank oil from an undisclosed source. Both crude oils are derived from sandstone formations with reservoir temperatures of 70° and 101°C respectively.

Brines Analytical grade sodium chloride, purity 99.9% was obtained from BDH and used throughout the study. Water was purified by reverse osmosis, and deionised in a Milli-Q-Reagent system immediately prior to use.

Surfactants Information on the pure and commercial grade surfactants studied with regard to structures, contaminations and activities is detailed in Table I. All surfactants were used as received.

Cloud Point Measurements Cloud points were recorded by the visual observation of aqueous solutions containing 1% W/V surfactant. The measurement defines the temperature at which the system under test shows a characteristic transitional change from a clear solution to an opalescent or cloudy state. All cloud points were recorded in both ascending and descending temperature cycles to ensure data confidence. The influence of salt and/or oils on the cloud point were systematically evaluated.

Phase Inversion The phase inversion of brine/oil/surfactant systems was established routinely by measuring solution conductivity employing a Jenway PWA 1 meter and cell. The process identifies the range over which a large decrease in conductivity occurs as the sytem under test is converted from an oil in water emulsion to a water in oil emulsion. Phase

TABLE I SURFACTANT SAMPLES UNDER OBSERVATION

TRADE NAME	SUPPLIER	CHEMICAL FORMULA OF MAJOR SYNTHESISED PRODUCT	% SURFACTANT ACTIVITY	NATURE OF SURFACTANT COMPOSITION		INORGANIC SALT CONTENT/%	WATER CONTENT/%
				ANIONIC %	NONIONIC %		
T100	HOECHST W GERMANY	$Bu_3Ph(EO)_{10}OH$	100	-	100	-	-
A7	ICI ENGLAND	$C_{13-15}H_{27-31}(EO)_7OH$	100	-	100	-	-
NP6	ICI ENGLAND	$C_9H_{17}Ph(EO)_6OH$	100	-	100	-	-
POE10	SIGMA ENGLAND	$CH_3(CH_2)_7CH=CH(CH_2)_8(EO)_{10}OH$	100	-	100	-	-
D3620	HOECHST W GERMANY	$Bu_3Ph(EO)_4SO_3Na$	35	~26	~9	5	60
A3C	ICI ENGLAND	$C_{13-15}H_{27-31}(EO)_3OCH_2CO_2Na$	85	~60	~25	7	8
LEONOX I.O.S. (MOL WT = 375)	MITSUBISHI JAPAN	$CH_3(CH_2)_mCH=CH(CH_2)_nSO_3Na$	35	~35	-	4	*53
DIOCTYL SULPHOSUCCINATE	SIGMA ENGLAND	$(CH_3(CH_2)_3CH(C_2H_5)CH_2)_2C_2O_4C_4H_7SO_3Na$	100	~100	-	-	-

* 8% unsulphonated oil reported present in this formulation

inversion temperatures were measured in well stirred systems
undergoing a temperature change of typically 1 K min^{-1}. Cooling
profiles were also recorded and only when the conductivity
measurements in the heating and cooling cycles matched were the
data recorded. Salinity loadings necessary for phase inversion
at a specific temperature were also evaluated. Because of
potentially undesirable effects such as gel formation, test
temperatures of 40° or 60° C were selected for these experiments.
Aqueous solutions containing either 150 or 300 g dm^{-3} NaCl were
prepared and used as titrants to promote phase inversion in
oil/water/surfactant systems. Care was taken to maintain the
water to oil ratio as close to 1 as possible. The surfactant
loadings necessary to produce such phase inversions were related
to anticipated requirements for all pure oils and stock tank
oils. For the purpose of standardisation most tests were
performed on systems containing 5% W/V surfactant.

RESULTS

NONIONIC SURFACTANTS

Cloud Points The influence of added NaCl on the observed cloud
points of 1% W/V solutions of the four nonionic surfactants under
observation are given in Figure 1. Approximately linear
correlations were observed as the aqueous NaCl level was
increased, with negative coefficients recorded between 0.22 - 0.3
K.g^{-1}dm^3. Higher loadings of surfactant were found to increase
the cloud point. It was observed also that the inclusion of
small quantities of oils to surfactant solutions could either
elevate or depress the cloud point. The significance of this
fact will be developed later.

Phase Inversion Temperatures It was possible to determine the
Phase Inversion Temperature (PIT) for the system under study by
reference to the conductivity/temperature profile obtained
(Figure 2). Rapid declines were indicative of phase preference
changes and mid-points were conveniently identified as the
inversion point. The alkane series tended to yield PIT values
within several degrees of each other but the estimation of the
PIT for toluene occasionally proved difficult. Mole fraction
mixing rules were employed to assist in the prediction of such
PIT values. Toluene/decane blends were evaluated routinely for
convenience, as shown in Figure 3. The construction of PIT/EACN
profiles has yielded linear relationships, as did the mole
fraction oil blends (Figures 4 and 5). The compilation and
assessment of all experimental data enabled the significant
parameters, attributable to such surfactant formulations, to be
tabulated as in Table II.
    The PIT dependence profiles generated for the nonionic
surfactants in contact with alkanes are given in Figure 6. Their
linear correlations allow suitable coefficients to be extracted
from these data which may be used in later, derivable inter-
relationships. It was observed that variations in the water to
oil ratio (WOR) affected the recorded PIT (Figure 7). The

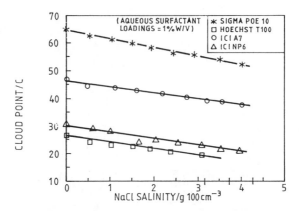

Figure 1. Cloud point variation for different salinities
(Aqueous surfactant loadings = 10 g/dm^3)

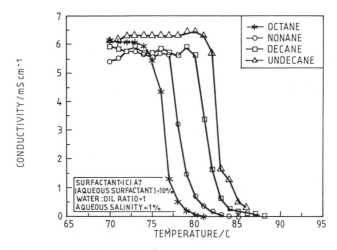

Figure 2.   Conductivity/temperature profiles for the
alkanes

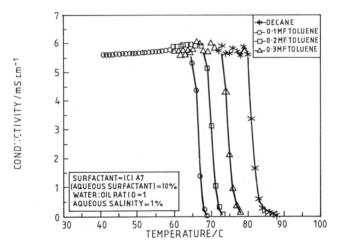

Figure 3. Conductivity/temperature profiles for
toluene/decane blends

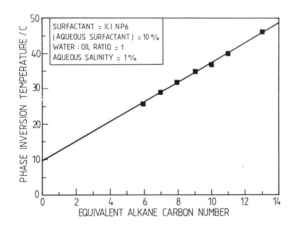

Figure 4. Phase inversion temperature variation for the
alkanes

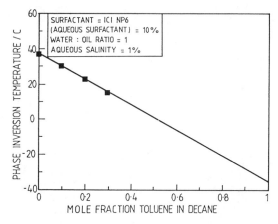

Figure 5.   Phase inversion temperature variation for toluene/decane blends

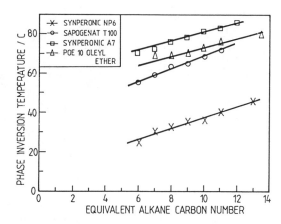

Figure 6.   Variation of phase inversion temperature with equivalent alkane carbon number

TABLE II SUMMARY OF PHASE INVERSION TEMPERATURE DATA AND APPROPRIATE DERIVED EQUIVALENTS

SURFACTANT	AQUEOUS PHASE NaCl SALINITY /g dm⁻³	$\dfrac{\text{d PIT}}{\text{d EACN}}$ /K(Unit ACN)⁻¹	PIT AT EACN=0 /°C	OIL PIT/°C (MEASURED OR EXTRAPOLATED)				OIL EACN FROM PIT			
				TOLUENE	CYCLO-HEXANE	GULLFAKS CRUDE	NINIAN CRUDE	TOLUENE	CYCLO-HEXANE	GULLFAKS CRUDE	NINIAN CRUDE
ICI NP6	10	2.7	10	- 35	-	23	20	- 16	-	4.8	3.7
ICI A7	10	2.2	59	32	-	74	74	- 12	-	6.8	6.8
HOECHST T100	10	3.4	35	9	-	61.5	60.5	- 7.6	-	7.8	7.5
SIGMA POE 10 OLEYL ETHER	50	2.3	50	14	-	65	65	- 16	-	6.5	6.5
ICI A3C	200	5.2	18	- 23	24	*62	*62	- 7.9	1.2	< 8.5	< 8.5

(Water to oil ratios were always 1 and the total surfactant loading per system was 5% W/V for all tests except for the ICI A3C system where 2.5% W/V was employed)

* Denotes values recorded on blends composed of 25% by volume crude oil plus 75% by volume decane

recorded PIT values were observed to increase as the WOR
declined. No mole fraction blending was available for stock tank
oils and percentage volume mixes were adopted for test purposes
(Figure 8). All nonionic formulations were capable of phase
inverting the two crude oil samples in their native state which
permitted a direct EACN value to be assigned to the oils from
previously derived standard PIT data shown in Table II. Both
reservoir crude oil samples showed somewhat variable EACN
values,within the range of 3.7 to 7.8.

Optimal Salinities Because the EACN values ascribed to oil
systems are usually derived from salinity scanning, it was
considered appropriate to evaluate the salinity tolerance of ICI
NP6, and to determine the EACN value for toluene so that a
comparison could be made between the two experimental techniques
employed in this study. Inspection of Table III reveals an EACN
value for toluene of -10.3 which can be compared to the value of
-16 determined from PIT data. Such large negative values
indicate the importance of the hydrophilic group in determining
alkane equivalences and it was desirable to probe how ionic
groupings, which also contain oxyethylene linkages as integral
parts of their hydrophilic segment, would confer EACN values on
toluene and indeed on stock tank oils.

ANIONIC SURFACTANTS

Phase Inversion Temperatures Because of the high solubility and
salt tolerance of carboxymethylates it was considered more
appropriate to establish PIT values for the alkane series rather
than determine extremely high optimal salinities. The estimation
of such data permitted the role of temperature to be assessed on
an ionic formulation containing oxyethylene groupings.
Temperature tolerance was much improved over the native nonionic
surfactant hydrophiles (see Figure 9 and Table II) although
difficulties were experienced in establishing a PIT value for
both stock tank oils. Experimentally, only phase mixtures
containing 25% by volume of crude oil in contact with 75% decane
were capable of inverting before difficulties arose. Toluene was
again determined to have a value less than zero (Figure 10).

Optimal Salinities Phase inversions at optimal salinity were
assessed routinely by salt titrations into systems maintained at
constant temperature. For the Leonox IOS surfactant system,
increasing levels of salinity were necessary to cause the
emulsion state to phase invert as the alkane molecular weight
increased (Figure 11). The initial conductivity value at the
condition where zero salt had been added may in part reflect the
salt contamination naturally present within the supplied
formulation. The internal olefin sulphonate species again
revealed a linear relationship between EACN and optimal salinity
as did all ionic formulations under test (see Figures 12 and 13,
plus Table III). The estimation of EACN values for both toluene

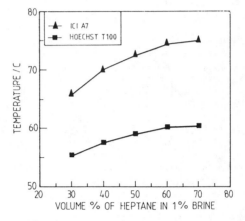

Figure 7. Influence of oil and water volume ratio on the observed PIT

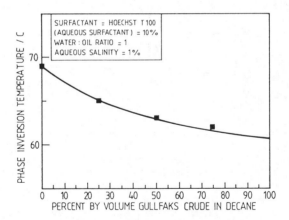

Figure 8. Phase inversion temperature variation for decane/Gullfaks crude blends

TABLE III   SUMMARY OF OPTIMAL SALINITY DATA AND APPROPRIATE DERIVED EQUIVALENTS

SURFACTANT	TEST TEMPERATURE /°C	$\frac{dS^*}{dEACN}$ /g dm⁻³ (Unit ACN)⁻¹	S* AT EACN=0 /g dm⁻³	S* VALUE FOR OILS/g dm⁻³ (MEASURED OR EXTRAPOLATED)				OIL EACN FROM S*			
				TOLUENE	CYCLO-HEXANE	GULLFAKS CRUDE	NINIAN CRUDE	TOLUENE	CYCLO-HEXANE	GULLFAKS CRUDE	NINIAN CRUDE
LECNOX IOS	60	6.1	-1	-2	11	*55	*58	-0.2	2.0	< 9.2	< 9.7
ICI NP6	40	17.8	-142	-325	-153	-	-	-10.3	-0.6	-	-
HOECHST D3620	60	8.0	-32	-35	-9	-	-	-0.4	2.9	-	-
SIGMA DIOCTYL SULPHOSUCCINATE	40	0.68	0	-5.2	1.7	**3.5	**3.5	-7.6	2.5	< 5.2	< 5.2

(Water to oil ratios were all initially 1 prior to titration, and the initial surfactant loading per system was 5% W/V. Salinity values relate exclusively to aqueous phase NaCl solutions).

* Denotes values recorded on blends composed of 25% by volume crude oil plus 75% by volume decane
** Denotes values recorded on blends composed of 50% by volume crude oil plus 50% by volume decane

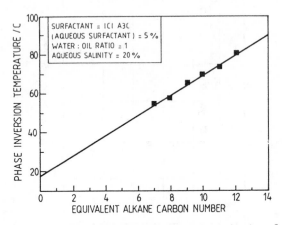

Figure 9.   Phase inversion temperature variation for the alkanes

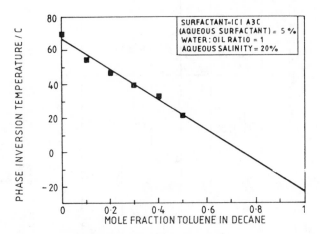

Figure 10.   Phase inversion temperature variation for decane/toluene blends

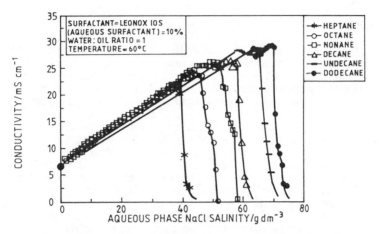

Figure 11.   Conductivity titration profiles for the alkanes

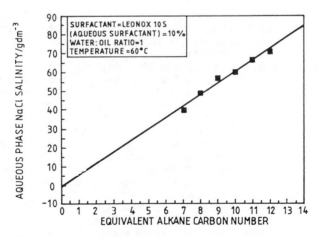

Figure 12.   Optimal salinity variation for the alkanes

and cyclohexane (see Figures 14 and 15) was undertaken with all ionics under test. Sulphonate types gave values not too far removed from those proposed by Wade et al (6) whilst other hydrophilic head group surfactants conferred other equivalences. Difficulties were again observed in the production of phase inversion with stock tank oils.

DISCUSSION

Nonionics The selection of nonionic surfactant candidates for test purposes was influenced by a number of requirements which included (a) commercial availability, (b) structural variety, (c) availability of anionic derivatives, and (d) some degree of conformance to the desirable structural requirements proposed by Graciaa et al (7).

Surfactant blends of interest will exhibit clouding phenomena in aqueous solutions undergoing a phase transition from a one phase system to a two phase system at a discrete and characteristic temperature, referred to as the Cloud Point (CP). This value indicates the temperature at which sufficient dehydration of the oxyethylene portion of the surfactant molecule has occurred and this results in its "displacement" from solution. The addition of lyotropic salts will depress the CP, presumably due to the promotion of localised ordering of water molecules near the hydrophilic sheath of the surfactant molecule (8). Furthermore, the addition of different oils to surfactant solutions can induce either an elevation or a depression of the recorded CP and can be used to qualitatively predict the PIT (8,9).

The ability of various oils to be solubilised within micellar interiors or within the palisade layer will undoubtedly influence the conditions for favourable phase behaviour. Any solubilisation that interferes with oxyethylene hydration may cause substantial reductions in the CP. Thus the combined effect of temperature, oil and salinity can control the performance potential of nonionics. It is believed that the alkane series under study in the present work were solubilised into the main interior of the micellar core, but the inclusion of toluene could involve some degree of penetration into the palisade layer. Mole fraction mixing rules were adopted to facilitate the evaluation of the PIT value for toluene and cyclohexane (see Figures 3 and 5) which could then be ascribed EACN values derived from previous alkane/PIT data (Table II). The low negative value recorded for toluene could reflect surfactant hydration interference. Alternatively, deep and effective penetration of toluene into the micellar core could afford such effecient phase transitions by expansion of the average hydrophobe assembly. Reservoir crudes will obviously contain a multiplicity of components, all of which may have either the ability to promote or reduce oxyethylene hydration. The low molecular weight aromatics will induce stronger influences than those observed with their higher alkyl homologues. The PIT values determined for the two reservoir crudes under study yielded EACN values between 3.7-7.8 which may, in part, reflect an equivalent polarity range. No difficulties

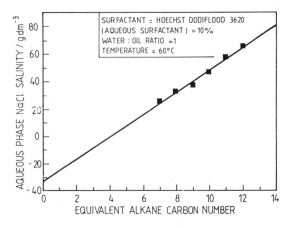

Figure 13.   Optimal salinity variation for the alkanes

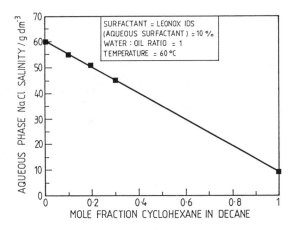

Figure 14.   Optimal salinity variation for
decane/cyclohexane blends

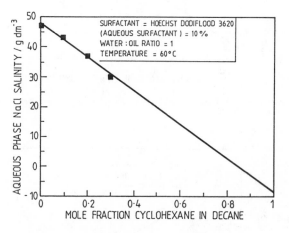

Figure 15. Optimal salinity variation for
decane/cyclohexane blends

were experienced in attaining phase inversion conditions with reservoir crudes and adequate solubilisation parameters were available.   Variations in the water to oil ratio were shown to affect th recorded PIT (Figure 7).   Increased loadings of oleic phase produced increased PIT values over the accessible range of study.   Shinoda and Arai (8,10) had previously observed that little variation could be expected for alkanes but that pronounced effects were dominant with aromatics.

Salinity Effects in the Inversion Process   It has been shown for anionics that the Salager (11) equation could relate salt and alcohol effects to phase behaviour according to:-

$$\text{ln } S^* = k \text{ (ACN)} + f \text{ (A)} + \dots \qquad (1)$$

where S^* = aqueous phase optimal salinity
             k = a constant related to the nature of the hydrophilic group (for sulphonates k = 0.16)

and f (A) = function of alcohol type and concentration

    In the absence of alcohol the salinity/alkane sensitivity will be dominated by the nature of the surfactant hydrophile. A linear response was observed between optimal salinity and alkane chain length for the ICI NP6 surfactant with a recorded coefficient of $ds*/d$ EACN = 17.8gdm^{-3}/unit EACN.   This high salt tolerance was expected for nonionics and can reflect an equivalence between the ability of both salt and temperature to effect a phase inversion change per unit EACN, via cloud point adjustment.   As expected, toluene displayed a remarkable effect on the extrapolated optimal salinity level necessary for phase inversion.   The disruption of oxyethylene hydration being far more dominant than any effects derived from limited salt addition.   The $[\text{EACN}]_S^*$ value of -10.3 is at variance to that calculated earlier from PIT data and may reflect a non-linear correlation between salt and temperature influences, particularly over the theoretical range necessary here.   Cyclohexane was also assessed (see Table III) and ascribed an $[\text{EACN}]_S^*$ value of -0.6. On the basis of these results, the inclusion of such oils as replacements for separator gas components would be inappropriat for pure nonionic systems.
    Of interest here is the question relating to the value for the slope coefficient, k, from equation (1), when surfactant structures incorporating both ionic (say sulphonate) and nonionic moieties are included together.   The changes in electric double layer effects imparted from salt addition might dominate the packing constraints and therefore the phase inversion process, or perhaps oxyethylene dehydration effects from the presence of toluene could also play a role.

ANIONICS

Phase Inversion Temperature   The carboxymethylate surfactant sample available for test purposes exhibited excellent salinity tolerance, in fact too high for practical sea water flooding

requirements. Optimal salinity studies were impracticable but
the role of the oxyethylene groups could be established by PIT
scanning in the presence of 20% W/V NaCl (see Figures 9 and 10).
The dual combination of hydrophilic groups permitted a better
temperature tolerance than that usually found for native
nonionics (Table II). It is to be expected that changes in
salinity loadings will strongly affect charged headgroups while
temperature variations will alter oxyethylene hydration. The
presence of added toluene was found to strongly reduce the PIT,
yielding an [EACN]$_{PIT}$ value of -7.6 which approaches values
recorded for nonionic surfactants. Such negative deviations are
in general accord with other reported data for this generic class
of surfactants (12). Cyclohexane showed an EACN of 1.2 which
indicates a less dramatic influence for this saturated cyclic
species. Thus oil polarity, oil molar volume and therefore its
location within micellar structures will influence hydration,
molecular assembly packing and, consequently, the PIT.
Difficulties were experienced when stock tank oils were
introduced, but this may be a consequence of reduced
solubilisation parameter values normally found for reservoir
crudes.

Optimal Salinities The phase inversion process may be considered
to reflect the balanced nature of the adsorbed surfactant species
at the oil/water interface. Simple geometric packing
considerations can be used to define the relative areas occupied
by hydrophilic and hydrophobic groupings which thereby determine
the direction of any preferred curvature. The ability to
transform surfactant molecules from one preferred curvature state
to another is a necessary requirement for the phase inversion
process to occur. Variations in aqueous salinity and
temperature, plus the inclusion of various oleic and cosolvent
phases, can all induce effective changes in either the
hydrophilic or hydrophobic segment and promote phase inversion
(13).
        Linear responses were evaluated experimentally in optimal
salinity/EACN scanning for all ionics under study. Of practical
interest is the natural presence of inorganic salts in the
commercial grade anionic formulations. For example, concentrates
containing 35% W/W surfactant and 5% W/W inorganic salt (along
with other components) will yield a 7.1g dm^{-3} salt loading in a
prepared 5% W/V surfactant solution. If the inorganic salt is
sodium sulphate then the equivalent NaCl level will be 7.1g dm$_{-3}$
simply in ionic strength terms. Thus variations in surfactant
loading will naturally induce variable salinity levels and make
grid point phase diagrams difficult to unravel. If the inherent
salt levels were calculated and some appropriate adjustment
invoked to equate sodium sulphate to NaCl, then an approximately
parallel plot (of optimal salinity against EACN) to the one
recorded here, would be observed. The data recorded herein
relates solely to optimal levels derived from added salts.
Inspection of Figures 12-15 and Table III reveals sulphonate
performance characteristics in contact with alkanes, toluene and
cyclohexane. Both sulphonates exhibited some degree of salinity
tolerance with the ethoxylated sulphonate being superior. The

inclusion of oxyethylene groups to sulphonate surfactants has been reported to reduce any co-solvent requirements and promote salinity tolerance; but as a consequence solubilisation parameters may be impaired (3). The EACN values determined for toluene (~0) and cyclohexane (between 2 and 3) are in general agreement with values proposed by Wade and co-workers (5). Since the presence or absence of oxyethylene groups appears to have little influence on $[EACN]_S^*$ for toluene and cyclohexane it must be concluded that the ionic sulphonate group dominates the phase evolution process during salt addition. Electric double layer effects may thus control preferred interfacial surfactant curvature, whilst the oil molar volume controls the extent of oil inclusion and micellar hydrophobe expansion. Thus, temperature changes mainly effect the oxyethylene portion of the surfactant. No estimates are available for stock tank oil EACN values, and this may be due to inadequate surfactant loading to facilitate complete solubilisation at the phase inversion condition. All the optimal salinity values recorded in this study were determined by salt titration experiments from an initial condition of WOR = 1. Typically the WOR had shifted to ~1.2 before inversion was recorded; titrations necessitating WOR values 1.4 were never recorded. Dioctyl sulphosuccinate has been reported (14) to be sensitive to the relative volume of oil and water contacted and care was taken therefore to minimise such effects. This surfactant was relatively sensitive to salt loadings and served as the only pure sample under test. The extrapolated EACN values for toluene and cyclohexane were -7.6 and 2.5 respectively, which reveals the importance of the structural nature of surfactants.

CONCLUSIONS

The following conclusions based on the experimental work reported in this paper are :

- Favourable phase inversion conditions, as monitored by conductivity, were established for all surfactant blends in contact with alkanes.

- If favourable crude oil inversion conditions are observed it is possible to calculate alkane equivalences based on either optimal salinity or PIT data.

- Commercial sulphonate formulations behave in a manner qualitatively similar to that expected from pure components during optimal salinity evaluation.

- Optimal salinity values directly influence the nature of the ionic groups, while temperature variations (PIT tests) strongly effect the oxyethylene linkages.

- Inherent salt loadings in commercial grade anionic surfactants will influence observed phase evolution processes.

- The EACN values for toluene were found to vary depending upon the nature of the surfactant molecular structure, but sulphonate systems confirm an EACN equivalence of ˜0.

LITERATURE CITED

1. DOE P.H., WADE W.H., SCHECHTER R.S., J.Coll.Int.Sci., 1977, 59, 3, 525-31.

2. DOE P.H., EL-EMARY M., WADE W.H., SCHECHTER R.S., J.Am.Oil Chem.Soc., 1978, 55, 505-12.

3. CARMONA I., SCHECHTER R.S., WADE W.H., WEERASOORIYA U., SPE No 11771, 1985.

4. ABE M., SCHECHTER D., SCHECHTER R.S., WADE W.H., WEERASOORIYA U., YIV S., J.Coll.Int.Sci., 1986, 114 2,342-56.

5. CAYIAS J.L., SCHECHTER R.S., WADE W.H., J.Coll.Int.Sci., 1977, 59, 1, 31-8.

6. CASH L., CAYIAS J.L., FOURNIER G., MACALLISTER D., SCHARRES T., SCHECHTER R.S., WADE W.H., J. Coll.Int.Sci. 1977, 59, 1, 39-44.

7. GRACIAA A., FORTNEY L., SCHECHTER R.S., WADE W.H., YIV S., SPE/DOE NO.9815, 1981.

8. SHINODA K., ARAI H., J. PHYS.CHEM., 1964, 68, 12, 3485-90.

9. AVEYARD R., LAWLESS T.A., J.CHEM.SOC. Farad Trans I., 1986, 82, 2951-63.

10. SHINODA K., ARAI H., J.Coll.Int.Sci., 1967, 25, 429-31.

11. SALAGER J.L., MORGAN J.C., SCHECHTER R.S., WADE W.H., YIV S., J.Coll.Int.Sci., 1982, 89, 1, 217-25.

12. OLSEN D.K., JOSEPHSON C.B., Report for U.S. Department of Energy, July 1987.

13. AVEYARD R., Chem.Ind., 1987, 474-8.

14. AVEYARD R., BINKS B.P., CLARK S., MEAD J., J.Chem.Soc.Farad. Trans.I, 1986, 82, 125-142.

RECEIVED November 28, 1988

Chapter 17

Interfacial Tension of Heavy Oil–Aqueous Systems at Elevated Temperatures

E. Eddy Isaacs, J. Darol Maunder, and Li Jian[1]

Alberta Research Council, Oil Sands and Hydrocarbon Recovery Department, P.O. Box 8330, Postal Station F, Edmonton, Alberta T6H 5X2, Canada

Oil/water interfacial tensions were measured for a number of heavy crude oils at temperatures up to 200°C using the spinning drop technique. The influences of spinning rate, surfactant type and concentration, NaCl and $CaCl_2$ concentrations, and temperature were studied. The heavy oil type and pH (in the presence of surfactant) had little effect on interfacial tensions. Instead, interfacial tensions depended strongly on the surfactant type, temperature, and NaCl and $CaCl_2$ concentrations. Low interfacial tensions (<0.1 mN/m) were difficult to achieve at elevated temperatures.

At a given NaCl concentration, an increase in temperature resulted in an increase in interfacial tension. In contrast, for a narrow range of $CaCl_2$ concentrations, interfacial tensions decreased with increasing temperatures. Changes of the amphiphile at the oil/water interface accounted for some of the experimental observations. Since the extent of oil desaturation is dependent on interfacial tension, the tension data could be used to assess the ability of surfactants to reduce oil saturations in the reservoir for application of surfactants and foams to thermal recovery processes.

The use of surfactants to achieve low ($<10^{-1}$ mN/m) interfacial tensions between oil and water as a means of enhancing recovery from partially depleted conventional reservoirs is well recognized [1]. In steam injection processes

[1]Current address: Research Institute of Petroleum Exploration and Development, P.O. Box 910, Beijing, Peoples Republic of China

for recovering heavy oil from underground deposits, surfactants
are used mainly to stabilize the formation of a foam with steam
or a non-condensible gas, which acts to divert the steam from
depleted zones thereby improving reservoir conformance [2]. A
major concern with foam processes is the detrimental effect of
residual oil in the swept zone on both foam formation and
propagation [3-6]. Reductions in residual saturations over and
above that obtained by steam injection are desirable and, in many
heavy oil reservoirs, essential to the application of steam-foam
processes. However, the extent of heavy oil desaturation is
dependent on the reduction in interfacial tension between oil and
water [7-9]. Thus, foam-forming surfactants can improve their
own cause by reducing interfacial tensions at steam temperatures.
Interfacial tension data combined with an understanding of the
factors responsible for the interfacial tension behavior at the
oil/water interface is, therefore, pertinent to developing
rational injection strategies for the application of surfactants
and foams to heavy oil recovery.

The interfacial tension behavior of alkane-aqueous NaCl
systems containing both pure and commercial surfactant mixtures
has been extensively studied. Under narrow ranges of
experimental conditions, very low (<0.01 mN/m) interfacial
tensions are obtained. Puig et al. [10,11], Franses et al. [12]
and Hall [13] have shown that very low tensions are not caused by
monolayer adsorption but by a film of a surfactant-rich phase or
liquid-crystalline layer at the oleic-water interface. A
pictorial representation of some of the microstructures
(interpreted from the work of Winsor [14]) that can form in the
aqueous, oleic, and interfacial region are shown in Figure 1.
Thus, the tension minima are a consequence of changes in
microstructures as a function of surfactant concentration,
salinity and temperature. However, Chan and Shah [15], Shinoda
et al. [16] and , more recently, Aveyard et al.[17] have
suggested that very low interfacial tension minima are not
necessarily the consequence of a phase change but rather the
distribution of surfactant between oil and water and the degree
of dissociation of surfactant in the micelle and the oil-water
interface.

The interfacial tension behavior between a crude oil (as
opposed to pure hydrocarbon) and an aqueous surfactant phase as a
function of temperature has not been extensively studied.
Burkowsky and Marx [18] observed interfacial tension minima at
temperatures between 50 and 80°C for crude oils with some
surfactant formulations, whereas interfacial tensions for other
formulations were not affected by temperature changes. Handy et
al. [19] observed little or no temperature dependence (25-180°C)
for interfacial tensions between California crude and aqueous
petroleum sulfonate surfactants at various NaCl concentrations.
In contrast, for a pure hydrocarbon or mineral oil and the same
surfactant systems, an abrupt decrease in interfacial tension was
observed at temperatures in excess of 120°C [20]. Nonionic
surfactants showed sharp minima of interfacial tension for crude

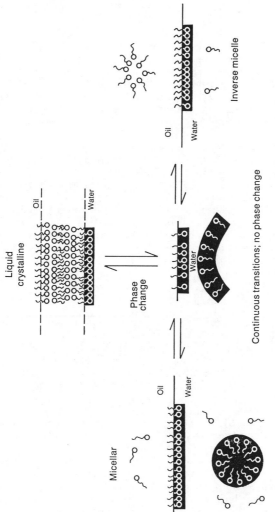

Figure 1: Schematic illustration of possible changes in the microstructure of amphiphiles at the oil/water interface.

oil systems at a temperature corresponding to the cloud point
[19].

Isaacs and Smolek [21] observed that low tensions obtained
for an Athabasca bitumen/brine-sulfonate surfactant system were
likely associated with the formation of a surfactant-rich film
lying between the oil and water, which can be hindered by an
increase in temperature. Babu et al. [22] obtained little effect
of temperature on interfacial tensions; however, values of about
0.02 mN/m were obtained for a light crude (39°API), and were
about an order of magnitude lower than those observed for a heavy
crude (14°API) with the same aqueous surfactant formulations.
For pure hydrocarbon phases and ambient conditions, it is well
established that the interfacial tension behavior is dependent on
the oleic phase [15,23]. In general, interfacial tension values
of crude oil-containing systems are considerably higher than the
equivalent values observed with pure hydrocarbons.

In the present paper, interfacial tensions were measured for
a number of heavy crude oils at temperatures up to 200°C using
the spinning drop technique. However, reliable data cannot be
obtained by this or any other drop shape method because of the
small density difference between heavy crudes and water which,
moreover, tends to decrease as the temperature increases. This
problem was overcome by using aqueous D_2O instead of H_2O as has
been previously described [5,8,21]. The influence of surfactant
type and concentration, mono- and divalent cation concentrations,
and pH on the attainment of low interfacial tensions are reported
and discussed.

EXPERIMENTAL DESCRIPTION

Materials. Samples of dewatered crude oils were obtained from:
the Athabasca oil sands of the McMurray formation by extraction
using the commercial hot water process (Suncor Inc.); the
Bluesky-Bullhead formation at Peace River, Alberta by solvent
extraction of produced fluids; the Clearwater formation at Cold
Lake, Alberta by solvent extraction of core material; and the
Karamay formation in Xing-Jiang, China. A summary of the
physical and chemical properties of the crude oils, including
chemical composition, and density-temperature and viscosity-
temperature relationships, is given in Table I.

Surfactant mixtures were used as obtained and are listed with
their properties in Table II. Sodium chloride and calcium
chloride were Fisher reagent grade. Deuterium oxide was Aldrich
Gold Label and had a surface tension of 70.4 mN/m at 23°C
measured with a Wilhelmy plate tensiometer.

Equipment and Procedures. Crude oil/aqueous interfacial tensions
were measured using a spinning drop tensiometer built at the
Alberta Research Council and designed for operation at elevated
temperatures [21]. The main difficulty in operating at elevated
temperatures was the wear on the bearing; improved operation was

Table I: Physical and chemical properties of heavy oils

Measurement	Temperature °C	Athabasca	Peace River	Clearwater	Karamay
Gravity, °API	15	8.9	7.5	10.4	19.7
Density, kg/dm^3	50	0.989	0.994	0.976	0.912
	100	0.958	0.961	0.946	0.876
	150	0.927	0.927	0.916	0.841
	200	0.897	0.899		
Viscosity, mPa.s	50	3350	9136	3172	369
	60	1700	3578	-	-
	100	-	137	133	31
Elemental Analysis					
Carbon		83.3	82.9	84.3	85.6
Hydrogen		10.6	10.7	10.8	12.5
Nitrogen		0.4	0.4	0.4	0.6
Oxygen		-	1.6	1.1	1.2
Sulfur		4.8	5.8	4.4	0.1
Fractional Analysis					
Saturates		24.6	17.0	20.1	48.0
Aromatics		26.2	20.4	11.4	11.5
Polars I		32.5	45.8	16.5	9.2
Polars II		32.5	45.8	7.0	1.2
Polars III		32.5	45.8	23.1	18.9
Asphaltenes		14.7	18.4	16.6	2.0
Acid No. mg of KOH per g		3.6	3.6	1.4	5.0

Table II: Properties of surfactants

Surfactant	Type	Source	% Active	Equivalent Weight
Chaser SD1000	sulfonate dimmer	Chevron	42	-
Enordet C$_{16}$-C$_{18}$	α-olefin sulfonate	Shell	30	356
Enordet LTS-18	alkylaryl sulfonate	Shell	12	457
SunTech IV	alkylaryl sulfonate	Sun	15	418
TRS 10-80	petroleum sulfonate	Witco	85	420

achieved by using precision bearings from RHP Canada Inc (No. R8/15) which could be oiled during experiments. Shafts were drilled into the aluminum block used to heat the sample in the spinning drop tube, to allow for oiling of the bearing during operation.

Densities were measured using a Paar DMA 60 meter equipped with DMA 512 and DMA 601 HP external cells. Values in the 50-150°C range were interpolated from measured data (3-5 points); values above 150°C were extrapolated and are less accurate. Interfacial tension measurements at the minimum density difference encountered (0.05 g/cm^3) could be in error by as much as 10%, which is within the repeatability of measurements with heavy crude oil samples (see below).

Unless stated otherwise, values of interfacial tension were obtained using D_2O as the aqueous phase. Although the physical properties of H_2O and D_2O are nearly identical, studies have shown that critical micelle concentrations (CMC) of several surfactants [24] and micellar aggregation numbers [25] are higher in D_2O than H_2O. For nonionic surfactants, no significant differences have been observed in the aggregation number [26], but the cloud point is lower in D_2O than H_2O [27]. It has been suggested [26,27] that the greater strength of the O-D•••O compared to the O-H•••H bond results in differences in intermicellar interactions. For the purpose of the present study, any differences in values measured in D_2O and H_2O have been shown to be small [21]; trends in interfacial tension behavior are expected to be the same.

The aqueous phases were prepared by dispersing surfactant in D_2O or in formation water using magnetic stirring. The solutions were then diluted to the appropriate concentration by the addition of NaCl or $CaCl_2$, or NaOH or HCl concentrates in D_2O. All concentrations refer to the active surfactant concentration at room temperature.

Oil drops of 2-5 μL were introduced into 0.4 cm i.d. capillary tubes containing the aqueous phase. The more viscous heavy oils were heated for a short period to facilitate this addition. The tubes were then sealed with a tightly fitting silicon-rubber septum. A teflon screw was used to apply pressure on the septum after the capillary tube was inserted into the shaft of the tensiometer. In this manner, temperatures up to 200°C were achieved without loss of liquid.

The 0.4 cm i.d. capillary tubes were used instead of the recommended 0.2 cm i.d. [28] in order to facilitate the addition of highly viscous oils. Figure 2 shows the effect of spinning rate on the interfacial tension of an n-butanol/deionized water system using two tube sizes. The dashed line represents the best fit for 15 data points measured using a 0.2 cm i.d. tube, where the mean interfacial tension is 1.76 mN/m (standard deviation of 0.02) with a range of 1.73 to 1.80 mN/m for speeds ranging from

4,050 to 13,460 rev/min. These values agree well with literature [28,29]. Data for the 0.4 cm i.d. tube are in good agreement only below about 8,000 rev/min, increasing gradually thereafter with increasing speed.

It is apparent that with the larger diameter tubes, at high frequencies the drop diameter lags behind the rotational speed of the tube causing an apparent increase in tension. With a more viscous oleic phase as in this study, smaller diameter drops, and lower tension systems, lagging should be more severe and result in higher apparent tensions. As an example, Figure 3 shows that the choice of spinning rate is important in determining the apparent tension for the Clearwater bitumen/surfactant-in-brine system. Generally, for heavy oils and tensions below 1 mN/m, speeds below 6,000 rev/min are preferable. The lower the tension, the lower the preferred speed while still maintaining a length to diameter ratio of about 4. Because of the small density difference between heavy oil and D_2O, the falloff of tensions due to buoyancy effects will occur at a much lower frequency than 4,000 rev/min as reported for n-butanol/H_2O and alkane/H_2O systems [28,29].

All measurements were carried out without prior equilibration of the aqueous and hydrocarbon phases. The values reported were obtained at least one hour after steady state had been reached at a given temperature. For calculating the interfacial tensions, the working equations described by Manning et al. [30] were used. Three replicate experiments using an Alberta heavy oil in produced water containing surfactant (LTS-18) as a function of temperature are shown in Figure 4. The same trend was observed for the three experiments with an interfacial tension minimum at 75°C. However, the apparent tension values of one set of measurements varied substantially from the other two, and considerably more than the reproducibility of the method (about 6 %). This variation is probably due to the inhomogeneity of the heavy crudes samples because of their colloidal nature and the differences in the chemical species present with each drop of oil selected.

This type of difficulty associated with measurements using chemically ill-defined substrates was also observed during sessile drop measurements carried out on Athabasca bitumen in D_2O [31]. Values in the range of 15-20 mN/m were obtained for measurements with several drops of bitumen, while interfacial tensions for other pure aqueous and oleic systems were accurate to ±0.5 mN/m.

RESULTS AND DISCUSSION

Effect of NaCl Concentration. The presence of surfactant in brine can have a dramatic effect on crude oil-aqueous surfactant tensions even at elevated temperatures [5,21]. Figure 5 shows that the effect of sodium chloride concentration on Athabasca bitumen-D_2O interfacial tensions measured at constant surfactant

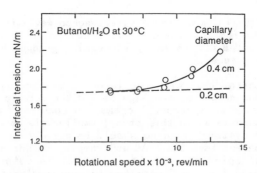

Figure 2: Effect of spinning rate on the interfacial tension of *n*-butanol/water system at 30°C.

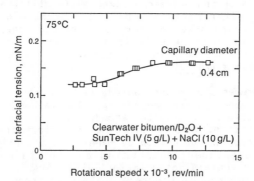

Figure 3: Effect of spinning rate on the interfacial tension of Clearwater bitumen/D_2O, and Sun Tech IV (5 g/L) and NaCl (10 g/L) system at 75°C.

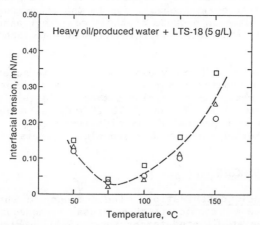

Figure 4: Effect of temperature on the interfacial tension of an Alberta heavy oil in produced water containing LTS-18 surfactant. Data are from three separate replicate experiments conducted under the same conditions.

concentration (2 g/L) and temperature (100°C) depends on the
surfactant type. In the case of Enordet C_{16-18}, interfacial
tensions were only slightly affected by salinity, decreasing
almost linearly with increasing NaCl concentration. Both the
behavior and the range of interfacial tension values were similar
to that exhibited [21] by sodium dodecyl sulfate (SDS), a pure
association colloid. The decrease in interfacial tension may be
due to changes in size, shape and aggregation number of the
micelles [14] resulting from changes in the ionic environment
(increased NaCl concentration) which do not involve a phase
change (see Figure 1).

In contrast to Enordet C_{16-18}, both TRS 10-80 and Sun Tech IV
displayed a tension minimum where the interfacial tension was
reduced by 2-3 orders of magnitude (Figure 5). The NaCl
concentration at which the tension was at a minimum and the
extent of the reduction varied between the two surfactants. At
NaCl concentrations much higher than that at which the minimum
tension occurred, surfactant precipitation caused non-equilibrium
shapes and no measurements were possible. In contrast, Enordet
C_{16-18} is extremely brine tolerant and data were measured up to
160 g/L NaCl.

In n-octane/aqueous systems at 27°C, TRS 10-80 has been shown
to form a surfactant-rich third phase, or a thin film of liquid
crystals (see Figure 1), with a sharp interfacial tension minimum
of about 5×10^{-4} mN/m at 15 g/L NaCl concentration [13].
Similarly, in this study the bitumen/aqueous tension behavior of
TRS 10-80 and Sun Tech IV appeared not to be related to monolayer
coverage at the interface (as in the case of Enordet C_{16-18}) but
rather was indicative of a surfactant-rich third phase between
oil and water. The higher values for minimum interfacial tension
observed for a heavy oil compared to a pure n-alkane were
probably due to natural surfactants in the crude oil which
somewhat hindered the formation of the surfactant-rich phase.
This hypothesis needs to be tested, but the effect is not unlike
that of the addition of SDS (which does not form liquid crystals)
in partially solubilizing the third phase formed by TRS 10-80 or
Aerosol OT at the alkane/brine interface [11,12].

Effect of Temperature. In the absence of surfactant, interfacial
tensions of the Athabasca [21], Karamay [5], and other heavy oils
[32] show little or no dependence on temperature. For
surfactant-containing systems, Figure 6 shows an example of the
effect of temperature (50-200°C) on interfacial tensions for the
Athabasca, Clearwater and Peace River bitumens in Sun Tech IV
solutions containing 0 and 10 g/L NaCl. The interfacial tension
behavior for the three bitumens was very similar. At a given
temperature, the presence of brine caused a reduction in
interfacial tension by one to two orders of magnitude. The
tensions were seen to increase substantially with temperature.
For the case of no added NaCl, the values approached those
observed [21] in the absence of surfactant.

At first glance, the tension-temperature behavior may be interpreted as that expected for surfactant concentrations at or below the CMC, the concentration at which monolayer coverage is complete. Since the CMC for anionic surfactants increases with temperature [17,33], the surface coverage decreases resulting in a decrease in tension. However, the temperature behavior was very different for Enordet C_{16-18} (2 g/L) a surfactant which does not form a surfactant-rich film, and NaCl (0-160 g/L; data given in Table III) solutions. The tensions were not drastically affected by temperature. Moreover, measurements (Table IV) for the Clearwater sample at a higher surfactant concentration (5 g/L) exhibited a trend similar to that observed in Figure 6 with 2 g/L Sun Tech IV. The main difference at the higher concentration was that tension values were 3-5 times lower below 100°C but not significantly different at 150° and 200°C. It thus appears that the interfacial tension-temperature behavior observed for Sun Tech IV and the three bitumens is consistent with a liquid crystalline phase at low temperatures, whose formation is hindered as the temperature increases. As shown previously [5,21], to maintain low interfacial tensions at elevated temperatures required increasing surfactant or electrolyte concentration.

Effect of pH. Interfacial tensions between heavy crude oils and alkaline solutions were measured at temperatures up to 180°C by Mehdizadeh and Handy [34]. They observed that tensions increased with an increase in temperature. However, recovery efficiencies obtained at high temperatures were comparable to those obtained at lower temperatures, apparently because the ease of emulsification at high temperatures counteracted the increase in tension.

To our knowledge, no data on the effect of pH and temperature has been measured in the presence of added surfactant. Figure 7 shows such pH dependence data as measured in this study for Athabasca bitumen in D_2O containing Sun Tech IV (2 g/L) at a constant ionic content of 10^{-2} M (adjusted using HCl, NaOH and NaCl) at 50 and 150°C. Data at 100°C (not shown for clarity) fell between those at 50 and 150°C. The dashed curve represents previously reported data [21] at 50°C in the absence of added surfactant, and without maintaining ionic strength. In marked contrast to the no-surfactant case, the interfacial tensions showed essentially no dependence in the 2-12 pH range at 100 and 150°C. The results at 50°C showed no dependence in the 2-9 pH range with a small tension minimum at about pH 11. However, at a given pH, tensions increased with increasing temperature. These results suggest that the surfactant preferentially adsorbs at the oil/water interface and, at high pH, hinders the formation and/or adsorption of natural surfactants at the interface. Further studies are needed to substantiate the results at lower interfacial tension regimes and to investigate the possible implications for a caustic-surfactant process.

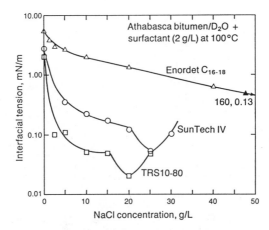

Figure 5: Effect of NaCl concentration on the Athabasca bitumen/D₂0 interfacial tension for Enordet C₁₆₋₁₈, Sun Tech IV and TRS 10-80. Closed triangle represents the data measured for Enordet C₁₆₋₁₈ at concentrations up to 160 g/L NaCl.

Table III: Interfacial tension data for the Athabasca bitumen/D₂0 and Enordet C₁₆₋₁₈ (2 g/L) system as a function of NaCl concentration and temperature

NaCl Conc., g/L	Interfacial Tension, mN/m		
	50°C	100°C	150°C
0	4.5	6.7	3.5
1.25	2.8	4.0	3.3
2.5	2.1	3.0	3.1
5.0	1.9	2.7	2.7
10.0	1.4	2.0	2.2
20.0	0.66	1.3	1.6
40.0	0.44	6.63	0.86
80.0	0.33	0.42	0.49
100.0	0.21	0.26	0.37
120.0	0.23	0.22	0.22
140.0	0.21	0.21	0.21
160.0	0.17	0.13	0.13

Table IV: Interfacial tension data for the Clearwater bitumen/D_2O
and Sun Tech IV system as a function of
surfactant and NaCl concentration and temperature

		Interfacial Tension, mN/m	
Temperature, °C	NaCl Conc., g/L	2 g/L Surfactant	5 g/L Surfactant
25	0	–	0.33
50	0	2.1	0.45
75	0	–	0.75
100	0	3.8	0.95
125	0	–	2.6
150	0	5.3	4.3
200	0	5.9	–
25	10	–	0.04
50	10	0.07	0.07
75	10	–	0.10
100	10	0.25	0.14
125	10	–	0.25
150	10	0.44	0.33
200	10	0.28	0.49

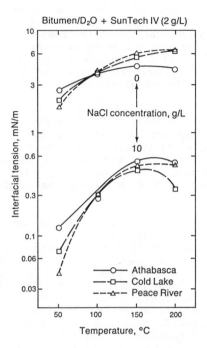

Figure 6: Interfacial tensions for Athabasca, Cold Lake and Peace River bitumen/D_2O systems as a function of temperature and NaCl (0 and 10 g/L) concentration.

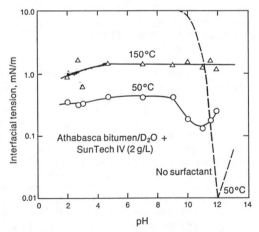

Figure 7: Variation of interfacial tension between Athabasca bitumen and D_2O containing Sun Tech IV (2 g/L) as a function of pH and temperature at constant ionic strength of 10^{-2} M. The dashed line represents data from reference [21] at 50°C in the absence of added surfactant or brine.

Effect of Ca^{2+}. In many reservoirs the connate waters contain substantial quantities of divalent ions (mostly Ca^{+2}). In alkaline flooding applications at low temperatures, the presence of divalent ions leads to a drastic increase in tensions [35,36]. Kumar et al. [37] also found that Ca^{+2} and Mg^{+2} ions are detrimental to the interfacial tensions of sulfonate surfactant systems. Detailed studies at elevated temperatures appear to be non-existent.

Figure 8 shows the effect of Ca^{+2} on the interfacial tensions of two oils (Karamay and Clearwater) in Sun Tech IV (5 g/L) and NaCl (10 g/L) solutions at 150°C. The interfacial tension values for the two oils were very similar with as much as an 8-fold reduction, depending on concentration. Interfacial tension minima in the range of 0.06 to 0.1 mN/m were observed at 0.05 and 0.5 g/L CaCl$_2$ for both oils.

It has been reported by Celik and Somasundaran [38] that the interaction of divalent (and trivalent) cations with sulfonate surfactants causes surfactant precipitation followed by dissolution of the precipitate at higher concentrations. The precipitate redissolution phenomenon is not observed with monovalent ions. Indeed, some surfactant precipitation in the spinning drop tube was observed above concentrations corresponding to the first minimum of Figure 8; it is not known whether redissolution took place at higher concentrations resulting in the second tension minimum.

The interfacial tension-temperature relationships at various CaCl$_2$ concentrations for Karamay crude in a Sun Tech IV (5 g/L) and NaCl (10 g/L) solution are shown in Figure 9. For 0, 0.025 and 0.1 g/L Ca^{+2}, an increase in interfacial tension with temperature was observed. The interfacial tension values above 150°C were about the same for these concentrations. At temperatures below 100°C, the effect of Ca^{+2} was to increase interfacial tension, probably by hindering the formation of a surfactant-rich phase. This is consistent with the detrimental effect on light oil/brine interfacial tensions (increase from about 10^{-3} to about 10^{-1}) reported by Kumar et al. [37].

For 0.5 g/L Ca^{+2}, the interfacial tension-temperature behavior was reversed compared to that observed at lower concentrations (Figure 9), showing a substantial decrease with temperature. The extent of the reduction was about 3 orders of magnitude compared to the interfacial tension in the absence of surfactant (about 10 mN/m at 180°C). As shown in Figure 10, the same trend was observed for the interfacial tension-temperature behavior of Karamay crude in formation water containing Sun Tech IV. This is not surprising considering that the formation water contained a high amount of divalent (0.25 g/L) and some monovalent (2 g/L) cations. While the mechanism for interfacial tension reduction with increasing temperature remains to be elucidated, this is a positive outcome in terms of reduction in

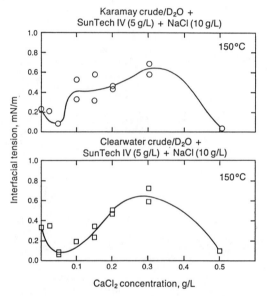

Figure 8: Effect of $CaCl_2$ concentration on the interfacial tension of oil/D_2O systems containing Sun Tech IV and NaCl for Karamay and Clearwater crudes at 150°C.

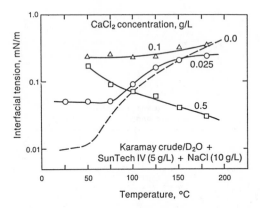

Figure 9: Effect of CaCl$_2$ concentration and temperature on
the interfacial tension of the Karamay crude/D$_2$0 system
containing Sun Tech IV and NaCl. Dashed line represents data
in the absence of CaCl$_2$ (reference [5]).

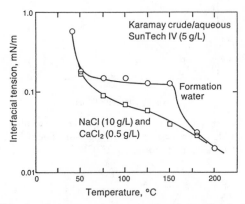

Figure 10: Effect of temperature on the interfacial tension of
Karamay crude in reservoir formation water and synthetic
brine.

residual saturations which may be crucial for reservoir application of foams at steam temperatures.

SUMMARY AND CONCLUSIONS

The interfacial tension results reported in this paper are part of a study to examine the benefits of using commercial foam-forming surfactants with steam-based processes for obtaining additional oil recovery. Low interfacial tension at elevated temperatures is needed to reduce residual oil saturation and to allow foams to form, or enhance their performance.

Modifications of the conventional spinning drop tensiometer were required for operating at temperatures up to 200°C. Measurements carried out with heavy oil samples required the use of D_2O instead of H_2O to maintain a sufficient density difference between oil and water. For accurate measurements, considerable care must be used to ensure that heavy oil drops do not lag behind the rotation of the capillary tube in the tensiometer. Also, repeatability of measurements conducted with chemically ill-defined substances may be hampered by the inhomogeneity of the oil drops.

In the presence of Sun Tech IV surfactant, the interfacial tension of an Athabasca/brine system showed little or no dependence on pH. However, at a given pH, tension values increased with temperature (50-150°C).

For results where comparisons could be made, the interfacial tension behavior was practically independent of the type of heavy oil used. Interfacial tensions strongly depended on the surfactant type, temperature, and NaCl and $CaCl_2$ concentrations. Changes in the structure of the amphiphile at the oil/water interface is affected by these variables and accounted for some of the experimental observations.

Low interfacial tensions (<0.1 mN/m) were difficult to maintain at elevated temperatures. At a given NaCl concentration, an increase in temperature (25-200°C) resulted in an increase in interfacial tension. In contrast, in the presence of high amounts of $CaCl_2$ (0.2 and 0.5 g/L) interfacial tensions decreased with increasing temperatures. Thus, consideration of the range of temperature and water chemistry to be encountered in a given reservoir is important in designing surfactant and foam enhanced recovery processes.

ACKNOWLEDGMENTS

This work was carried out as part of the AOSTRA and Alberta Research Council Core Research Program. Our thanks to Dr. Karel Smolek for assistance with measurements, to Wendy Zwickel for editing, and to Yvonne Mariacci for assistance in typing the manuscript.

LITERATURE CITED

1. Fundamentals of Enhanced Oil Recovery, H.K. van Poollen and Associates Inc., Penn Well Books, Tulsa, 1980.

2. Islam, M.R.; Selby, R.J.; Farouq Ali, S.M. Meeting of the Petroleum Society of CIM, Calgary, June 7-10, 1987 (CIM 87-38-77).

3. Holm, L.W. Soc. Pet. Eng. J., 1968, 359-369.

4. Friedmann, F.; Jensen, J.A. 56th California Regional Meeting of the SPE of AIME, Oakland, April 2-4, 1986 (APE 15087).

5. Isaacs, E.E.; Li, J.; Green, M.K.; McCarthy, F.C.; Maunder, J.D. AOSTRA J. Res., 1988, in press.

6. McPhee, C.A.; Tehrani, A.D.H.; Jolly, R.P.S. SPE/DOE Enhanced Oil Recovery Symposium, Tulsa, April 17-20, 1988 (SPE/DOE 17360).

7. Gopalakrishan, P.; Bories, S.A.; Combarnous, M. Report of the Group d'Etude IFP-IMF, Toulouse, 1978, (SPE 7109).

8. Isaacs, E.E.; McCarthy, F.C.; Smolek, K.F. 2nd European Conference on EOR, Paris, November, 1982.

9. Isaacs, E.E.; Prowse, D.R.; Rankin, J. J. Can. Pet. Tech., 1982, 21, 33.

10. Puig, J.E.; Franses, E.I.; Davis, H.T.; Miller, W.G.; Scriven, L.E. Soc. Pet. Eng. J., 1979, 19, 71.

11. Puig, J.E.; Mares, M.T.; Miller, W.G.; Franses, E.I. Colloids and Surfaces, 1985, 16, 139.

12. Franses, E.I.; Puig, J.E.; Talmon, Y.; Miller, W.G.; Scriven, L.E.; Davis, H.T. J. Phys. Chem., 1980, 84, 1547.

13. Hall, A.C. Colloids and Surfaces, 1980, 1, 209.

14. Winsor, P.A. Chem. Rev., 1968, 68, 1.

15. Chan, K.S.; Shah, D.O. J. Disp. Sci. Technol., 1980, 1, 55.

16. Shinoda, K.; Hanrein, M.; Kunieda, H.; Saito, H. Colloids and Surfaces, 1981, 2, 301.

17. Aveyard, R.; Binks, B.P.; Clark, S.; Mead, J. J. Chem. Soc., Faraday Trans. 1, 1986, 82, 125.

18. Burkowsky, M.; Marx, C. Tenside Deterg., 1978, 15. 247.

19.   Handy, L.L.; Amaefule, J.O.; Zeigler, V.M.; Ershaghi, I.   4th Symposium on Oilfield and Geothermal Chemistry, SPE of AIME, Houston, 1979, (APE 7867).

20.   Handy, L.L.; El-Gassier, M.; Ershaghi, I.   5th Symposium on Oilfield and Geothermal Chemistry, SPE of AIME, Stanford, 1980, (SPE 9003).

21.   Isaacs, E.E.; Smolek, K.F.   Can. J. Chem. Eng., 1983, 61, 233.

22.   Babu, D.R.; Hornof, V.; Neale, G.   Can. J. Chem. Eng., 1984, 62, 156.

23.   Cayias, J.L.; Schechter, R.S.; Wade, W.H.   J. Colloid Interface Sci., 1977, 59, 31.

24.   Emerson, M.F.; Holtzer, A.   J. Phys. Chem., 1967, 70, 783.

25.   Chang, N.; Kaler, E.   J. Phys. Chem., 1985, 89, 2996.

26.   Binana-Limbele' W.; Zana, R.   J. Colloid Interface Sci., 1988, 121, 81.

27.   Pandit, N.K.; Caronia, J.   J. Colloid Interface Sci., 1988 122, 100.

28.   Manning, C.D.; Scriven, L.E.   Rev. Sci. Inst., 1977, 48, 1699.

29.   Currie, P.K.; Van Nieuwkoop, J.   J. Colloid Interface Sci., 1982, 87, 301.

30.   Manning, C.D.; Pesheck, C.V.; Puig, J.E.; Seeto, Y.; Davis, H.T.   United States Department of Energy Report, 1983, (DOE/BC/10116-12).

31.   Isaacs, E.E.; Morrison, D.N.   AOSTRA J. Res., 1985, 2, 113.

32.   Flock, D.L.; Le, T.H.; Gibeau, I.P.   J. Can. Pet. Tech., 1986, 72.

33.   Fennell Evans, D.; Wightman, P.J.   J. Colloid Interface Sci., 1982, 82, 515.

34.   Mehdizadeh, A.; Handy, L.L.   SPE 13072, 59th Annual Technical Conference, SPE of AIME, Houston, 1984, (SPE 13072).

35.   Jennings, H.T., Jr.; Johnson, C.E.; McAuiliffe, C.D.   J. Pet. Tech., 1974, 26, 1344.

36.   Cooke, C.E., Jr.; Williams, R.E.; Kolodzie, P.A.   J. Pet. Tech., 1974, 26, 1365.

37.   Kumar, A.; Neale, G.; Hornof, V.   J. Can. Pet. Tech., 1984,
      37.

38.   Celik, M.S.; Somasundaran, P.   J. of Colloid Interface Sci.,
      1988, 122, 163.

RECEIVED November 28, 1988

Chapter 18

Oil Recovery with Multiple Micellar Slugs

S. Thomas,[1] S. B. Supon,[2] and S. M. Farouq Ali[1]

[1]Department of Mining, Metallurgical and Petroleum Engineering, The
University of Alberta, Edmonton, Alberta T6G 2G6, Canada
[2]Petroleum and Natural Gas Engineering, Pennsylvania State University,
University Park, PA 16802

Micellar flooding process has been shown to be
an effective tertiary oil recovery technique
for watered-out light oil reservoirs, based
upon many field tests. However, under the
present oil prices it has not been economic in
most cases. In other instances, recovery
methods such as carbon dioxide flooding may be
more attractive. The present research was
aimed at improving the efficiency of the
micellar flooding process through the use of
multiple micellar slugs. The slugs were
selected on the basis of the phase behaviour
of various ternary systems (crude, water,
surfactant). Equilibrium micellar phases
chosen from the single phase region or,
micellar compositions formed along the
dilution path of a suitable micellar slug in
the single phase region were used to form the
composite slugs. Oil-rich and water-rich
slugs were compared for their effectiveness to
recover tertiary oil. Multiplicity of slugs
on recovery efficiency was examined by
injecting portions of oil-rich and water-rich
slugs alternately until the desired slug
volume was reached. The use of graded
composite slugs - starting with an oil-rich
slug and ending with a water-rich slug - was
successful in improving process efficiency.
(Process efficiency, in this work, is defined
as the tertiary recovery per unit volume of
slug injected). For example, process
efficiency increased five fold when a single
slug was replaced by a graded composite slug
of equivalent size.

0097–6156/89/0396–0345$06.00/0
© 1989 American Chemical Society

The works of various investigators such as Gogarty and
Tosch (1), Healy and Reed (2), and Davis and Jones (3),
have shown that the micellar flooding process can be used
effectively to mobilize residual oil in watered-out light
oil reservoirs. Many field tests conducted in the U.S.
have further proved its effectiveness. However, the
economics of the process remain unattractive for
implementing the process for tertiary oil recovery.
Recent works by Holm (4), Pope (5), Sayyouh and Farouq
Ali (6), Enedy and Farouq Ali (7) made significant
contributions towards improving the efficiency of the
process. The primary objective of this research was to
devise a micellar flooding process that is economically
as well as technically attractive to the industry,
through the use of multiple micellar slugs.

Process Description

Process efficiency, in this study, is defined as the
tertiary oil recovery per unit volume of the slug
injected. This refers to the efficiency of an oil-rich
slug. Economic recovery efficiency varies from slug to
slug due to variations in the surfactant content. It
should be noted that the micellar slugs were formulated
with an effort to keep the cost a minimum.

Basic Process. Hydrocarbon, water/brine and surfactant
are the basic components of a micellar solution. A small
amount of alcohol is usually added to improve the
solution stability, to adjust the viscosity, and to
reduce the surfactant loss due to adsorption on the
reservoir rock. A suitably formulated micellar slug
miscibly displaces the residual oil, leading to a very
small final oil saturation. For economic reasons, a
small pore volume (2-5%) micellar slug is injected, and
it is effectively propagated through the reservoir by the
mobility buffer. The mobility buffer protects the slug
from dilution by the drive water too early in the
process. The efficiency of the process can be augmented
by using a suitable composite slug configuration. A
composite slug consists of oil-rich and water-rich slugs,
injected alternately in two or more steps, until the
desired pore volume is reached. The economics of the
process can be further enhanced by using a graded
composite slug. Such slugs can reduce the surfactant
requirement to a minimum, and maximize the tertiary
recovery efficiency.

Phase Behaviour Studies and Slug Formulation

The efficiency of a micellar slug to recover tertiary oil
is largely dependent on its ability to remain a single
phase during the flooding process so that the oil may be
displaced "miscibly" and hence, completely. However,

various factors such as slug dilution by reservoir
fluids, surfactant loss to the reservoir rock, reservoir
salinity etc. influence the phase behaviour adversely.
Once slug breakdown occurs, the displacement becomes
immiscible and some of the resident oil can be left
trapped in the pores. Tertiary recovery can be
substantially improved if the slug composition has a
minimal multiphase region to prolong the miscible
displacement regime, and a low interfacial tension in the
multiphase region to enhance the immiscible displacement.
Another factor contributing to oil displacement is the
relatively larger oleic phase volume at slug breakdown.
With these ends in view, a number of phase behaviour
studies were made to formulate efficient micellar slugs
for the tertiary recovery of three light oils - Bonnie
Glen, Provost, and Bradford crudes.

Phase Behaviour Characteristics. Studies of ternary
systems, representing crude oil, water/brine and
surfactant systems were completed prior to the selection
of micellar slugs. The brine used in this study was
2%(wt./vol.) sodium chloride in distilled water. It was
found that the phase boundaries depend, to a great
extent, on the water/oil ratio in the initial system.
Because of the multicomponent nature of the micellar
solutions, the phase behaviour of the system is much more
complex than a simple three component system. Therefore,
a pseudoternary system could only be used to
qualitatively correlate the observations on the phase
behaviour of the micellar solutions. Due to economic
reasons and viscosity requirements, the surfactant and
oil concentrations were limited to 5-15% and <50%,
respectively. The type and amount of alcohol used were
determined by trial and error to obtain adequate
viscosity control and solution stability. Having
determined the general area of composition for a suitable
micellar solution, several solutions were prepared at a
given water/oil ratio and varying surfactant
concentrations (1-15%). When more than one surfactant
were used, they were blended together in a known
proportion and treated as a single component in the
ternary system. The solutions were prepared by
dispersing the surfactant/cosurfactant mixture in the
hydrocarbon phase and gradually adding the aqueous phase
while maintaining moderate agitation with a magnetic
stirrer, until the solution was homogeneous (~5-10 min.).
The solutions were allowed to equilibrate in graduated
cylinders for 2-4 weeks at the desired temperature. The
phase volumes were measured and the type of solution
(Type II-, Type II+ or Type III) was determined with the
help of IR (infrared) analysis for sulfonate content.
The composition points, the tie-lines and the phase
envelope showing the multiphase regions were then plotted
on the ternary diagram. Micellar solutions were selected

on the basis of their ability to solubilize brine and oil, as well as the compatibility with polymer.

Multiple Micellar Slugs

The efficiency of the displacement process can be further improved by using a multiple micellar slug - a combination of oil-rich and water-rich slugs - instead of a conventional single slug. In this method, a slug that is more compatible with oil (oil-rich) is followed by a slug that is more compatible with water (water-rich). These slugs may be divided into portions and injected alternately in several steps as a 'composite micellar slug'. For example, a 5% pv composite slug injected in three steps can be thought of as (5/3)% pv oil-rich slug followed by (5/3)% pv water-rich slug followed again by (5/3)% pv of the oil-rich slug. The choice of the component slugs is based on the equilibrium micellar phases determined from the phase diagram. Such a combination enhances the stability, and therefore, the performance of the micellar slugs in recovering the tertiary oil. Another approach to prolong the integrity of the micellar slug is to use a 'graded composite slug'. Here, the component slugs are injected such that a gradation is formed with respect to the oil and/or the surfactant concentration of the individual slugs, i.e. the grading starts with oil-rich slug and ends with water-rich slug. Compositions that fall on the same dilution path in the phase diagram are in equilibrium with each other and hence, have favourable phase behaviour. This helps to delay the slug breakdown and thus, improve the displacement efficiency.

Objectives

The primary objectives of this research were :
1.   Formulate efficient micellar slugs for the tertiary recovery of three light oils viz. Bradford crude, Bonnie Glen crude, and Provost crude;
2.   Improve the efficiency of the micellar flooding process through the use of multiple micellar slugs;
3.   Devise methods to reduce the economic disadvantages inherent in the process by suitable selection of slugtype, size and combination, such as substituting a single slug with a graded composite slug of equivalent size.

Experimental Procedure and Apparatus

Apparatus. A constant rate displacement pump charged with mercury was used to displace the fluid of interest from steel cylinders to the core. A pressure transducer connected to the chart recorder provided the pressure history of each core flood. An automatic sampler with timer was used to collect the effluent Schematic of the experimental set up is shown in Figure 1.

Procedure. Core floods were carried out in horizontally mounted Berea sandstone cores of length 61 cm and diameter 5 cm. Porosity varied from 18 to 25% and brine permeability from 100 to 800 μm^2 . The cores were coated with a thin layer of epoxy and cast in stainless steel core holders using molten Cerrobend alloy (melting point 70°C). The ends of the cores were machined flush with the core holder and flanges were bolted on. Pore volume was determined by vacuum followed by imbibition of brine. Absolute permeability and porosity were determined. The cores were initially saturated with brine (2% NaCl). An oil flood was then started at a rate of 10m/day until an irreducible water saturation (26-38%) was established. The cores were then waterflooded with brine (2% NaCl) to obtain residual oil saturation.

A micellar flood was then started with the injection of the micellar slug, polymer buffer, and the drive water in succession, at a rate of 1.3 m/day. Two types of polymers - polyacrylamide polymer (Dow Pusher 700) and Xanthan Gum polymer (Kelzan XC) - were used as the polymer buffers. Sodium chloride brine (1%) was used as the drive water. Effluent was collected and analyzed for surfactant content using the IR and UV techniques. All experiments were conducted at room temperature (22-24°C).

Analysis of the Produced Fluids

The oil phase of the effluent was analyzed using the IR method for sulfonate content. Petrostep B100 and Petrostep B110 absorb IR radiation at 1008 cm-1 and 1176 cm-1 respectively. In the case of slug L1, it is assumed that the proportion of sulfonates (a blend of Petrostep B100 and Petrostep B110) is identical in the injected and the produced fluids.) The surfactant content of each sample and the total amount of sulfonate produced were then calculated. Sulfonate produced in the aqueous phase was determined using the UV method. Absorbance was measured at 206 nm using 10 mm couvettes. It was found that the bulk of the sulfonate recovered was in the oleic phase of the effluent. Five to ten percent of the total sulfonate injected went into the aqueous phase. The loss of surfactant is calculated as milliequivalents per 100 gram rock.

Accuracy. All fluids properties measured were within ±1%. The volumes injected and produced were in agreement within ±0.1%. The reproducibility of the core floods was between 0.002-8%.

<u>Crude Oils and Connate Water.</u> The multiple micellar slug
process was developed for the tertiary recovery of three
light oils viz. Bradford crude, Bonnie Glen crude and
Provost crude. The viscosities and densities of the
three crude oils used are listed in Table I.

Table I. Viscosities and Densities of Crude Oils

Crude Oil	Viscosity, mPa.s	Density, g/ml
Bonnie Glen	5	0.836
Bradford	5	0.817
Provost	11	0.866

     A 2% (w/v) sodium chloride solution was used as the
connate water in all corefloods. It should be mentioned
that it is desirable to incorporate divalent cations,
such as Ca^{++} and Mg^{++} , in the slug formulations as well
as in the connate water to simulate an actual reservoir.

Micellar Solutions

Micellar slug B4 was formulated for the tertiary recovery
of Bonnie Glen crude. Figure 2 shows the ternary
diagram for this crude oil, distilled water, and
surfactant system. Solutions a, b, c, and d separated
into oil-rich and water-rich microemulsions in
equilibrium with each other. For example, solution a
equilibrated into a water-rich microemulsion a' and an
oil-rich microemulsion a''. Similarly, solutions b, c,
and d equilibrated to form b', b''; c', c''; and d', d''
respectively. The tie lines shown connect the solutions
in equilibrium. Point   P refers to the plait point,
where the oil-rich and water-rich solutions become
miscible and hence, a single phase. Micellar solution B4
lies in the single phase region, just outside the binodal
curve. Solution B4 was diluted with distilled water once
to form solution B4W and, twice to form solution B4X.
Solutions B4, B4W, and B4X were used in combination to
form the composite slug for the recovery of Bonnie Glen
crude.
     Micellar slug L1 was formulated for the recovery of
the Provost crude. Figure 3 shows the ternary phase
behaviour diagram for this crude, distilled water and
surfactant system. Petroleum sulfonates Petrostep B100
and Petrostep B110 were the surfactants used in these
micellar solutions. Solutions A, B, C, and D separated
into water-rich microemulsions in equilibrium with excess
oil. Micellar solution L1 lies in the single phase
region. Solutions L1W1 and L1W2 were formed by diluting
solution L1 with distilled water. When solutions L1W1
and L1W2 were allowed to equilibrate, they separated into
two phases, the compositions of which are given by the

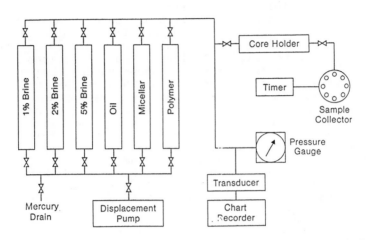

Figure 1. Schematic of Core Flooding Apparatus

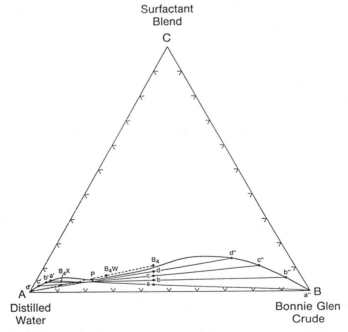

Figure 2. Phase Behaviour or Bonnie Glen Crude, Distilled Water, and Surfactant System

ends of the tie lines passing through the overall
composition points.
        Micellar slugs for the recovery of Bradford crude
were formulated using a different approach. Upon the
completion of the phase diagram, equilibrium micellar
phases lying along the binodal curve were chosen as the
oil-rich and water-rich solutions. These solutions have
similar surfactant/cosurfactant concentration, but differ
in oil/brine ratios. Figures 4 and 5 show the ternary
phase behaviour diagrams for these systems. Since the
solutions lie along the same tie line, they are in
equilibrium with each other, but have opposing affinities
for oil and brine. Two oil-rich slugs were formulated,
using two different surfactants, for the tertiary
recovery of Bradford crude were P1 and T1. The
corresponding water-rich slugs were P2 and T2. Penn
State surfactants PRL-5A and TRS-10B were used in these
slugs. Table II shows the composition and viscosities of
the various micellar slugs.

Polymers

The success of the micellar flooding process for
recovering tertiary oil is largely dependent on the
mobility buffer which serves to delay viscous fingering
of the drive water into the slug and hence, its dilution.
Solutions of polyacrylamide polymer and Xanthan gum
polymer were used as mobility controlling agents in these
experiments. Choice of the type and concentration of the
polymer is made depending on the extent to which it is
degraded by brine and the rock minerals, resistance
factor of the polymer solution, micellar-polymer
interaction and the thermal stability of the polymer
solution. It was found that polymer degradation by the
slug is directly dependent on the surfactant
concentration of the slug. The presence of alcohol in
the slug improves the micellar-polymer compatibility.
Polymer degradation by brine is determined by the ionic
strength of the brine. Divalent cations have grater
effect in reducing the polymer viscosity than monovalent
cations.

Discussion of Results

Single Slugs. Minimum slug size required for the
efficient recovery of the tertiary crude was determined
by carrying out a series of single slug runs. The slug
sizes varied from 2-15% pv. Table III shows the results
for a single micellar slug driven by 500 ppm polymer
solution for three crude oils.

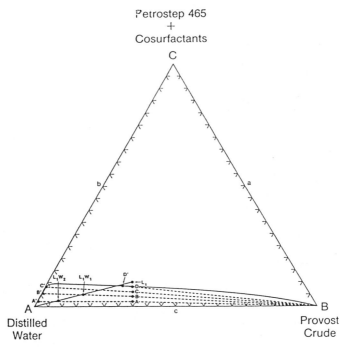

Figure 3. Phase Behaviour of Provost Crude, Distilled Water and Surfactant System

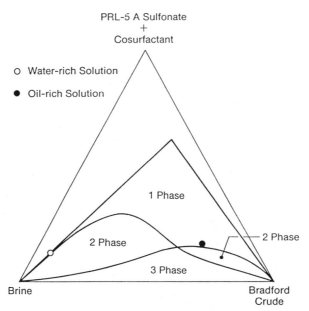

Figure 4. Phase Behaviour of Bradford Crude, Brine, and Surfactant (PRL-5A) System

Table II.   Composition and Viscosities of Micellar Slugs

Components	Composition, % vol.								Composition, % w	
	B4	B4W	B4X	L1	L1W1	L1W2	P1	P2	T1	T2
Crude oil*	39.00	19.50	9.75	30.05	15.00	7.50	63.50	5.50	85.00	2.00
Distilled water	50.00	75.00	87.50	60.09	80.00	90.00				92.50
Brine (2% w/v NaCl)	9.00	4.50	2.25				20.10	84.00	5.00	
Petrostep 465				5.54	2.81	1.40				
Petrostep B110				3.07	1.56	0.78				
Petrostep B100										
PRL-5							11.99	7.63		
TRS-10									8.01	4.00
IPA	1.50	0.75	0.38	1.25	0.63	0.32				
NBA	0.50	0.25	0.13							
Isoamyl Alcohol							4.51	2.87		
Cyclohexanol									1.49	1.00
Viscosity, m.Pa.s	45.50	2.50	1.50	17.74	2.40	1.60	10.00	8.70	1.50	21.20

* B4, B4W, B4X : Bonnie Glen crude
  L1, L1W1, L1W2 : Provost crude
  P1, P2, T1, T2 :Bradford crude

Table III. Tertiary Recovery with Single Micellar Slugs

Crude Oil	Slug Type	Slug Size % pv	Tertiary Recovery % oil in place
Bonnie Glen	B4(oil-rich)	2.0	16.9
		5.0	61.4
		6.0	73.6
		10.0	79.0
		15.0	82.4
Provost Crude	L1(oil-rich)	2.0	64.0
		5.0	73.4
Bradford Crude	P1(oil-rich)	2.5	24.4
		5.0	27.3
		10.0	36.7,
			41.2
	P2(water-rich)	2.5	3.7
		5.0	16.5
		10.0	44.3
	T1(oil-rich)	2.5	19.9
		5.0	23.6
		10.0	47.5
	T2(water-rich)	2.5	7.1
		5.0	15.7
		10.0	30.9

The results show that the largest increment in tertiary recovery occurs around a micellar slug size 5% pv for all the slugs used. Figure 6 shows the results for Bonnie Glen crude oil, indicating that the relative increase in tertiary recovery reaches a maximum at slug sizes between 4 and 8% pv. Further increase in slug size gives only minimal increment in tertiary recovery and, greatly reduces the cost-effectiveness of the slug. Process efficiency peaked at 5% pv slug size for micellar slug B4 in the case of Bonnie Glen crude oil. This is shown in Figure 7. Tertiary recovery of Bradford crude using single and composite slugs of various sizes is shown in Figure 8.

Composite Slugs. A single micellar slug was replaced by an equivalent volume of composite slugs to examine the effect of composite slugs on tertiary recovery. An oil-rich slug was followed by a water-rich slug in a 50:50 volume ratio. Selected results are shown in Table IV. Slug compositions T1, T2 and B4 and B4W are given in Table II and shown in Figures 2, 3, 4 and 5, respectively.
The results show that for a given slug size, a composite slug recovered more crude than a single slug of equivalent size.

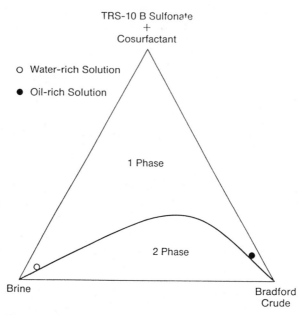

Figure 5.   Phase Behaviour of Bradford Crude, Brine,
and Surfactant (TRS-10B) System

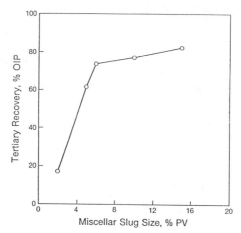

Figure 6.   Effect of Micellar Slug Size on Tertiary
Recovery

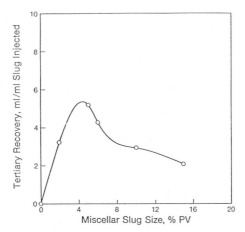

Figure 7. Effect of Micellar Slug Size on the Oil Recovery Efficiency of Micellar Slug B4

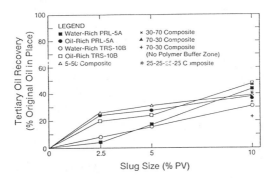

Figure 8. Tertiary Recovery of Bradford Crude Using Oil-Rich, Water-Rich and Composite Slugs

Table IV. Tertiary Recovery Using Paired (Oil-Rich and Water-Rich) Micellar Slugs

Micellar Slug	Total Slug Size, % pv	Tertiary Recovery, % oil in place		
		Single Slug oil-rich	water-rich	Composite Slug
T1/T2	2.5	19.9	7.1	25.0
	5.0	23.6	15.7	31.6
B4/B4W	5.0	62.92	–	80.9
	10.0	77.02	–	94.7

Next, the proportions of oil-rich and water-rich slugs were varied for a given slug size (10% pv) and the tertiary recoveries were compared. As seen in Table V, oil-rich slugs were more effective

Table V. Effect of Oil-Rich/Water-Rich Slug Volume Ratio on Tertiary Recovery

Micellar Slug	oil-rich : water-rich ratio (vol.)	Tertiary Recovery % oil in place
T1/T2	100 : 00	47.5
	70 : 30	42.2
	50 : 50	39.9
	30 : 70	34.5
	00 : 100	30.9

than water-rich slugs in recovering tertiary oil. The efficiency of the process can be improved by following the oil-rich slug with water-rich slug. For example, a 70:30 composite slug recovered as much crude as a single slug of equivalent size.

Multiplicity of Slugs. The effect of multiplicity of slug on recovery efficiency was examined by alternating the oil-rich and water-rich slugs in 2, 3 and 4 steps, for a given slug size (5% pv). In each case, tertiary recovery was higer than for 2 steps. (Slug size "5/3" stands for a total slug size of 5% pv split into 3 equal parts.) The order of injection o/w/o represents oil-rich slug followed by water-rich slug followed by oil-rich slug. Table VI shows selected results for a 5% slug in several steps. It is clear that the three slug configuration gave higher oil recovery than a single slug.

Table VI. Comparison of Tertiary Recovery Using Single and Composite Micellar Slugs

Slug Type	Slug Size % pv	Order of Injection	Tertiary Recovery % oil in place
B4	5	o	62.93
B4	5/2*	o/w	80.85
B4	5/3	o/w/o	87.02
L1	5	o	73.35
L1	5/3	o/w/o	81.91

* denotes a 5% pv slug, consisting of two equal parts, being oil-rich, water-rich, respectively.

Graded Composite Slugs. In graded micellar slugs, two or more slugs are injected so that a gradation starts with oil-rich slug and ends with a water-rich slug. Micellar compositions formed along the dilution path of an oil-rich slug are excellent in forming a gradation and are efficient in mobilizing the tertiary oil. Table VII shows the results of single, composite, and graded composite slugs of B4 and L1, in recovering Bonnie Glen crude and Provost crude, respectively.

Table VII. Comparison of Tertiary Recovery with Single, Composite, and Graded Composite Micellar Slugs

Slug Type	Slug Size % pv	Order of Injection	Tertiary Recovery, % oil in place
B4	5	o	62.93
	5/3*	o/w/o	87.02
	5/3	o/w/+w	97.94
L1	5	o	73.35
	5/3	o/w/o	81.91
	5/3	o/w/+w	96.87

o : oil-rich   w : water-rich,   +w : more water-rich
* denotes a 5% pv slug, consisting of three equal parts, being oil-rich, water-rich and oil-rich, respectively.

Smaller Slug Size. Efficiency of graded composite slugs over a single slug was high enough to reduce the slug volume requirement to less than one-half and still obtain the same tertiary recovery. This is shown in Table VIII.

Each volume of a composite slug injected recovered 3 to 15 times its volume in tertiary oil. Sulfonate loss in graded composite slug runs was considerably lower than in single slug runs. This shows that slug dissipation due to sulfonate loss is lower in graded composite slug runs, and hence the residual oil was mobilized and recovered more efficiently. Typical oil recovery curves for 5% pv single slug runs and multiple slug runs using micellar slug B4 are shown in Figures 9 and 10. Similar curves for micellar slug L1 are shown in Figures 11 and 12, respectively.

Conclusions

Based upon over 50 micellar floods carried out on sandstone cores, the following conclusions are reached:
1. Oil-rich slugs recovered more crude oil than water-richslugs.
2. Small micellar slugs (2-5% pv) can recover 3 to 15 times their volume of waterflood residual oil. Final oil saturation was reduced to 0.02% pv for 2% pv slug size and,0.01% pv for 5% pv slug size.
3. Oil displacement efficiency can be improved considerably when a composite slug is used in place

Table VIII. Tertiary Recovery with Smaller Slug Size

Slug Type	Slug Size % pv	Tertiary Recovery % oil in place	Sulf. Recovery meq/100g rock	Process Efficiency*
B4(s)	5	62.93	0.085	5.02
B4(g)	5	97.94	0.028	6.26
B4(s)	2	16.86	0.040	3.01
B4(g)	2	92.36	0.017	15.39
L1(s)	5	73.35	0.037	5.51
L1(g)	5	96.87	0.018	6.65
L1(s)	2	63.99	0.022	11.34
L1(g)	2	70.26	0.006	12.27

s : single slug   g : graded composite slug
* vol. of oil recovered/Total slug volume

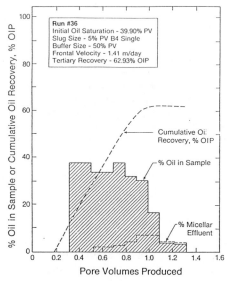

Figure 9. Production History of 5% PV Micellar Slug B4
and 50% PV Buffer Injected at 1.41 m/day

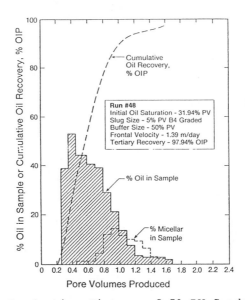

Figure 10.   Production History of 5% PV Graded
Composite Slug of B4 and 50% PV Buffer Injected at 1.35
m/day

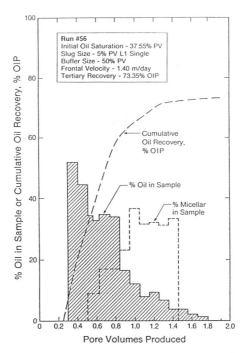

Figure 11.   Production History of 5% PV Micellar Slug
L1 and 50% PV Buffer Injected at 1.40 m/day

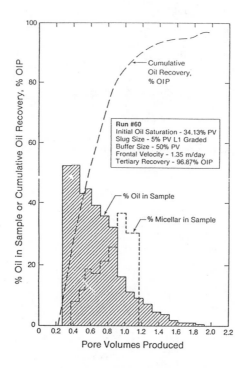

Figure 12.   Production History of 5% PV Graded
Composite Slug of L1 Injected at 1.35 m/day

of a single slug. Process efficiency increased substantially (12-15%) for a 2% pv slug size when a graded composite slug was used instead of a single slug of equivalent size.

4. Oil recovery increases with an increase in micellar slug size. But the process efficiency drops and the surfactant loss increases.

5. The slug requirement, and therefore, the cost of production, can be reduced by using a graded composite slug in place of a conventional single slug. For example, a 5% pv of a graded composite slug B4 recovered 97.74% oil in place, whereas a 2% pv of the same slug recovered 92.36% oil in place, increasing the process efficiency from 6 to 15.

Literature Cited

1. Gogarty, W.B. and Tosch, W.C.: "Miscible-Type Waterflooding : Oil Recovery with Micellar Solutions", J. Pet. Tech. (Dec. 1968) 1407-1415.

2. Healy, R.H. and Reed R.L.: "Physicochemical Aspects of Microemulsion Flooding", Soc. Pet. Eng. J. (Oct.1974) 14, 491-501.

3. Davis, J.A. and Jones, S.C.: "Displacement Mechanisms of Micellar Solution", J. Pet. Tech. (Dec. 1968), 1415-1428.

4. Holm, L.W.: "Mobilization of Waterflood Residual Oil by Miscible Fluids", SPE Res. Eng. (July 1986) 354-362.

5. Pope, G.A, Tsaur, K., Schechter, R.S. and Wang, B.: "The Effect of Several Polymers on the Phase Behaviour of Micellar Fluids", Soc. Pet. Eng. J. (Dec. 1982) 816-830.

6. Sayyouh, M.H.M., Farouq Ali, S.M. and Stahl, C.D.: "Rate Effects in Tertiary Micellar Flooding of Bradford Crude Oil", Soc. Pet. Eng. J. (Aug. 1981) 469-479.

7. Enedy, S.L., Farouq Ali, S.M. and Stahl, C.D.: "Competing Roles of Interfacial Tension and Surfactant Equivalent Weight in the Development of a Chemical Flood", Soc. Pet. Eng. J. (Aug. 1982) 472-480.

RECEIVED February 7, 1989

Chapter 19

Effect of Demulsifiers on Interfacial Properties Governing Crude Oil Demulsification

Surajit Mukherjee[1] and Arnold P. Kushnick

Exxon Chemical Company (ECTD), Houston, TX 77029

Effectiveness of a crude oil demulsifier is correlated with the lowering of shear viscosity and dynamic tension gradient of the oil-water interface. Using the pulsed drop technique, the interfacial dilational modulii with different demulsifiers have been measured. The interfacial tension relaxation occurs faster with an effective demulsifier. Electron spin resonance with labeled demulsifiers indicate that the demulsifiers form 'reverse micelle' like clusters in bulk oil. The slow unclustering of the demulsifier at the interface appears to be the rate determining step in the tension relaxation process.

Crude oil is almost always produced as persistent water-in-oil emulsions which must be resolved into two separate phases before the crude oil can be accepted for transportation. The water droplets are sterically stabilized by the asphaltene and resin fractions of the crude oil. These are condensed aromatic rings containing saturated carbon chains and napthenic rings as substituents, along with a distribution of heteroatoms (S, O and N) and metals (Ni, V). It has been suggested (1) that a stabilizing film protecting the water drops from coalescence is created by hydrogen bonding of the N-, O-, and S- containing groups at the water drop--oil interface.

Although, many other methods (e.g. electrostatic separation, heating, centrifugation, etc.) may be used to separate the oil and water phases, chemical demulsification is the most inexpensive and widely used technique to resolve crude oil emulsions. The demulsifiers are oil-soluble water-dispersible non-ionic polymeric

[1]Current address: Lever Research, Inc., Edgewater, NJ 07020

(molecular weight 2,000-100,000) surfactants. They are added to the crude oil in very small (10-400 ppm) amounts. One of the most commonly used demulsifier is the oxyalkylated alkyl phenol formalehyde resin, the alkyl group may be butyl, amyl or nonyl group. The interfacial activity is controlled by the amounts of ethylene and propylene oxides attached to the resin.

The major problem in demulsifying crude oil emulsions is the extreme sensitivity to demulsifier composition. There have been attempts (2, 3) to correlate demulsifier effectiveness with some of the physical properties governing emulsion stability. However, our understanding in this area is still limited. Consequently, demulsifier selection has been traditionally based on a 'trial and error' method with hundreds of chemicals in the field.

Our goal is to develop a property-performance relationship for different types of demulsifiers. The important interfacial properties governing water-in-oil emulsion stability are shear viscosity, dynamic tension and dilational elasticity. We have studied the relative importance of these parameters in demulsification. In this paper, some of the results of our study are presented. In particular, we have found that to be effective, a demulsifier must lower the dynamic interfacial tension gradient and its ability to do so depends on the rate of unclustering of the ethylene oxide groups at the oil-water interface.

Experimental Techniques

The oil-water dynamic interfacial tensions are measured by the pulsed drop (4) technique. The experimental equipment consists of a syringe pump to pump oil, with the demulsifier dissolved in it, through a capillary tip in a thermostated glass cell containing brine or water. The interfacial tension is calculated by measuring the pressure inside a small oil drop formed at the tip of the capillary. In this technique, the syringe pump is stopped at the maximum bubble pressure and the oil-water interface is allowed to expand rapidly till the oil comes out to form a small drop at the capillary tip. Because of the sudden expansion, the interface is initially at a nonequilibrium state. As it approaches equilibrium, the pressure, $\Delta P(t)$, inside the drop decays. The excess pressure is continuously measured by a sensitive pressure transducer. The dynamic tension at time t, is calculated from the 'Young-Laplace' equation

$$\Delta P(t) = \frac{2\sigma(t)}{R_d} \tag{1}$$

where R_d is the radius of the drop. The interfacial dilational modulas is then calculated by a Fourier Transformation of the transient interfacial tension data (see later).

The interfacial shear viscosities are measured by the deep channel viscous traction surface viscometer (5) at the Illinois Institute of Technology. The oil-water equilibrium tensions are measured by either the spinning drop or the du Nouy ring (6) method.

Electron Spin Resonances are measured at 9.5 GHz at room temperature. Demulsifiers are labeled by reacting the terminal OH groups with the spin-label 3-chloroformyl 2,2,5,5 tetramethyl pyrroline 1-oxyl (Figure 5a). This is done by the Schotten-Baumann reaction (7). The carboxylic form of the spin-label (obtained from Eastman Kodak Company) is dissolved in a benzene/pyridine mixture and is reacted, in situ, with thionyl chloride (also from Eastman Kodak Company). After 15 minutes, vacuum-dried demulsifier is added to the mixture. The solution is mixed and left overnight. All the reactions are carried under a nitrogen blanket. Excess benzene is then added and the insoluble part is eliminated. The tagged demulsifier is separated from the free label either by precipitating it with methanol or by size exclusion chromatography.

Results and Discussion

Interfacial Shear Viscosity

The demulsification process involves coalescence of smaller water droplets into larger ones. During this process, the oil in the liquid film between the droplets drains out, thereby thinning the film and finally rupturing it. The faster the film thins, the greater is the demulsification effectiveness. The liquid drainage rate depends, among other factors, on the interfacial shear viscosity. A high (>1 surface poise) interfacial shear viscosity significantly slows down the liquid drainage. Consequently, the emulsion is stable. But the reverse is not true. Emulsions have been found to be stable (8) even with a low interfacial viscosity.

Our results suggest that the lowering of interfacial shear viscosity, although necessary, is not a sufficient criterion for effective demulsification. In addition, a demulsifier must also rapidly dampen any fluctuations in the oil-water interfacial tension. The demulsification data with four different demulsifiers for a crude oil-water system (Table I) support this conclusion. Structurally, the demulsifier P1 and RO are of moderate (MW = 2,000-5,000) molecular weights, whereas P1 and P2 are large (MW >50,000) three dimensional structures.

The results show that although all the demulsifiers lower the shear viscosity, they differ widely in their demulsification effectiveness, as measured by the residual bottom sediment and water content (Figure 1) (BS and W%) of the dehydrated oil. For example, the demulsifier OP1, although it lowers both the equilibrium interfacial tension (Figure 2) and the shear viscosity (Table I), nevertheless is ineffective. This is because it takes a much longer time for the oil-water interfacial tension to reach equilibrium with OP1 than with P1 or P2 (see later).

In general, there is no correlation between the tension and the shear viscosity of an oil-water interface. However, for systems containing demulsifiers, a low interfacial tension (IFT) often leads to a lowering of the shear viscosity. Demulsifiers, in general, are large disordered molecules and when they are present at the interface they create a mobile, low viscosity zone. However, a low IFT is not a necessary condition for a low viscosity interface. A large demulsifier such as P1, although not very surface active, can still lower the shear viscosity to a very low value (Table I).

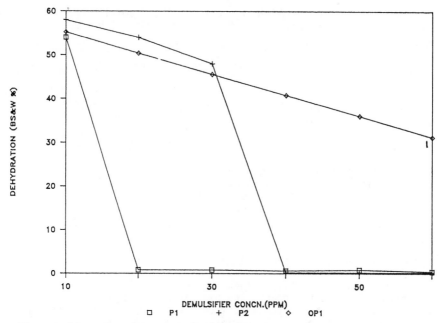

Figure 1:   Residual   water   content   of   a   brine-in-crude oil
emulsion with the demulsifiers P1, P2 and OP1.

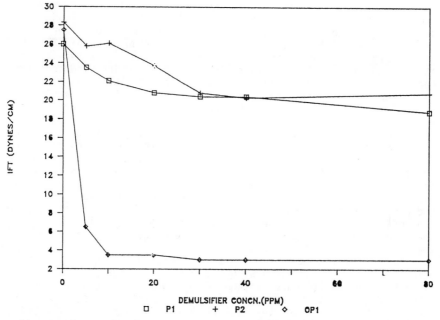

Figure   2:   Interfacial   tension of brine vs the crude oil with
P1, P2 and OP1.

Table I. Comparison of Effectiveness and Interfacial Properties of
Different Demulsifiers in a Crude Oil-Water System

Demulsifier	Concn. (ppm)	BS&W (vol%)	IFT (Dynes/cm)	Shear Viscosity (Surface Poise)
NONE	-	50	28	2.0×10^{-1}
RO	200	16	9	8×10^{-2}
P1	60	0.2	21	1×10^{-4}
P2	60	0.6	20	1×10^{-2}
OP1	40	50	4	1×10^{-2}

RO: Oxyalkylated nonyl phenol resin
P1:   Oxyalkylated cross-linked polypropylene glycol
P2:   Cross-linked polypropylene glycol
OP1: Oxyalkylated trimethylol propane

Dynamic Interfacial Tension Gradient

When the interfacial film between two droplets thins, the liquid in
the film flows out towards the plataeu border region. Such a flow
removes some of the surfactants from the interface thus creating an
uneven concentration of the surfactant along the interface (9, 10).
Furthermore, the interfacial film often thins unevenly thus creating
locally thin and thick regions (11). A local thinning implies an
increase in the interfacial area and hence a decrease in the
surfactant concentration. On the other hand, the surfactant
concentration increases in the thick region. This nonuniform
surfactant concentration at the interface leads to local variations
in the interfacial tensions which produces a flow of liquid from the
high to the low tension regions. This is known as the
'Gibbs-Marangoni' effect. This interfacial tension induced flow
opposes the outward drainage in the film. It also helps to 'heal'
the film to its original uniform thickness. The net result is a
reduced rate of film thinning and consequently a more stable
emulsion.
     In order to be effective, a demulsifier has to reduce the
'Gibbs-Marangoni' resistive liquid flow. It does so by migrating
from the interior to the interface, equalizing the interfacial
surfactant concentration and bringing the interfacial tension of the
film to its equilibrium value. As pointed out by Ross and Haak
(12), if this process is faster than the surfactant migration
induced flow along the interface, the local thin spots in the film
are not 'healed' and the liquid drainage in the film continues
unabated. Consequently, the drop coalescence rate is enhanced and
the emulsion is rapidly destabilized.
     The importance of rapid relaxation in demulsification
effectiveness can be seen with the crude oil-water dynamic tension
results with P2 (Figure 3) and OP1 (Figure 4). As can be seen, it
takes only about 60 seconds for the interface to reach its
equilibrium state with the effective demulsifier P2, whereas with
less effective demulsifier OP1, the equilibrium is reached only
after 800 seconds.

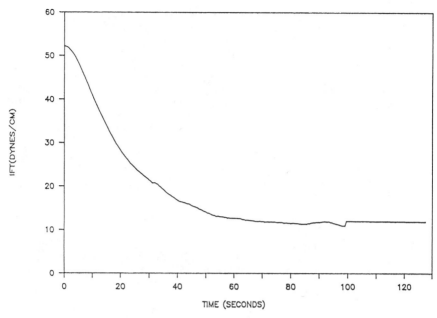

Figure 3:   Dynamic interfacial tension of the crude oil-brine
interface as a function of time, with 40 ppm P2.

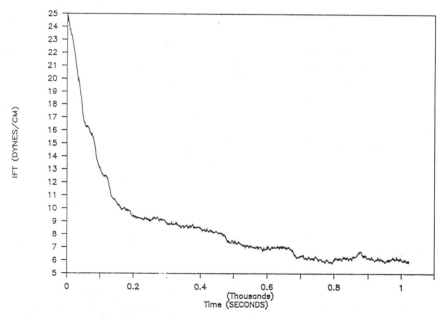

Figure 4:   Dynamic interfacial tension of the crude oil-brine
interface as a function of time, with 40 ppm OP1.

The time it takes for an oil-water interface to reach equilibrium depends strongly on the ethylene oxide content of the demulsifier molecule. The interfacial tension decay process can often be an order of magnitude longer than what is observed in systems where the surfactant diffusion from bulk to interface is the rate determining step. Demulsifiers form 'reverse- micelle' like structures in the bulk oil with the ethylene oxide groups clustering together to minimize their interactions with the oil phase. We believe, the unclustering of the ethylene oxide groups as well as rearrangement of the demulsifier molecules at the interface are the rate determining steps in the tension relaxation process. Similar interfacial tension behavior has been observed (13) with large oil-soluble water-insoluble surfactant such as Cholesterol.

Electron Spin Resonance

We have studied demulsifier association by the electron spin resonance (ESR) technique. The spin label is covalently attached (Figure 5a) to the demulsifier. Normally, the ESR spectrum of a freely tumbling nitroxyl radical consists of three sharp peaks (Figure 5b). However, the spectrum for a tagged ethoxylated nonyl phenol resin (Figure 6a or 6b) shows only a single broad peak.
The results may be explained in the following manner. The spin labels are attached next to the ethylene oxide groups. Clustering of the ethylene oxide groups brings the nitroxide radicals in close proximity with other. As the orbitals containing unpaired electron from different labels overlap, there is a spin exchange which leads to a single resonance peak. Similar exchange broadening has been observed (14) in lipid films in which spin probes are excluded to form aggregates. With stronger interaction, the single resonance peak narrows.
The ethylene oxide clustering depends on the polarity of the medium. The peak-to-trough separation gives an indication of the clustering strength. A smaller separation implies stronger clustering. We have found that in toluene (Figure 6a) the separation is 15 gauss whereas in isopropyl alcohol (Figure 6b) it increases to 25 gauss. This suggests a weaker clustering of the ethylene oxide groups with increasing polarity of the medium.

Interfacial Dilation Modulus

The complex interfacial dilational modulus (ϵ*) is a key fundamental property governing foam and emulsion stability. It is defined as the interfacial tension increment ($d\sigma$) per unit fractional interfacial area change (dA/A) i.e.,

$$\epsilon^*(f) = \epsilon'(f) + i\ \epsilon''(f) = \frac{d\sigma}{\frac{(d\ A)}{A}} = \frac{d\sigma}{d\ \ln A} \qquad (2)$$

where $\epsilon'(f)$ and $\epsilon''(f)$ are the real and the imaginary components at a frequency f. We have used a procedure (4), where the complete frequency spectrum is obtained by a Fourier Transform of the dynamic interfacial tension data. The relevant relationships are:

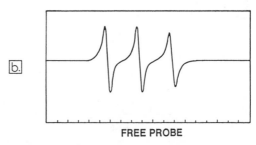

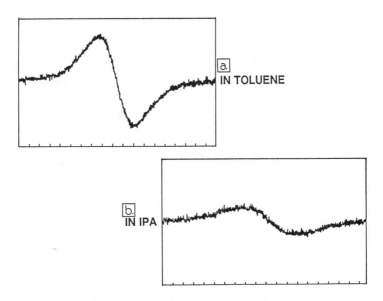

Figure 5:   a) Chemical structure of a nonyl phenol resin and the spin label 3-chloroformyl 2,2,5,5 tetramethyl pyrroline 1-oxyl; b) ESR of a free label in toluene.

Figure 6:   ESR of the labeled nonyl phenol resin in a) toluene and in b) isopropyl alcohol.

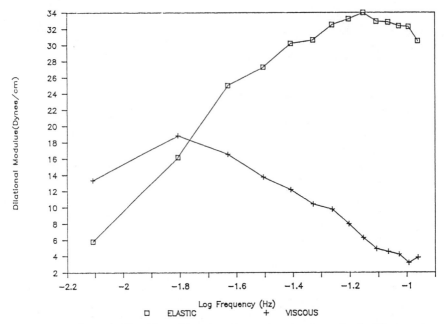

Figure 7:   Elastic (in-phase) and viscous (out-of-phase)
components of crude oil-brine interfacial dilational modulus
with 40 ppm P2.

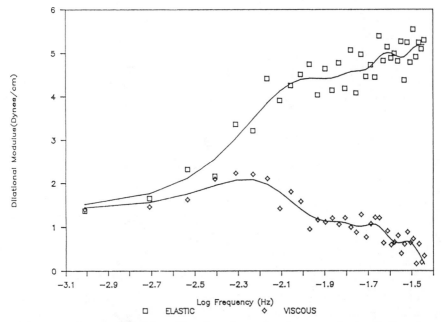

Figure 8:   Elastic (in-phase) and viscous (out-of-phase)
components of crude oil-brine interfacial dilational modulus
with 40 ppm OP1.

$$\varepsilon^*(f) = \frac{\int_0^\infty \Delta\sigma(t)e^{-2\pi ift}dt}{\int_0^\infty \Delta\ln A(t)e^{-2\pi ift}dt} \qquad (3)$$

$$\varepsilon'(f) = \frac{2\pi f}{\Delta\ln A}\ \text{Im FT }(\Delta\sigma) \qquad (4)$$

$$\varepsilon''(f) = \frac{2\pi f}{\Delta\ln A}\ \text{Real FT }(\Delta\sigma) \qquad (5)$$

The real component, $\varepsilon'(f)$, is the dilational elastic modulus (Figure 7). It is the interfacial tension gradient that is in phase with the area change. At a low frequency, the interfacial film behaves as a soluble monolayer. Interfacial concentration and the tension are then governed by the bulk concentration and so remain invariant to a change in area. On the other hand, at a very high frequency the interfacial film behave as a completely insoluble monolayer and the variation in interfacial tension resulting from a local change in area is virtually instantaneous.

The imaginary component, $\varepsilon''(f)$, is the dilational viscosity modulus. This arises when the demulsifier in the monolayer is sufficiently soluble in the bulk liquid, so that the tension gradient created by an area compression/expansion can be short circuited by a transfer of demulsifiers to and from the surface. It is 90° out of phase with the area change.

The shape of the curves for the dilational modulus (Figures 7 and 8) suggests a single relaxation mechanism, probably the unfolding of the demulsifier molecules at the interface. The frequency peak in the $\varepsilon''(f)$ plot is a measure of the characteristic relaxation time. A shorter relaxation time, by inducing faster film drainage, increases demulsification efficiency.

The dynamic response data for P1 and P2 (Figure 7) are similar. They are, however, quite different from that of OP1 (Figure 8). The characteristic relaxation times for P1 and P2 are 50 and 69 seconds respectively, whereas with OP1 it is 158 seconds. This indicates that with P1 and P2, the oil-water interface will have much shorter response time leading to an improved demulsification effectiveness.

Summary

For effective demulsification of a water-in-oil emulsion, both shear viscosity as well as dynamic tension gradient of the water-oil interface have to be lowered. The interfacial dilational modulus data indicate that the interfacial relaxation process occurs faster with an effective demulsifier. The electron spin resonance with labeled demulsifiers suggests that demulsifiers form clusters in the bulk oil. The unclustering and rearrangement of the demulsifier at the interface may affect the interfacial relaxation process.

Acknowledgment

The authors wish to thank the Exxon Chemical Company for permission to publish this paper. We thank Dr. Darsh T. Wasan and Mr. Chandrashekhar Shetty at IIT, Chicago for measuring some of the interfacial shear viscosities. A special thanks to Ms. Layce Gebhard of Exxon Research and Engineering for measuring the ESR spectra. Finally, a special thanks to Ms. Rose Mary Rangel of Energy Chemicals (Exxon Chemical Company) for the preparation of the manuscript.

Literature Cited

1. Oren, J. J.; Mackay, D. M. Fuel. 1977, 56, 382.

2. Burger, P. D.; Hsu, C.; Arendele, J. P. SPE 16285, SPE International Symposium on Oilfield Chemistry. San Antonio, Texas. February 4-6, 1987; p 457.

3. Blair, C. M. Chem. Ind. 1960, 538.

4. Clint, J. H.; Neustadter, E. L.; Jones, T. J. Proc. 3rd. Eur. Symp. on Enhanced Oil Recovery, Bournemouth, U.K., 1981, p 135.

5. Wasan, D. T.; Gupta, L.; Vora, M. K. AICHE J. 1971, 17(6), 1287.

6. Adamson, A. W. Physical Chemistry of Surfaces, 3rd Ed.; J. Wiley and Sons.: New York, 1976; Chapter 1.

7. Tormala, P.; Lattila, H.; Lindberg, J. J. Polymer. 1973, 14, 481.

8. Neustadter, E. L.; Whittingham, K. P.; Graham, D. E. In Surface Phenomena in Enhanced Oil Recovery; Shah, D. O., Ed.; Plenum Press: New York, 1981; p 307.

9. Zapryanov, Z.; Malhotra, A. K.; Adrengi, N.; Wasan, D. T. Int. J. Multiphase Flow. 1983, 9, 105.

10. Ivanov, I. B.; Jain, R. K. In Dynamics and Instability of Fluid Interfaces; Sorensen, T. S. Ed.; Lecture Notes in Physics Series No. 105; Springer Verlag: Berlin, W. Germany, 1979; p 120.

11. Jain, R. K.; Ivanov, I. B.; Maldarelli, C.; Ruckenstein, E. Ibid; p 140.

12. Ross, S.; Haak, R. M. J. Phys. Chem. 1969, 73, 2828.

13. Van Hunsel, J.; Bleys, G.; Joos, P. J. Colloid Interface Sci. 1986, 114, 432.

14. Smith, I. C. P.; Butler, K. W. In Spin Labeling: Theory and Practice, Berliner, L. J., Ed.; Academic Press: New York, 1976; Vol. 1, p 423.

RECEIVED December 21, 1988

CRUDE OIL CHEMISTRY

Chapter 20

Useful Surfactants from Polar Fractions of Petroleum and Shale Oil

Kazem M. Sadeghi, Mohammad-Ali Sadeghi, Dawood Momeni, Wen Hui Wu,[1] and Teh Fu Yen

School of Engineering, University of Southern California, Los Angeles, CA 90089-0231

Fossil fuel derived liquids from two different origins, shale oil and petroleum crude oil, were subjected to solvent fractionation through a silica gel column. The solvent system for fractions consisted of n-hexane, toluene, and toluene/methanol (for polar fractions). The polar fraction of the samples were subfractioned by ion exchange chromatography. The columns used were anion exchange resin, cation exchange resin, and clay-$FeCl_3$ to obtain the acid, base, and neutral fractions, respectively. The polarity increased for each column as more polar solvents were used. In order to compare surface activity of subfractions derived from shale oil and crude oil, the interfacial tension (IFT) of each subfraction was measured against aqueous solutions with different amounts of sodium silicate concentrations. It was proven that representative samples obtained from shale oil fractionation led to much lower interfacial tensions compared to the ones obtained with crude oil fractionation samples. It was also shown that the most polar fraction of the anion exchange column was from shale oil.

Surface active agents, more commonly known as "surfactants," are the groups of chemical compounds that in the most common form constitute an ionic or polar portion (hydrophilic head) and a hydrocarbon portion (hydrophobic tail). The ionic or polar portion interacts strongly with the water via dipole-dipole or ion-dipole interactions and

[1]Current address: Research Institute of Petroleum Processing, P.O. Box 914, Beijing, Peoples Republic of China

is solvated. On the other hand, the strong interactions between the water molecules arising from dispersion forces and hydrogen bonding of the chain act cooperatively to squeeze the hydrocarbon out of water. Furthermore, the hydrophobic moiety is a single or double hydrocarbon chain and the hydrophilic moiety is either an anionic, cationic, nonionic or zwitterionic polar group (1). The unique property of surface active materials makes them able to react strongly at various interfaces (e.g., air-water, oil-water, water-solid, oil-solid, etc.) and to lower the interfacial surface energy. Surfactants in solution tend to accumulate and adsorb at interfaces between their solution and adjacent phases. The orientation of these molecules as well as molecular interaction and molecular packing result in an interfacial behavior different from that in bulk phases. The present state-of-the-art enhanced oil recovery processes reveals that of the potential oil reserves, about 60 percent are estimated to be compatible to chemical flooding with surfactants (2). The surfactant selection for a tertiary oil recovery process is made on the basis of ultralow interfacial tension between the oil and the aqueous phase. Melrose and Brander (3) and Taber (4) have shown that successful immiscible oil displacement depends on the existance of a very low interfacial tension, between the oil and water phases. A value of about 10^{-3} dyne/cm or less is required to mobilize the oil. It is shown that the recovery of residual oil from laboratory test cases is greatly improved for systems with ultra low interfacial tension (3). The achievement and maintenance of low interfacial tensions during chemical flooding therefore seems essential.

Numerous methodologies have been developed for separation of polar compounds from crude oils (5-7). Among these, the most well-known is the scheme developed by the Bureau of Mines in American Petroleum Institute Research Project 60 (5,6) which involves ion-exchange chromatographic and ferric chloride complexation techniques for removal of acids, bases, and neutral nitrogen compounds. However, this procedure is rather complex and tedious. The definition of acids or bases by ion-exchange methods is in terms of the hydrogen donating or accepting tendency of the molecule. Since many polar compounds are amphoteric, their definitions as acids or bases depend on the analytical sequence employed. Seifert et al. (8-10) have extracted acids from crude oil and showed that carboxylic acids are primarily responsible for the observed surface activity. Some long-chain acids in the crude oil, however, exist as natural esters, amides, and other acid-base complexes. The presence of these compounds in crude oil has been identified by many investigators including Snyder (11,12) and McKay et al. (13).

The current research objective is to evaluate the surface activity of the subfractions obtained from the solvent fractionated crude oil and shale oil samples as they are passed through the separation process developed for this work. The columns used are anion exchange resin, cation exchange resin, and clay-$FeCl_3$ to obtain the acid, base, and neutral fractions, respectively. The clay-$FeCl_3$ complexation technique alone could specifically concentrate nitrogen and oxygen-containing compounds in shale oil. Although petroleum sulfonates (i.e., sodium salts of sulfonated crude oil) are known as main candidates in practical surfactant flood systems (14) they are usually very hard to structurally characterize and need to be combined with co-surfactants and blocking agents to enhance or protect the main surfactant. Besides having a broad range of molecular weights which make them more complex, their production is often cumbersome and very costly. The surface active compounds obtained from crude oil and especially from shale oil samples in this work, are shown to be very stable and produce very low interfacial tension in an alkaline system. Structural characterization studies of these compounds are also discussed.

Experimental

Sample. Petroleum crude oil sample from Long Beach Field (TUMS Well C-331, API° 20), California, and shale oil obtained by retorting at 500°C the Green River Oil Shale (Anvil Point Mine) were studied. About 20 g of shale oil was dissolved in 200 ml of THF and then filtered. The sample was recovered by a rotary evaporator. Although the same procedure was done on the other samples, the percent ash was different for each sample. All the samples were evaporated to a constant weight in a vacuum oven at 50°C.

Silica Gel Chromatography. The ratio of sample to absorbent was about 1:35. The columns were exhaustively eluted with n-hexane, toluene, 4:1 toluene/methanol, and 2:1 toluene/methanol volume ratios, to get Fractions I, II, III, and IV, respectively (Figure 1). Fraction III was then sub-fractionated futher by ion exchange chromatography. In order to separate polar fractions from the samples, as discussed in detail in one of our works (Sadeghi, K.M.; Sadeghi, M.-A.; Wu, W.H.; Yen, T.F. Fuel, in press.), a column was slurry packed with the Baker analyzed reagent grade silica gel (60-200 mesh) in n-hexane and topped with a layer of sand before each experiment. Silica gel had neutral activity and was thermally activated before use.

Ion Exchange Chromatography. The polar oil (Fraction III) obtained from silica gel chromatography was mixed with cyclohexane. The slurry was then passed through an anion exchange resin column packed with Amberlyst A-27 (Aldrich Chemical Co.), a strongly basic, macroreticular resin. In order to obtain the acidic fractions from the anion exchange resin column, an eluting scheme based on the increasing polarity of the solvent system was employed. This scheme included the use of cyclohexane, toluene, a mixture of 3:2 toluene/methanol, and a mixture of 3:2 toluene/methanol saturated with CO_2, for extraction into the four fractions. The one obtained from elution with cyclohexane was further fractionated by passing it through a cation exchange resin column packed with an Amberlyst 15 (Aldrich Chemical Co.), which is a macroreticular resin and strongly acidic in its nature. The solution obtained was exposed to the same elution scheme in the first stage with the exception of substituting the toluene methanol/CO_2 mixture with a mixture of 5.4:3.6:1.0 toluene/methanol/isopropylamine in the last phase. Three basic fractions resulted from these fractionations. A column packed with clay-$FeCl_3$ (Engelhard Minerals and Chemicals) was employed to further subfractionate the extracted sample obtained from cyclohexane eluting in the second stage. The solvent system used was cyclohexane and dichloromethane. The ratio of sample to absorbent was about 1:10 for the three absorbents used. Two neutral fractions were obtained at this last stage of separation process for each sample tested. The detailed flowchart of the separation procedure is shown in Figure 2.

Interfacial Tension (IFT) Measurements. All IFT measurements were done using a University of Texas Model 300 Spinning Drop Interfacial Tensiometer. The basic principle is to introduce a drop (about 2 $\mu\ell$) of an oil sample into a glass capillary tube (1.5 mm I.D., 78 mm long) filled with the aqueous medium. The tube is then spun about its main axis. The oil drop will elongate to a length determined by the IFT value of the system. Details of the theory and application can be found elsewhere ([15,16]). According to the equipment manufacturer, the formula used to calculate IFT value is:

$$\text{IFT (dyne/cm)} = [1.234(\Delta d)^3 \Delta\rho]/p^2$$

where
Δd = the thickness of the elongated oil drop in cm;
$\Delta\rho$ = density difference between the oil and the aqueous phase in gm/ml;
p = period of spinning in seconds.

RESULTS AND DISCUSSION

The weight percentage breakdown of fractions and subfractions obtained from fractionation of both the crude oil and shale oil samples are shown in Figure 3 and 4, respectively. The percentage recoveries of Fraction III from the crude oil and shale oil samples were 16.5% and 24.1%, respectively. To investigate the interfacial activity of these subfractions upon reaction with alkali, IFT measurements were carried out with a 1% solution of each fraction in toluene against aqueous

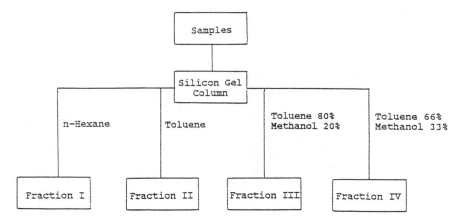

Figure 1. Solvent Fractionation Scheme Using Silicon Gel Column for Crude Oil and Shale Oil.

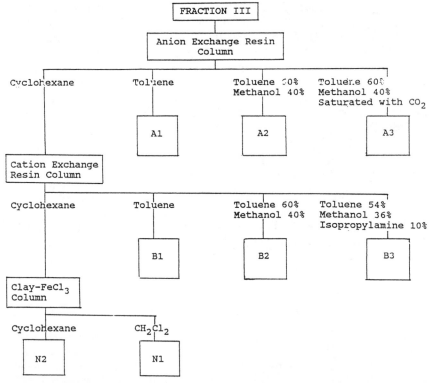

Figure 2. Separation of Fraction III (see Figure 1) to Subfractions by Ion Exchange Chromatography (A = Acid, B = Base, and N = Neutral Fractions).

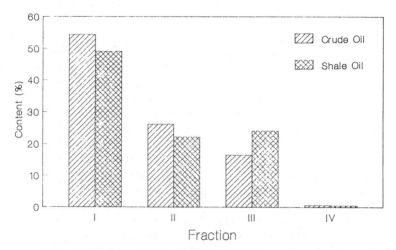

Figure 3. Content of Fractions of Crude Oil and Shale Oil by Silicon Gel Column.

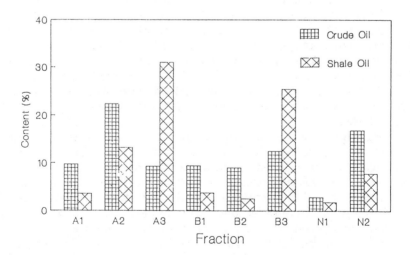

Figure 4. Content of Subfractions of Crude Oil and Shale Oil by Ion Exchange Chromatography (A = Acid, B = Base, and N = Neutral Fractions).

solutions of different sodium silicate contents. The results of these investigations are shown in Figures 5 through 7. As these results clearly indicate, the fractions show considerable interfacial activity under alkaline conditions. The separated shale oil fractions result in lower IFT values compared to the ones obtained from crude oil fractions. The lowest IFT and therefore highest surface activity is achieved for fraction A3 separated from shale oil with the lowest concentration of sodium silicate in the aqueous solution. This is very significant when considering that this fraction (A3) has the highest yield of recovery (31% of Fraction III, equal to 7.4% of the original shale oil sample) among the others separated from both the crude oil and shale oil samples. It is interesting to note that the same conditions were applied to crude oil; and the fraction A3 obtained from this sample exhibits a low value of IFT but it is more than 40 times higher than the one separated from the shale oil. The highest yield in this category, however, belongs to the fraction eluted from the anion exchange resin column with the toluene/methanol solvent system. Although many studies have been done in the past (17-20) to explain the interaction of alkali and acids in the oil for the lowering of IFT, the results of this study do not imply a direct relationship between acidity of the fraction and its surface activity. Jang et al. (21) reported that non-reactive, naturally occurring esters, amides and acid-base complexes were present in sufficient quantities when fractions of crude oil were isolated and characterized.

The results of experiments with crude oil fractions in this study also suggest that several species were present in reaction interface. There are mainly long chain carboxylic acids. The difference in size and structure is expected to give them different pka values. As a result, different surface activity (i.e., IFT value) is obtained with different levels of alkali concentration. Crude oil fractions with lower surface activity only yield surface inactive salts that may appear as precipitates at the interface.

The highest surface activity of fraction A3 extracted from shale oil needs to be explored in detail in order to understand this very unique phenomena. The benchmark experiments performed by Lee et al. (22) in studies of dissociation phenomena of Stuart oil shale in an alkaline environment proved the formation of carboxylic acids as it was verified from GC results. In another study by Lee et al. (23), it was shown that the hydroxyl ions from an alkaline solution could decompose the silicate and aluminasilicate structures in oil shale samples, provided that ultrasonic radiation and electrolytic current were simultaneously applied.

A hypothetical structural model developed by Yen (24) represented the organic components of Green River oil shale. The major components were isoprenoids, steroids, terpenoids and cartenoids. The common bridges consisted of disulfide, ether, ester, heterocyclic and alkadiene. Elemental analysis of typical oil shale samples has shown

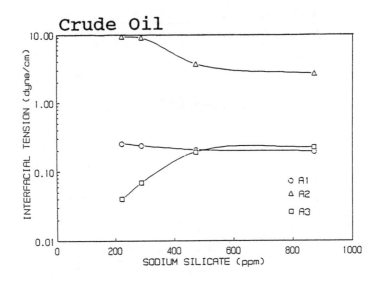

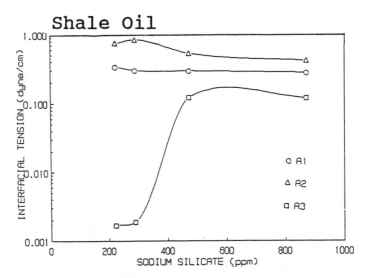

Figure 5. Interfacial Tension versus Alkali Concentrations for the Acid (A1 to A3) Fractions of   Crude Oil and   Shale Oil.

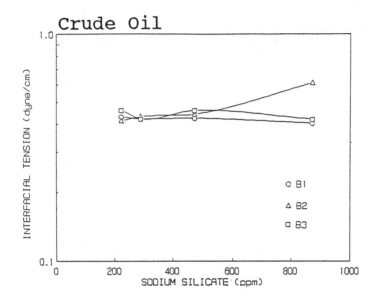

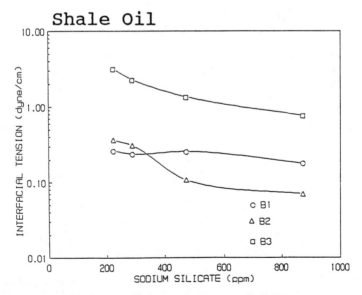

Figure 6. Interfacial Tension versus Alkali Concentrations for the Base (B1 to B3) Fractions of   Crude Oil and   Shale Oil.

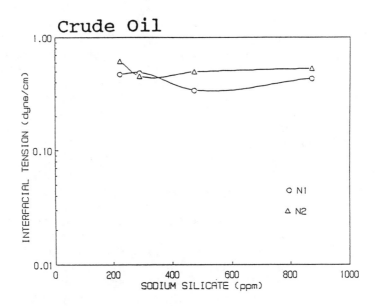

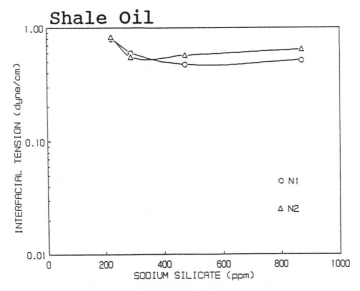

Figure 7. Interfacial Tension versus Alkali Concentrations for the Neutral (N1 and N2) Fractions of   Crude Oil and   Shale Oil.

that they are very rich in nitrogen compounds (30-84%). Poulson and co-workers (25) reported that pyridines and pyrroles were the two major types of nitrogen compounds in shale oil. Cyclic amides and anilides were proposed as possible additional types of nitrogen compounds in shale oil (26). The concentrating of the nitrogen-containing compounds in shale oil and related fossil fuels by clay-$FeCl_3$ complexation chromatography techniques was reported previously by Yen et al. (27,28). Based on these studies and our research in progress, we expect that enriched nitrogen compounds in oil shale could facilitate the acid extraction, mainly carboxylic acids, and subsequently increase the surface activity of the acid fraction (A3) extracted from the shale oil sample. A hypothetical model is proposed (Figure 8) to show the association of molecules at the oil-water interface. It is postulated that the surfactant molecules will be oriented in such a way that each molecule lies on the oil side of the interface and each nitrogen compound lies on the aqueous side. When the interfacial tension falls in the alkaline environment the surface pressure increases and the hydrocarbon chains of the surfactant molecules are prevented from moving close together because of the width of the nitrogen compounds. More research work is essential to structurally characterize this surfactant, but these results would clearly indicate that methods are to be developed to aid in the in-situ generation of bio-surfactants in oil shale processing which could significantly reduce the interfacial tension and lead to the ultimate recovery of shale oil.

CONCLUSION

Surface active compounds were extracted from both the crude oil and shale oil samples through a separation scheme developed for this work. These substances could effectively decrease interfacial tension between them and aqueous alkaline phase one hundred thousand fold (Figure 5). The surface activity of fractions derived from shale oil samples were much higher than the ones obtained from crude oil. The abundance of nitrogen compounds in shale oil samples was considered to be the main reason for reducing the interfacial tension to its lowest value. A hypothetical model was developed to describe the interfacial activity of the acidic fraction derived from shale oil. Experiments with this fraction led to the lowest IFT value compared to the other fractions. Since the lowering of interfacial tension is the first major step in enhancement of heavy oil recovery from both the shale oil and petroleum reservoir, it is strongly believed that application of those such surfactants, with their low cost and high stability, could lead to optimum recovery of residue oils from these reservoirs in an alkaline environment. Employment of such new methods as proposed by Lee et al. (22) ultrasonic radiation and electrolytic dissociation processes,

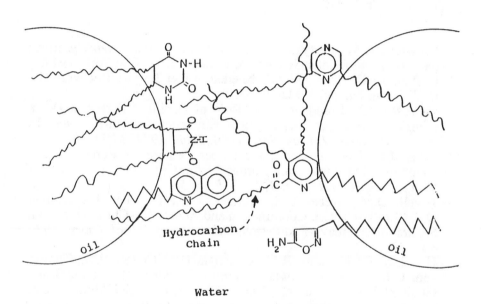

Figure 8. A Hypothetical Model of Shale Oil at the Oil-Water Interface.

could further enhance the generation of these surfactants and lead to the ultimate reserves recovery.

ACKNOWLEDGMENTS

The authors thank Western Extraction Technology, Inc. and EER Labs, Inc. for their partial financial support. We also would like to thank Dr. H. L. Wong and Mr. Leon Lemons for the preparation of this manuscript.

LITERATURE CITED

1. Schwartz, A.M.; Perry, J.W. *Surface Active Agents, Their Chemistry and Technology*; Inter-Science Publishers, Inc.: New York, 1949.
2. Geffen, T.M. "Here's What's Needed to Get Tertiary Recovery Going," *World Oil* 1975, March, 53.
3. Melrose, J.C.; Brandner, C.F. "Role of Capillary Forces in Determining Microscopic Displacement Efficiency for Oil-Recovery by Water Flooding," *J. Canadian Petrol. Tech.* 1974, *13(1)*, 13.
4. Taber, J.J. "Dynamic and Static Forces Required to Remove a Discontinuous Oil Phase from Porous Media Containing Both Oil and Water," *Soc. Petrol. Eng. J.* 1969, *9(1)*, 3.
5. Jewell, D.M.; Weber, J.H.; Bunger, J.W.; Plancher, H.; Latham, D.R. "Ion-Exchange, Coordination and Adsorption Chromatographic Separation of Heavy-End Petroleum Distillates," *Anal. Chem.* 1972, *44*, 1391.
6. Hirsch, D.E.; Hopkins, R.C.; Coleman, H.J.; Cotton, F.O.; Thompson, C.J. "Separation of High-Boiling Petroleum Distillates Using Gradient Elution Through Dual-Packed (Silica Gel-Alumina Gel) Adsorption Columns," *Anal. Chem.* 1972, *44*, 915.
7. Suatoni, J.C.; Swab, R.E. "Preparative Hydrocarbon Compound Type Analysis by High Performance Liquid Chromatography," *J. Chromatogr. Sci.* 1976, *14*, 535.
8. Seifert, W.K.; Howells, W.G. "Interfacially Active Acids in a California Crude Oil," *Anal. Chem.* 1969, *41*, 554.
9. Seifert, W.K. "Effect of Phenols on the Interfacial Activity of Crude Oil (California) Carboxylic Acids and the Identification of Carbazoles and Indoles," *Anal. Chem.* 1969, *41*, 562.
10. Seifert, W.K.; Teeter, R.M. "Preparative Thin-layer Chromatography and High Resolution Mass Spectrometry of Crude Oil Carboxylic Acids," *Anal. Chem.* 1969, *41*, 786.
11. Snyder, L.R. "Nitrogen and Oxygen Compound Types in Petroleum. Total Analysis of a 400-700∘F Distillate from a California Crude Oil," *Anal. Chem.* 1969, *41*, 314.

12. Snyder, L.R. "Nitrogen and Oxygen Compound Types in Petroleum. Total Analysis of an 850-1000°F Distillate from a California Crude Oil," *Anal. Chem.* 1969, *41*, 1084.
13. McKay, J.F.; Cogswell, T.E.; Latham, D.R. "Analytical Methods for the Analysis of Acids in High-Boiling Petroleum Distillates," *Prepr., Am. Chem. Soc., Div. Pet. Chem.*, 1974, *19*, 25.
14. Shah, D.O.; Schechter, R.S., Eds. *Improved Oil Recovery by Surfactant and Polymer Flooding*; Academic Press, Inc.: New York 1977.
15. Cayias, J.L.; Schechter, R.S.; Wade, N.H. *Adsorption at Interfaces*, ACS Symp. Series No. 8, 1975, 234.
16. Chan, M. "Interfacial Activity in Alkaline Flooding Enhanced Oil Recovery," Ph.D. Thesis, USC, Los Angeles, 1980.
17. Dunning, H.N.; Moore, J.W.; Denekas, M.O. "Interfacial Activities and Porphyrin Contents of Petroleum Extracts," *Ind. Eng. Chem.* 1953, *45*, 1759.
18. Bansal, V.K.; Chan, K.S.; McCallough, R.; Shah, D.O. "The Effect of Caustic Concentration on Interfacial Charge, Interfacial Tension and Droplet Size: A Simple Test for Optimum Caustic Concentration for Crude Oils," *J. Canadian Petrol. Tech.* 1978, *17(1)*, 69.
19. Chan, M.; Sharma, M.M.; Yen, T.F. "Generation of Surface Active Acids in Crude Oil for Caustic Flooding Enhanced Oil Recovery," *I & EC Process Des. Dev.* 1982, *21*, 580.
20. Chan, M.; Yen, T.F. "A Chemical Equilibrium Model for Interfacial Activity of Crude Oil in Aqueous Alkaline Solution: The Effects of pH, Alkali and Salt," *Canadian J. Chem. Eng.* 1982, *60*, 305.
21. Jang, L.K.; Sharma, M.M.; Chang, Y.I.; Chan, M.; Yen, T.F. "Correlation of Petroleum Component Properties for Caustic Flooding," In *Interfacial Phenomena in Enhanced Oil Recovery*; Wasan, D., Payatakes, A., Eds.; AIChE Symposium Series No. 212, 1982, *78*, 97.
22. Lee, A.S.; Sadeghi, M.-A.; Yen, T.F. "Characterization of the Stuart Oil Shale System. 1. New Method of Releasing Organic Matter," *Energy & Fuels* 1988, *2*, 88.
23. Lee, A.S.; Lian, H.J.; Yen, T.F. "Dissociation of Organic and Mineral Matrix of Maoming Oil Shale at Low-Temperature and Ambient-Atmosphere," *Proc. Int. Conf. on Oil Shale and Shale Oil*, May 16-19, 1988, Chemical Industry Press, Beijing, China, 112.
24. Yen, T.F. "Structural Investigations on Green River Oil Shale Kerogen," In *Science and Technology of Oil Shale*; Yen, T.F., Ed.; Ann Arbor Science, Michigan, 1976, 193.
25. Poulson, R.E.; Frost, C.M.; Jensen, H.B. "Characteristics of Synthetic Crude from Crude Shale Oil Produced by In-situ Combustion Retorting," In *Shale Oil, Tar Sands, and Related Fuel Sources*; Yen, T.F., Ed; ACS Adv. Chem. Ser. No. 151, 1976, 1.

26. Poulson, R.E.; Jensen, H.B.; Cook, G.L. "Nitrogen Bases in a Shale-Oil Light Distillate," *Prepr., Am. Chem. Soc. Div. Pet. Chem.,* 1971, *16(1)*, A49.
27. Yen, T.F.; Shue, F.F.; Wu, W.H.; Tzeng, D. "Ferric Chloride-Clay Complexation Method. Removal of Nitrogen-Containing Components from Shale Oil and Related Fossil Fuels," In *Geochemistry and Chemistry of Oil Shales*; Miknis, F.P., McKay, J.F., Eds.; ACS Adv. Sym. Ser. No. 230; 1983, 457.
28. Shue, F.F.; Yen, T.F. "Concentration and Selective Identification of Nitrogen- and Oxygen-Containing Compounds in Shale Oil," *Anal. Chem.* 1981, *53*, 2081.

RECEIVED January 26, 1989

Chapter 21

Microscopic Studies of Surfactant Vesicles Formed During Tar Sand Recovery

Mohammad-Ali Sadeghi, Kazem M. Sadeghi, Dawood Momeni, and Teh Fu Yen

School of Engineering, University of Southern California, Los Angeles, CA 90089–0231

By applying ultrasonic energy and using an alkaline solution, self-propagating surfactants are formed. The principle of membrane-mimetic chemistry via the process mechanism is being explored in this work by the thorough examination of the photomicrographs. Giant-sized multi-lamellar surfactant vesicles are analyzed under both transmittance and reflectance light during their initial formation and, after a few hours, four months, and three years following the completion of the process. From these studies, the optimum time period needed for the vesicles to stabilize is determined. The vesicles are proven to be very stable under the slow reaction condition of six-hour process time. With free radical initiator added (e.g., H_2O_2), the reaction takes minutes to complete. Based on the surfactant vesicles characterization derived from current investigations, it is suggested that these surfactants could retain their effectiveness necessary for bitumen recovery in the reservoir environment for a number of years.

Tar sands, also called bituminous sands or oil sands, are essentially siliceous materials such as sands, sandstones or diatomaceous earth deposits impregnated with 5 to 20 percent by weight of a dense, viscous, low gravity bitumen. Deposits of tar sands exist and are widely distributed throughout the world. Wen et al. ([1]) pointed out that bitumens, increasing in the order of the size and complexities, consist of the following fractions: (1) gas oil, (2) resin, (3) asphaltene, (4) carbene, and (5) carboid. The chemical and physical properties of tar sands depend significantly on the relative amounts of each fraction and their properties. The values of molecular weight, number of condensed aromatic rings, number of heteroatoms (N, O, S) and concentration of

0097–6156/89/0396–0391$06.00/0
© 1989 American Chemical Society

metals (V, Ni, Ti, etc.) all generally increase from the gas oil fraction to the preasphaltene fraction (carbene and carboid). Generally, there are three types of hydrocarbons present in bitumens; paraffinic, naphthenic, and aromatic, although many of the molecules are combinations of the three types, especially the heavier fractions.

Self-generated surfactants (produced from fossil fuels by a chemical/physical process) or natural surfactants (exiting in fossil fuels) are derived from the inherent organic acids and replaceable acidic protons which are present in crude oils or bitumens (e.g., mercaptans). Yen and Farmanian (2) isolated native petroleum fractions that form surfactants and contain hydrogen displacable components including one, two, three, or four of the following types:

$$R-SH$$

$$R-NH(R_1)$$

$$R-OH$$

$$R-COOH$$

in which R is a hydrocarbon of 3 to 20 carbon atoms, R_1 is hydrogen or a hydrocarbon of 3 to 20 carbon atoms, and R and R_1 are either separate or cyclically combined. These acids, when coming into contact with alkaline solutions, yield in-situ surface active materials or self-generated surfactants of various forms of liquid crystals (e.g., giant micelles, large vesicles, etc.). These self-generated surfactants reduce the interfacial tension between the crude oil hydrocarbon and water. In the tar sand production processes in operation in Alberta, Canada, caustic additives (e.g., NaOH) are used to improve recovery rates by about 15%. Bowman (3) isolated some of the dissolved organics by foam fractionation techniques and showed that the materials were surfactants. A similar discovery was found by Ali (4), where a tentative mechanism was given as the following:

$$R - S - R' + O_2 \longrightarrow R-SO - R'$$
$$\downarrow O_2 + H_2O$$
$$RSO_xH + R'COOH$$

(x = 2 or 3; sulphenic or sulphonic acids). He concluded from infrared studies that the two functional groups of carboxylic acids and sulphonic acids were on the same molecule, in other words, that the parent sulphide was a cyclic one. The study on the reactions initiated by ultrasonic energy in the U.S.A. dates back to 1927, which is about the same period for the research on the reaction initiated by ionizing radiation (5). The mechanisms, responsible for the observed increases in rates in transport and unit operation processes utilizing ultrasonic energy, can be divided into two categories; (1) First-order effects of fluid particles (acceleration, displacement, and velocity) and (2) Second-order phenomena (cavitation, radiation pressure, acoustic streaming, and interfacial instabilities). Mostly, it is one or more of the second-order effects which are responsible for the enhancements in the transport process (6). The use of ultrasonic energy in conjunction with self-generated surfactants to recover bitumen from tar sand is a new extraction technique having a 95% recovery and reaction times that are in minutes (7,8). Chan et al. (9,10) proposed a simple equilibrium chemical model for the surface reaction at heavy crude oil-caustic interfaces. Each acid species dissociates at certain pH value, referred to as the onset pH or pK_a value, and yields the surface active anion. The dissociation constant of each species depends on its molecular size and structure. For crude oil, being such a complex mixture of organic acids, the pK_a values range from 8 to 11 (11). Some other components in crude oil besides fatty acids have been reported as being interfacially active (12,13). It was reported that compounds of zinc, copper, nickel, titanium, calcium, and magnesium were found to be adsorbed selectively to crude oil-water interfaces. All these metals occur in petroleum as porphyrin-metal chelates, or other types of complexes or even possibily form non-nitrogen binding complexes. The interfacial films were thought to consist of waxes and resins with stabilizing porphyrin-metal complexes, free porphyrins, porphyrin oxidation products and protein-metal salts or complexes.

The concept of differing pK_a values for the different acidic species has been used to suggest a method of enhancement for generation of self-propagating surfactants formed during the bitumen extraction process (Yen, T.F., USC, Unpublished data). By the application of ultrasonic energy and using an alkaline solution, self-propagating surfactant vesicles are formed from the natural surfactant vesicles which have onion-like multicompartment structures. Prolonged ultrasonication of these results in the formation of submicroscopic spheroidal vesicles consisting usually of a fairly uniformed single-compartment bilayer surfactant vesicle surrounding a solvent core (14). The reaction can be reversible. Many unilamellar vesicles can coalesce together and form a giant multilamellar vesicle in which the liquid-crystal feature can be observed in microscopic range using a polarized light source and analyzer. To verify the existence of self-generated surfactants during tar sand recovery process and to analyze their structural behavior, the vesicles are studied in this work under transmittance and reflectance light of the microscope. The principles of membranemimetic chemistry are employed to describe the process phenomena. The vesicles' stability for a prolonged period of time is carefully studied and the duration for these surfactants which can withstand the severity of the

reservoir environmental conditions to effectively enhance the bitumen
recovery efficiency from tar sand deposits is evaluated.

EXPERIMENTAL

Tar sand samples mined from Athabasca, in Alberta, Canada, were
High Grade Athabasca Tar Sand, Sample No. 1981HG. The bitumen
present in the tar sand was 14.5% by weight as determined by Soxhlet extraction (15). A quantity of 100 grams of tar sands was added
to a 800 ml of alkaline solution (20:1 by volume distilled water to
sodium silicate) and sonicated for several hours until all the bitumen
had dissipated in the solution (sand was bitumen-free). The sodium silicate reagent (Na_2SiO_3) was BJ-120 (wt. ratio $SiO_2/Na_2O = 1.80$)
from the PQ Corporation, Huntington Beach, CA. A 10-gallon trans-
ducerized tank (Branson Model ATH610-6) and a companion generator
(Branson Model EMA 0-6) having six piezo- electric transducers with
a 40 kHz frequency was used. We have reported the detailed experimental procedures in previous publications (7,8,15,16). As reported
before (8), the reaction takes minutes to recover bitumen with the aid
of radicals (e.g., hydrogen peroxide, benzoyl peroxide). The slow reaction kinetics (no radicals added) were used to isolate the micelles
formed during the process, for analysis under the microscope.

The surfactants formed in the tar sand recovery were examined using light microscopic techniques. A representative sample was
pipeted during sonication and immediately taken for observation under
a Leitz SM-Lux-Pol Cross Polarizing Microscope. One drop of each
sample to be examined was placed over a clean glass slide and re-
strained with a cover lid. Photomicrographs were then taken (equipped
with a high speed camera and color film using WILD MP515 Semipho-
tomat) of both reflection polarized light and transmittance of the inci-
dent light upon the sample. Air tight amber vials were used to store
the processed solution for the given durations of twelve hours, four
months and three years, before examining the surfactants under the
microscope.

RESULTS AND DISCUSSION

The heavy-end portions (usually called heavy fractions) of bitumen
(e.g. asphaltenes, preasphaltenes) can exist both in a random oriented
particle aggregate form or in an ordered micelle form, peptized with
resin molecules (16,17). In their natural state, asphaltenes exists in
an oil-external (Winsor's terminology) or reversed micelle. The polar
groups are oriented toward the center, which can be water, silica (or
clay), or metals (V, Ni, Fe, etc.). The driving force of the polar groups

assembled toward the center originates from either hydrogen-bonding, or charge transfer, or even acid-base salt formation.

In crude oil, asphaltene micelles are present as a discrete or colloidally dispersed particles in the oily phase. As the various intermediate and low boiling fractions are removed during the distilling or evaporation process, the particles of asphaltene micelles are massed together to form the larger colloidal sizes (Yen, T. F., "Asphaltic Materials," Encyclopedia of Polymer Science & Engineering, 2nd Ed., John Wiley, in press). The so-called precipitation or aggregation processes are colloidal in nature (Figure 1). Single sheet, usually the pi-system with saturated substitutes, is associated into stacks or piles of 4-5 layers. These "stacks" are commonly referred to as "asphaltenes", which are precursors of micelles (18,19) (either reverse, random-oriented or normal) and the supermicelle. The accompanied transformation of liquid crystal-gel stage or the floc stage is widely observed in industrial practices. Actually, investigators have isolated supermicelles, particles, flocs, etc. and performed many experimental determinations, including molecular weights which range from a few thousand up to hundreds of millions. They all attribute this discrepancy to "asphaltene", which may be misleading for other investigators. One knows for certain that the single sheet of asphaltene is under 1000 in molecular weight. Often people believe this is the controversy of Yen's model (20) of asphaltene, yet clearly all findings, including the most recent ones (21,22), can always be interpreted by that simple model.

It is essential to state that the heavy fractions such as asphaltene and preasphaltene do contain large numbers of polar molecules (23,24). These polar molecules behave exactly as surfactants or amphiphiles (asphaltene usually contains a long-chain substituent (25)). We again have to emphasize that it is almost not possible to create a colloidal micelle from pure hydrocarbon and water without any surfactant. Hence, we conclude to say that asphaltene or asphaltene-like molecules (asphaltics) will participate in a manner according to membrane-mimetic chemistry.

Surfactant molecules can be considered as building blocks for certain forms of geometry in colloidal chemistry. Various forms of association molecules can be obtained as the concentration of surfactant in water is increased and/or physicochemical conditions are changed (e.g. CMC, Craft-point, etc.). Figure 2 schematically shows the most likely structural configurations and assemblages of surfactants association in an aqueous system (26). Upon addition of oil and a short-chain alcohol, for example, one can convert the oil-in-water micelles into water-in-oil microemulsions. It is therefore possible to induce a transition from one structure to another by changing the physicochemical conditions such as temperature, pH and addition of mono or di-valent cations to the surfactant solution. It should be also noted that the sur-

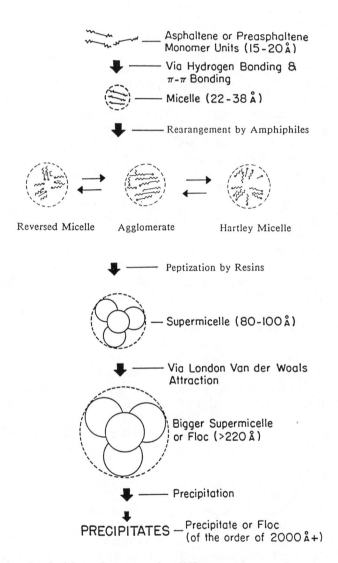

Figure 1. Asphaltic substances in different stages of a process. The size and weight depend on the environment [modified from Ref. 27 and others].

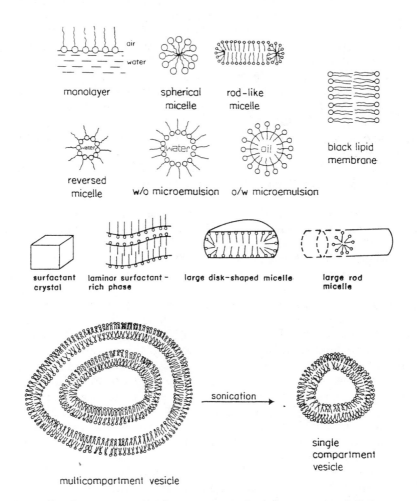

Figure 2. Some organized structures of surfactants in different media and the effect of sonication to vesicles [modified from Ref. 26 and others].

factant may skip several phase transitions depending upon its structure and the physicochemical conditions.

In an alkaline environment (pH greater than 10), for a membrane-mimetic system, the interaction of cations (e.g., sodium, potassium) with the peptized resin molecules acts in a membrane mimetic fashion. The cation reacts with the acid-bearing functional group to develop a host-guest cavity in a cylindrical shape. The counter anion (e.g., hydroxide, silicate) is activated and dissolves in the oil phase. In this manner, the resin molecule containing the heterocycle center is dissociated and any ionizable proton such as in -COOH, -SH, or -NH is replaced with the cation. The self-generated surfactant molecules are produced. When this self-generated surfactant migrates into the naturally reversed micelles (oil external-water internal) of asphaltene, it disrupts the polar structure to form a Hartley micelle (polar-external) as shown in Figure 3. This is termed as the "micellar inversion process." The counter anions emulsify the oil and the micellar structure becomes a microemulsion stabilized by the self-generated surfactant molecules (7,15).

In the micellar inversion process, the heavy fractions (asphaltenes and preasphaltenes) are separated from the bitumen. This separation results, as the micelles containing the less polar or lighter hydrocarbon components and the individual molecules breaking away from the micelle, moving to the surface of the aqueous phase and coalescing to form an oil layer. After the separation of lighter components, the higher polar molecules form asphaltenes and preasphaltenes micelles strongly reassociated and complexed with metals to form colloidal-size flocs and precipitates. As Briggs et al. (27) indicated that the preasphaltene fraction contains an aliphatic to aromatic hydrogen ratio of 3.6, and the presence of alkyl side chains and the larger molecules by weight would attribute to the large size of colloidal particles. These particles are formed by inter-molecular association of 3 to 5 polyaromatic molecules (single sheet) of asphaltene and/or preasphaltene. Most of these colloidal particles are spherical with diameters in the range of 22-38 angstroms. The ultrasonic vibration energy effect is expected to accelerate the aggregation of these colloidal particles through the orthokinetic flocculation mechanism. Four of the 22-38 angstroms particles could agglomerate to form a supermicelle or floc with a diameter in excess of 220 angstroms (see Figure 1). According to many indications, such as x-ray small-angle scattering data (27-29), supermicelle structures with dimensions in excess of 1000 angstroms would precipitate from the solution.

The purpose of employing various light incidents when using a microscope was to distinguish between different crystals such as mineral crystals (e.g., clays, silicons, NaOH, etc.) from surfactant crystals (30-33). It was found that under transmittant light only bituminous hydrocarbons were non-transparent (absorb light) and appeared either

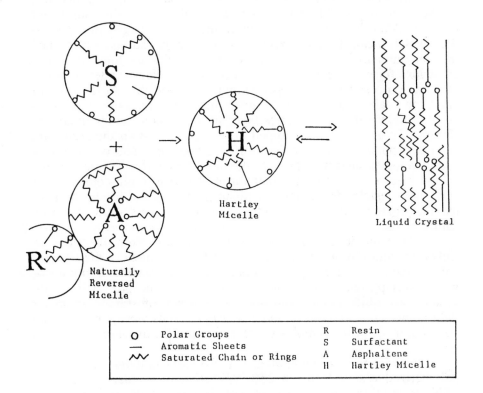

O	Polar Groups	R	Resin
—	Aromatic Sheets	S	Surfactant
Ꮗ	Saturated Chain or Rings	A	Asphaltene
		H	Hartley Micelle

Figure 3. Dynamic equilibrium of asphaltene micelle inversion process.

black, orange and/or yellow in photomicrographs. This is the stage of reaction in which the removal of hydrocarbon contaminants is suspected to take place. As shown in Figures 4a (transmittance) and 4b (reflectance), asphaltic micelles (isotropic black circles) are going through the micellar inversion process (34).

At the completion of above reaction (after the 6th hour of sonication) another batch of solution was withdrawn for observation. Figure 4c shows the emulsions that have formed during the reaction. These emulsions were all optically active under polarized light (reflectant) as shown in Figure 4d. This was interpreted as a sign for surfactant activity. Also there were no asphaltic micelles observed in the solution, which meant total removal of contaminants had taken place and the contaminants had precipitated as solid asphaltic agglomerates. The emulsions formed during the reaction started to coalesce a few minutes after the processing had stopped (Figures 5a and 5b). At the time of emulsion coalescence, many surfactant molecules rearranged in an orderly fashion and formed a thick layer surrounding some oil emulsions (Figures 5c and 5d). The trapped oil emulsions within the surfactant layer continuously burst and their surfactant molecules moved outwards adding another surfactant layer to the outer boundary. This is the suspected mechanistic sequence that forms the large multi-compartment surfactant vesicles from single-compartment vesicles during the sonication process. The optical activity of these layers shows the true nature of the surfactant molecules being highly organized.

These membrane-mimetic agents are also osmotically active as shown in Figure 6 (4 months later). Notice the centers of the vesicles are completely covered with the upgraded bitumen without any air or water gaps present as compared to Figures 5c and 5d. Also, these vesicles are quite stable even after three years of storage in an air tight amber vial (Figures 7a and 7b). Amber vials were used to avoid any photolysis of the samples which might initiate unsuspected radical reactions.

In order to determine whether these surfactant vesicles were of polymerized vesicle forms, a 25% V/V ethanol (standard grade) was added to the three year old sample solution. Alcohols are known (34) to destroy surfactant vesicles derived from natural phospholipids, however, synthetically prepared polymerized vesicles are stable in as much as 25% (V/V) alcohol addition. Photomicrographs shown in Figures 7c and 7d indicate that these vesicles partially retain their stability (being mesomorphic) and therefore are suspected to be polymerized surfactants. Whether surfactant molecules of these vesicles are single or multipla bonds in tail, or in head groups remains to be seen.

As indicated by Puig et al. (35), surfactant retention and attendant pressure buildup in the rock can be greatly reduced if the surfactant dispersion is converted into the liquid crystalline state. Unilameller vesicles are preferred in the field work rather than the multilamellar

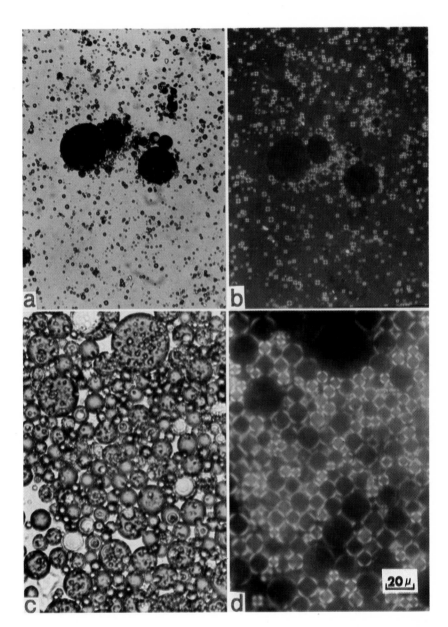

Figure 4.   Bitumen removal during 3rd/4th hour of the processing [(a) transmittance and (b) reflectance]. Emulsification phenomena after completion (6th hour) of the processing [(c) transmittance and (d) reflectance].

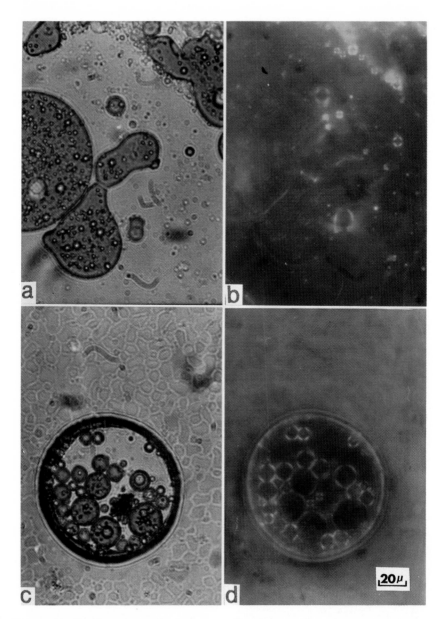

Figure 5. Coalescence phenomena a few hours after completion of the processing [(a) transmittance and (b) reflectance]. Multilamellar surfactant vesicle development phenomena twelve hours after the completion of the processing [(c) transmittance and (d) reflectance].

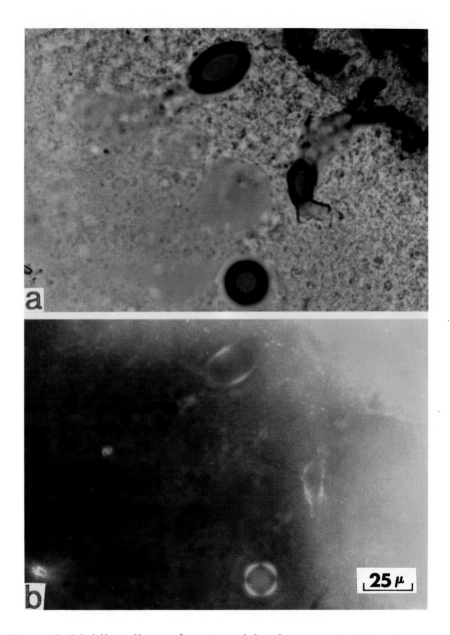

Figure 6. Multilamellar surfactant vesicles four months following the processing [(a) transmittance and (b) reflectance].

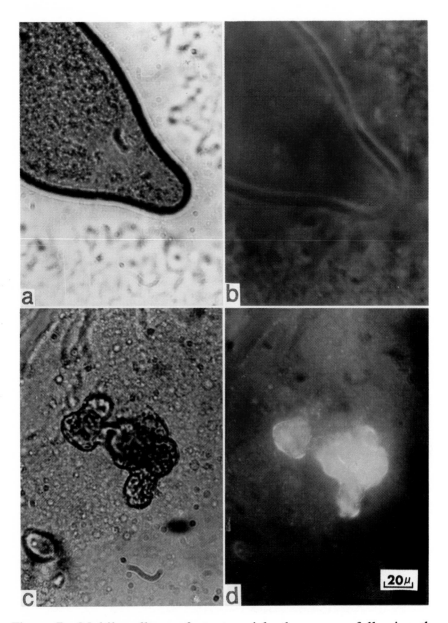

Figure 7. Multilamellar surfactant vesicle three years following the processing [(a) transmittance and (b) reflectance]. Partial disruption of vesicle in 7a and 7b, upon addition of 25% V/V of ethanol [(c) transmittance and (d) reflectance].

vesicles. Other than the use of sonication, different chemical additives can be equally efficient in the application of vesicles as surfactant delivery agents for enhanced oil recovery or decontamination in soil.

CONCLUSION

Benchmark studies demonstrated that the application of ultrasonic energy in an alkaline solution media (i.e. pH > 10) will induce generation of self-propagating surfactants which could effectively extract bitumen from tar sand deposits. The photomicrographs taken by the application of both the transmittance and reflection of polarized light were carefully analyzed and it was concluded that the surfactant vesicles generated during and after the process completion were quite stable even after three years of storage. This indicates that these surfactant vesicles were polymerized during the processing due to the presence of free radicals, whether generated by sonication or added as initiators (7,8). The membrane-mimetic chemistry (34) concepts were applied to thoroughly examine the reaction pathways of the process by the microscopic studies. The overall studies did indicate clearly that the development of such a process in a tar sand reservoir environment could similarly lead to generation of multiple-layered or multi-compartment surfactant vesicles and subsequently to the ultimate in-situ bitumen recovery. In a similar manner, emulsification, transportation, upgrading, production, recovery, etc., of tar sand and related heavy oil industries can be benefited.

ACKNOWLEDGMENTS

The authors thank Energy and Environment Research Laboratories, Inc. for partial financial support. We want to acknowledge Dr. Jih-Fen Kuo of Groundwater Technology, Inc. and Dr. John G. Reynolds of Lawrence Livermore National Lab for technical discussions. We also thank Dr. H.L. Wong and Mr. Leon Lemons for the preparation of this manuscript.

LITERATURE CITED

1. Wen, C. S.; Chilingarian, G. V.; Yen, T. F. "Properties and Structure of Bitumens," *Bitumens, Asphalts, and Tar Sands*; Chilingarian, G. V.; Yen, T. F., Eds.; Elsevier Publishing Company: New York, 1978, 7, p 155.
2. Yen, T. F.; Farmanian, P. A. "Native Petroleum Surfactants," U.S. Patent 4 411 816, 1983.

3. Bowman, C. W. "Molecular and Interfacial Properties of Athabasca Tar Sands," *Proceedings of the Seventh World Petroleum Congress*; Elsevier Publishing Company: New York, 1967, *3*, p 583.

4. Ali, L. H. "Surface-Active Agents in the Aqueous Phase of the Hot-Water Flotation Process for Oil Sands," *Fuel*, 1978, *57*, 357.

5. Chendke, P. K.; Fogler, H. S. "Second-Order Sonochemical Phenomena. Extensions of Previous Work and Applications in Industrial Process," *Chem. Eng. J.* 1974, *8*, 165.

6. Frederick, J. R. *Ultrasonic Engineering*; John Wiley and Sons: New York, 1965.

7. Sadeghi, M.-A.; Sadeghi, K. M.; Kuo, J. F.; Jang, L. K.; Yen, T. F. "Treatment of Carbonaceous Materials," U.S. Patent 4 765 885, 1988.

8. Kuo, J. F.; Sadeghi, K. M.; Jang, L. K.; Sadeghi, M.-A.; Yen, T. F. "Enhancement of Bitumen Separation From Tar Sand by Radicals in Ultrasonic Irradiation," *Appl. Phys. Comm.* 1986, *6(2)*, 205.

9. Chan, M.; Sharma, M. M.; Yen, T.F. "Generation of Surface Active Acids in Crude Oil for Caustic Flooding Enhanced Oil Recovery," *Ind. Eng. Chem. Process Des. Dev.* 1982, *21*, 580.

10. Chan, M. "Interfacial Activity in Alkaline Flooding Enhanced Oil Recovery," Ph.D. Dissertation, USC, Los Angeles, 1980.

11. Chan, M.; Yen, T.F. "Role of Sodium Chloride in the Lowering of Interfacial Tension Between Crude Oil and Alkaline Aqueous Solution," *Fuel*, 1981, *60*, 552.

12. Dunning, H. N.; Moore, J. W.; Denekas, M. O. "Interfacial Activities and Porphyrin Contents of Petroleum Extracts," *Ind. Eng. Chem.* 1953, *45*, 1759.

13. Dodd, C. G.; Moore, J. W.; Denekas, M. O. "Metallifeorus Substances Adsorbed at Crude Petroleum-Water Interfaces," *Ind. Eng. Chem.*, 1952, *44*, 2585.

14. Huang, C. "Studies of Phosphatidylcholine Vesicles. Formation and Physical Characteristics," *Biochemistry*, 1969, *8*, 344.

15. Sadeghi, K. M.; Sadeghi, M.-A.; Kuo, J. F.; Jang, L. K.; Yen, T. F. "Self-Propogated Surfactants Formed During Separation of Bitumen From Tar Sands Using an Alkaline Solution and Sonication," Presented at ACS, 1987 Pacific Conference on Chemistry and Spectroscopy, October 28-30, 1987.

16. Sadeghi, M.-A.; Jang, L. K.; Kuo, J. F.; Sadeghi, K. M.; Palmer, R. B.; Yen, T. F. "A New Extraction Technology for Tar Sand Production," *The 3rd UNITAR/UNDP Inernational Conference on Heavy Crude and Tar Sand*; AOSTRA: Alberta, 1988, p 739.

17. Yen, T. F. "The Role of Asphaltene in Heavy Crude and Tar Sands," *The Future of Heavy Crude and Tar Sands*; McGraw Hill: New York, 1981, p 174.

18. Dickie, J. P.; Haller, M. N.; Yen, T. F. "Electron Microscopic Investigations on the Nature of Petroleum Asphaltics," *J. Colloid Interface Sci.*, 1969, *29(3)*, 475.

19. Yen, T. F. "Structural Differences Between Asphaltenes Isolated from Petroleum and from Coal Liquid," *Chemistry of Asphaltenes*; Bunger, J. W.; Li, N. C., Eds.; Advances in Chemistry Series No. 195; American Chemical Society: Washington, D.C., 1981, p 39.
20. Yen, T. F. "Present States of the Structure of Petroleum Heavy Ends and Its Significance to Various Technical Applications," *ACS, Div. Petrol. Chem., Prepr.* 1972, *17(4)*, F102.
21. Ravey, J. C.; Duconret, G.; Espinat, D. "Asphaltene Macrostructure by Small Angle Neutron Scattering," *Fuel*, 1988, *67*, 1560.
22. Kotlyar, L. S.; Ripmerster, J. A.; Sparks, B. D.; Woods, J. "Comparative Study of Organic Matter Derived from Utah and Althabasca Oil Sands," *Fuel*, 1988, *67*, 1529.
23. Tissot, B. P.; Welte, O. H. *Petroleum Formation and Occurrence*; Springer Verlag: New York, 1984, 2nd Ed.
24. Yen, T. F. "Chemical Aspects of Interfuel Conversion," *Energy Sources*, 1973, *1*, 117.
25. Yen, T. F. "Long-Chain Alkyl Substituents in Native Asphaltic Molecules," *Nature, Phy. Sci.*, 1971, *233*, 36.
26. Fendler, J. H., "Microemulsions, Micelles, and Vesicles for Membrane Mimetic Photochemistry," *J. Phys. Chem.*, 1980, *84*, 1485.
27. Briggs, D. E.; Addongton, D. V.; Mckeen, J. A. *Coal IV*, CEP Technical Manual, 1978.
28. Pollack, S. S.; Yen, T. F. "Structural Studies of Asphaltics by X-Ray Small Angle Scattering," *Anal. Chem.*, 1970, *42*, 623.
29. Dwiggins, C. W., Jr. "A Small Angle X-Ray Scattering Study of the Colloidal Nature of Petroleum," *J. Phys. Chem.*, 1965, *69*, 3500.
30. Hartshorne, N. H.; Stewart, A. *Crystals and the Polarizing Microscope*; Arnold: London, 1970.
31. Hartshorne, N. H. *The Microscopy of Liquid Crystal*; Microscope Publications Ltd: London, 1974.
32. Tiddy, G. J. T. "Surfactant-Water Liquid Crystal Phases," *Physics Reports (Review Section of Physics Letters)* 1980, *57(1)*, 1.
33. Franses, E. I.; Talmon, Y.; Scriven, L. E.; Davis, H. T.; Miller, W. G. "Vesicle Formation and Stability in the Surfactant Sodium 4-(1'- Heptylnonyl) benzenesulfonate," *J. Colloid Interface Sci.*,1982, *86(2)*, 449.
34. Fendler, J. H. *Membrane Mimetic Chemistry*; Johy Wiley & Sons: New York, 1982.
35. Puig, J. E.; Franses, E. I.; Yeshayahu, T.; Davis, H. T.; Miller, W. G.; Scriven, L. E. "Vesicular Dispersion Delivery Systems and Surfactant Waterflooding," *Soc. Petrol. Eng. J.*, 1982, February, 37.

RECEIVED January 26, 1989

Chapter 22

Modifications in the Composition of Crude Oils During In Situ Combustion

A. Audibert and J. Roucaché

Institut Français du Pétrole, B.P. 311, 92506 Rueil-Malmaison, Cedex, France

During enhanced oil recovery by in-situ combustion, crude oil undergoes chemical and physical changes. In in-situ combustion laboratory tests, air injection was stopped to interrupt the reactions. The organic matter sampled ahead of the burnt zone was analyzed using an analytical procedure specifically designed to characterize the evolution of the composition of the crude oil. The coke deposit was characterized by Infrared Spectroscopy and Oil Show Analyzer. The residual oil and the produced oil samples were characterized by SARA Analysis. The interpretation of tests involving crude oils with different geochemical compositions shows the possible influence of the crude oil composition on the amount of coke deposit and on its ability to undergo in-situ combustion. The results which provide valuable information for numerical simulation of in-situ combustion, concern not only the coke deposit (amount, composition, oxygen reactivity) but also the organic matter sampled ahead of the combustion zone (composition, coke precursors) and the produced oil.

During enhanced oil recovery by in-situ combustion, a crude oil undergoes chemical changes (pyrolysis reactions) and physical changes (dilution by the cracking products, vaporization and condensation of some fractions). Both phenomena are important for oil production :
- easier and higher recovery due to the change in oil viscosity,
- influence of the amount of coke deposit on the propagation of the combustion process.

0097–6156/89/0396–0408$06.00/0
© 1989 American Chemical Society

The cracking and the low-temperature oxidation of crude oils have been studied previously in order to simulate the thermal transformations of oil to gas and coke during enhanced oil recovery (1-6). Other authors characterize the thermal modifications of oil in the presence of a vapor phase (7).

The objective of this work is to study the possible influence of the crude oil composition on the amount of coke deposit and on its ability to undergo in-situ combustion. Thus, the results would provide valuable information not only for numerical simulation of in-situ combustion but also to define better its field of application. With this aim, five crude oils with different compositions were used in specific laboratory tests that were carried out to characterize the evolution of the crude oil composition. During tests carried out in a porous medium representative of a reservoir rock, air injection was stopped to interrupt the reactions. A preliminary investigation has been described previously (8).

EXPERIMENTAL

Oils properties.

Five crude oils with different geochemical compositions have been studied. The different properties of the oils are listed in Table I. These oils can be classified under different categories :
  . Oil A (Paris Basin) – from a marine origin which is already altered and will not be further transformed.
  . Oil B (Rumania) – from the class of naphtenic-aromatic crude oils. This oil will need only a small quantity of energy to be chemically modified to form lighter hydrocarbons.
  . Oils C (Boscan), D (Cerro Negro), E (Athabasca) – from the class of asphaltenic-aromatic crude oils, have a high sulfur content. The first two oils coming from carbonate source rocks contain polar compounds consisting of very stable polycyclic aromatics. On the other hand, the last oil contains aromatics which are less condensed and more reactive.

Procedure.

The combustion cell, which is 2.1 m long and 20 cm in diameter and the procedure have been described previously (9).

The fluids produced are regularly sampled during the propagation of the combustion front. The oil samples are separated and analyzed according to the analytical procedure detailed in the next paragraph. The air injection is generally stopped when the combustion front has propagated along the first half of the length of the porous medium. After a complete cooling of the combustion cell under rotation, samples are taken at different points of the porous medium, generally at each thermocouple and at half distance between two successive thermocouples. All the coke zone which is observed downstream from the combustion front is sampled.

TABLE I – OIL PROPERTIES

Oil	ρ kg/m^3	μ mPa.S	Saturates %	Aromatics %	Resins %	Asphaltenes %	Sulfur % wt
A	890 (15°C)	40 (30°C)	55.2	34	10.7	0.1	0.3
B	960 (20°C)	2000 (18°C)	47.4	23.6	28.8	0.3	0.2
C	1005 (15°C)	1300 (60°C)	15.5 (210$^+$)	43.0	29.5	12.0	5.3
D	1007 (15°C)	7092 (60°C)	17.3	32.1	36.8	13.8	4
E	1000 (20°C)	2150 (50°C)	21.5	36.5	30.8	8.1	4.3

Analytical procedure.

An analytical procedure (Fig. 1) has been improved to characterize both the residual organic matter in the sands and the oil recovered during the tests. For example, the analysis of samples by the Oil Show Analyzer (OSA, IFP Fina process, commercialized by Delsi Instruments-France) gives the organic content of rocks (10) and permits the study of the evolution in the oil composition. This analysis (Fig. 2) is based on a two-step procedure involving successively (a) pyrolysis from ambient temperature up to 600°C, and (b) combustion in air at 600°C. The first peak corresponds to the detection of the gaseous hydrocarbons up to C_7, volatilized at about 90°C during 2 minutes (gas amount S0 in g/100 g of rock), the following to the liquid hydrocarbons, volatilized at 300°C during 3 minutes (oil amount S1); the hydrocarbons released during the cracking of the residual organic matter from 300°C to 600°C form the peak S2. The detection of the CO_2 produced by combustion in air at 600°C during 5 minutes of the previously pyrolyzed sample leads to the peak S4 and allows the residual organic carbon content to be determined. The different product contents are generally given in g per 100 g of rock. Here, the different product contents are also given versus the total carbon content (for example S0/S0+S1+S2+S4) to eliminate the effect of oil saturation.

The oil extracted from the sand samples or the produced oil is studied by thin layer chromatography, which gives the distribution of the different structural groups : saturates, aromatics, resins, asphaltenes (SARA). If changes are observed in the oil composition, the different fractions are further analyzed by gas chromatography on capillary columns with a specific detector for each group (Flame ionisation detector for aromatics, flame photometry detector for thiophenic compounds and thermoionic specific detector for nitrogen compounds). After extraction of the liquid oil phase, the residual organic matter is isolated from the mineral phase by acid attack (HCl/HF), in a nitrogen atmosphere at 70°C (11). The residual organic matter is then characterized by elemental analysis and infrared spectroscopy.

RESULTS

The porous medium properties for the different tests are presented in Table II. The tests are carried out in a porous medium constituted of silica sand, kaolinite (4%), mixed with heavy crude oil and water.

Analysis of the residual organic matter by Oil Show Analyzer.

The evolution in the composition of the residual organic matter is studied by OSA in the coke zone and in all the zone ahead of the combustion front. Whatever the oil, in each sand sample, the amount of gaseous hydrocarbons (S0) is low because all the light hydrocarbons have been stripped by the gas flow.

For oil A, ahead of the coke zone, the amount of organic carbon in S0+S1+S2 has decreased from 6 g/100 g of rock (initial amount in the porous medium) to 3 g/100 g. The higher decrease can

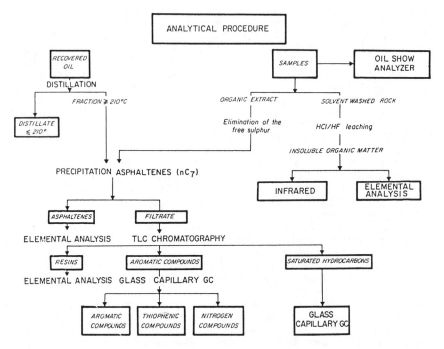

Figure 1. Analytical procedure. (Reproduced with permission from ref. 8. Copyright 1984 Institut Français du Petrole.)

TABLE II – POROUS MEDIUM PROPERTIES

Oil	A	B	C	D	E
So %	57.1	50.5	57.8	54.4	58.4
Sw %	35.4	20.1	26.5	25.1	26.7
Oil content kg/cm^3	167.4	172.7	207.3	203.4	209.5
Air requirement Nm^3/m^3	212	298	393	427	376
Front temperature °C	450–470	450–500	510–750	635–750	660–750
% burned	10.2	11.8	19.5	19.3	16.7

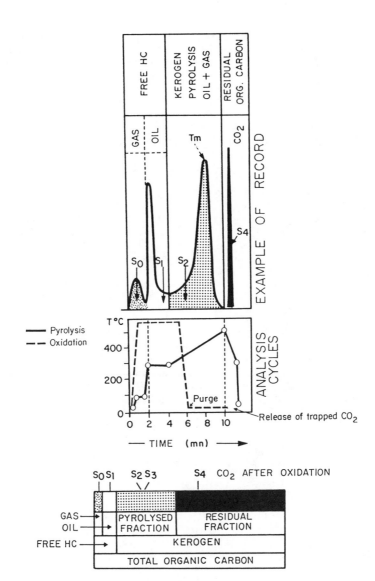

Figure 2. OSA diagram. (Reproduced with permission from ref. 8. Copyright 1984 Institut Français du Petrole.)

be observed for S2 which corresponds to the content of heavy
components released from 300 to 600°C. Just ahead of the coke zone,
there is an increase in the ratio S1/S2 with respect to the initial
oil. This is due to the presence of liquid hydrocarbons which have
been formed by cracking of heavier hydrocarbons or have been swept
by the combustion gases. The same cracking effect is observed for
the other oils B to E but on the contrary, the amount of organic
carbon in S0+S1+S2 is equal or even higher than the original one.
Thus, all the zone ahead of the coke zone has been enriched by the
cracking products but the heavy product content (S2) remains also
high. An example of the progressive changes of the organic matter
properties due to the combustion can be observed on Fig. 3 for OIL
E. The coke zone and the zone ahead of the coke where the changes
are noticeable have been specially studied.

Analysis of the coke zone

The amount of the residual organic matter is given in Table III for
different samples studied.

The material balance is consistent with the results obtained
by OSA (S2+S4 in g/100 g). For oil A, the coke zone is very narrow
and the coke content is very low (Table III). On the contrary, for
all the other oils, the coke content reaches higher values such as
4.3 g/ 100 g (oil B), 2.3 g/100 g (oil C), 2.5 g/100 g (oil D),
2.4/100 g (oil E). These organic residues have been studied by
infrared spectroscopy and elemental analysis to compare their
compositions. The areas of the bands characteristic of C-H bands
$(3000-2720 \text{ cm}^{-1})$, C=C bands $(1820-1500 \text{ cm}^{-1})$ have been measured.
Examples of results are given in Fig. 4 and 5 for oils A and B. An
increase of the temperature in the porous medium induces a decrease
in the atomic H/C ratio, which is always lower than 1.1, whatever
the oil (Table III). Similar values have been obtained in pyrolysis
studies (4). Simultaneously to the H/C ratio decrease, the bands
characteristics of CH_2 and CH_3 groups progressively disappear. The
absorbance of the aromatic C-H bands also decreases. This reflects
the transformation by pyrolysis of the heavy residue into an
aromatic product which becomes more and more condensed. Depending
on the oxygen consumption at the combustion front, the atomic O/C
ratio may be comprised between 0.1 and 0.3.

Analysis of the zone ahead of the coke zone.

In this zone, the quantity of extracted oil is generally sufficient
to obtain the distribution of the different structural groups (SARA
analysis) except for oil A (Fig. 6 to 9). For oil B (Fig. 6), for
the first two samples, the amount of extracted products is too low
and the analysis is uncertain. It can only be noticed that the
asphaltene content is null. On the contrary, just beyond the coke
zone (samples III-IV), the asphaltene content respectively reaches
12.9 and 5.4% whereas the asphaltene content of the initial oil is
only 0.3%. This effect is also observed for oil C (10% versus 6.3%)
(Fig. 7), D (24% versus 13.8%) (Fig. 8), E (24.4% versus 8.1%)
(Fig. 9). For all the oils, the amount of resins+asphaltenes
generally remains constant and the amount of saturates increases

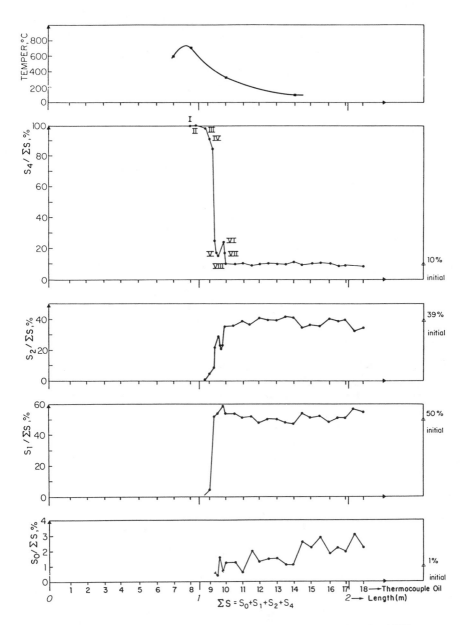

Figure 3. Analysis by OSA of samples taken in the porous medium for Oil E.

TABLE III – COKE PROPERTIES

Sample nb	A (1)	A (2)	B (1)	B (2)	C (1)	C (2)	D (1)	D (2)	E (1)	E (2)
I	1.02	0.75	0.103	–	2.1	0.34	2.51	0.31	1.96	0.32
II	0.79	0.55	3.52	0.5	2.27	0.35	2.37	0.4	1.9	0.42
III	0.7	0.7	4.27	0.65	1.59	0.4	2.17	0.58	2.39	0.5
IV	0.4	1.15	2.06	0.92	0.15	1	1.49	0.6	1.48	0.63
V	0.35	0.95	0.134	–	0.22	1.1	0.048	–	–	–
VI	–	–	0.096	–	0.194	0.8	0.094	–	–	–
VII	–	–	–	–	0.082	1.05	–	–	–	–

(1) amount of residual organic matter in g/100 g of rock.
(2) H/C ratio determined by elemental analysis.

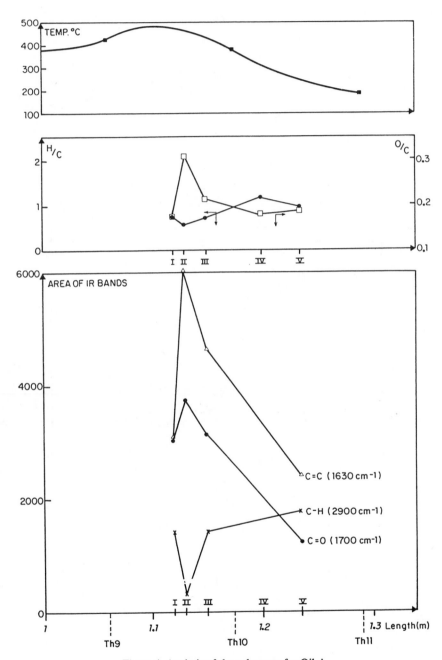

Figure 4. Analysis of the coke zone for Oil A.

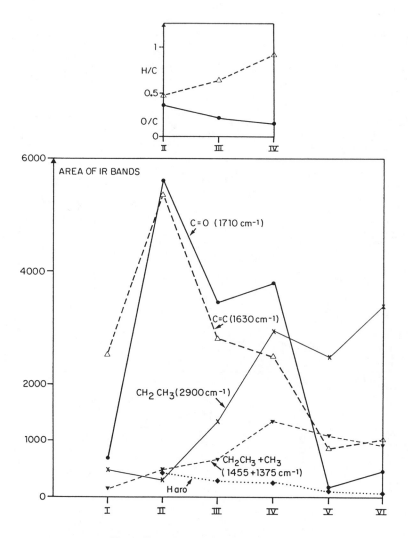

Figure 5. Analysis of the coke zone for Oil B.

COMPOSITION OF THE EXTRACT   OIL B

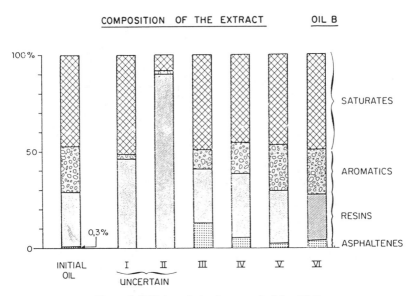

Figure 6. SARA analysis of extracted oil for Oil B.

COMPOSITION OF THE EXTRACT   OIL C

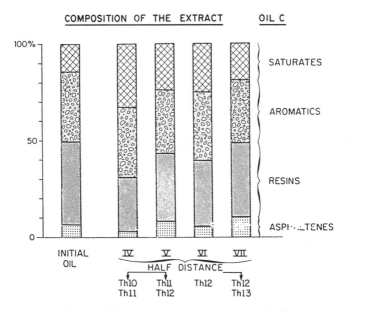

Figure 7. SARA analysis of extracted oil for Oil C.

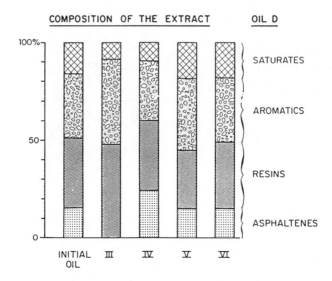

Figure 8. SARA analysis of extracted oil for Oil D.

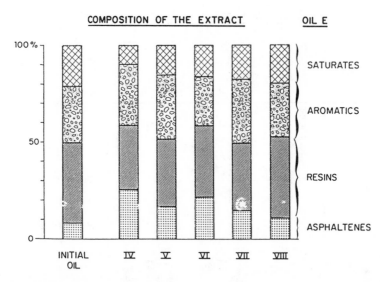

Figure 9. SARA analysis of extracted oil for Oil E. (Reproduced with permission from ref. 8. Copyright 1984 Institut Français du Petrole.)

whereas there is a slight decrease in the aromatic content. For example, for oil C, the C_{15}-C_{20} saturate content increases from 37.9% (initial value) to 58.7% (sample IV). Ahead of this zone, the oil extracted from the porous medium progressively recovers its initial properties.

Analysis of the recovered oil.

The oil recovered is first distilled. For oil A, the fraction having a boiling point lower than 210°C at atmospheric pressure (210⁻ or C_{12}^- fraction) represents 2.6 to 11% of the recovered oil, whereas the initial oil contains 2.5% of this fraction. The chemical changes affecting the produced oil are consistent with the evolution of its physical properties (Fig. 10). The same effect can be observed for a heavier crude. For example, for oil E, the 210⁻ fraction represents 10 to 15% of the recovered oil even though this fraction doesn't exist in the initial crude oil.

The variations in the composition of the fraction having a boiling point higher than 210°C (210⁺) depend strongly on the initial composition of the oil.

For oil A, slight differences in composition exist; the aromatic and resin fractions hardly decrease to form lighter saturate compounds. The effects are quite similar to those on oil B whose global composition does not change. But in the saturate fraction, the amount of n and iso alkanes is three times higher in the recovered samples than in the initial one (Fig. 11).

For crude oils C and D, some lighter hydrocarbons are formed during the cracking reactions but the composition of the 210⁺ fraction is hardly modified. In particular, it can be noticed that the asphaltene contents of both of the recovered oils remain high.

On the contrary, for oil E the quantity of asphaltenes decreases from 8.1% for the initial crude oil to 4.1% for the sample produced at the end of the test (Fig. 12). Moreover, the amounts of resins + asphaltenes decreases whereas the amounts of saturates and aromatics increase (51.4% in the initial oil, 72.4% for a sample recovered at t = 24 h). The analysis by GC shows that each oil fraction is enriched in components with molecular chains ranging from 15 to 30 carbons which don't exist in the initial oil (n-alkanes, aromatics C_{20}-C_{30} which are less complex than the initial ones, thiophenic compounds C_{15}-C_{25}). The elemental composition of the asphaltene fraction of the samples recovered during the test are shown in Table IV. There is an high increase in the oxygen/carbon ratio compared to the slight decrease in the hydrogen/carbon ratio. The results of the analysis of the resin fraction are quite similar. The thermal cracking of the oil induces the formation of lighter molecular weight compounds and of a polar denser residue. Those results are consistent with the observations of a field study (12).

CONCLUSIONS

Elemental analysis and infrared spectroscopy give a good characterization of the coke deposit. From one crude oil to

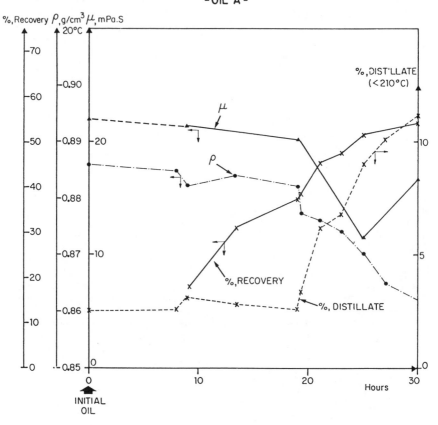

Figure 10. Characterization of the recovered oil for Oil A.

GC OF SATURATES

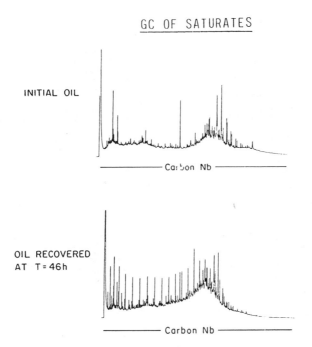

INITIAL OIL

Carbon Nb

OIL RECOVERED
AT T = 46 h

Carbon Nb

Figure 11. Saturates chromatography for a sample recovered at 46 h of test, compared to the initial one for Oil B.

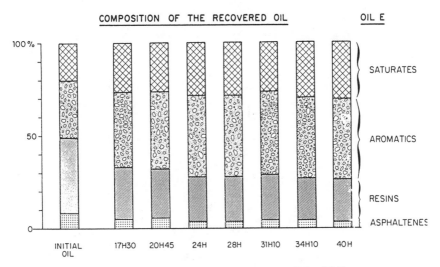

COMPOSITION OF THE RECOVERED OIL   OIL E

SATURATES

AROMATICS

RESINS
ASPHALTENES

INITIAL OIL   17H30   20H45   24H   28H   31H10   34H10   40H

Figure 12. SARA analysis of the recovered oil for Oil E.

TABLE IV - ELEMENTAL ANALYSIS OF ASPHALTENES
IN THE RECOVERED SAMPLES - OIL E -

	H/C	O/C
Initial	1.21	0.028
Sample recovered at t=		
17h30	1.12	0.144
20h45	1.15	0.137
24h	1.16	0.054
28h	1.16	0.133
31h10	1.17	0.053
34h10	1.16	0.052
40h	1.12	0.048

another, the amount of coke is quite different, the highest amounts
of coke being obtained from the heaviest crude oils, and especially
those which contain more reactive compounds. For all the oils, the
evolution of the coke composition versus the temperature evolution
is similar; for example, the H/C ratio is generally comprised
between 0.5 and 0.6 for a temperature equal to 400°C. The lowest
H/C values (i.e. lower than 0.4) are observed for the highest
temperatures of the combustion front (i.e. higher than 500°C). It
means that the coke composition can be related to the temperature
reached in the porous medium and its amount to the oil thermal
reactivity which is influenced by the oil geochemical composition.

As the sampling moves downstream from the coke zone, it may
be noticed that whatever the oil composition,
- the first samples contain the highest amount of residual
carbon and no asphaltenes in the extracted fraction of the sample,
- in the following samples having a lower residual carbon
content, the extracted oil contains a higher asphaltene content
than the initial oil. This effect is observed even if the initial
asphaltene content of the oil is quite null. It seems that the coke
formation can be written as the following balance :
Aromatics $\longrightarrow$ Resins $\longrightarrow$ Asphaltenes $\longrightarrow$ coke
where the resin + asphaltene content remains constant and
asphaltenes are the main precursors of coke. The same observations
have been made in low-temperature oxidation experiments (6).

Whatever the oil composition, the cracking reactions enhance
the amount of the 210^- fraction in the recovered oil. The
composition of the 210^+ fraction depends on the initial oil
composition : it is not modified if the oil is already altered or
if the oil contains very stable compounds.

This experimental work gives data that help us to understand
the mechanisms of coke formation during in-situ combustion.
Moreover, in case of a field application, a study of the oil
produced and the organic matter from cores taken behind the
combustion front, related to the analysis of the initial oil could
provide information on the propagation of the combustion front.

REFERENCES

1. Fassihi, M.R., Meyers, K.O., Weisbrod, K.R., "Thermal Alteration of Viscous Crude Oils", Annu. Fall Meeting of Soc. Petroleum Engrs, SPE paper No 14225 (Sept. 1985).
2. Ungerer, P., Béhar, F., Villalba, M., Heum, O.R., Audibert, A., "Kinetic Modelling of Oil Cracking". 13th Internation. Meeting on Organic Geochem., Venice (Sept. 1987).
3. Béhar, F., Audibert, A., Villalba, M., "Secondary Cracking of Crude oils : Experimental Study" - Submitted to Energy and Fuel (1988).
4. Abu Khamsin, S.A., Brigham, W.E., Ramey, H.J., "Reaction Kinetics of fuel formation for in-situ combustion", Fith Soc Petroleum Engrs Middle East Show, SPE paper No 15736, (March 1987).
5. Millour, J.P., Moore, R.G., Bennion, D.W., Ursenbach, M.G., Gie, D.N., "A simple Implicit Model for Thermal Cracking of Crude Oils", Annu. Fall Meeting of Soc. Petroleum Engrs, SPE Paper No 14226 (Sept. 1985).
6. Millour, J.P., Moore, R.G., Bennion, D.W., Ursenbach, M.G., Gie, D.N., "An Expanded Compositional Model for Low Temperature Oxidation of Athabasca Bitumen". J. Canad. Petroleum Technol., (May-June 1987), p. 24-32.
7. Monin, J.C., Audibert, A., "Thermal Cracking of Heavy Oil/Mineral Matrix Systems", Soc. Petroleum Engrs, Internation. Symp. on Oilfield Chemistry, SPE paper No 16269, (Feb. 1987).
8. Audibert, A., Roucaché, J., "Evolution of the Composition of a Crude Oil During In-situ Combustion", Characterization of Heavy Crude Oils and Petroleum Residues. Ed. Technip, Paris, p 135-139, (1984).
9. Burger, J., Sourieau, P., Combarnous, M., "Thermal Methods of Oil Recovery", Ed. Technip, Paris and Gulf Publishing Company, Houston (1985).
10. Espitalié, J., Marquis, F., Barsony, J., "Geochemical Logging by the Oil Show Analyzer" Analytical Pyrolysis - Techniques et Applications - Voorhees, Ed. Butterworth, Londres, (1984).
11. Durand, B., "Kerogen", Ed. Technip, Paris (1980).
12. Bojes, J.M., Wright, G.B., "Application of Fluid Analyses to the Operation of an in-situ Combustion Pilot". Annu. Technical Meeting of Petroleum Soc. of CIM, Paper No 86-37-61, (June 1986).

RECEIVED November 28, 1988

Chapter 23

Kinetics and Energetics of Oxidation of Bitumen in Relation to In Situ Combustion Processes

Leslie Barta, Andrew W. Hakin, and Loren G. Hepler

Department of Chemistry, University of Alberta, Edmonton, Alberta T6G 2G2, Canada

Using a "home made" aneroid calorimeter, we have measured rates of production of heat and thence rates of oxidation of Athabasca bitumen under nearly isothermal conditions in the temperature range 155-320°C. Results of these kinetic measurements, supported by chemical analyses, mass balances, and fuel-energy relationships, indicate that there are two principal classes of oxidation reactions in the specified temperature region. At temperatures much lc.jer than 285°C, the principal reactions of oxygen with Athabasca bitumen lead to deposition of "fuel" or coke. At temperatures much higher than 285°C, the principal oxidation reactions lead to formation of carbon oxides and water. We have fitted an overall mathematical model (related to the factorial design of the experiments) to the kinetic results, and have also developed a "two reaction chemical model". Subsequent measurements in which water vapor has been introduced along with oxygen have led to modified kinetics and also a modified chemical model for wet oxidation of Athabasca bitumen.

In situ combustion processes involve injection of air or oxygen into the formation to burn part of the bitumen, thereby producing heat that raises the temperature of the reservoir. At low temperatures oxidation is incomplete and leads to formation of "fuel" or partly oxidized bitumen that has higher viscosity and lower heating value than the original bitumen. These incomplete oxidation reactions, collectively called "low temperature oxidation" or "LTO", predominate during the ignition delay period of in situ combustion and also occur ahead of the combustion front when oxygen is available. It is therefore important to have knowledge of the kinetics and energetics of oxidation of bitumen in the temperature range in which LTO occurs.

0097–6156/89/0396–0426$06.00/0

In this paper we summarize some of the results of our measurements of rates of dry oxidation. Results of chemical analyses of residues produced by heating in flowing nitrogen atmosphere (distillation) are also reported and combined with our kinetic data to obtain values of kinetic parameters. Preliminary results of measurements of rates of wet oxidation are presented.

Experimental

The focus of our investigations of the kinetics of oxidation of Athabasca bitumen has been on the use of an aneroid calorimeter (1) for measuring rates of heat production under nearly isothermal (ΔT < 1.2°C in each experiment) conditions. Initial attention was given to just two of the variables that affect the kinetics of oxidation: (i) temperature and (ii) pressure of oxygen. Preliminary studies into a third variable, the partial pressure of water vapor in the system, are discussed in Part 3 of the Results and Calculations section. Each calorimetric sample ($\approx$1 g, 13.47 mass % bitumen) came from a large sample of "reconstructed" oil sand consisting of Athabasca bitumen loaded onto a chemically inert solid support material (60/80 mesh acid washed Chromosorb W) of well-defined particle size.

We represent the overall oxidation reaction by

$$\text{Bitumen} + O_2 \rightarrow \text{chemical products} + \text{heat} \qquad (1)$$

and express the rate of reaction in terms of the rate of production of heat as in

$$\text{Rate} = dQ/dt = k\,[p(O_2)]^r \qquad (2)$$

where Q is the instantaneous heat produced per gram of bitumen, t is time, $p(O_2)$ is the pressure of oxygen, k is a specific rate constant, and r is the reaction order with respect to pressure of oxygen. Because the stoichiometry of the process represented by Equation 1 is not known and because fuel quality changes with extent of reaction, we have used the method of initial rates for evaluation of r and k.

Elemental compositions and masses of organic residues produced by distillation and by oxidation were determined as described previously (2).

Results and Calculations

Part 1. Kinetics and Energetics of Dry Oxidation. The simplest approach to data analysis is to assume that only a single class of oxidation reactions is important and to make the related assumption that the temperature dependence of the single rate constant k can be represented by an Arrhenius equation. In this way we obtain

$$\ln W_i = \ln A - E_a/RT + r \ln [p(O_2)_i] \qquad (3)$$

in which W_i is the initial rate of heat production, A is the Arrhenius pre-exponential factor, and E_a is the average activation

energy. According to this simple model, both E_a and r are
independent of temperature. Consideration of a factorial design
model showed that we require the results of five experiments
(different temperatures and pressures of oxygen) to test Equation
3. Results of such experiments and related mathematical analysis
(2) have shown that the "first order mathematical model" or "one
class of reaction chemical model" represented by Equation 3 is
inadequate.

Because earlier experimental results and data analyses (3-10)
had led us to anticipate the inadequacy of the simple approach
considered above, we also planned and carried out (2) a second
order factorial design of experiments and related data analysis.
Mathematical analysis (of the results of 11 experiments) based on
the second order model showed that all of these results could be
represented satisfactorily by an equation of the form

$$\ln W_i = b_0 + (b_1+b_3/T)(1/T) + (b_2+b_4/T) \ln [p(O_2)_i] \qquad (4)$$

Comparison of Equation 4 with Equation 3 shows that

$$\ln k = b_0 + (b_1+b_3/T)(1/T) \qquad (5)$$

and

$$r = b_2 + b_4/T \qquad (6)$$

Because the dependence of kinetic parameters k and r on
temperature as described by Equations 5 and 6 is constrained by the
factorial design, it is possible that these equations and therefore
Equation 4 may not give the best overall representation of the
kinetic data. We have therefore carried out four more experiments
and analyzed these results along with those mentioned earlier to
obtain (2) the best overall mathematical representation that is
summarized by the following equations:

$$\ln k = a_0 + a_1T + a_2/(595-T)^2 \qquad (7)$$

$$r = a_3 + a_4/(600-T) \qquad (8)$$

$$\ln W_i = a_0 + a_1T + a_2/(595-T)^2 + a_3 \ln [p(O_2)_i] + a_4 \ln [p(O_2)_i]/(600-T) \qquad (9)$$

The composite rate constant k summarized by Equation 7 has a
maximum near 302°C. The reaction order r is nearly constant at low
temperatures and increases dramatically at higher temperatures.

The kinetic results and related analysis (2) summarized above
indicate that there is a change in the predominant class of oxidation reaction with increasing temperature, which led to the
expectation that the total heat developed in the overall oxidation
also depends on temperature. Because the measurements that led to
kinetic data based on initial rates were continued nearly iso-
thermally until oxidation was complete, it has also been possible
to establish (2) that the total heat developed increased by nearly
ten-fold over the range 155 to 320°C.

On the basis of our kinetic results, our heats of oxidation
mentioned above, independent heats of total combustion (Yan, H-k.;
Hepler, L.G. to be published), our chemical analyses, and the
results of earlier investigations by others (11-15), we have
developed a chemical model or picture of the dry oxidation process
as follows.

We propose that the complicated dry oxidation of bitumen can
be represented as the sum of contributions from two classes of
oxidation reaction. One class of reactions is the partial oxidation that leads to deposition of coke and formation of "oxygenated
bitumen", with very little production of carbon oxides and water.
This class of reactions is concisely summarized by

$$\text{Bitumen} + O_2 \rightarrow \text{Coke} + \text{(other chemical products)} + Q_D \qquad (10)$$

in which Q_D represents the heat production associated with deposition of fuel.

The second class of reactions is similar to conventional
combustion or burning reactions that yield mostly carbon oxides and
water vapor, as summarized by

$$\text{Bitumen} + \text{Fuel} + O_2 \rightarrow \text{Carbon Oxides} + \text{Water Vapor} + Q_B \qquad (11)$$

in which Q_B represents the heat produced by the "burning
reactions".

The "two classes of chemical reactions model" represented by
Equations 10 and 11 can be represented mathematically by the
following:

$$\ln W_D = \ln A_D - E_{a(D)}/RT + r_D \ln [p(O_2)] \qquad (12)$$

$$\ln W_B = \ln A_B - E_{a(B)}/RT + r_B \ln [p(O_2)] \qquad (13)$$

Not all of the bitumen initially present in the calorimetric
sample is available for oxidation, due to distillation during
heating to the temperature of the experiment. As a preliminary to
evaluating the kinetic parameters appearing in Equations 12 and 13,
we have performed distillation experiments as described in Part 2
to identify the mass fraction of the original bitumen that is
available for oxidation at each temperature. Kinetic parameters
based on the initial rate of heat production per gram of bitumen
available are listed in Table I. The temperature at which deposition and burning contribute equally to the total rate of heat
production is ~270°C. Deposition continues to be important to a
little over 300°C. The ratio of the rate of heat production by
burning to that by deposition is <0.001 at 155°C, indicating that
deposition is the sole process of interest at temperatures <155°C.

Table I. Values of Kinetic Parameters for Low Temperature
Oxidation of Athabasca Bitumen

Equation[a]	ln A	E_a (kJ/mol)	r
$\ln W_D = \ln A_D - E_{a(D)}/RT + r_D \ln p(O_2)$	17.3	51	0.55
$\ln W_B = \ln A_B - E_{a(B)}/RT + r_B \ln p(O_2)$	34.5	143	1.15

[a]Correlation coefficients are -0.995 for deposition (subscript D)
and -0.996 for burning (subscript B).

The enthalpy of the depositional process described by Equation
10 is obtained by dividing the measured heat produced at 155°C by
the mass of bitumen converted to coke, yielding $\Delta H_D = -2$ kJ g^{-1}.

By combining the total heat produced, the available mass of
organic material (see Part 2), and the fraction of the available
carbon that is converted to CO_2, ΔH_B can be calculated from the
relationship

$$Q \text{ (kJ/g bit avail)} = f_D \Delta H_D + f_B \Delta H_B \qquad (14)$$

where f_D and f_B are the mass fractions of the available bitumen
involved in each of the two processes. In a new experiment on dry
oxidation at 285°C and 210 kPa O_2, 870 kJ was produced by 0.053 g
of available bitumen and 36% of the available carbon was recovered
as CO_2. Since the fraction of available carbon is a reliable
estimate of the fraction of available bitumen (Part 2), $f_B = 0.36$
and therefore $f_D = 0.64$. Combining these values of the appropriate
mass fractions with the value of ΔH_D cited above leads to $\Delta H_B =$
-42 kJ g^{-1}. Results of a similar experiment at 317°C lead to $\Delta H_B =$
-45 kJ g^{-1}. The combined values of ΔH_D and ΔH_B yield an average
value of ~-45 kJ g^{-1} for the heat of combustion of bitumen, within
6% of the value obtained by bomb calorimetry (Yan, H-k.; Hepler,
L.G. to be published). In view of the much higher precision of
heat measurement with the bomb calorimeter as compared with our
aneroid calorimeter, the value $\Delta H_B = -41$ kJ g^{-1} that may be derived
from the measured heat of combustion is preferable to that obtained
from Equation 14, although the example given here serves as an
independent check on our value of ΔH_B.

In principle both ΔH_D and ΔH_B can be obtained from Equation
14, using data for the conversion of available carbon to CO_2 at two
different temperatures. In practice, however, this calculation
does not yield reliable values because of the very small range of
carbon conversion available under our experimental conditions and
because of the very large difference in magnitude of the two
enthalpies.

Part 2. Distillation Experiments. Distillation experiments were
performed to determine the fraction of the original mass of bitumen
in a calorimetric sample that is available for oxidation at each
temperature in the range 155-320°C, to determine the quality of
this "fuel" as expressed by the molar ratio of hydrogen to carbon,

and to provide some of the information needed to obtain a mass balance for carbon in the oxidation process.

Samples of bitumen on Chromosorb W (loading factor = 12.20 ± 0.05 mass %) were heated in flowing nitrogen at a pressure of 210 kPa absolute (flow rate = 13 mL min^{-1}) to 155, 180, 230, 275, or 330°C and allowed to equilibrate overnight. A cold trap was placed in the gas train at the outlet of the calorimeter to collect the volatile fractions that distilled at each temperature. The heated samples were cooled to room temperature under flowing nitrogen, removed from the calorimeter, analyzed for carbon, hydrogen, nitrogen, and sulfur, and ashed to determine the total mass of organic residue.

Table II. Mass % of Total Bitumen Remaining After Heating and Elemental Composition of the Residue Produced

Temperature (°C)	L.F.[a]	Mass %				
		Bitumen Remaining	C	H	S	N
25	12.20	100.00	82.49	9.64	5.45	0.41
155	11.30	83.90	83.75	9.32	6.60	0.71
182	8.77	71.90	82.45	9.29	7.80	0.46
229	7.74	63.40	80.78	8.58	8.71	0.13
275	6.16	50.50	81.69	7.67	7.52	0.16
330	5.18	42.50	80.34	7.58	10.28	0.10

[a]Loading factor (%).

Table II lists the elemental compositions of the organic residues produced at the various temperatures. The molar H/C ratio of the residue produced by heating decreases slightly with increasing temperature, from 1.3 at 155°C to 1.1 at 330°C. Carbon content of the organic residue produced by heating is approximately independent of temperature in the range 155-330°C. Table III lists the mass fractions, expressed as %, of original bitumen and of original C, H, and S remaining as residue. The mass % of whole bitumen and mass % of original carbon remaining at each temperature are approximately the same. Sulfur is not lost by heating until the temperature exceeds 230°C, and appreciable S (more than half the original mass present) remains even at 330°C.

Table III. Mass % of Total Bitumen and of C, H, and S Remaining as Residue After Heating

Temperature (°C)	Bitumen	C	H	S
25	100.0	100.0	100.0	100.0
155	83.9	85.2	81.1	101
182	71.9	76.2	69.3	103
187	71.9	74.9	-	-
229	63.4	62.1	56.4	101
275	50.5	50.0	40.2	69.7
330	42.5	41.4	33.4	80.2

No condensate was found in the cold trap (ice or dry ice/acetone) after any of the experiments, nor was any material recovered from the gas train by flushing with toluene and ethanol. However, the quantity of condensate expected on the basis of ashing of the heated samples is gravimetrically significant in all cases (15 mg expected at 155°C, 56 mg expected at 330°C).

Table IV. Mass Balance of Carbon in Dry Oxidation, Expressed as Mass % Relative to Original Mass of Carbon (Distillation) and Available Mass of Carbon (Final Products)

Temperature (°C)	Distillation		Final Products of Oxidation		
	$C_{\Delta(s)}$	$C_{\Delta(v)}$	$C_{ox(s)}$	$C_{ox(v)}$	Heat (kJ/g bit avail)
155	82.4	17.6	77.3	22.7	1.9
174	75.0	25.0	(69.7)	(30.3)	4.0
225	59.5	40.5	49.7	50.3	12.5
285	48.7	51.3	29.8	70.2	16.4
317	44.6	55.4	0.0	100.0	18.1

The mass balance for carbon during dry oxidation of bitumen in our calorimeter can be calculated from the above results in combination with some of our chemical results for oxidized samples reported elsewhere (2). This mass balance is summarized in Table IV, where the following relationships have been used:

$$C_{total} = C_{\Delta(s)} + C_{\Delta(v)} \tag{15}$$

$$C_{\Delta(s)} = C_{ox(s)} + C_{ox(v)} \tag{16}$$

Here "Δ" indicates products produced by thermostatting the calorimeter and its contents prior to oxidation and "ox" indicates products produced from the thermostatted samples by dry oxidation. The subscripts s and v refer to the phase (solid or vapor) in which the product remains or to which the product is transferred. Thus $C_{\Delta(s)}$ represents the amount of carbon actually available as fuel at each temperature, $C_{ox(s)}$ represents the amount of carbon remaining as oxygenated bitumen, $C_{\Delta(v)}$ represents the amount of carbon distilled during heating, and $C_{ox(v)}$ represents the amount of additional carbon converted by oxidation to volatile hydrocarbons and CO_2. The values listed in columns 1 through 4 of Table IV were obtained using Equations 15 and 16 rewritten (for convenient interpretation of the results) as Equations 17 through 20:

$$C_{\Delta(s)} = b_r c_r / b_o \tag{17}$$

$$C_{\Delta(v)} = 100 - C_{\Delta(s)} \tag{18}$$

$$C_{ox(s)} = c_c b_c / b_r \tag{19}$$

$$C_{ox(v)} = 100 - C_{ox(s)} \tag{20}$$

where b_r = mass of bitumen remaining in the heated sample, g
        c_r = carbon content of bitumen remaining, %
        b_o = original mass of bitumen, g
        c_c = carbon content of residual coke, %
        b_c = mass of residual coke, g

Values of $C_{ox(s)}$ at 155, 225, and 317°C were calculated from the results of the distillation experiments described above, using data (2) for residual cokes. The value of $C_{ox(s)}$ at 285°C was calculated from data from a new dry oxidation at 285°C. An estimate of the value of $C_{ox(s)}$ at 174°C was obtained by fitting a line to the values for $C_{ox(s)}$ at 155, 225, and 285°C (correlation coefficient = -0.999):

$$C_{ox(s)} = 133.4 - 0.336 \ t \tag{21}$$

where t is temperature in °C. Smoothed values of $C_{\Delta(s)}$ at each temperature were calculated from the fitted equation

$$\text{mass \% bitumen remaining} = 8.11 + 1.1817 \times 10^4 /t \tag{22}$$

Values of $C_{ox(s)}$, and thence $C_{ox(v)}$, for 174°C calculated as described here are shown in parentheses in Table IV.

The values in the first two columns of Table IV show the distribution of original carbon in products of distillation; values in the next two columns show the distribution of available carbon in products of oxidation. The large difference between the value of $C_{ox(s)}$ at 317°C predicted by Equation 21, 17.4%, and the observed value of 0% underscores the validity of our proposed change in mechanism near 285°C. Additional evidence for this change is provided by the carbon contents of the residual cokes: 82% at room temperature, 72% at 155°C, 56% at 225°C, and 55% at 285°C. The levelling-off of carbon content of residual cokes between 225 and 285°C suggests that the depositional reaction begins to compete with a second type of reaction. That this second type of reaction involves burning of coke and other hydrocarbons is apparent in the much lower yield of residual coke at 285°C and higher yield of CO_2 (see discussion of Table IV below) at 285°C as compared to the yield at 225°C. The last column lists the total heat produced per gram of bitumen available and shows that the total heat relative to the mass of material from which it is produced increases an order of magnitude over the temperature range 155–320°C.

On the basis of the analysis presented in Tables II, III, and IV and measurements of the mass of CO_2 evolved during oxidation, Figure 1 was constructed to display the fraction of original carbon mobilized by heating, the fraction of the remaining (available) carbon mobilized as incompletely oxidized hydrocarbon by oxidation, and the fraction of available carbon deposited as coke by oxidation. The distribution of available carbon between the mobile and non-mobile products of oxidation lends additional support to our proposed "two-reactions" mechanism.

Part 3. Effect of Water Vapor on Kinetics and Energetics of
Oxidation. A saturator containing distilled water at a known and
controlled temperature (and hence a known vapor pressure of water)
was inserted in the gas train between the oxygen cylinder and the
inlet to the calorimeter. The saturator can be thoroughly flushed
with oxygen and brought to the desired pressure of the oxidation
experiment prior to injection of oxygen to the calorimeter. To
prevent condensation of water vapor, the gas train between the oven
and the saturator was wrapped with resistance tape which could be
heated to a temperature above the temperature of the saturator.
Pressure of oxygen was obtained as the difference between the
measured total pressure and the vapor pressure of water at the
measured temperature of water in the saturator. To avert
complications due to corrosion, the aluminum combustion tube was
replaced with a 316 stainless steel tube of the same outside
diameter but having slightly thinner walls to aid in heat transfer.
The dry oxidation experiments described in Part 2 were used to
verify that neither the values of the calibration constants nor the
values of initial rates of heat production and total heats produced
were affected by the change in properties of the combusion tube.

Preliminary experiments to refine the operation of the
calorimeter during wet oxidation have been performed at 225°C.
Vapor pressure of water was approximately 13 kPa, corresponding to
a saturator temperature of about 55°C. Pressures of oxygen were
varied from 117 kPa to 368 kPa, with three experiments at $p(O_2)$ =
210 kPa to determine reproducibility of the results.

Initial rates of oxidation were determined as described in
Reference 1. Mass of residual coke was determined by ashing, and
carbon and hydrogen content of the residual coke was determined by
microanalysis. Because the relative error in the total heat
produced by oxidation of the bitumen exceeded 50%, the heats were
considered unreliable and are not reported here.

Experimental conditions and initial rates of oxidation are
summarized in Table V. For comparison, initial rates of dry
oxidation at the same temperature and pressure of oxygen predicted
by Equation 9 are included in parentheses. The predicted dry rate,
measured dry rate, and measured wet rates are compared in Figure
2. The logarithms of the initial rates of heat production during
wet oxidation increase approximately linearly (correlation
coefficient = 0.92) with the logarithm of the partial pressure of
oxygen and lead to values of ln k = 2.5 and r = 0.9, as compared
with values of ln k = 4.8 and r = 0.6 for dry oxidation at this
temperature.

The results of the chemical analyses are summarized and
compared in Table VI with similar results from dry oxidation. In
comparison with dry oxidation at 225°C, wet oxidation at this same
temperature leads to less residual coke. As shown by the molar H/C
ratio, this residual coke is enriched in carbon. Less of the
available carbon is converted to residual coke.

The conclusions that may be drawn from these exploratory
kinetic and chemical data for wet oxidation are "sketchy" at
best. Nevertheless, some observations and suggestions relevant to
our continuing investigation are presented here.

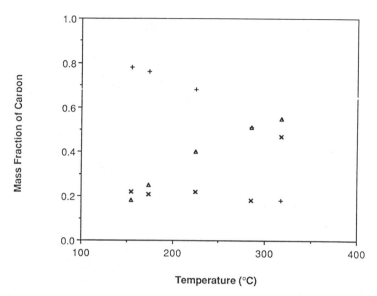

Figure 1. Distribution of available carbon. Symbols are:
+, carbon desposited as coke by oxidation, g/g available; x,
mobile carbon (minus carbon dioxide) produced by oxidation,
g/g available; Δ, mobile carbon produced by heating in inert
atmosphere, g/g original.

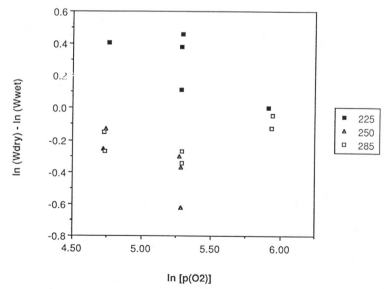

Figure 2. Rate of heat production by wet oxidation relative
to dry oxidation at the same temperature and pressure of
oxygen.

Table V. Initial Rates of Wet Oxidation at
225°C and 13 kPa H_2O [a]

$p(O_2)$ (kPa)	$p(H_2O)$ (kPa)	Initial Rate (W/g Bitumen)
198.2	12.3	1.2 (1.9)
197.4	13.1	1.3 (1.9)
197.4	13.1	1.7 (1.9)
368.4	13.6	2.8 (2.8)
117.0	13.6	1.0 (1.5)

[a]Loading factor of calorimetric samples was 12.20%. Numbers in parentheses are the predicted dry rates.

Table VI. Residual Coke and Conversion of Available Carbon During
Wet Oxidation of Athabasca Bitumen at 225°C [a]

LF After Heating (%)	LF After Oxidation (%)	$p(O_2)$ (kPa)	$p(H_2O)$ (kPa)	Residual Coke (g/g Bit Avail)	Molar H/C of Coke	$C_{ox(s)}$ [b] (%)
8.62	6.18	210.7	–	0.70	0.77	49.7
8.42	5.11	198.1	12.3	0.64	0.37	42.2
8.42	5.21	197.4	13.1	0.65	0.32	42.2
8.42	5.30	197.4	13.1	0.66	0.33	42.7
8.42	5.53	386.4	13.6	0.69	0.68	44.6

[a]Data for dry oxidation are included for comparison. Loading factor (LF) of the original sample used in the dry oxidation experiments is 13.47 mass % bitumen. Loading factor of the original sample used in the wet oxidation experiments is 12.20 mass % bitumen. Molar H/C of unoxidized residue produced by heating to 225°C = 1.27.
[b](Mass of carbon converted to coke/mass of carbon available) x 100.

First, the rate of heat production is again related to the sum of the rates of depositional and burning processes, and if the predominant factor affecting the overall rate is temperature, then it does not seem likely that the specific effect of water vapor on the oxidation reported here is chemical catalysis, since a lowering of activation energy for either process would result in an increase in the overall rate relative to dry oxidation.

Second, the difference between the rates of wet and dry oxidation is dependent on $p(O_2)$, being largest at low $p(O_2)$ and essentially negligible at high $p(O_2)$. Since $p(H_2O)$ was held constant in these experiments, this suggests a "concentration" effect; i.e., it appears that water vapor may interfere with some of the pathways by which the various components of bitumen react with oxygen. Millour et al. (16) and others (3-10, 11-15) have proposed that deposition of coke by low temperature oxidation of bitumen can be represented in terms of pseudocomponent reactions that involve conversion of original maltenes to new asphaltenes and resins and conversion of original asphaltenes to new resins. The results reported in Table VI suggest that water vapor induces selective removal of saturates, possibly by steam distillation of some of the polar intermediate compounds produced by oxidation of the original saturate fraction. Thus we observe a lower yield of coke, a lower molar H/C ratio of the residual coke, and a lower extent of conversion of total available carbon to coke. Such selective removal of saturates would be greatest at low $p(O_2)$ for a given $p(H_2O)$. Finally, the amount of residual coke and the extent of conversion of carbon to coke would decrease with increasing $p(H_2O)$ for a given $p(O_2)$.

What would be the corresponding effect of water vapor on the overall rate of heat production? The rate of heat production must now be considered as the sum of the rates of heat evolution by deposition and burning plus the rate of heat absorption by distillation; i.e., the overall rate of heat production must be smaller than the dry rate:

$$W_{wet} = W_{dry} - W_d \qquad (23)$$

where $W_{dry} = W_D + W_B$ and the subscript d refers to water-enhanced distillation.

Tentative values of a reaction order, x, and rate constant, k_d, for the proposed distillation reaction can be calculated from the data in Table V using Equation 23 and the relationship

$$\ln W_d = \ln k_d + x \ln [p(H_2O)/P_{total}] \qquad (24)$$

The result is $\ln k_d = 9.2 \pm 0.2$ and $x = 1.4 \pm 0.1$. Representing the rate constant k_d with an Arrhenius equation and approximating the activation energy, $E_{a(d)}$, with an average heat of vaporization equal to ~80 kJ mol^{-1}, a value of $\ln A_d = 28.7 \pm 0.2$ is obtained.

Further to our preliminary studies at 225°C subsequent wet oxidation experiments have been carried out at temperatures of 250 and 285°C. Partial pressures of oxygen were varied from 19 to 289 kPa whilst the partial pressure of water vapor in the calorimetric system was maintained at approximately 15 kPa.

As for the distillation experiments (Part 2) no condensate was found in the cold trap (dry ice/acetone) at the end of each wet oxidation experiment. On exiting the cold trap the flowing gases were directed through a drying tube containing phosphorus pentoxide supported on glass wool. It was on this boundary that most distillate was recovered. Apparently the distillate is too volatile to be condensed by dry ice/acetone and is only recovered when it undergoes a chemical reaction with the P_2O_5. The oily liquid produced at this boundary, which is found to be soluble in ethanol and water, is under investigation. In an attempt to

simplify this process it is planned to collect a sample of the
volatile distillate and subject it to a mass spectrophotometric
investigation. The results of these studies will be discussed in a
future publication.

Table VII. Initial Rates of Wet Oxidation at 285°C and 250°C[a]

p(O_2) (kPa)	p(H_2O) (kPa)	Temperature (°C)	Initial Rate (W/g bitumen)
113.00	15.15	284.98	12.84 (11.06)
113.92	15.00	284.33	14.19 (10.80)
198.81	15.00	283.03	19.12 (14.62)
197.95	17.14	283.82	21.31 (15.11)
381.17	14.72	284.82	25.42 (24.30)
377.51	15.87	283.92	26.17 (23.09)
114.37	15.58	249.93	3.59 (3.15)
112.82	16.50	248.55	3.86 (3.00)
194.57	17.14	248.44	5.47 (4.05)
197.47	15.00	249.78	7.95 (4.26)
197.76	16.66	250.05	6.24 (4.30)

[a]Loading factor of calorimetric samples was 12.20%. Numbers in
parentheses are the predicted dry rates calculated from Equation 9.

Initial rates of heat production are recorded together with
the experimental conditions in Table VII. Comparison with the
initial rates of heat production for dry oxidation, estimated from
Equation 9, shows that at both 250 and 285°C the rate of heat
production for the wet oxidation process is faster than that for
the dry oxidation process at the same temperature. For experiments
conducted at 285°C the relationship between the logarithms of the
initial rates of heat production for wet oxidation and the
logarithm of the partial pressure of oxygen in the system are found
to be linear. A linear least squares fit to these data gave
estimates of the kinetic parameters ln k and r which are contained
in Table VIII together with estimates of the same parameters for
the dry oxidation process, calculated from Equation 9. The
reaction order with respect to oxygen is decreased for the wet
oxidation process relative to the dry process at the same
temperature. However, this is over-compensated by increases in the
rate constant for the wet oxidation process, leading to an overall
increase in the initial rate of heat production compared to the dry
oxidation process at the same temperature. There are not yet
enough data points for the wet oxidation process at 250°C to
justify a fit to the data, but initial trends in the plot of the
logarithm of the initial rate of heat production against the
logarithm of the partial pressure of oxygen suggest a similar
situation as discussed for the wet oxidation process at 285°C. On
comparison with our wet oxidation results at 225°C it appears that
by raising the temperature only 25°C we have gone from the
situation in which the initial rate of heat production for the wet
oxidation process is slower than the rate of the dry oxidation to a
situation in which the rate of heat production for the wet
oxidation process is faster than that of the corresponding dry

process. Figure 2 clearly identifies the situation described above by showing a plot of the logarithm of the ratio of the dry to the wet initial rates of heat production against the logarithm of the partial pressure of oxygen in the system.

Table VIII. Values of Kinetic Parameters for the Wet Oxidation
of Athabasca Bitumen at 285°C

Equation[a]	ln k	r
$\ln W_{w285} = \ln k + r \ln p(O_2)$	7.03±0.35	0.53±0.07
$\ln W_{d285} = \ln k + r \ln p(O_2)$	6.22	0.65

[a]The correlation coefficient for wet oxidation at 285°C (subscript w285) is -0.97. Subscript d285 refers to the dry oxidation process at 285°C. Kinetic parameters for the dry oxidation process were calculated with the aid of Equation 9.

Equation 23 can no longer be used to describe the initial rate of heat production for the wet oxidation process at these elevated temperatures. However, it was expected that an equation of the form shown below would suffice:

$$W_{wet} = W_{dry} + W_x \tag{25}$$

Here W_x represents the initial rate of heat production for some process (or processes) that is dependent on the partial pressure of the water vapor in the calorimetric system and gives a positive contribution to the initial rate of heat production by wet oxidation (unlike the steam distillation process envisaged for wet oxidation at 225°C, which aids in the retardation of the initial rate of heat production). If this relationship were valid, then a plot of the logarithm of the initial rate W_x against the logarithm of the ratio of the partial pressure of water vapor in the system to the total pressure in the system would be linear. However, as demonstrated by Figure 3, no such simple relationship is found. This procedure illustrates the difficulties involved in interpreting experimental results when there are three variables, instead of two variables as in dry oxidation.

Table IX. Mass % Composition of Cokes Produced by the
Wet Oxidation Process at 285°C

Average Oxygen Pressure in System (kPa)	Mass %					
	C	H	S	N	O[a]	Molar H/C[c]
113.46	35.0	0.3	–	–	64.7	0.10
198.38	51.5	2.0	–	–	46.3	0.46
379.34	40.0	1.5	–	–	58.5	0.45
Unoxidized[b]	81.4	8.9	6.0	trace	3.8	1.31

[a]Mass composition of oxygen calculated by difference.
[b]Mass % composition of oil sand prior to oxidation experiment.
[c]The molar H/C ratio for the dry oxidation process at 285°C is estimated at 0.62.

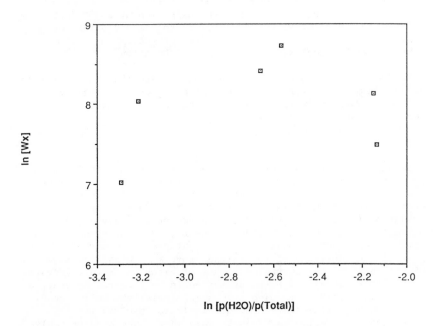

Figure 3. Variation of the difference between the wet and
the dry initial rates of heat production as a function of the
partial pressure of water vapor in the system at 285°C.

The elemental compositions of the oxidized oil sands produced from our calorimetric experiments at 285°C have been determined. The results, corrected for the elemental composition of the support material, are reported in Table IX. The elemental composition of the unoxidized bitumen is included for comparison. As predicted from our dry oxidation experiments, all of the sulfur present in the bitumen at the start of the wet oxidation experiments has been oxidized to volatile products and subsequently removed. The mass percents of carbon and hydrogen present in the analyzed cokes follow the pattern described by the initial rate of heat production for the process described by W_x, i.e., the mass percents of carbon and hydrogen in the cokes initially increase with an increase in oxygen pressure, reach a maximum value at some intermediate oxygen pressure, and then decrease with increasing oxygen pressure. The molar H/C ratio is very small at low oxygen pressures; however, it can be seen to increase with an increase of the oxygen pressure.

The results of ashing experiments on the samples from wet oxidation experiments conducted at 285°C are contained in Table X. The loading factors of the oxidized oil samples are dramatically reduced compared to those for samples oxidized at 225°C (see Table VI). The amount of residual coke per gram of bitumen available in each sample can also be seen to have undergone a dramatic reduction as have the estimates of $C_{ox(s)}$, the amount of available carbon converted to coke, compared to the lower temperature wet oxidation study.

Table X. Residual Coke and Conversion of Available Carbon During Wet Oxidation of Athabasca Bitumen at 285°C

LF After Heating[b] (%)	LF After Oxidation (%)	$p(O_2)$ (kPa)	$p(H_2O)$ (kPa)	Residual Coke (g/g Bit Avail)	$C_{ox(s)}$[a] (%)
6.05	1.16	113.00	15.15	0.14	5.5
6.05	1.23	113.92	15.00	0.14	6.9
6.05	1.62	198.81	15.00	0.20	14.1
6.05	1.52	197.95	17.14	0.20	11.3
6.05	1.00	381.17	14.72	0.14	7.1
6.05	1.52	377.51	15.87	0.18	8.2

[a](Mass of carbon converted to coke/mass of carbon available) x 100. For dry oxidation at 285°C $C_{ox(s)}$ is estimated at 29.8%.
[b]Estimated from Equation 22.

Both of the above chemical studies point towards the increased importance of the burning process at 285°C in determining the initial rate of heat production. The role of water as yet remains undefined other than at the higher temperature of 285°C it appears to have the opposite effect on the bitumen sample compared to the process at 225°C; i.e., it appears that water vapor encourages pathways by which the various components of bitumen react with oxygen. Preliminary calculations of the total heats evolved during the wet oxidation of bitumen sands indicate that they are independent of the partial pressure of oxygen in the system at

285°C, a conclusion that is supported by the loading factors of the residual cokes shown in Table X.

Our earlier (2) and present results pertaining to the dry oxidation of Athabasca bitumen have provided a clear picture of the dependence of rate and energetics on temperature and on pressure of oxygen. These results have led to an overall mathematical model and also to a "two reaction chemical model" that can be represented by two relatively simple equations. Introducing the additional variable of partial pressure of water vapor that is pertinent to wet oxidation processes also introduces many complications and ambiguities so that the present results have led to specific information but no clear overall picture. Continuing measurements of rates of wet oxidation, supplemented with chemical investigations, are providing more specific information from which we hope to develop an overall mathematical model and also a chemical model, as already done for dry oxidation.

LITERATURE CITED

1. Zhang, Z.-l.; Barta, L.; Hepler, L. G. AOSTRA J. Research 1987, 3, 249.
2. Barta, L.; Hepler, L. G. Energy and Fuels 1988, 2, 309.
3. Dabbous, M. K.; Fulton, B. F. Soc. Petrol. Eng. J. 1974, 14, 253.
4. Fassihi, R. Z.; Brigham, W. E.; Ramey, H.J., Jr. Soc. Petrol. Eng. J. 1984, 24, 399 and 408.
5. Phillips, C. R.; Hsieh, I. C. Fuel 1985, 64, 985.
6. Yoshiki, K. S.; Phillips, C. R. Fuel 1985, 64, 1591.
7. Verkoczy, B.; Jha, K. N. J. Can. Petrol. Techn. 1986, 25, 47.
8. Vossoughi, S.; Bartlett, G. W.; Willhite, G. P. Soc. Petrol. Eng. J. 1985, 25, 656.
9. Kharrat, R.; Vossoughi, S. J. Petrol. Techn. 1985, 37, 1441.
10. Burger, J.; Sahuquet, B. Rev. Inst. Fr. Petrol. 1977, 32, 141.
11. Ciajolo, A.; Barbella, R. Fuel 1984, 63, 657.
12. Moschopedis, S. E.; Speight, J. G. J. Mat. Sci. 1977, 12, 990.
13. Moschopedis, S. E.; Speight, J. G. Fuel 1975, 54, 210.
14. Moschopedis, S. E.; Speight, J. G. Fuel 1973, 52, 83.
15. Noureldin, N. A.; Lee, D. G.; Mourits, F. M.; Jha, K. N. AOSTRA J. Research 1987, 3, 155.
16. Millour, J. P.; Moore, R. G.; Bennion, D. W.; Ursenback, M. G.; Gie, D. N. An Expanded Compositional Model for Low Temperature Oxidation of Athabasca Bitumen, paper no. 86-32-41, Petroleum Society of CIM. June 8-11, 1986.

RECEIVED February 7, 1989

Chapter 24

Thermodynamic and Colloidal Models of Asphaltene Flocculation

S. Kawanaka, K. J. Leontaritis, S. J. Park, and G. A. Mansoori

Department of Chemical Engineering (M/C 110), University of Illinois, Chicago, IL 60680

This paper reviews the experiences of the oil industry in regard to asphaltene flocculation and presents justifications and a descriptive account for the development of two different models for this phenomenon. In one of the models we consider the asphaltenes to be dissolved in the oil in a true liquid state and dwell upon statistical thermodynamic techniques of multicomponent mixtures to predict their phase behavior. In the other model we consider asphaltenes to exist in oil in a colloidal state, as minute suspended particles, and utilize colloidal science techniques to predict their phase behavior. Experimental work over the last 40 years suggests that asphaltenes possess a wide molecular weight distribution and they may exist in both colloidal and dissolved states in the crude oil.

The key to solving many of the technical problems that face the fossil fuel industries in our modern technological society today lies heavily in understanding the thermodynamic and transport aspects of these problems. Most of the irreplenishable energy resources available are mainly mixtures of gases, liquids and solids of varying physical and chemical properties contained in the crust of the earth in a variety of geological formations. Knowledge of the fluid phase equilibrium thermodynamic and transport characteristics of these mixtures is a primary requirement for the design and operation of the systems which recover, produce and process such mixtures.

There has been extensive progress made in the past several years in the formulation of statistical thermodynamics of mixtures and transport phenomena modeling of multiphase flow in composite media. This knowledge may now be applied to the understanding and prediction of the phase and transport behavior of reservoir fluids and other

0097–6156/89/0396–0443$06.00/0

hydrocarbon mixtures. The present report is designed with the purpose of describing the role of modern theoretical and experimental techniques of statistical thermodynamics and transport and electrokinetic phenomena to develop methods that will predict asphaltene and asphalt flocculation during the production, transportation, and processing of petroleum.

The mechanisms of gas injection and oil recovery involved with miscible gas flooding are basically of three kinds [1,2]: (i) The first-contact miscible gas drive, (ii) The condensing gas drive (or the enriched gas drive), and (iii) The vaporizing gas drive (or the high pressure gas drive) processes. The first and second processes are based on the injection of hydrocarbons that are soluble in the residual oil, while the third process involves injection of a high density gas, such as high-pressure nitrogen or carbon dioxide. In the case of the first-contact miscible process, a typical injection fluid is propane, which is soluble in oil. For the condensing gas drive process the injection fluid could be natural gas containing relatively high concentration of intermediate hydrocarbons, such as ethane, propane, and butane.

Miscible flooding of petroleum reservoirs by carbon dioxide, natural gas, and other injection fluids has become an economically viable technique for petroleum production (1,2). The most common problem in petroleum recovery is poor reservoir volumetric sweep efficiency, which is due to channeling and viscous fingering because of the large difference between mobilities of the displacing and displaced fluids. Introduction of a miscible fluid in the petroleum reservoirs in general will produce a number of alterations in the flow behavior, phase equilibrium properties, and the reservoir rock characteristics. One such alteration is asphaltene and wax precipitation, which is expected to affect productivity of a reservoir in the course of oil recovery from the reservoir. (3,4,5). In most of the instances observed asphaltene and wax precipitation may result in plugging of or wettability reversal in the reservoir. Effect of asphaltene deposition could be positive (like prevention of viscous fingering) or negative (complete plugging of the porous media) depending on whether it could be controlled and predicted before it occurs.

The parameters that govern precipitation of asphaltene and wax appear to be composition of crude and injection fluid, pressure, and temperature of the reservoir. With alterations in these parameters the nature of asphaltene and wax substances which precipitate will vary. Also, precipitation of asphaltene is generally followed with polymerization or flocculation of the resulting precipitate, which produces an insoluble material in the original reservoir fluid (6,7,8). Because of the complexity of the nature of asphaltic and wax substances the phenomena of precipitation and flocculation of these substances are not well understood. Also in view of the complexity of the petroleum reservoirs, study and understanding of the in situ precipitation of asphaltene and wax seems to be a challenging and timely task. Such an understanding will help to design a more profitable route for miscible gas flooding projects.

In part II of the present report the nature and molecular characteristics of asphaltene and wax deposits from petroleum crudes are discussed. The field experiences with asphaltene and wax deposition and their related problems are discussed in part III.   In order to predict the phenomena of asphaltene deposition one has to consider the use of the molecular thermodynamics of fluid phase equilibria and the theory of colloidal suspensions.   In part IV of this report predictive approaches of the behavior of reservoir fluids and asphaltene depositions are reviewed from a fundamental point of view.   This includes correlation and prediction of the effects of temperature, pressure, composition and flow characteristics of the miscible gas and crude on: (i) Onset of asphaltene deposition; (ii) Mechanism of asphaltene flocculation.   The in situ precipitation and flocculation of asphaltene is expected to be quite different from the controlled laboratory experiments.   This is primarily due to the multiphase flow through the reservoir porous media, streaming potential effects in pipes and conduits, and the interactions of the precipitates and the other in situ material presnet. In part V of the present report the conclusions are stated and the requirements for the development of successful predictive models for the asphaltene deposition and flocculation are discussed.

Nature and Properties of Asphaltenes

The classic definition of asphaltenes is based on the solution properties of petroleum residua in various solvents. The word asphaltene was coined in France by J.B. Boussingault in 1837. Boussingault described the constituents of some bitumens (asphalts) found at that time in eastern France and in Peru. He named the alcohol insoluble, essence of turpentine soluble solid obtained from the distillation residue "asphaltene", since it resembled the original asphalt.

In modern terms, asphaltene is conceptually defined as the normal-pentane-insoluble and benzene-soluble fraction whether it is derived from coal or from petroleum. The generalized concept has been extended to fractions derived from other carbonaceous sources, such as coal and oil shale (8,9). With this extension there has been much effort to define asphaltenes in terms of chemical structure and elemental analysis as well as by the carbonaceous source. It was demonstrated that the elemental compositions of asphaltene fractions precipitated by different solvents from various sources of petroleum vary considerably (see Table I). Figure 1 presents hypothetical structures for asphaltenes derived from oils produced in different regions of the world. Other investigators (10,11) based on a number of analytical methods, such as NMR, GPC, etc., have suggested the hypothetical structure shown in Figure 2.

It has been shown (9) that asphaltenes contain a broad distribution of polarities and molecular weights. According to these studies, the concept of asphaltenes is based on the solubility behavior of high-boiling hydrocarbonaceous materials in benzene and low-molecular weight n-paraffin hydrocarbons. This solubility behavior is a result of physical effects that are caused by a spectrum of chemical properties. Long also

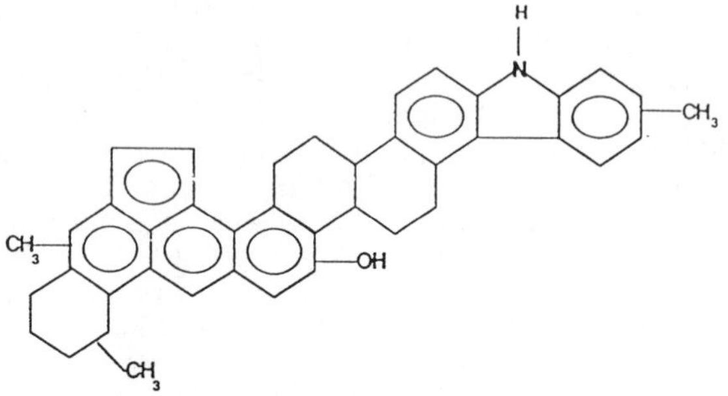

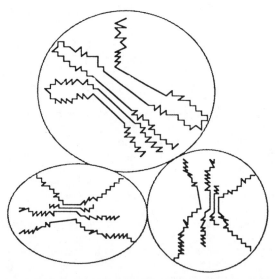

Figure 1. An example of a hypotetical structure of asphaltene, among the many suggested, showing their aromatic character.

Figure 2. Asphaltene structure deduced from microscopic and macroscopic analysis, showing their micro- and macro-molecular bonding. T. F. Yen, 1972, first suggested this type of structure.

explained that by considering molecular weight and molecular polarity as separate properties of molecules, the solvent-precipitation behavior of materials derived from various carbonaceous sources can be understood. Future quantification of Long's approach probably can be achieved by developing a polarity scale based on solubility parameter.

According to Marcusson (11) there is a close relation between asphaltenes, resins, and high molecular weight polycyclic aromatic hydrocarbons which may exist in petroleum. The heavy polycyclic aromatics of petroleum, on oxidation gradually form neutral resins (and probably asphaltogenic acids,); asphaltenes are made as a result of further oxidation of neutral resins. On the contrary, the hydrogenation of asphaltic products containing neutral resins and asphaltenes produces heavy hydrocarbon oils, i.e., neutral resins and asphaltenes are hydrogenated into high molecular weight polycyclic aromatic structures. Asphaltenes are very similar to neutral resins in the ultimate analysis. The process of transformation of neutral resins into asphaltenes is very simple even at low temperatures. Specially neutral resins can easily be converted to asphaltenes in the presence of air or oxygen at elevated temperatures. Thus neutral resins and asphaltenes are similar in their chemical structure.

The physical and physico-chemical properties of asphaltenes are different with those of neutral resins. The molecular weight of asphaltenes is very high. Published data (9,12,13,14,15,16) for the molecular weight of petroleum asphaltenes range from approximately 1,000 to 2,000,000. The reported molecular weight of asphaltene varies considerably depending upon the method of measurement. A major concern in reporting molecular weights is the association of asphaltenes which can exist at the conditions of the method of measurement. Vapor pressure osmometry (VPO) has become the most prevalent method for determining asphaltene molecular weights. However, the value of the molecular weight from VPO must be weighed carefully since, in general, the measured value of molecular weight is a function of the solvent and its dielectric constant. Reported molecular weights from ultracentrifuge and electron microscope studies are high. To the contrary, those from solution viscosity and cryoscopic methods are low. Asphaltenes are lyophilic with respect to aromatics, in which they form highly scattered colloidal solutions. Specifically, asphaltenes of low molecular weight are lyophobic with respect to paraffins like pentanes and petroleum oils.

Thus the degree of dispersion of asphaltenes in petroleum oils depends upon the chemical composition of the oil. In heavy highly aromatic oils the asphaltenes are colloidally dispersed; but in the presence of an excess of petroleum ether and similar paraffinic hydrocarbons they are coagulated and precipitated. The coagulated and precipitated asphaltenes can be re-peptized (colloidally dispersed) by the addition of aromatics. Neutral resins are particularly effective as peptizers. Such components as resins or high molecular weight aromatics are readily adsorbed by asphaltenes and act as protective layers, isolating the colloidal particles from the coagulative action of lyophobic constituents of petroleum oils. If the proportion of the peptizing constituents in a petroleum oil is sufficient, the asphaltenes form stable suspensions.

Sachanen (11) explained that the solution of asphaltenes in an aromatic solvent is preceded by swelling of the powdered asphaltenes accompanied by evolution of heat. Nellensteyn discovered that the peptizing or precipitating properties of different liquids with respect to asphaltenes are closely related to the surface tension. Flocculation occurs when the solvent has a surface tension below 24 dynes/ cm at 25°C. Total peptization takes place when the surface tension exceeds 26 dynes/ cm. In the intermediate zone between 24 and 26 dynes/ cm, either flocculation or peptization may occur, depending on the properties of the asphaltenes. Asphaltenes in solution are generally recognized as colloidal systems. Several investigators have observed that colloidal solutions can be precipitated by flow through capillaries and porous media (4,7). The precipitation is believed to result from the electrical interactions with the walls which disturb the stabilizing electrical forces around the colloidal particles and the particles agglomerate to a precipitate. The asphaltene particles are electrically charged and thus can be precipitated by application of an electrical potential or by flow of the asphaltene-containing product through the sand, due apparently to electrical effects resulting from the flow. On the contrary, application of a counter-potential may prevent precipitation of asphaltenes from the crude oil flowing through a porous material.

Asphaltenes are not crystallized and cannot be separated into individual components or narrow fractions. Thus, the ultimate analysis is not very significant, particularly taking into consideration that the neutral resins are strongly adsorbed by asphaltenes and probably cannot be effectively separated from them. Not enough is known of the chemical properties of asphaltenes. On heating, they are not melted, but decompose, forming carbon and volatile products above 300-400 °C. They react with sulfuric acid forming sulfonic acids, as might be expected on the basis of the polyaromatic structure of these components. The color of dissolved asphaltenes is deep red at very low concentration in benzene as 0.0003 per cent makes the solution distinctly yellowish. The color of crude oils and residues is due to the combined effect of neutral resins and asphaltenes. The black color of some crude oils and residues is related to the presence of asphaltenes which are not properly peptized.

Field Experiences

Asphaltene deposition during oil production and processing is a very serious problem in many areas throughout the world. In certain oil fields (12,13,14) there have been wells that, especially at the start of production, would completely cease flowing in a matter of a few days after an initial production rate of up to 3,000 BPD. The economic implications of this problem are tremendous considering the fact that a problem well workover cost could get as high as a quarter of a million dollars. In another field the formation of asphaltic sludges after shutting in a well temporarily and/or after stimulation treatment by acid has resulted in partial or complete plugging of the well (7). Still in another field, deposit of asphaltenes in the tubing was a very serious production problem and necessitated frequent tubing washings or scrapings to maintain production (17). Asphaltenes have played a significant role in the production history and economics of the deep horizons of some oil reservoirs (18). For example in one case asphaltene problems have ranged from asphaltene deposition during early oil production to asphaltene flocculation and deposition resulting from well acidizing and carbon dioxide injection for enhanced oil recovery.

Table I. Elemental compositions of asphaltenes precipitated by different flocculants from various sources (16)

Source	Flocculant	C	H	N	O	S	H/C	N/C	O/C	S/C
		Elemental Composition(wt%)					AtomicRatios			
Canada	n-pentane	79.5	8.0	1.2	3.8	7.5	1.21	.013	.036	.035
	n-heptane	78.4	7.6	1.4	4.6	8.0	1.16	.015	.044	.038
Iran	n-pentane	83.8	7.5	1.4	2.3	5.0	1.07	.014	.021	.022
	n-heptane	84.2	7.0	1.6	1.4	5.8	1.00	.016	.012	.026
Iraq	n-pentane	81.7	7.9	0.8	1.1	8.5	1.16	.008	.010	.039
	n-heptane	80.7	7.1	0.9	1.5	9.8	1.06	.010	.014	.046
Kuwait	n-pentane	82.4	7.9	0.9	1.4	7.4	1.14	.009	.014	.034
	n-heptane	82.0	7.3	1.0	1.9	7.8	1.07	.010	.017	.036

Even for reservoirs in which asphaltene deposition was not reported previously during the primary and secondary recovery, it was reported that asphaltene deposits were found in the production tubing during carbon dioxide injection enhanced oil recovery projects (18).

Asphaltene precipitation, in many instances, carries from the well tubing to the flow lines, production separators, and other downstream equipment. It has also been reported (19) that asphaltic bitumen granules occured in the oil and gas separator with oil being produced from certain oil fields.

The downtime, cleaning, and maintenance costs are a sizable factor in the economics of producing a field prone to asphaltene deposition. Considering the trend of the oil industry towards deeper reservoirs, heavier and as a result asphaltic crudes, and the increased utilization of miscible gas injection techniques for recovering oil, the role of asphaltene deposition in the economic development of asphaltene containing oil discoveries will be important and crucial.

Modeling of Asphaltene Flocculation, and its Interaction with Oil

Solution of the asphaltene problem calls for detailed   analyses of asphaltene containing systems from the statistical mechanical standpoint and development of molecular models which could describe the behavior of asphaltenes in hydrocarbon mixtures. From the available laboratory and field data it is proven that the asphaltene which exists in oil consists of very many particles having molecular weights ranging from one thousand to several hundred thousands. As a result distribution-function curves are used to report their molecular weights. The wide range of asphaltene size distribution suggests that asphaltenes may be partly disolved and partly in colloidal state (in suspension) peptized (or stabilized) primarily by resin molecules that are adsorbed on asphaltene surface (12,13,14,19,20). As a result, a realistic model for the interaction of asphaltene and oil should take into account both the solubility in oil of one segment and suspension characteristic (due to resins) of another segment of the molecular weight distribution curve of asphaltene. We have proposed two different models (12,13,21,22) which are based on statistical mechanics of particles (monomers and polymers) dissolved or suspended in oil. A combination of the two models is general enough to predict the asphaltene-oil interaction problems (phase behavior or flocculation) wherever it may occur during oil production and processing.

Such a model may be constructed by joining the concepts of continuous thermodynamic theory of liquid-solid phase transition, fractal aggregation theory of colloidal growth, and steric colloidal collapse and deposition models.

Solubility Model of Interaction of Asphaltene and Oil : It is generally assumed that two factors are responsible for maintaining the mutual solubility of the compounds in a complex mixture such as the petroleum crude: These are the ratio of polar to nonpolar molecules and the ratio of the high molecular weight to low molecular weight molecules in the mixture.   Of course, polar and nonpolar compounds are basically immiscible, and light and heavy molecules of the same kind are partially miscible depending on the differences between their molecular weights. However, in the complex mixture of petroleum crude or coal liquids and the like all these compounds are probably mutually soluble so long as a certain ratio of each kind of molecule is maintained in the mixture. By introduction of a solvent into the mixture this ratio is altered. Then the heavy and/or polar molecules separate from the mixture either in the form of another liquid phase or to a solid precipitate. Hydrogen bonding and the sulfur and/or the nitrogen containing segments of the separated molecules could start to aggregate (or polymerize) and as a result produce the irreversible asphaltene deposits which are insoluble in solvents.   In order to formulate the necessary model for prediction of the "onset of deposition" of asphaltene we have taken advantage of the theories of polymer solutions (22,23,24). Both asphaltenes and asphaltene-free crude consist of mixtures of molecules with a virtually continuous molecular weight distributions in order to formulate the theory of interaction of oil and asphaltene systems we can utilize the concept of continuous mixture (16,21) joined with the thermodynamic theory of heterogeneous polymer solutions (22,24). We have already formulated the necessary continuous mixture model for the prediction of the "onset of deposition" of asphaltene due to injection of a miscible solvent. In the course of our ongoing research we intend to extend our model to varieties of crude oils and miscible solvents of practical interest **(Figure 3)**.

Suspension Model of Interaction of Asphaltene and Oil : This model is based upon the concept that asphaltenes exist as particles suspended in oil. Their suspension is assisted by resins (heavy and mostly aromatic molecules) adsorbed to the surface of asphaltenes and keeping them afloat because of the repulsive forces between resin molecules in the solution and the adsorbed resins on the asphaltene surface (see Figure 4). Stability of such a suspension is considered to be a function of the concentration of resins in solution, the fraction of asphaltene surface sites occupied by resin molecules, and the equilibrium conditions between the resins in solution and on the asphaltene surface. Utilization of this model requires the following (12): 1. Resin chemical potential calculation based on the statistical mechanical theory of polymer solutions. 2. Studies regarding resin adsorption on asphaltene particle surface and measurement of the related Langmuir constants. 3. Calculation of streaming potentials generated during flow of charged asphaltene particles. 4. Development and use of asphaltene colloidal and aggregation models for estimating the amount of asphalt which may be irreversibly aggregated and flocculated out. The amount of resins adsorbed is primarily a function of their concentration in the liquid state (the oil). So, for a given system (i.e., fixing the type and amount of oil and asphaltenes) changing the concentration of resins in the oil will cause the amount of resins adsorbed on the surface to change accordingly. This means that we may drop the concentration of resins in the oil to a point at which the amount of resins adsorbed is not high enough to cover the entire surface of asphaltenes. This may then permit the asphaltene particles to come together (irreversible aggregation), grow in size, and flocculate.

One major question of interest is how much asphaltene will flocculate out under certain conditions. Since the system under study consist generally of a mixture of oil, aromatics, resins, and asphaltenes it may be possible to consider each of the constituents of this system as a continuous or discrete mixture (depending on the number of its components) interacting with each other as pseudo-pure-components. The theory of continuous mixtures (24), and the statistical mechanical theory of monomer/polymer solutions, and the theory of colloidal aggregations and solutions are utilized in our laboratories to analyze and predict the phase behavior and other properties of this system.

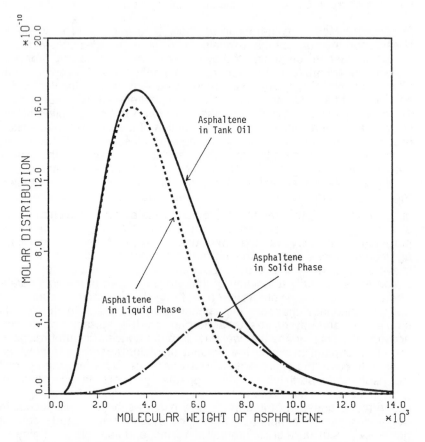

Figure 3. Molecular weight distributions of asphaltenes before and after flocculation predicted by our continuous mixture model.

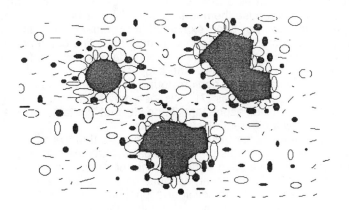

NOTES

1.0 ◯ represents resin molecules

2.0 ● represents aromatic molecules

3.0 ⌒ represents oil molecules of different size and paraffinic nature

4.0 ◼ represents asphaltene particles of different sizes and shapes

Figure 4. Asphaltene particle peptization effected by adsorbed resin molecules. This physical model is the basis of our asphaltene Thermodynamic-Colloidal Model.

Discussions

It has been shown experimentally (4,7,12,13,14) beyond any reasonable doubt that the electrical charge of colloidal asphaltenes is a very important property and, regardless of the charge sign, it seems possible to device colloidal asphaltene deposition preventive measures by controlling the electrical effects attributed to the charge of asphaltenes. The primary electrokinetic phenomenon in effect is the "streaming potential" generated by the movement of the electrically charged colloidal asphaltene particles due to the flow of oil. This streaming potential seems to neutralize the similar charge of the colloidal asphaltene particles and cause them to flocculate. The electric charge of colloidal asphaltenes has not been explained yet, primarily because of the complexity of the composition of asphaltic materials. The difference in charge (+or-) displayed by asphaltene particles derived from different crudes has not been explained either. One suggestion has been that the large quantities of nickel and vanadium found in asphaltene deposits may hold the key to these charges (7). This idea may be investigated by analyzing metal contents of asphaltene deposits that contain colloidal asphaltene particles with different electric charge.

One thing that appears to be universally accepted is that resins in the petroleum act as peptizing agents of the colloidal asphaltene particles. A number of experiments have been performed that point out the peptizing role of resins (4,7,12,19,25). However, because of the significance of the resins as peptizing agents of the colloidal asphaltene particles and the fact that based on current experimental information they appear to be the main hope for combating the colloidal asphaltene problem in the field, more experiments must be performed to establish clearly and beyond any doubt the resin role in the asphaltene deposition problem and generate enough thermodynamic properties of the resins to be utilized in modeling efforts to the problem.

Experimental evidence (4,25) suggests that for an oil mixture there is a critical concentration of resins below which the colloidal asphaltene particles may flocculate and above which they cannot flocculate regardless of how much the oil mixture is agitated, heated, or pressurized sort of changing its composition. The authors believe that there should be certain unique geological conditions which favor the formation of an oil whose actual resin concentration is less than its critical resin concentration. Such conditions are responsible for the transformations the hydrocarbon deposits undergo. These geological conditions could conceivably be identified and established after sufficient experimental data on actual resin concentration and critical resin concentration are generated for different crude oils around the world. As a result, geological conditions alone could provide clues for predicting potential asphaltene problems for oils that have not yet been produced.

The postulated and sufficiently proven notion that asphaltenes are oxidation products of resins and that resins are oxidation products of oil (11) sort of makes the probability of finding oils whose actual resin concentration is less than their critical resin concentration small. In other words, an oil deposit must have a substantial amount of resins before starting to form asphaltenes during the natural geological transformation process. Table II seems to corroborate this statement. Thus, it would be interesting to find out what kind of geological conditions are suitable for generating oils whose actual resin concentration is less than their critical resin concentration. A model which is based on the notion of the resin critical concentration described above and predicts the phase behavior of asphaltenes in oil mixtures was published recently by the authors (12). Since asphaltene deposition takes place during primary, secondary and tertiary oil recovery, and in the refinery, injection of peptizing agents (i.e.: resins) in proper amounts and places could prevent or at least control the colloidal asphaltene deposition problem. Furthermore, experiments could be performed (i.e.: of the coreflood type) where peptizing agents are injected to study their effect on inhibition of asphaltene deposition or permeability reductions.

Table II. Resin and Asphaltene Content of Crude Oils (11)

Crude Oil	Sp.gr. (60°/60°F)	Resins (% by wt)	Asphaltenes (% by wt)
Pennsylvania	0.805	1.5	0.0
Oklahoma, Tonkawa	0.821	2.5	0.2
Oklahoma, Okla. City	0.835	5.0	0.1
Oklahoma, Davenport	0.796	1.3	0.0
Texas, Hould	0.936	12.0	0.5
Texas, Mexia	0.845	5.0	1.3
Louisiana, Rodessa	0.807	3.5	0.0
Calif., Huntington Beach	0.897	19.0	4.0
Mexico, Panuco	0.988	26.0	12.5
Russia, Surachany	0.850	4.0	0.0
Russia, Balachany	0.867	6.0	0.5
Russia, Bibi-Eibat	0.865	9.0	0.3
Russia, Dossor	0.862	2.5	0.0
Russia, Kaluga	0.955	20.0	0.5
Asia, Iraq (Kirkuk)	0.844	15.5	1.3
Mississippi, Baxterville[18]	0.959	8.9	17.2

One interesting question posed by previous researchers (14,19) is why there was asphaltic bitumen deposited at the bottom of the well considering that no phase change or any substantial temperature or pressure changes had taken place. The conclusion was that the question could only be answered after considerable light was thrown upon the nature of the asphaltic bitumen prior to its separation from the crude oil in the well. There were a few efforts to try to determine the size and nature of asphaltene particles while they still are in the original oil (15,19). Katz and Beu (19) did not see any asphaltene particles in the original oil of size 65 Å or larger, but they did see these particles after mixing the crude with solvents. They concluded that the particles, if they do exist, must be smaller than 65 Å. Witherspoon et al. (15), using ultracentrifuge techniques, found that the particles that eluded Katz and Beu do exist and are of the 35-40 Å range. By careful work of electron micrography with rapid lyophilization, the size of asphaltene is found to be 20-30 Å (27). In native oil or solutions, the asphaltene particle size can be doubled (28,29).

Conclusions

Because the asphaltene problem is so elusive, it seems that, before one can formulate a comprehensive model describing the problem, the true asphaltene deposition mechanism(s) must be clearly understood and backed by field and experimental data; then an accurate and representative model can be formulated. In this report we have tried to address this basic question. The mere fact that there are several basic schools of thought with regards to the asphaltene deposition problem points out that we are still far from formulating a universally accepted model for describing the behavior of asphaltenes in crude oil.Establishing the state of the asphaltene particles in the original crude oil seems to be a basic building block in the scientific quest to find a solution to the asphaltene deposition problem. More experiments must be done to duplicate Witherspoon et al.s' ultracentrifuge work for different oils and possibly utilize other contemporary experimental techniques to establish the state of asphaltenes in crude oils. Meanwhile, a model that describes the phase behavior of asphaltenes in oil must take into account the lack of positive information on the structure of asphaltenes in the original oil and rely as little as possible on the concept of an asphaltene molecule or of specific properties of asphaltenes. This was the philosophy that we followed in our models, mentioned earlier, predicting the behavior of asphaltenes in petroleum.

Acknowledgments

This research is supported in part by the National Science Foundation Grant CBT-8706655 and in part by the Shell Oil Company.

Literature Cited

1. Mansoori, G.A.; Jiang, T.S. Proc. of the 3rd European Conference on Enhanced Oil Recovery, Rome, Italy, April,1985.
2. Stalkup, F.I. Miscible Displacement: Society of Petroleum Engineers Monograph, June 1983; Chapters 1,2,3

3. David, A., Asphaltenes Flocculation During Solvent Simulation of Heavy Oils. American Institute of Chemical Engineers, Symposium Series 1973, 69 (no. 127), 56-8.
4. Preckshot, C.W.; Dehisle, N.G.; Cottrell, C.E.; Katz, D.L., Asphaltic Substances in Crude Oil Trans. AIME 1943, 151, 188.
5. Shelton, D.A.; Yarborough, L. J. of Petroleum Technology 1977,1171.
6. Cole, R.J.; Jessen, F.W. Oil and Gas Journal 1960, 58, 87.
7. Lichaa, P.M. and Herrera, L. Society of Petroleum Engineers Journal, paper no. 5304, 1975, 107.
8. Speight, J.G.; Long, R.G.; Trawbridge, T.D. Fuel 1984, 63, no. 5, 616.
9. Long, R.B. In Cemistry of Asphaltenes; Bunger, J.W.; Li, N.C., Eds.; Advances in Chemistry Series No. 195; Washington, DC, 1981; p 17.
10. Yen, T.F., Present Status of the Structure of Petroleum Heavy Ends and its Significance to Various Technical Applications. Preprints ACS, Div. Pet. Chem. 1972, 17, No. 4, 102.
11.Sachanen, A.N. The chemical constituents of petroleum; Reinhold Publishing Corp., 1945.
12. Leontaritis, K.J.; Mansoori, G.A. SPE Paper#16258; Proceedings of the 1987 SPE Symposium on Oil Field Chemistry, Society of Petroleum Engineers, Richardson, TX, 1987.
13. Leontaritis, K.J.; Mansoori, G.A.; Jiang, T.S., Asphaltene Deposition in Oil Recovery: a Survey of Field Experiences and Research Approaches J. of Petr. Sci. and Eng.,1988 (to appear).
14. Adalialis, S. M.Sc. thesis, Petroleum Engineering Department, Imperial College of the University of London, London, 1982.
15. Ray, B.R.; Witherspoon, P.A.; Grim, R.E. J. Phys. Chem. 1957, 61, 1296,.
16. Speight, J.G.; Moschopedis, B.C. In Cemistry of Asphaltenes; Bunger, J.W.; Li, N.C., Eds.; Advances in Chemistry Series No. 195; Washington, DC, 1981; p 115.
17. Haskett, C.E.; Tartera, M. Journal of Petroleum Technology 1965, 387-91.
18. Tuttle, R.N. Journal of Petroleum Technology 1983, 1192.
19. Katz, D.L.; Beu, K.E. Ind. and Eng. Chem. 1945, 37, 195.
20. Koots, J.A.; Speight, J.C., Fuel, 54 ,1975, p179.
21. Mansoori, G.A.; Jiang. T.S.; Kawanaka, S. Arabian J. of Sci. & Eng. 1988, 13, No. 1, 17.
22. Du, P.C.; Mansoori, G.A. SPE Paper #15082, Proc. of the 1986 California Regional Meeting of SPE, Society of Petroleum Engineers, Richardson, TX, 1986.
23.Hirschberg, A. ; de Jong, L.N.J. ; Schipper, B.A.; Meijers, J.G., Society of Petroleum Engineers Journal 1984, 24, No 3, 283-293.
24. Scott, R.L.; Magat, M. J. Chem. Phys. 1945, 13, 172; Scott, R.L. J. Chem. Phys. 1945, 13, 178.
25. Swanson, J. Phys. Chem. 1942, 46, 141.

26. Kawanaka, S.; Park, S.J.; and Mansoori, G.A.   SPE/DOE Paper
    #17376; Proc. 1988 SPE/DOE Symposium on Enhanced Oil
    Recovery, p 617, Society of Petroleum Engineer, Tulsa, OK, 1988.
27. Dickie, J.P.; Haller, M.N.; Yen, T.F. J. Coll. Interface Sci. 1969, 29,
    475.
28. Dwiggens, C.W. J. Phys. Chem. 1965, 69, 3500.
29. Pollack, S.S.; Yen, T.F. Anal. Chem. 1970, 42, 623.

RECEIVED March 8, 1989

FOAM PROPERTIES IN POROUS MEDIA

Chapter 25

Dynamic Stability of Foam Lamellae Flowing Through a Periodically Constricted Pore

A. I. Jiménez and C. J. Radke

Department of Chemical Engineering, University of California, Berkeley, CA 94720

The stability threshold or critical capillary pressure of foam flowing in porous media depends on the flow rate, with higher velocities breaking the foam. We present a hydrodynamic theory, based on a single lamella flowing through a periodically constricted cylindrical pore, to predict how the critical capillary pressure varies with velocity. As the lamella is stretched and squeezed by the pore wall, wetting liquid from surrounding pores fills or drains the moving film depending on the difference between the conjoining/disjoining pressure and the porous-medium capillary pressure. The interplay between stretching/squeezing and draining/filling ascertains the critical velocity at which the film breaks in a given porous medium of fixed wetting-liquid saturation.

Comparison of the proposed dynamic stability theory for the critical capillary pressure shows acceptable agreement to experimental data on $100\text{-}\mu\text{m}^2$ permeability sandpacks at reservoir rates and with a commercial α-olefin sulfonate surfactant. The importance of the conjoining/disjoining pressure isotherm and its implications on surfactant formulation (i.e., chemical structure, concentration, and physical properties) is discussed in terms of the Derjaguin-Landau-Verwey-Overbeek (DLVO) theory of classic colloid science.

Foam generated in porous media consists of a gas (or a liquid) dispersed in a second interconnected wetting liquid phase, usually an aqueous surfactant solution (1). Figure 1 shows a micrograph of foam flowing in a two-dimensional etched-glass porous medium micromodel (replicated from a Kuparuk sandstone, Prudhoe Bay, Alaska (2)). Observe that the dispersion microstructure is not that of bulk foam. Rather discontinuous

0097–6156/89/0396–0460$06.00/0
© 1989 American Chemical Society

gas bubbles essentially fill the pore space. They are separated
from each other by thin surfactant-stabilized lamellae and from
the pore walls by liquid channels and films. Each lamella joins
the pore wall through curved regions known as Plateau borders
(3). The lamellae can either block flow paths or flow as
connected bubble trains (4). Not all of the gas need be discontinuous (1). Some can transport as a continuous phase.

The primary factor controlling how much gas is in the form of
discontinuous bubbles is the lamellae stability. As lamellae
rupture, the bubble size or texture increases. Indeed, if bubble
coalescence is very rapid, then most all of the gas phase will be
continuous and the effectiveness of foam as a mobility-control
fluid will be lost. This paper addresses the fundamental mechanisms underlying foam stability in oil-free porous media.

Lamellae longevity in porous media is dominated by the mean
capillary pressure difference between the nonwetting foam phase
and the continuous wetting aqueous phase. Consider Figure 2 which
reports the transient pressure drop across the 7-cm long etched-
glass micromodel of Figure 1 upon injection of an 80% quality foam
stabilized by a 0.5 wt % active commercial α-olefin sulfonate
surfactant of carbon number between 14 and 16 (ENORDET AOS 1416).
The foam is steadily generated in an identical upstream micromodel
and flows into the second downstream micromodel where the pressure
behavior is recorded. The two pressure histories labelled wet and
dry refer, respectively, to the downstream micromodel being completely aqueous surfactant saturated or being completely dry.
For the wet case, the foam enters and achieves steady state after
several pore volumes. A mobility reduction compared to water of
about 90% ensues. However, for the dry case, there is about a one
pore-volume time lag before the pressure responds. During this
time, visual observations into the micromodel indicate a catas-
tropic collapse of the foam at the inlet face. The liquid surfactant solution released upon collapse imbibes into the smaller
pores of the medium. Once the water saturation rises to slightly
above connate (ca 30%), foam enters and eventually achieves the
same mobility as that injected into the wet medium.

These observations are consistent with the concept of a
critical capillary pressure for foam to exist in porous media.
The dry micromodel porous medium exerted essentially an infinite
capillary suction pressure and the foam lamellae drained to
breakage at the inlet. After enough wetting liquid filled the
smaller pores to decrease the capillary pressure below a critical
value, P_c^*, foam could enter. Recently Khatib, Hirasaki and Falls
(5) directly measured the critical capillary pressure of foam
coalescence for bead packs of permeabilities ranging from 72 to
8970 μm^2 and for various surfactant formulations. These authors
found that with all other variables held constant, P_c^* decreases
with increasing flow velocity (see Figure 8 of (5)). That is, for
a porous medium with a given permeability and wetting-phase
saturation (and corresponding capillary pressure) there is a
critical gas velocity above which lamellae rupture. Similarly, if
the capillary pressure is increased at a given velocity of flow,
the foam will break. These findings suggest that for higher
velocities the wetting-liquid content of the medium must increase

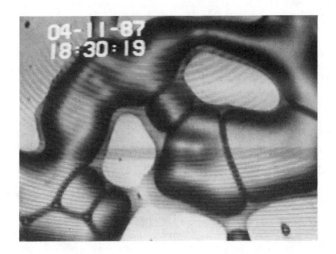

Figure 1. Micrograph of foam in a 1.1 μm^2, two dimensional etched-glass micromodel of a Kuparuk sandstone. Bright areas reflect the solid matrix while grey areas correspond to wetting aqueous surfactant solution next to the pore walls. Pore throats are about 30 to 70 μm in size. Gas bubbles separated by lamellae (dark lines) are seen as the nonwetting "foam" phase.

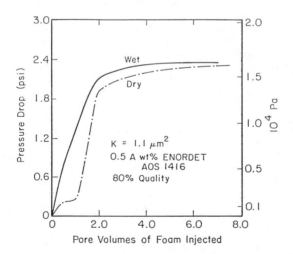

Figure 2. Transient pressure drop across the porous-medium micromodel of Figure 1 for foam pregenerated in an identical upstream medium. The foam frontal advance rate is 186 m/d. In the wet case, foam advanced into the downstream micromodel which was completely saturated with aqueous surfactant solution. In the dry case, the downstream micromodel contained only air.

proportionately in order to reduce P_c^* enough for foam to remain stable.

The purpose of this paper is to explain quantitatively the origin of a critical capillary pressure for foam existence in porous media. We first address the value of P_c^* for static foam that is not flowing and show that it depends on the medium wetting-liquid saturation and absolute permeability in addition to surfactant formulation. We then focus on dynamic or flowing foam stability. A hydrodynamic theory is introduced to explain how P_c^* depends on velocity by analyzing the stabilty of foam lamellae moving through a periodically constricted sinusoidal pore.

Stability of Static Foam in Porous Media

To explain the role of the medium capillary pressure upon foam coalescence, consider a flat, cylindrical, stationary foam lamella of thickness, 2h, circa 1000 Å, and radius, R (i.e., 50 to 100 μm), subject to a capillary pressure, P_c, at the film meniscus or Plateau border, as shown in Figure 3. The liquid pressure at the film meniscus is $(P_g - P_c)$, where P_g is the gas pressure.

Inside the lamella, there is an excess force or conjoining/disjoining pressure, Π, as introduced and tested experimentally by Derjaguin et al. ($\underline{6}$,$\underline{7}$). The conjoining/disjoining pressure is a function of the film thickness, 2h. A typical isotherm for Π(h) is shown in dimensionless form in Figure 4. In this case the very short range molecular contributions ($\underline{8}$), thought to be of structural origin, are not shown because such ultrathin films seem unlikely for foam application in porous media. The particular form of $\hat{\Pi}$ $(\hat{h})$ in Figure 4 is calculated from the constant potential and weak overlap subcase of the DLVO theory ($\underline{9}$):

$$\hat{\Pi} = - \frac{1}{\theta \, \hat{h}^3} + \exp(-\hat{h}) , \qquad (1)$$

where $\hat{\Pi} = \Pi/B$, $\hat{h} = 2\kappa h$, and $\theta = B/\kappa^3 A_H$. The first term on the right side of this expression reflects attractive dispersion forces while the second term corresponds to repulsive electrostatic double-layer forces. B is a known function of the surface potential, ionic strength, surfactant concentration, and temperature ($\underline{9}$), A_H is a Hamaker constant, and $1/\kappa$ is the Debye length. θ is a measure of the ratio of repulsive to attractive forces whose value in Figure 4 is set at 5. Later we shall also employ the constant and low surface charge density (i.e., q) form of the DLVO theory ($\underline{9}$):

$$\hat{\Pi} = - \frac{1}{\theta \, \hat{h}^3} + \text{csch} \ (\hat{h}/2) , \qquad (2)$$

where $\hat{\Pi} = \epsilon\Pi/2\pi q^2$, $\theta = 2\pi q^2/\epsilon\kappa^3 A_H$, and ϵ is the bulk permittivity of the surfactant solution.

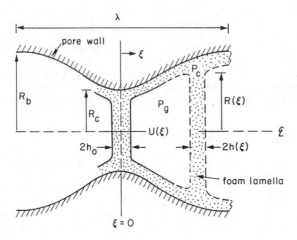

Figure 3. Schematic of flowing lamellae in a periodic constricted pore. The porous medium imposes a capillary pressure, P_c, at each Plateau border.

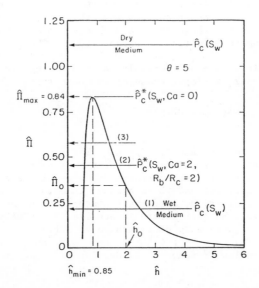

Figure 4. A conjoining/disjoining pressure isotherm for the constant- potential and weak overlap electrostatic model.

The driving force for exchange of fluid into and out of the film is $(P_c - \Pi)$. When P_c is greater than Π the pressure difference $(P_c - \Pi)$, drives liquid out of the lamella and vice versa. If it happens that $P_c = \Pi$, then an equilibrium metastable film is possible. Whether a particular intersection of P_c and Π yields a metastable or an unstable foam film depends on the specific shape of the conjoining/disjoining curve (10-12).

For a nonthinning, unbounded film, Vrij (10) showed via a thermodynamic analysis (i.e., a surface energy minimization) that when $\partial\Pi/\partial h$ is greater than zero the film is unstable. Thus, in Figure 4 the critical thickness limit for metastable films, $\hat{h}_{min}$, is given by the maximum of the conjoining/disjoining curve, Π_{max}, or equivalently when $\partial\Pi/\partial h = 0$ for any value of θ. Unbounded films thinner than $\hat{h}_{min}$ are unstable to infinitesimal perturbations; films thicker than $\hat{h}_{min}$ are stable to such perturbations.

It is now possible to explain the origin of a critical capillary pressure for the existence of foam in a porous medium. For strongly water-wet permeable media, the aqueous phase is everywhere contiguous via liquid films and channels (see Figure 1). Hence, the local capillary pressure exerted at the Plateau borders of the foam lamellae is approximately equal to the mean capillary pressure of the medium. Consider now a relatively dry medium for which the corresponding capillary pressure in a dimensionless sense is greater than $\hat{\Pi}_{max}$, as shown in Figure 4. Given sufficient time any foam lamella present in such a medium will thin down to a thickness less than $\hat{h}_{min}$ and rupture. Hence, the entire foam breaks. This explains why foam cannot enter a completely dry porous medium. Conversely, for a relatively wet medium with $\hat{P}_c < \hat{\Pi}_{max}$ the lamellae will establish a metastable equilibrium state with $\hat{\Pi} = \hat{P}_c$ and $\hat{h} > \hat{h}_{min}$. In this case, unless subjected to large disturbances (i.e., those with amplitude of order h), the foam lamellae remain intact and the foam is metastable. Clearly for a static foam, a critical capillary pressure at $\hat{P}_c^* = \hat{\Pi}_{max}$ demarks the boundary between metastable and unstable foam lamellae.

In a given porous medium at a fixed water saturation (i.e., with $P_c(S_w)$ set), static foam stability depends solely on the value of Π_{max}. If $P_c(S_w)$ is greater than Π_{max}, the foam must break. Π_{max} in turn is determined by the surfactant formulation, since the shape of the conjoining/disjoining pressure isotherm reflects the surfactant charge, size, structure, and concentration as well as the ionic strength and hardness of the aqueous solvent.

A given surfactant package characterized by a very high value of Π_{max} will produce a highly stable foam.

Since, in general, lower permeability media exhibit higher capillary-pressure suction, we argue that it is more difficult to stabilize foam when the permeability is low. Indeed the concept of a critical capillary pressure for foam longevity can be translated into a critical permeability through use of the universal Leverett capillary-pressure J-function (13) and, by way of example, the constant-charge model in Equation 2 for Π:

$$K^* = \phi \left[\frac{J\sigma\epsilon}{2\pi \ q^2 \ \hat{\Pi}_{max}} \right]^2 , \tag{3}$$

where ϕ is the porosity, σ is the surfactant solution equilibrium surface tension, and J is a known function of the wetting-phase saturation, S_w (13). For absolute permeabilities below K^* the foam is unstable, while above K^*, it is metastable. Figure 5 shows some typical curves generated from Equation 3 for K^* as a function of the liquid saturation, S_w, for two indifferent ionic concentrations. As expected, lower water saturations require a more permeable medium for foam stability. With the simple constant-charge electrostatic model, there is also a strikingly strong dependence of the critical permeability on the ionic concentration. At low salt concentrations foam can survive in low permeability media. According to DLVO theory, increasing significantly the ionic concentration for a given absolute permeability and water saturation is detrimental to stability. We can conclude that the stability of static foam in porous media depends on the medium permeability and wetting-phase saturation (i.e., through the capillary pressure) in addition to the surfactant formulation. More importantly, these effects can be quantified once the conjoining/disjoining pressure isotherm is known either experimentally (8) or theoretically (9). Our focus now shifts to the velocity dependence of P_c^* (and K^*).

Stability of Dynamic Foam in Porous Media

To explain coalescence of dynamic foam in porous media consider a single lamella flowing through a periodically constricted sinusoidal pore, as shown in Figure 3. As noted above, the capillary pressure of the medium imposed on the lamella at the pore wall is assumed constant and is set by the local wetting-liquid saturation. Assuming constant volumetric flow, Q, the lamella transports from pore constriction, of radius R_c, to pore body, of radius R_b, with a local interstitial velocity $U(\xi)$ which varies according to the square of the pore radius given by the periodic function, $\hat{R} = R(\xi)/R_c$:

$$\hat{R}(\xi) = (1+\underline{a}) + \underline{a} \ \cos[\pi(1+2\xi/\lambda)] , \tag{4}$$

where ξ is the axial distance measured from the pore constriction, and λ is the wavelength of the periodic pore. <u>a</u> characterizes the pore structure and is given by $(1/2)$ $(R_b/R_c$ $-1)$. Thus, for a perfectly straight pore <u>a</u> is equal to zero.

Upon moving from the pore constriction $(\xi=0)$ to the pore body $(\xi = \lambda/2)$, the lamella is stretched as it conforms to the wall. To achieve the requisite volume rearrangement a radial pressure differential is induced which thins the film but results in no net fluid efflux into the Plateau borders. The converse occurs when the film is squeezed upon moving from a pore body to a pore constriction. If R_b/R_c, or equivalently <u>a</u>, is large enough to thin the lamella at the pore body to about h_{min} in Figure 4, then rupture is imminent. Since eventual encounter with a large pore body is virtually assured, this reasoning predicts the collapse of all foam lamellae independent of the flow velocity.

Conversely, whenever the stretched lamella of local thickness h exerts a conjoining/disjoining pressure $\Pi(h)$ which lies above the value of P_c, then fluid transports out of the pendular wetting liquid surrounding the sand grains and into the thin lamellae to prevent the thickness from falling below h_{min}. Unfortunately, the thin lamella resists instantaneous fluid exchange. According to the simple Reynolds parallel-film model (<u>14</u>,<u>15</u>) adopted below, this hydrodynamic resistance is inversely proportional to h^3 (we ignore any resistance to liquid flow along the channels lining the pore walls and in the Plateau borders (<u>16</u>)). When the foam flow rate Q is low, there is sufficient time to heal the thinning of a lamella at the pore body. When Q is raised sufficiently, however, healing cannot occur quickly enough and the film thickness drops below h_{min} initiating breakage. Accordingly, for a given capillary pressure in a porous medium there will be a specific foam flow rate at which the lamellae rupture. This is the proposed origin of the flow-rate dependence of P_c^* measured by Khatib, Hirasaki and Falls (<u>5</u>).

To quantify how P_c^* varies with Q it is first necessary to compare the time scales of lamella transport, film drainage/ filling, and film collapse. The characteristic time for a lamella to transport through one constriction, $\lambda/<U>$, is estimated to be between 0.1 and 10 s by taking a typical value of 100 μm for λ and a range for the average intersticial velocity of 10^{-5} to 10^{-3} m/s. Likewise, the characteristic time for liquid film drainage according to Reynolds theory, $3\mu R^2/4h^2P_c$, is estimated to be between 0.1 and 1 s if the viscosity of the surfactant solution, μ, is taken as 1 mPa·s, the lamella initial thickness is 0.1 μm, the lamella radius is 100 μm, and the capillary pressure is 1 to 10 kPa. These two time constants overlap assuring that film liquid exchange by capillary pumping or suction can keep pace with the pore-wall stretching and squeezing of the film during flow through several constrictions. On the other hand, based on a linear stability analysis summarized in Appendix A for a free,

nonthinning film including both disjoining and conjoining forces
(17), the time for film rupture is $1.5\mu h_{min}^5 \sigma/A_H^2$ which for $\sigma = 30$
mN/m and $h_{min} = 10$ nm gives values between 0.5 μs and 5 ms.
Hence, film rupture is much faster than either film transport or
film drainage and filling. In this initial analysis we therefore
neglect any time lag for the rupture event and assess breakage as
instantaneous whenever h_{min} is reached. This approximation also
neglects any effects of drainage velocity (16,18-20) or stretching
rate (17) on rupture time.
    Based on the picture above, we combine the expressions for
the rate of change of the film thickness due to wall conformity
and capillary- pressure-driven influx or efflux into a dimensionless evolution equation (17):

$$\frac{\partial \hat{h}}{\partial \hat{\xi}} = \frac{2\pi a\hat{h}\,\sin\left[\pi(1+\hat{\xi})\right]}{1+\underline{a} + \underline{a}\,\cos\left[\pi(1+\hat{\xi})\right]} - \frac{\theta\hat{h}^3}{Ca}\left(\hat{P}_c - \hat{\Pi}\right)\left(1 - \alpha_s\hat{R}^2/\hat{h}\right)^{-1} \quad (5)$$

where

$$\hat{P}_c = \begin{cases} P_c/B & \text{constant potential,} & (6a) \\[2mm] \epsilon P_c/2\pi q^2 & \text{constant charge,} & (6b) \end{cases}$$

$$Ca = \frac{3\mu Q}{4\pi\lambda\kappa A_H} \quad , \quad\quad\quad\quad\quad\quad (6c)$$

$$\alpha_s = 6\mu\kappa R_c^2/5\mu_s \quad , \quad\quad\quad\quad\quad\quad (6d)$$

with $\hat{\xi} = 2\xi/\lambda$ and with μ_s defining the Newtonian viscosity of the
interface. $\hat{\Pi}$ follows from either Equation 1 or 2, depending
whether the constant-potential or constant-charge model is
invoked, and $\hat{R}$ varies with ξ according to Equation 4.
    The first term on the right side of Equation 5 corresponds to
the rate of stretching/squeezing and is sensitive to pore
geometry. The sharper the constriction (i.e., high R_b/R_c), the
larger is the contribution of this term to the total rate of
thinning of the lamella. The second term represents the rate of
film drainage/filling caused by the difference between the
capillary pressure and the instantaneous conjoining/disjoining
pressure. Note that if $\hat{P}_c$ is always less than $\hat{\Pi}(\xi)$ this term
causes film thickening and vice versa.
    Following Ivanov et al. (21-23) a factor involving the
parameter α_s is included in the drainage/filling term to account
for surfactant viscous dissipation at the gas/liquid interface in
the limit of high surfactant concentrations. When $\alpha_s = 0$ the

lamella exhibits an infinitely viscous interface obeying the no-slip condition. For finite values of α_s the lamella exhibits less resistance to fluid influx or efflux.

The scaling factor appearing in the denominator of the draining/ filling term is a modified capillary number, defined by Equation 6c, which parameterizes how fast the lamella moves through the constriction. In the limit of Ca → ∞ the second term is negligible compared to the first term indicating that the time of flow through the constriction is too short to allow liquid to move into or out of the lamella under the influence of the porous-medium capillary pressure. In the limit of Ca → 0 the lamella is at rest, and, regardless of the shape of the constriction, the capillary pressure will drive liquid into or out of the lamella until a metastable equilibrium or an unstable state is reached. In this case, as stated earlier, the limiting capillary pressure above which lamellae cannot exist, P_c^*, is identical to Π_{max}.

For a finite flow velocity both the stretching/squeezing and the drainage/filling rates play important roles. Figure 6 reports the numerical marching solution of Equation 5 for the parameters listed and with the conjoining/disjoining curve of Figure 4 (17). In curve 1 of Figure 6 with $\hat{P}_c$ = 0.225, the film thickness generally increases from $\hat{h}_o$ = 2 to oscillate periodically from pore to pore about $\hat{h}_e$ ~ 2.4, where the preset capillary pressure intersects the conjoining/disjoining pressure curve. The film evolution may also be traced in Figure 4 by a thickening along the $\hat{\Pi}$ curve followed by periodic swings about $\hat{P}_c$ = 0.225. The maximum and minimum film thicknesses do not occur precisely at the pore-throat and pore-body locations because of time lag due to capillary pumping and suction. Without large scale disturbances, the film of curve 1 can transport indefinitely without breaking.

In curve 3 with $\hat{P}_c$ = 0.575, reflecting a drier porous medium, the film thickness generally diminishes toward the periodic steady state. However, near the first pore body the lamella thickness falls below $\hat{h}_{min}$ = 0.85 corresponding to $\hat{\Pi}_{max}$ of Figure 4. At this point the film becomes unstable and ruptures.

Curve 2 in Figure 6 shows the film evolution when the minimum film thickness just reaches $\hat{h}_{min}$. This curve defines the critical capillary pressure, $\hat{P}_c^*$. Thus upon changing the pertinent parameters, the solution to Equation 5 traces a locus of critical capillary pressures, $\hat{P}_c^*$ as a function of Ca, $\hat{h}_o$, $\underline{a}$, θ, and α_s. Conversely, given the wetting-liquid saturation, or equivalently P_c of the permeable medium, the locus of marginally stable states specifies a critical Ca^*, $\underline{a}^*$, etc. We investigate briefly the

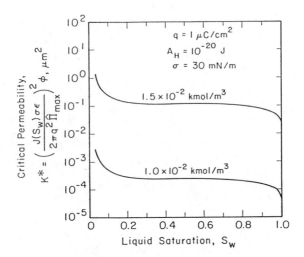

Figure 5. The critical absolute permeability necessary to sustain the stability of a static foam as a function of liquid saturation. Calculations are for the constant-charge electrostatic model.

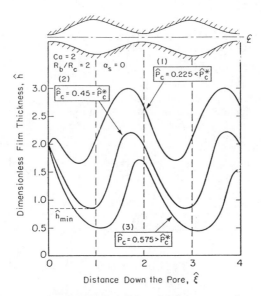

Figure 6. Evolution of the lamella thickness as it transports down the periodic pore for the conjoining/disjoining pressure isotherm of Figure 4. Three capillary pressures are considered in curves 1 through 3. These capillary-pressure values are also labelled in Figure 4. Curve 2 defines the critical or marginally stable capillary pressure, $\hat{P}_c^*$.

behavior of $\hat{P}_c^*$ with Ca (flow rate), α_s (surface viscosity), and $\underline{a}$ (pore geometry).

Figure 7 reports calculations of the effect of flow velocity on the critical capillary pressure for the constant-charge electrostatic model and for different initial film thicknesses. Values of $\hat{P}_c$ and Ca lying above the curves reflect unstable lamellae while values below the curves yield metastable foams. At capillary numbers less than 10^{-4}, $\hat{P}_c^*$ asymptotically approaches the static stability limit of $\hat{\Pi}_{max}$, as discussed in the previous section. Higher capillary numbers demand lower critical capillary pressures for metastability, or equivalently, higher water saturations in the porous medium. The reason is that capillary pumping into the lamella is required to stabilize the film against rupture due to the pore-wall stretching. Lower capillary pressures are necessary at higher velocity because of the finite fluid resistance. At high capillary numbers all curves eventually result in a zero $\hat{P}_c^*$. This means that conceptually, a completely water-saturated medium would be required to support the foam lamellae. How the critical capillary pressure approaches zero depends on the initial film thickness (for R_b/R_c fixed). Thinner initial films clearly rupture more readily. For the particular parameters listed, once $\hat{h}_o$ is greater than about 5 all curves collapse onto a single curve. This is because at large initial thicknesses, capillary suction quickly depletes fluid from the lamellae.

For Ca < 0.1 in Figure 7 the critical capillary pressure is also independent of the initial film thickness. In this case, the hydrodynamic resistance to fluid filling or draining is small enough that the film reaches the periodic steady state in less than half a pore length. Figure 7 confirms the trend observed by Khatib, Hirasaki and Falls that P_c^* falls with increasing flow rate (5).

The role of surface viscosity is demonstrated in Figure 8 with the constant-potential and weak overlap electrostatic model. We note that the underlying theory is restricted to large values of the surface viscosity, μ_s (21-23). As $\alpha_s = 6\mu\kappa R_c^2/5\mu_s$ increases, the gas/liquid interface varies from a no-slip or inextensible boundary toward a stress-free boundary. Hence, there is less fluid resistance to capillary pumping or suction and the lamellae can withstand a drier porous medium. The critical capillary pressure, accordingly, first increases monotonically up to $\alpha_s \sim 0.5$. Thereafter, the critical capillary pressure decreases, indicating the need for a wetter porous medium to support the foam lamellae. For α_s approaching unity in Figure 8, the local radius factor multiplying α_s in Equation 5 dominates and forces a greater resistance for healing inflow when the lamella

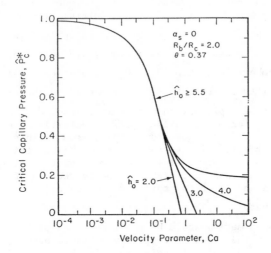

Figure 7.   The effect of gas velocity on the critical capillary
pressure.

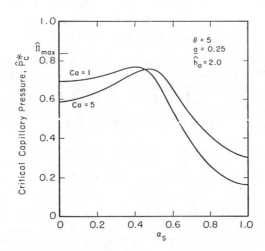

Figure 8.   The effect of surface viscosity on the critical
capillary pressure.

thickness is near $\hat{h}_{min}$ (i.e., when $\hat{R}$ is near R_b/R_c) compared to that for $\alpha_s = 0$. The higher hydrodynamic resistance for lamella thickening permits $\hat{h}$ to fall below $\hat{h}_{min}$ unless $\hat{P}_c$ is lowered. Since the correction factor for surface viscosity in Equation 5 is not rigorous, the results in Figure 8 for α_s greater than about 0.4 should be viewed with caution.

Figure 9 reports the effect of the pore-body to pore-throat radius ratio, R_b/R_c, on the critical capillary pressure for various values of the capillary number and for the constant-potential and weak overlap electrostatic model. Given a constant value of Ca and a wetting-phase saturation (i.e., given $\hat{P}_c$), Figure 9 teaches that large pore-body to pore-throat radii ratios lead to a more unstable foam. The effect is more dramatic for higher capillary numbers.

In a given porous medium there will be a distribution of pore-body and pore-throat sizes. If a lamella exits a constriction into a pore body whose ratio, R_b/R_c, is less than that set by the appropriate curve in Figure 9, the lamella survives. Conversely, if it exits into a pore body whose value of R_b/R_c is greater than that in Figure 9, that lamella must break. Thus, particular combinations of pore throats and pore bodies (i.e., those with R_b/R_c greater than critical) can be classified as termination sites, in analogy with the germination sites for foam generation (1). The overall rate of foam coalescence in a porous medium then depends directly on the number density of termination sites. This number density in turn is a function, among other variables, of the wetting-phase saturation. Given the pore-throat and pore-body size distribution and their probability of interconnection, the termination site density appears calculable. Hence, our periodic-pore theory in Figure 9 bears directly on foam coalescence kinetics in porous media (24).

Comparison to Experimental Results

Huh, Cochrane and Kovarik (25) recently studied the behavior of aqueous surfactant CO_2 foams in etched-glass porous-medium micromodels. They report qualitative visual observations of lamellae stretching and breaking during transport through pore bodies of the medium. This confirms the basic tenet of the proposed dynamic foam stability theory.

Figure 10 reports as dark circles the critical-capillary-pressure measurements of Khatib, Hirasaki and Falls (5) for foam stabilized by an α-olefin sulfonate of 16 to 18 carbon number (ENORDET AOS 1618) in an 81- μm^2 permeability sandpack. Using the parameters listed and the constant- charge electrostatic model for the conjoining/disjoining pressure isotherm, the data are rescaled and plotted as $\hat{P}_c^*$ versus Ca on a logarithmic abscissa. The

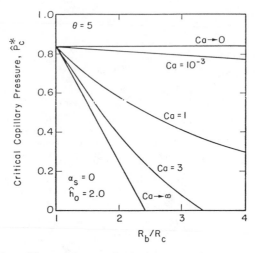

Figure 9. The effect of pore-body to pore-throat radius ratio on the critical capillary pressure.

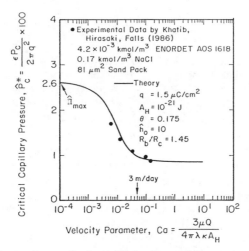

Figure 10. Comparison of the critical-capillary-pressure data of Khatib, Hirasaki and Falls (5) (darkened circles) to the proposed dynamic foam stability theory (solid line). Best fitting parameters for the constant-charge electrostatic model are listed.

experimental data clearly show the decrease of the critical
capillary pressure at higher flow rates.

The solid line in Figure 10 is best eye fit using the suite
of parameter values labelled on the figure. These values are not
unique, but they are physically reasonable. Under this constraint
the theory proves acceptable. We find about a 3-fold reduction in
P_c^* for capillary numbers relevant to reservoir displacement rates.
More definitive evaluation of the proposed theory must await
direct measurement of the conjoining/disjoining curve for the
specific surfactant formulation under investigation.

Conclusions

The stability of flowing lamellae controls the longevity of foam
in porous media. By analyzing a single lamella as it percolates
through a periodically constricted tube, we have developed a
hydrodynamic stability theory that explains the physical phenomena
governing foam coalescence in porous media. The conjoining/
disjoining pressure isotherm proves to be the crucial physical
property of the surfactant system. It determines the maximum
capillary pressure that foam can sustain at rest in a porous
medium. This static critical capillary pressure can be associated
with a critical permeability or a critical wetting-liquid
saturation for a given medium through the Leverett J-function.

For transporting foam, the critical capillary pressure is
reduced as lamellae thin under the influence of both capillary
suction and stretching by the pore walls. For a given gas
superficial velocity, foam cannot exist if the capillary pressure
and the pore-body to pore-throat radii ratio exceed a critical
value. The dynamic foam stability theory introduced here proves
to be in good agreement with direct measurements of the critical
capillary pressure in high permeability sandpacks.

Acknowledgment

This work was supported by a Department of Energy Grant DE-AC-
76SF00098 to the Lawrence Berkeley Laboratories of the University
of California.

Legend of Symbols

$\underline{a}$	=	$1/2\ (R_b/R_c-1)$ dimensionless sinusoidal pore geometric factor
A_H	=	Hamaker constant, J
B	=	pre-exponential scale factor of the electrostatic disjoining pressure for constant potential and small degrees of overlap, Pa
Ca	=	$3\mu Q/4\pi\lambda\kappa A_H$, capillary number
E	=	dimensionless disturbance amplitude
h	=	lamella half thickness, m
$\hat{h}$	=	$2\kappa h$, dimensionless lamella thickness

H   =   dimensionless deviation in film thickness

i   = $\sqrt{-1}$

J   =   Leverett capillary-pressure function

K   =   absolute permeability, m^2

P_c =   capillary pressure, Pa

$\hat{P}_c$ =   dimensionless capillary pressure defined by Equation 6

P_g =   gas pressure, Pa

q   =   lamella surface charge density, C/m^2

Q   =   volumetric flow rate, m^3/s

R   =   local pore radius, m

R_b =   radius of pore body, m

R_c =   radius of pore constriction, m

$\hat{R}$ = R/R_c, dimensionless local pore radius

S_w =   wetting-phase saturation

t   =   time, s

U   =   interstitial velocity, m/s

x   =   linear distance tangential to a lamella, m

α =   dimensionless disturbance wavenumber

α_s = $6\mu\kappa R_c^2/5\mu_s$, surface-viscosity parameter

β =   dimensionless equilibrium film thickness, defined in Equation A4

γ = $\theta/3$

ϵ =   bulk permittivity of the surfactant solution, F/m

θ =   dimensionless ratio of electrostatic repulsive to Hamaker attractive forces in the conjoining/disjoining pressure isotherm, defined under Equations 1 and 2

κ =   inverse Debye length, m^{-1}

λ =   wavelength of periodic pore, m

μ =   Newtonian viscosity of the surfactant solution, Pa·s

μ_s =   Newtonian surface viscosity, g/s

$\hat{\xi}$ =   axial distance from pore constriction, m

ξ = $2\xi/\lambda$, dimensionless axial distance from pore constriction

$\hat{\Pi}$ =   conjoining/disjoining pressure, Pa

Π =   dimensionless conjoining/disjoining pressure, defined under Equations 1 and 2

σ =   equilibrium surface tension of surfactant solution, N/m

ϕ =   porosity

ω =   dimensionless disturbance growth rate

Subscripts

o   =   initial

e   =   equilibrium

max   =   maximum

min   =   minimum

Superscripts

^   =   dimensionless

*   =   critical

Literature Cited

1. Ransohoff, T.C.; Radke, C.J. SPE Reservoir Engineering 1988, 3(2), 573-585.
2. Manlowe, D.S. M.S. Thesis, University of California, Berkeley, 1988.
3. Bikerman, J.J. Foams; Springer-Verlag: New York, 1973; Chapter 1.
4. Ginley, G.M.; Radke, C.J. this series, 1989.
5. Khatib, Z.I.; Hirasaki, G.J.; Falls, A.H. SPE Reservoir Engineering 1988, 3(3), 919-926.
6. Derjaguin, B.V. Acta Physicochim. 1939, 10(1), 25-44.
7. Derjaguin, B.V.; Titievskaja, A.S. Proc. 2nd Int. Congr. Surface Activity; Butterworths, London, 1957, Vol. 1, pp. 211-219.
8. Exerowa, D.; Kolarov, T.; Khristov, K.H.R. Colloids and Surfaces 1987, 22, 171-185.
9. Verwey, E.J.W.; Overbeek, J.T.G. The Theory of Lyophobic Colloids; Elsevier: Amsterdam, 1948.
10. Vrij, A. Disc. Faraday Soc. 1966, 42, 23-33.
11. Ivanov, I.B.; Radoev, B.; Manev, E.D.; Scheludko, A. Trans Faraday Sci. 1970, 66, 1262-1273.
12. Scheludko, A. Adv. Colloid Interface Sci. 1967, 1, 391-464.
13. Leverett, M.C. Trans. AIME. 1941, 142, 152-169.
14. Reynolds, O. Phil. Trans. Roy. Soc. (London) 1886, A177, 157-234.
15. Rao, A.A.; Wasan, D.T.; Manev, E.D. Chem. Eng. Commun. 1982, 15, 63-81.
16. Malhotra, A.K.; Wasan, D.T. Chem. Eng. Commun. 1986, 48, 35-56.,
17. Jimenez, A.I. Ph.D. Thesis, University of California, Berkeley, in preparation, 1989.

18. Gumerman, R.J.; Homsy, G.M. Chem. Eng. Commun. 1975, 2, 27-36.
19. Sharma, A; Ruckenstein, E. J. Coll. Int. Sci. 1987, 119, 1-13; 14-29.
20. Sharma, A.; Ruckenstein, E. Langmuir 1987, 3, 760-768.
21. Rodoev, B.P.; Dimitrov, D.S.; Ivanov, I.B. Colloid and Polymer Sci. 1974, 252, 50-55.
22. Ivanov, I.B.; Dimitrov, D.S. Colloid and Polymer Sci. 1974, 252, 982-990.
23. Ivanov, I.B.; Dimitrov, D.S.; Somasundaran, P.; Jain, R.K. Chem. Eng. Sci. 1985, 40(1), 137-150.
24. Friedmann, F.; Chen, W.H.; Gauglitz, P.A. SPE 17357, presented at the SPE/DOE Enhanced Oil Recovery Symposium, Tulsa, OK, April 17-20, 1988.

25. Huh, D.G.; Cochrane, T.D.; Kovarik, F.S. SPE 17359,
     presented at the SPE/DOE Enhanced Oil Recovery Symposium,
     Tulsa, OK, April 17-20, 1988.
26. Drazing, P.G.; Reid, W.H. Hydrodynamic Stability; Cambridge,
     1981.
27. Chandrasekhar, S. Hydrodynamic and Hydromagnetic Stability;
     Dover, 1961.
28. Williams, M.B.; Davis, S.H. Journal of Coll. and Int. Sci.,
     1982, 90(1), 220-228.
29. Gallez, D.; Prevost, M. Chemical Hydrodynamics, 1985,
     6(5/6), 731-745.

Appendix A - Linear Stability Analysis of a Free Lamella

To establish the characteristic time for rupture of a free
(unbounded), nondraining lamella, we perform a simple, normal
modes linear stability analysis(26,27) including both an
attractive Hamaker conjoining force and a repulsive electrostatic
disjoining force. Consider a planar film, initially of equilibrium half thickness h_e, that undergoes a gentle symmetric disturbance in thickness. Following Williams and Davis (28) and Gallez
and Prevost (29), the dimensionless film thickness $\hat{h} = h/h_e$
evolves in dimensionless time, $\hat{t}$, and space, $\hat{x}$, according to

$$\hat{h}_{\hat{t}} + \left[\hat{h}^{-1}\hat{h}_{\hat{x}} + \hat{h}^3\hat{h}_{\hat{x}\hat{x}\hat{x}} - \gamma\beta^4\hat{h}^3\hat{h}_{\hat{x}} \exp\left(-\beta\hat{h}\right)\right]_{\hat{x}} = 0 \ , \qquad (A1)$$

where subscripts indicate differentiation. The last term on the
left of Equation A1 corresponds to the constant surface-potential
form of the weak overlapping double-layer disjoining force given
in Equation 1. The nondimensional variables and parameters in
Equation A1 are

$$\hat{t} \equiv t/\left[\mu\sigma h_e^5/6A_H^2\right] \ , \qquad (A2)$$

$$\hat{x} \equiv x/h_e^2 \sqrt{\sigma/3A_H} \ , \qquad (A3)$$

$$\beta \equiv 2\kappa h_e \ , \qquad (A4)$$

and
$$\gamma \equiv B/3\kappa^3 A_H \ . \qquad (A5)$$

For an infinitesimal disturbance with $\hat{h} = 1+H$ and $H \ll 1$, Equation
A1 may be linearized out to terms of $O(H^2)$ as

$$\frac{\partial H}{\partial \hat{t}} + \frac{\partial^4 H}{\partial \hat{x}^4} + \left[1 - \gamma \beta^4 \exp(-\beta)\right] \frac{\partial^2 H}{\partial \hat{x}^2} = 0 \ . \qquad (A6)$$

According to normal modes, the infinitesimal disturbance is taken as sinusoidal:

$$H = E \exp (\hat{\omega}t + i\hat{\alpha}x) \; , \tag{A7}$$

where ω is the dimensionless growth rate, α is the dimensionless wavenumber, and E ($\ll 1$) is the disturbance amplitude.

Substitution of Equation A7 into A6 gives the dispersion equation relating the disturbance growth rate and wavelength:

$$\omega = \alpha^2 [1 - \gamma\beta^4 \exp(-\beta) - \alpha^2] \; . \tag{A8}$$

We conclude that H grows ($\omega > 0$) in time if $\alpha^2 < 1 - \gamma\beta^4 \exp(-\beta)$ but decays ($\omega < 0$) to zero if $\alpha^2 > 1 - \gamma\beta^4 \exp(-\beta)$. Thus, when $1-\gamma\beta^4 \exp(-\beta)$ is negative corresponding to $\partial\Pi/\partial h < 0$, the film is metastable. Conversely, when $1-\gamma\beta^4 \exp(-\beta)$ is positive corresponding to $\partial\Pi/\partial h > 0$, the lamella is unstable to the longer wavelength disturbances. In this latter case, the fastest growing mode is characterized by

$$\omega_{max} \equiv \frac{1}{4} \left[1 - \gamma\beta^4 \exp(-\beta)\right]^2 \; , \tag{A9}$$

with the corresponding wavenumber

$$\alpha_{max} \equiv \sqrt{\frac{1}{2} \left[1-\gamma\beta^4 \exp(-\beta)\right]} \; . \tag{10}$$

Finally, the rupture time follows from Equation A7 by setting H = - 1 (i.e., $\hat{h}$ = 0) and by using the expressions above for α_{max}, ω_{max} and $x = \pi/\alpha_{max}$:

$$t = \frac{\ln E^{-4}}{\left[1 - \gamma\beta^4 \exp(-\beta)\right]} \left[\frac{\mu\sigma h_e^5}{6A_H^2}\right] \; . \tag{A11}$$

Upon setting E = 0.1 and γ = 0, we recover the characteristic rupture time listed in the main text.

RECEIVED April 3, 1989

Chapter 26

Influence of Soluble Surfactants on the Flow of Long Bubbles Through a Cylindrical Capillary

G. M. Ginley[1] and C. J. Radke

Department of Chemical Engineering, University of California, Berkeley, CA 94720

Flow of trains of surfactant-laden gas bubbles through capillaries is an important ingredient of foam transport in porous media. To understand the role of surfactants in bubble flow, we present a regular perturbation expansion in large adsorption rates within the low capillary-number, singular perturbation hydrodynamic theory of Bretherton. Upon addition of soluble surfactant to the continuous liquid phase, the pressure drop across the bubble increases with the elasticity number while the deposited thin film thickness decreases slightly with the elasticity number. Both pressure drop and thin film thickness retain their 2/3 power dependence on the capillary number found by Bretherton for surfactant-free bubbles. Comparison of the proposed theory to available and new experimental data at capillary numbers less than 10^{-2} and for anionic aqueous surfactant above the critical micelle concentration shows good agreement with the 2/3 power prediction on capillary number and confirms the significant impact of soluble surfactants on bubble-flow resistance. Finally, scaling arguments extend the single-bubble theory to predict the effective viscosity of the flowing bubble regime in porous media. Again, comparison of the effective-viscosity prediction to available pressure-drop data in Berea sandstone demonstrates good agreement.

---

[1]Current address: Marathon Oil Company, Littleton, CO 80160

0097–6156/89/0396–0480$06.50/0
© 1989 American Chemical Society

Foam is a promising fluid for achieving mobility control in underground enhanced oil recovery (1-3). Widespread application of this technology to, for example, steam, CO_2, enriched hydrocarbon, or surfactant flooding requires quantitative understanding of foam flow properties in porous media. Because foam in porous media is a complicated dispersion of gas (or liquid) in an aqueous surfactant phase (4), the pressure drop-flow rate relationship depends critically on the pore-level microstructure or texture (i.e., on the bubble size and/or bubble-size distribution). Foam texture in turn depends on the dynamic interaction of the generation and breakage mechanisms (5), both of which are strong functions of pore geometry, and surfactant type and concentration.

Numerous visual micromodel studies of foam generated and shaped in oil-free, water-wet porous media with robust stabilizing surfactants, show that the bubble size is variable but generally is on the order of one to several pore-body volumes (4,6-11). As shown in the schematic of Figure 1, the bubbles ride over thin-film cushions of the wetting phase adjacent to the rock surfaces. They are separated from one another by surfactant-stabilized lamellae which terminate in Plateau borders (12). The curvature in the Plateau borders is set primarily by the saturation of the continuous wetting phase which occupies the smallest pores (13,14). To a reasonable approximation the mean capillary suction pressure of the wetting phase is applied to each lamella. When flowing, the lamellae transport as a contiguous bubble train which snakes through available pores not occupied by the wetting phase.

The foam microstructure depicted in Figure 1 suggests that important aspects of the hydrodynamic resistance of flowing bubble trains in porous media can be captured in studies of bubbles in capillaries (7). This work considers the hydrodynamic behavior of a single gas bubble translating in a cylindrical capillary whose radius is smaller than that of the undeformed bubble. The continuous liquid phase contains a surfactant whose concentration is near to or above the critical micelle concentration. Specifically, we extend Bretherton's analysis for a clean gas bubble (15) to include the effects of a soluble surfactant which is kinetically hindered from attaining local equilibrium at the gas/liquid interface. The shape of the bubble and the resulting pressure drop across the bubble are obtained numerically for small deviations in surfactant adsorption from equilibrium. Given the dynamic pressure drop across a single bubble, we briefly show how foam-flow behavior in porous media may be predicted using scaling arguments similar to those adopted for non-Newtonian polymer solutions (16).

Previous Work

In 1961 Bretherton solved the problem of a long gas bubble, uncontaminated by surface-active impurities, flowing in a cylindrical tube at low capillary numbers, $Ca \equiv \mu U/\sigma_o$ (μ is the

Newtonian liquid viscosity, U is the bubble velocity, and σ_o is
the equilibrium surface tension), where surface tension and
viscous forces dominate the bubble shape ($\underline{15}$). Using a
lubrication analysis, Bretherton established that the bubble
slides over a stationary, constant-thickness film whose
thickness divided by the radius of the tube, h_o/R_T, varies as the
capillary number to the 2/3 power. The dimensionless pressure
drop to drive the bubble, $(-\Delta P_B)R_T/\sigma_o$, is calculated from the
altered shape of the bubble at its two ends and also scales as the
capillary number to the 2/3 power. Experimental measurements by
Bretherton of the film thickness deposited by the bubble are in
good accord with his theory for capillary numbers ranging from
10^{-5} to 10^{-2}. To our knowledge no corresponding measurements have
been reported for the pressure drop of single, clean bubbles at
low capillary numbers.

Lawson and Hirasaki recently analyzed the case of single
bubbles immersed in a surfactant solution and flowing at low
capillary numbers through a narrow capillary ($\underline{7}$). These authors
consider the rate-limiting step in the transfer of surfactant to
and from the interface to be finite adsorption-desorption
kinetics. Surface and bulk diffusion resistances are neglected,
and surfactant depletion in the thin film region is shown to be
negligible. Following Levich ($\underline{17}$), the surfactant adsorption and
the surface tension along the bubble are assumed to deviate only
slightly from their equilibrium values. Lawson and Hirasaki
attribute all the effects of the surfactant to the constant film
thickness region of the bubble while the bubble ends are ignored.
Because of this approach, these authors are unable to obtain the
value of the deposited film thickness as part of their analysis.
By asserting that the film thickness remains proportional to the
2/3 power of the capillary number, they establish that the dynamic
pressure drop for surfactant-laden bubbles also varies with the
capillary number to the 2/3 power but with an unknown constant of
proportionality. New pressure-drop data for a 1 wt% commercial
surfactant, sodium dodecyl benzene sulfonate (Siponate DS-10), in
water, after correction for the liquid indices between the
bubbles, confirmed the 2/3 power dependence on Ca and revealed
significant increases over the Bretherton theory due to the
soluble surfactant.

Here we also consider sorption kinetics as the mass-
transfer barrier to surfactant migration to and from the
interface, and we follow the Levich framework. However, our
analysis does not confine all surface-tension gradients to the
constant thickness film. Rather, we treat the bubble shape and
the surfactant distribution along the interface in a consistent
fashion.

Problem Statement

Figure 2 portrays a schematic of a long bubble flowing in zero
gravity through a tube filled with a completely wetting

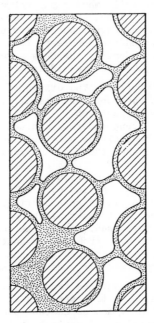

Figure 1. Schematic of the bubble-flow regime in porous media.
Open space corresponds to bubbles, dotted space is the aqueous
surfactant solution, and cross-hatched areas are sand grains.

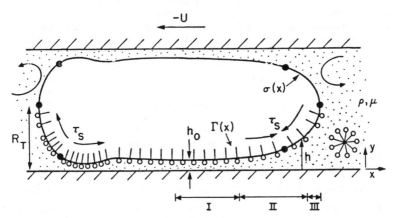

Figure 2. Flow of a single gas bubble through a liquid-filled
cylindrical capillary. The liquid contains a soluble
surfactant whose distribution along the bubble interface is
sketched.

surfactant solution. The reference frame is such that the bubble
is stationary with the walls moving past at a velocity of -U. The
height of the bubble interface, h, is measured from the tube wall.
In this analysis the undistorted bubble radius is always greater
than the tube radius, and the bubble is longer than at least twice
the tube radius. Consequently, there is a region of constant
liquid film thickness, h_o, in the middle of the bubble, even when

surfactants are present (18). The origin of the axial coordinate,
x, is placed near the constant thickness film, but its exact
location is initially unspecified.

The shape of the front and rear menisci change as a result
of the resistance to bubble flow. Calculation of this deviation
in bubble shape establishes the dynamic pressure drop across the
bubble.

The expected surfactant distribution is also portrayed
qualitatively in Figure 2. At low Ca, recirculation eddies in
the liquid phase lead to two stagnation rings around the bubble,
as shown by the two pairs of heavy black dots on the interface
(18,19). Near the bubble front, surfactant molecules are swept
along the interface and away from the stagnation perimeter. They
are not instantaneously replenished from the bulk solution.
Accordingly, a surface stress, τ_s, develops along the interface

directed from low to high surface tension (i.e., from high to low
surfactant adsorption). Surface stresses at the rear stagnation
perimeter also occur, only now surfactant accumulates near the
ring and there is a net flux of surfactant away from the
interface. Sufficiently far from the stagnation rings, the
surfactant achieves sorption equilibrium, and the surface tension
approaches the equilibrium value, σ_o. Thus a stress-free,

constant thickness film underlies the bubble (18). Equilibrium
surface tension in the constant thickness film portion of the
bubble implies that all resistance to flow occurs near the ends.
Thus, the bubble can be viewed as infinite in length, and the
front and back menisci may be treated separately.

If the supply of surfactant to and from the interface is
very fast compared to surface convection, then adsorption
equilibrium is attained along the entire bubble. In this case
the bubble achieves a constant surface tension, and the formal
results of Bretherton apply, only now for a bubble with an
equilibrium surface excess concentration of surfactant. The net
mass-transfer rate of surfactant to the interface is controlled
by the slower of the adsorption-desorption kinetics and the
diffusion of surfactant from the bulk solution. The characteristic time for diffusion is $\Gamma_o^2/(Dc_o^2)$ where D is the bulk surfactant
diffusion coefficient, c_o is the bulk concentration, and Γ_o is the

equilibrium adsorption at that concentration. Conversely, the characteristic time for adsorption is $\Gamma_o/(k_1 c_o \Gamma_{max})$ where k_1 is an adsorption rate constant and Γ_{max} corresponds to monolayer coverage. Hence, low surfactant concentrations likely lead to diffusion control whereas kinetic limitations may dominate at high concentrations. Since our focus is on concentrations above the critical micelle concentration, we take sorption kinetics to be rate limiting.

In examining either end of the bubble, the interface can be divided into three distinct regions (15,20), as shown in Figure 2. Region I is characterized by a constant film thickness, h_o. In region III near the tube center, viscous stresses scale by the tube radius and for small capillary numbers do not significantly distort the bubble shape from a spherical segment. Thus, even though surfactant collects near the front stagnation point (and depletes near the rear stagnation point), the bubble ends are treated as spherical caps at the equilibrium tension, σ_o. Region II provides a transition between the two asymptotic limits. Viscous stresses now scale by the local thickness of the film, h, and the bubble shape varies from the constant thickness film to the spherical segment. Here the surfactant distribution along the interface may be important. Fortunately, for small capillary numbers, dh/dx < 1 and the lubrication approximation may be used throughout. Region II is quantified below.

Problem Formulation

The axial velocity profile in region II follows from the lubrication form of the Navier-Stokes equations with the following boundary conditions. The liquid velocity at the tube wall equals -U, and the tangential stress at the surfactant-contaminated, gas-liquid interface is given by the negative gradient of the surface tension, -dσ/dx (17). Pressure in the liquid phase is replaced by a normal stress balance and the Young-Laplace equation, consistent with the lubrication approximation. Finally, macroscopic continuity demands that the average flow rate at any x position along the bubble must equal the flow rate in the constant thickness thin film region. Combination of the normal stress balance, the Navier-Stokes equation, and the macroscopic continuity balance results in a third-order differential equation for the position of the bubble interface as a function of x:

$$\frac{h^3}{3\mu} \sigma h_{xxx} + \frac{h^3}{3\mu} \sigma_x h_{xx} + \frac{h^2}{2\mu} \sigma_x - Uh + Uh_o = 0, \tag{1}$$

where the subscript x denotes differentiation. If $\sigma_x = 0$ (i.e., if $\sigma = \sigma_o$), this relationship reduces to that of Bretherton for a constant-tension bubble. The second term originates from the normal stress balance, and the third term emerges from the shear-stress boundary condition.

A second equation is needed to determine the surface tension as a function of axial position. We adopt the quasistatic assumption that σ is a unique equilibrium function of the surface excess concentration, Γ, even during dynamic events (17). A surface species continuity balance dictates how Γ varies along the interface. Upon neglect of surface diffusion and for $h_x < 1$, the steady state form of this balance is

$$(\Gamma U_s)_x = k \, c_o \Gamma_{max} \left[1 - \frac{\Gamma}{\Gamma_o} \right] , \tag{2a}$$

where

$$k \equiv k_1 / (1 + k_1 \Gamma_{max} / k_m) , \tag{2b}$$

and U_s is the bubble surface velocity so that ΓU_s is the convected flux of surfactant along the bubble interface. k_1 is the adsorption rate constant, and k_m is a constant mass-transfer coefficient. At high surfactant concentrations and with slow sorption kinetics, Equation 2 reduces to the Langmuir kinetic model. For low surfactant concentrations and with slow diffusion rates, Equation 2 reflects a Nernst constant diffusion-layer-thickness model (17).

With the surface-velocity expression known from the hydrodynamics, Equation 2 can be rewritten as

$$\left[\Gamma \left\{ \frac{\sigma h^2}{2\mu} h_{xxx} + \frac{h^2}{2\mu} \sigma_x h_{xx} + \frac{h}{\mu} \sigma_x - U \right\} \right]_x = k \, c_o \Gamma_{max} \left(1 - \frac{\Gamma}{\Gamma_o} \right) . \tag{3}$$

The above expression and the quasistatic adsorption assumption provide the additional information necessary to establish both the bubble profile, $h(x)$ from Equation 1, and the surfactant distribution, $\Gamma(x)$ from Equation 3.

To reduce Equations 1 and 3 to canonical form we adopt the scaling of Bretherton (15):

$$\eta \equiv \frac{h}{h_o^{(0)}} \tag{4a}$$

and

$$\xi \equiv \frac{x}{h_o^{(0)}} \left\{ \frac{3\mu U}{\sigma_o} \right\}^{1/3} \tag{4b}$$

Here, $h_o^{(0)}$ is the constant thin film thickness obtained by Bretherton for a constant-tension bubble. The value of the surface tension used in the capillary number is the equilibrium value, σ_o. It also proves convenient to express the surface

excess concentration as a deviation from equilibrium in the
following manner:

$$\theta = \left[\frac{\Gamma}{\Gamma_o} - 1\right] (3Ca)^{-2/3}.$$ (5)

Finally, the surface tension is expanded about the equilibrium
adsorption, Γ_o, and only the first term in the deviation from
equilibrium is retained.

These transformations, after elimination of terms that are
appropriately higher order in capillary number, yield the
following expressions (21):

$$\eta^3 \eta_{\xi\xi\xi} - \frac{3}{2}\alpha\eta^2 \theta_\xi - \eta + \eta_o = 0 ,$$ (6)

and

$$\frac{3}{2} \eta^2 \eta_{\xi\xi\xi\xi} + 3\eta\eta_\xi\eta_{\xi\xi\xi} - 3\alpha\eta_\xi\theta_\xi - 3\alpha\eta\theta_{\xi\xi} + \beta\theta = 0 ,$$ (7)

where again the subscript, ξ, refers to differentiation with
respect to ξ. Equations 6 and 7 quantify the interface shape and
surfactant distribution, respectively, in region II.

Two important parameters, α and β, arise which depend on the
equilibrium and kinetic properties of the surfactant. First, α
measures the fractional change in equilibrium surface tension with
a fractional change in surfactant adsorption:

$$\alpha \equiv \left[-\frac{d\sigma_o}{d\Gamma_o}\right] \left[\frac{\Gamma_o}{\sigma_o}\right] .$$ (8)

It is based on equilibrium properties and is directly related to
the Gibbs elasticity (17). In the present context α gauges how
strongly the surface tension depends on the surfactant distribution along the bubble interface. Second, β captures the kinetics
of the adsorption process and is defined by

$$\beta \equiv 3P^{(0)} \left[\frac{k_c c_o \Gamma_{max}}{\Gamma_o}\right] \left[\frac{R_T\mu}{\sigma_o}\right] .$$ (9)

Observe that β is a Damköhler number since it can be interpreted
as the ratio of a characteristic contact time for flow in a thin
film $(R_T\mu/\sigma_o)$, to a characteristic time for adsorption, $(\Gamma_o/[k_c c_o \Gamma_{max}])$. The constant $P^{(0)} = 0.643$ reflects the curvature of the
bubble front for the constant-tension Bretherton analysis.

The ratio of the two parameters, $\alpha/\beta = E$, is also important:

$$E = (-d\sigma_o/d\Gamma_o) \Gamma_o^2/(3P^{(0)}\mu R_T k_c c_o \Gamma_{max}) .$$ (10)

E is one of several elasticity numbers characterizing the stabilizing effect which adsorbed surfactant molecules have on an interface during mass-transfer processes (22). Note that E is inversely proportional to the capillary radius so that the effect of soluble surfactants on the bubble-flow resistance is larger for smaller capillary radii.

When β approaches infinity, Equation 7 reveals that θ equals zero, which corresponds to infinitely fast sorption kinetics and to an equilibrium surfactant distribution. In this case Equation 6 becomes that of Bretherton for a constant-tension bubble. Equation 6 also reduces to Bretherton's case when α approaches zero. However, $\alpha = 0$ means that the surface tension does not change its value with changes in surfactant adsorption, which is not highly likely. Typical values for α with aqueous surfactants near the critical micelle concentration are around unity (21).

Matching. Equations 6 and 7 demand boundary conditions. Near the constant thickness film region the interface position asymptotically approaches h_0, and the surface excess concentration limits to Γ_0. Likewise, Equations 6 and 7 for large values of ξ (large positive values for the bubble front and small negative values for the bubble rear) must meld into the static-cap region. However, the small-slope approximation inherent in Equations 6 and 7 does not apply in region III. Thus, the problem is singular and matching conditions are required.

We utilize the ad hoc procedure of Bretherton (15) which has been formally justified by Park and Homsy (20). Figure 3 displays the meridional circle characterizing the spherical cap of the bubble front. As shown, let $\hat{x}$ be a shifted axial coordinate measured from the origin of that circle and let $R_T - h_*$ be its radius. h_*/R_T is considered to be small. The translated $\hat{x}$ coordinate, after scaling according to Equation 4b or $\zeta \equiv \hat{x}\,(3Ca)^{1/3}/h_0^{(0)}$, is related to the ξ coordinate by $\zeta = \xi - \xi_*$. Matching between regions II and III is then stated mathematically by

$$\lim_{\xi \to \infty} \eta(\xi = \zeta + \xi_*) = \lim_{\hat{x}/R \to 0} \left\{ \frac{R_T}{h_0^{(0)}} - \sqrt{\left(\frac{R_T - h_*}{h_0^{(0)}}\right)^2 - (3Ca)^{-2/3}\,\zeta^2} \right\}. \quad (11)$$

From Equation 6 the outer limit of the inner solution is a parabola: $\eta(\xi \to \infty) = 1/2\,P\,\xi^2 + Q\,\xi + R$. The outer solution, corresponding to the right side of Equation 11, is a circle. From a Taylor expansion about $\hat{x} = 0$, its inner limit is also a polynomial. Following the proscription of Equation 11, equating terms of $O(\zeta^2)$ specifies the constant film thickness, h_o. Terms of $O(\zeta)$ locate the coordinate shift factor, ξ_*, and the constant terms yield h_*, all as expressions of the known constants P, Q, and R.

According to the Young-Laplace expression, the pressure difference across the bubble-front interface is $2\sigma_o/(R_T - h_*)$. Since $h_*/R_T < 1$, the dynamic pressure contribution in excess of the static value, $2\sigma_o/R_T$, is given by:

$$\frac{\Delta P \; R_T}{2\sigma_o} = \frac{h_*}{R_T} .$$ (12)

This completes the analysis for the bubble front.

Matching procedures for the bubble rear follow by analogy. In this case, however, the bubble aft translates over the constant thickness film deposited by the bubble front so that h_o is already fixed.

Regular Perturbation Solution. To effect an analytical expression for the bubble-flow resistance, we consider fast sorption kinetics or equivalently, small deviations from equilibrium surfactant coverage making β large. Hence, a regular perturbation expansion is performed in $1/\beta$ about the constant-tension case. The resulting equations for θ and η are to zero and first order in $1/\beta$ (21):

$$\underline{0[1]} \qquad\qquad \theta^{(0)} = 0 ,$$ (13)

$$[\eta^{(0)}]^3 \eta^{(0)}_{\xi\xi\xi} - \eta^{(0)} + 1 = 0 ,$$ (14)

$$\underline{0\left[\frac{1}{\beta}\right]}$$

$$\theta^{(1)} = -\frac{3}{2} \eta^{(0)}_{\xi} \frac{1}{[\eta^{(0)}]^2} ,$$ (15)

and

$$[\eta^{(0)}]^3 \eta^{(1)}_{\xi\xi\xi} + 3[\eta^{(0)}]^2 \eta^{(0)}_{\xi\xi\xi} \eta^{(1)} - \eta^{(1)}$$

$$= \frac{3}{2}\alpha \frac{1}{\eta^{(0)}} \left\{ 3[\eta^{(0)}_{\xi}]^2 - \frac{3}{2} \eta^{(0)} \eta^{(0)}_{\xi\xi} \right\} - \eta^{(1)}_o ,$$ (16)

where the superscripted variables refer to the order in $1/\beta$. Equations 13 and 14 represent the constant-tension equations with

an equilibrium surfactant distribution and have been solved by
Bretherton. Equations 15 and 16 can be solved to obtain the
first-order contribution to the surface position and surfactant
surface excess concentration due to the presence of a soluble
surfactant.

Once $\eta^{(0)}$ is known, the surfactant distribution $\theta^{(1)}$ is
available directly from Equation 15. Thereafter, the first-order
correction to the interface position $\eta^{(1)}$ may be calculated.
Following Bretherton, Equations 14 and 16 are restated as initial
value problems and solved by a numerical marching technique. We
use a 4th-order, Runge-Kutta procedure and calculate $\eta^{(0)}$ and $\eta^{(1)}$
for the front and back of the bubble separately ($\underline{21}$). The matching condition, Equation 11, is applied to each order in $1/\beta$.
Extensive details on the mathematical and numerical procedures are
available elsewhere ($\underline{21}$).

<u>Results</u>

Figures 4 and 5 depict the calculated surfactant distribution,
expressed as $\theta^{(1)}$, for the bubble front and rear, respectively.
For perspective, the locations of the stagnation rings are
indicated by arrows. As anticipated, near the front stagnation
perimeter there is a deficiency of surfactant because surface
convection is directed away from that point. The largest
gradients in surface velocity occur just aft of the stagnation
ring. Hence, in Figure 4, the surfactant adsorption attains a
minimum at $\xi \sim 10$.

Again, as expected, in Figure 5 there is an excess of surfactant near the rear stagnation ring due to surface convection
towards that point. Forward from that location, however, there is
also a depletion relative to equilibrium adsorption. This is
caused by the traveling wave in the rear bubble profile as
demonstrated in Figure 2 and in Figure 7 to follow.

The total dimensionless film thickness, $\eta = \eta^{(0)} + (1/\beta)$
$\eta^{(1)}$, is plotted on the ζ scale in Figures 6 and 7 for the bubble
front and rear, respectively. A range of elasticity numbers from
zero (the Bretherton equilibrium surfactant coverage case) to
unity is shown for comparison. The bubble-front profile in Figure
6 is only slightly altered even for E = 1. The effect of
surfactant on the rear profile is more interesting, as
demonstrated in Figure 7. Here additional and higher amplitude
oscillations appear as E is raised.

The main results of our first-order regular perturbation
analysis are the expressions for the constant thin film thickness,
h_o, and for the total hydrodynamic pressure drop across the entire
bubble (front and back), $-\Delta P_B$:

$$h_o/R_T = 1.34 \ [1 - 1.56(10^{-2})E] \ Ca^{2/3}, \tag{17}$$

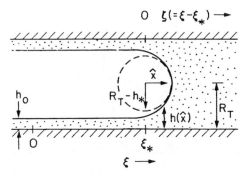

Figure 3. Schematic of matching to the spherical cap at the bubble front. The radius of the flow-altered sphere is $R_T - h_*$. For a static bubble, the bubble radius is R_T.

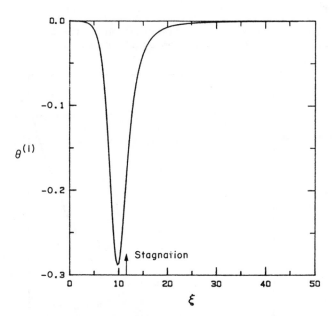

Figure 4. The surfactant distribution at the bubble front expressed as a deviation from equilibrium.

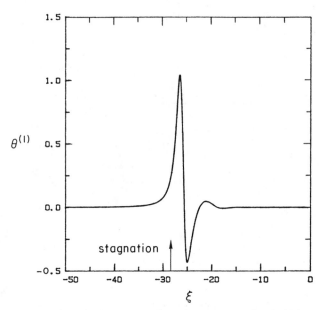

Figure 5.   The surfactant distribution at the bubble rear
expressed as a deviation from equilibrium.

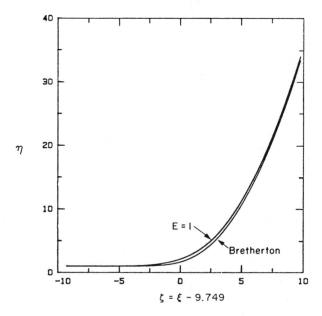

Figure 6.   The bubble shape at the front for the elasticity
number equal to 0 and 1.

and

$$\frac{(-\Delta P_B)R_T}{\sigma_o} = 9.40 \ [1 + 0.469E] \ Ca^{2/3} \ . \tag{18}$$

The first term in both Equations 17 and 18 is the constant surface-tension contribution and the second term gives the first-order contribution resulting from the presence of a soluble surfactant with finite sorption kinetics. A linear dependence on the surfactant elasticity number arises because only the first-order term in the regular perturbation expansion has been evaluated. The thin film thickness deviates negatively by only one percent from the constant-tension solution when $E = 1$, whereas the pressure drop across the bubble is significantly greater than the constant-tension value when $E = 1$.

Comparison to Experiment

Goldsmith and Mason (18) have experimentally observed the constant thickness thin film region underlying a gas bubble flowing in a tube. They report that its value does not differ significantly upon addition of surfactant to the continuous liquid phase. This is in accord with Equation 17 demonstrating a very small influence of surfactant on the film thickness.

Figure 8 gives ambient temperature pressure drop-flow rate data for trains of isolated bubbles in precision cylindrical capillaries. A dimensionless pressure drop per bubble is plotted as a function of the capillary number on logarithmic scales. Open triangles are the data of Hirasaki and Lawson (7) for a 1 wt% aqueous solution of a commercial sodium dodecyl benzene sulfonate (Siponate DS-10) in a 1 mm diameter tube. Open squares reflect preliminary data of Ginley (21) for a 1 wt% solution of sodium dodecyl sulfate in a 50 wt% mixture of glycerol and water. The capillary tube diameter is 2 mm both for these experiments and for those in pure water, shown as open circles. The lowest solid line, labelled by $E = 0$, corresponds to the constant-tension theory of Bretherton. The remaining two lines are best fit according to a 2/3 power dependence on the capillary number. Figure 8 reveals that the few data available for surfactant-laden bubbles do confirm the capillary-number dependence of the proposed theory in Equation 18. Careful examination of Figure 8, however, reveals that the regular perturbation analysis carried out to the linear dependence on the elasticity number is not adequate. More significant deviations are evident that cannot be predicted using only the linear term, especially for the SDBS surfactant. Clearly, more data are needed over wide ranges of capillary number and tube radius and for several more surfactant systems. Further, it will be necessary to obtain independent measurements of the surfactant properties that constitute the elasticity number before an adequate test of theory can be made. Finally, it is quite apparent that a more general solution of Equations 6 and 7 is needed, which is not restricted to small deviations of surfactant adsorption from equilibrium.

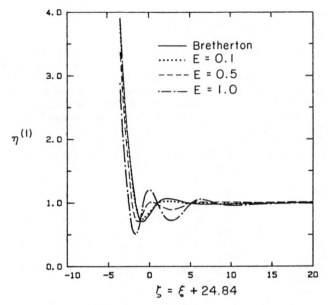

Figure 7. The bubble shape at the rear for the elasticity number equal to 0, 0.1, 0.5, and 1.0.

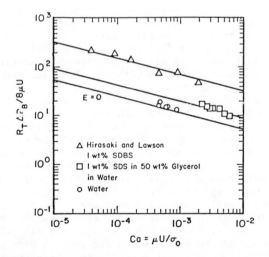

Figure 8. Experimental data of the dimensionless pressure drop per bubble as a function of capillary number for 1 and 2 mm diameter glass capillaries. The solid line denoted by $E = 0$ gives the theory of Bretherton.

Extension to Porous Media

In this section we illustrate how the proposed theory for single, surfactant-laden bubbles in a cylindrical tube can be extended to predict the hydrodynamic resistance of bubble trains flowing in porous media. Some of the basic ideas are known ($\underline{7}$, $\underline{23}$), so the present discussion is brief.

Refer again to Figure 1. In strongly water-wet media, capillarity forces the aqueous surfactant phase to occupy the smaller pores in addition to residing in lamellae and in thin films coating the solid grains. Thus, flowing bubbles trains must transport in the larger pore channels, making foam a "nonwetting" phase. Accordingly, Darcy's law can be written as

$$\underset{\sim n}{V} = - \frac{k_{rn} K}{\mu_e} \nabla P_n , \tag{19}$$

where $\underset{\sim n}{V}$ is the superficial velocity of the flowing foam, P_n is a volume-averaged pressure of the foam, K is the absolute permeability, and k_{rn} is the traditional relative permeability to the nonwetting phase evaluated at the saturation of the flowing bubble trains, S_n. Unfortunately, because the foam is neither a continuum fluid nor a bulk fluid, its effective or apparent viscosity, μ_e, is not a constant. Indeed, μ_e depends on S_n in addition to surfactant type and concentration, bubble texture, and flow rate. Equation 19 essentially serves as a definition of μ_e; progress can be made only when μ_e is known.

Three steps are necessary to transform Equation 18 into an expression for μ_e. First, the pressure drop over a unit capillary length is taken as directly proportional to the linear bubble density, n_B, and the pressure drop per bubble. Second, the bubbles form intervening lamellae rather than liquid indices. This requires a curvature correction from that of the capillary to that of the Plateau border ($\underline{7}$). Third, the scaling arguments of Blake and Kozeny are invoked in a manner similar to that done for non-Newtonian polymers in porous media ($\underline{16}$). The final result is ($\underline{23}$):

$$\mu_e = \left\{ \mu^{2/3} \sigma_o^{1/3} n_B \sqrt{k_{rn} K} \; a \; \left[1 + 0.25b \; \epsilon (r_1 + r_2) \; \sqrt{\phi S_n / 2 k_{rn} K} / r_n r_1 r_2 \right] \right.$$

$$\left\{ \left[\frac{2r_2}{r_1(r_1+r_2)} \left[1 + (r_1/r_2)^2 \right] \right] \right\} \Big/ 2 \sqrt{2} \; \tau_n^{1/3} S_n^{1/6} \phi^{1/6} v_n^{1/3} \; , \qquad (20)$$

where $a = 9.40$, $b = 0.469$, ϕ is the porosity, τ_n is the tortuosity of the nonwetting phase, and r_1 and r_2 are nondimensional radii of curvature of the Plateau border. These are known functions of the quantity, $J\tau_n \sqrt{k_{rn}/S_n}$, where J is the Leverett capillary-pressure function (24) evaluated at the saturation of the wetting phase. The role of the surfactant in controlling the foam effective viscosity is inherent in the parameter $\epsilon \equiv R_T E$, whose definition is available in Equation 10. Again, smaller radii pores, or equivalently lower permeability media, exhibit more sensitivity to the surfactant properties, as confirmed in Equation 20. Note the shear-thinning behavior of the bubble flow regime and the linear dependence on the texture through n_B. Equation 20 should be compared to other empirical expressions which are currently available (5,25).

Barring direct measurement of foam texture, we adopt the following reasoning. Because of the generation of foam bubbles by the snap-off and division mechanisms (4), bubble sizes are expected to be approximately that of pore bodies. Thus, the linear bubble density should scale roughly as $n_B = \delta/D_g$ where D_g is the grain diameter of the medium (obtained from the absolute permeability) and δ is a proportionality constant not significantly larger than unity.

Figure 9 compares Equation 20 with the recent pressure drop flow rate data of Friedmann, Chen, and Gauglitz (5) for a 1 wt% commercial sodium alkyl sulfonate dimer (Chaser SD-1000) stabilized foam in a Berea sandstone. These data are particularly useful because they have been corrected for foam blockage and therefore correctly reflect the flowing bubble regime. The solid line in Figure 9 is best fit according to Equation 20. Unfortunately, neither of the parameters ϵ or δ is available. Two sets of estimates are shown in Figure 9. When $\epsilon = 0$ (i.e., no surfactant effect) the bubble size is about 30% of a grain diameter. When $\epsilon = 0.1$ mm (i.e., a value characteristic of those in Figure 8) the bubble size is about 10 grain diameters. We assert that Equation 20 not only predicts the correct velocity behavior of foam but it does so with reasonable parameter values (23).

These results are very encouraging. They do, however, emphasize the need to measure flowing bubble texture and surfactant properties independently before rigorous experimental tests of detailed theories can be made.

Conclusions

The effect of a soluble surfactant on the flow of long bubbles in a cylindrical tube has been quantified when the surfactant

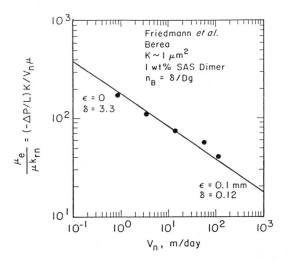

Figure 9. Experimental data for the effective viscosity of the foam bubble regime in Berea sandstone as a function of the foam superficial velocity. The solid line is drawn according to the scaling theory with values of the two sets of parameters ϵ and δ listed.

exhibits small deviations from equilibrium adsorption. A regular
perturbation expansion in large adsorption rates is constructed
about the low capillary-number, singular perturbation theory of
Bretherton. The pressure drop across the bubble increases upon
the addition of surfactant, whereas the thin film thickness
decreases slightly. Both the pressure drop and the thin film
thickness retain their 2/3 power dependence on the capillary
number found by Bretherton for surfactant-free bubbles. To first
order, a linear dependence on the surfactant elasticity number
arises. Comparison of the proposed theory to new and available
pressure-drop data for single bubbles immersed in anionic
surfactant solutions confirms the $Ca^{2/3}$ prediction but demonstrates that higher order terms in the elasticity number are
required.

Using Kozeny scaling arguments, the proposed single-bubble
theory is applied to predict the flow resistance of foam in
porous media. Comparison of the predicted foam effective
viscosity to experimental data at characteristic reservoir rates
in Berea sandstone shows good agreement. The scaling procedure,
therefore, provides a powerful tool for quantitative modeling of
foam flowing in the bubble regime through porous media.

<u>Acknowledgments</u>

This work was supported by a Department of Energy Grant DE-AC-76
SF00098 to the Lawrence Berkeley Laboratory of the University of
California. We thank John Newman for helpful discussions on
several mathematical issues.

<u>Legend of Symbols</u>

a,b	=	known constants
Ca	=	$\mu U/\sigma_o$, capillary number, ratio of viscous to surface tension forces
c_o	=	bulk concentration of surfactant, mol/m^3
D	=	diffusion coefficient of surfactant, m^2/s
D_g	=	grain diameter, m
E	=	elasticity number, α/β
h	=	local film thickness measured from the capillary wall, m
h_o	=	constant thin film thickness, m
h_*	=	distance from wall to the minimum of the spherical cap of the bubble ends, m
J	=	Leverett capillary-pressure function
K	=	absolute permeability, m^2
k	=	effective rate constant defined in Equation 2b, $m^3/mol\cdot s$
k_r	=	relative permeability
k_m	=	mass-transfer coefficient, m/s

k_1 = adsorption rate constant, $m^3/mol \cdot s$

L = porous-medium length, m

n_B = number of bubbles per unit length, m^{-1}

P, Q, R = second, first and zero-degree constants in the constant curvature parabola of region II

ΔP = pressure difference, Pa

ΔP_B = negative of the pressure drop across the entire bubble, Pa

R_T = capillary tube inner radius, m

$r_{1,2}$ = dimensionless radii of curvature of a Plateau border

S = saturation

U = bubble velocity, m/s

U_s = surface velocity of bubble, m/s

V_n = superficial velocity of nonwetting phase, m/s

x = axial position, m

$\hat{x}$ = shifted axial coordinate, m

α = $(-d\sigma_o/d\Gamma_o)(\Gamma_o/\sigma_o)$, fractional Gibbs elasticity

β = $1.929\ (k\ c_o \Gamma_{max}/\Gamma_o)(R_T \mu/\sigma_o)$, Damköhler number

Γ = surface excess concentration, mol/m^2

Γ_o = equilibrium surface excess concentration, mol/m^2

Γ_{max} = surface excess concentration corresponding to monolayer coverage, mol/m^2

δ = $D_g n_B$, proportionality constant

ϵ = $R_T E$, radius independent surfactant elasticity number, m

ζ = $\xi + \xi_*$, dimensionless shifted axial coordinate

η = $h/h_o^{(0)}$, dimensionless film thickness

η_o = $h_o/h_o^{(0)}$, dimensionless thin film thickness

θ = $(\Gamma/\Gamma_o - 1)(3Ca)^{-2/3}$, dimensionless relative surface excess concentration

μ = liquid viscosity, $mPa \cdot s$

μ_e = effective viscosity of foam in bubble-flow regime, $mPa \cdot s$

ξ = $x/h_o^{(0)}(3Ca)^{-1/3}$, dimensionless axial coordinate

ξ_* = origin of spherical bubble caps on the ξ scale

ρ = liquid density, kg/m^3

σ = surface tension, mN/m

σ_o = equilibrium surface tension, mN/m

τ =   tortuosity

τ_s =   shear stress at gas/liquid interface of bubble, Pa

ϕ =   porosity

Subscript

n   =   nonwetting

Superscripts

(0)   =   zero order in $1/\beta$
(1)   =   first order in $1/\beta$

<u>Literature Cited</u>

1. Dilgren, R.E.; Deemer, A.R.; Owens, K.B. SPE 10774,
    presented at the California Regional Meeting of SPE, San
    Francisco, CA, March 24-26, 1982.
2. Mohammadi, S.S.; Van Slyke, D.C. SPE 16736, presented at the
    62nd Annual Fall Meeting of SPE, Dallas, TX, September 27-30,
    1987.
3. Patzek, T.W., Konis, M.T. SPE 17380, presented at the SPE/
    DOE Enhanced Oil Recovery Symposium, Tulsa, OK, April 17-20,
    1988.
4. Ransohoff, T.C.; Radke, C.J. <u>SPE Reservoir Engineering</u> 1988,
    <u>3(2)</u>, 573-585.
5. Friedmann, F.; Chen, W.H.; Gauglitz, P.A. SPE 17357,
    presented at the SPE/DOE Enhanced Oil Recovery Symposium,
    Tulsa, OK, April 17-20, 1988.
6. Mast, R.F. SPE 3997, presented at the Annual Fall Meeting of
    SPE, New Orleans, LA, October 4-6, 1972.
7. Hirasaki, G.J.; Lawson, J.B. <u>SPEJ</u> 1985, <u>25</u>, 176-190.
8. Owete, O.S.; Brigham, W.E. <u>SPE Reservoir Engineering</u> 1987,
    <u>2(3)</u>, 315-323.
9. Kuhlman, M.I. SPE 17356, presented at the SPE/DOE Enhanced
    Oil Recovery Symposium, Tulsa, OK, April 17-20, 1988.
10. Huh, D.G.; Cochrane, T.D.; Kovarik, F.S. SPE 17359,
    presented at the SPE/DOE Enhanced Oil Recovery Symposium,
    Tulsa, OK, April 17-20, 1988.
11. Manlowe, D.S. M.S. Thesis, University of California,
    Berkeley, 1988.
12. Bikerman, J.J. <u>Foams</u>; Springer-Verlag: New York, 1973;
    Chapter 1.
13. Khatib, Z.I.; Hirasaki, G.J.; Falls, A.H. <u>SPE Reservoir
    Engineering</u> 1988, <u>3(3)</u>, 919-926.
14. Jimenez, A.I.; Radke, C.J., In <u>Advances in Oil Field
    Chemistry</u>; Borchardt, J.K., Yen, T.F., Eds; Chapter , in
    this book.
15. Bretherton, F.P. <u>J. Fluid Mech.</u> 1961, <u>10</u>, 166-188.
16. Christopher, R.H.; Middleman, S. <u>I&EC Fundamentals</u> 1965,
    <u>4(4)</u>, 423-426.

17. Levich, V.G. Physiochemical Hydrodynamics; Prentice-Hall: Englewood Cliffs, 1962; Chapters VII, VIII.
18. Goldsmith, H.L.; Mason, S.G. J. Coll. Sci. 1963, 18, 237-261.
19. Shen, E.I.; Udell, K.S. J Appl. Mech. 1985, 52, 253-256.
20. Park, C.W.; Homsy, G.M. J. Fluid Mech. 1984, 139, 291-308.
21. Ginley, G.M. M.S. Thesis, University of California, Berkeley, 1987.
22. Berg, J.C. In Recent Developments in Separation Science; Li, N.N., Ed.; CRC Press: Cleveland, 1972; Vol. 2, pp. 1-31.
23. Ettinger, R.A. M.S. Thesis, University of California, Berkeley, in progress, 1989.
24. Leverett, M.C. Trans. AIME 1941, 142, 152-169.
25. Marfoe, C.H.; Kazemi, H. SPE 16709, presented at the 62nd Annual Fall Meeting of SPE, Dallas, TX, September 27-30, 1987.

RECEIVED November 28, 1988

Chapter 27

Mobility of CO_2 and Surfactant Adsorption in Porous Rocks

Hae Ok Lee and John P. Heller

New Mexico Petroleum Recovery Research Center, New Mexico Institute of Mining and Technology, Socorro, NM 87801

High pressure equipment has been designed to measure
foam mobilities in porous rocks. Simultaneous flow
of dense CO_2 and surfactant solution was established
in core samples. The experimental condition of dense
CO_2 was above critical pressure but below critical
temperature. Steady-state CO_2-foam mobility measurements were carried out with three core samples. Rock
Creek sandstone was initially used to measure CO_2-
foam mobility. Thereafter, extensive further studies
have been made with Baker dolomite and Berea sandstone to study the effect of rock permeability.
Also, other dependent variables associated with CO_2-
foam mobility measurements, such as surfactant
concentrations and CO_2-foam fractions have been
investigated as well. The surfactants incorporated
in this experiment were carefully chosen from the
information obtained during the surfactant screening
test which was developed in the laboratory. In
addition to the mobility measurements, the dynamic
adsorption experiment was performed with Baker
dolomite. The amount of surfactant adsorbed per gram
of rock and the chromatographic time delay factor
were studied as a function of surfactant concentration at different flow rates.

For a miscible displacement at the required reservoir conditions,
carbon dioxide must exist as a dense fluid (in the range 0.5 to
0.8g/cc). Unfortunately, the viscosity of even dense CO_2 is in the
range of 0.03 to 0.08 cp, no more than one twentieth that of crude
oil. When CO_2 is used directly to displace the crude, the unfavorable viscosity ratio produces inefficient oil displacement by
causing fingering of the CO_2, due to frontal instability. In
addition, the unfavorable mobility ratio accentuates flow non-

uniformities due to permeability stratification or other heterogeneities.

The mobility of CO_2 in porous rock can be decreased by
containing it in a foam-like dispersion. Such CO_2-foams have been
proposed as a useful injection fluid in enhanced oil recovery
(1,2). A critical literature review on general foam rheology is
given elsewhere (3). The foam flooding method modifies the flow
mechanism by changing the structure of the displacing fluid at the
pore level. This method of decreasing the mobility of a low-
viscosity fluid in a porous rock requires the use of a surfactant
to stabilize a population of bubble films or lamellae within the
porespace of the rock (4). However, the degree of thickening
achieved apparently depends to a great extent on the properties of
the rock itself. These properties probably include both the
distance scale of the pore space and the wettability, and so can be
expected to differ from reservoir to reservoir, as well as to some
extent within a given field. Laboratory measurements of CO_2-foam
mobility as well as studies involving mobility control of CO_2/sur-
factant in core flooding have been investigated by several researchers recently (5-7). Special efforts have been made to
investigate the reservoir application of mobility control foams in
CO_2 floods (8), and the influence of reservoir depth on enhanced
oil recovery by CO_2 flooding (9). Furthermore, the economic model
of mobility methods for CO_2 flooding has been taken into consideration to determine the profitability of carbon dioxide
flooding in non-waterflooded fields, and of the use of thickening
agents for mobility control (10). Also, an actual field test was
conducted with CO_2-foam. The results of the field test of CO_2
mobility control at Rock Creek have been published (11).

In order to understand the nature and mechanisms of foam flow
in the reservoir, some investigators have examined the generation
of foam in glass bead packs (12). Porous micromodels have also been
used to represent actual porous rock in which the flow behavior of
bubble-films or lamellae have been observed (13,14). Furthermore,
since foaming agents often exhibit pseudo-plastic behavior in a
flow situation, the flow of non-Newtonian fluid in porous media has
been examined from a mathematical standpoint. However, representation of such flow in mathematical models has been reported to be
still inadequate (15). Theoretical approaches, with the goal of
computing the mobility of foam in a porous medium modelled by a
bead or sand pack, have been attempted as well (16,17).

The use of surfactant as a foaming agent to stabilize the CO_2-
foam motivated addtional subresearch areas related to foam flooding. Over the past years, continous efforts and a great deal of
time were invested to find the most suitable surfactant for CO_2
flooding (18,19). In addition to research to find the most
effective surface active agents, the use of surfactant causes
concern that, due to the inherent amphipatic nature of this type
of molecule, adsorption on the rock surfaces would reduce their
concentration to such an extent that the process would not be
successful. Consequently there has been much careful analysis of
adsorption, and several independent experiments have been carried
out to study this question (20-22).

The purpose of this paper is to present the result of two

kinds of relevant laboratory measurement. The first kind measures
the mobility of CO_2-foam at different flow rates, carrying out
steady-state experiments using real rock samples as the porous
media. The second experiment deals with the dynamic adsorption of
surfactant on porous rock. The mobility of CO_2-foam is greatly
dependent on the size of the porespace, as expressed in the
single-phase permeability of the rock sample. This is shown by
comparison of mobility measurements made with Baker dolomite and
with Berea sandstone. This result was supported by the previous
work with Rock Creek and Berea sandstones. Also the adsorption
results indicate that indeed additional surfactant is required due
to the permanent adsorption, and that the chromatographic delay
associated during the flow of surfactant through the core sample
cannot be ignored. In addition to presenting the results of these
measurements, the operation of the apparatus and the conduct of
both experiments are discussed.

Measurements of CO_2-Foam Mobility

In this section the laboratory measurements of CO_2-foam mobility
are presented along with the description of the experimental
procedure, the apparatus, and the evaluation of the mobility. The
mobility results are shown in the order of the effects of surfactant concentration, CO_2-foam fraction, and rock permeability. The
preparation of the surfactant solution is briefly mentioned in the
Effect of Surfactant Concentrations section. A zwitteronic
surfactant Varion CAS (ZS) from Sherex (23) and an anionic surfactant Enordet X2001 (AEGS) from Shell were used for this experimental study.

Experimental. A simplified representaion of the basic elements of
the flow system is shown in Figure 1 and the schematic of the CO_2-
foam mobility measurement experiment is presented in Figure 2. The
CO_2 flows through the capillary tube and the pressure drop across
the tube is measured by a Validyne differential pressure transducer. An Isco pump is used to pressurize the brine/surfactant
solution, which also flows through the foam generator and the core.
The foams are generated inside the short core used as a foam generator, where the mixing between CO_2 and surfactant solution occurs.
The mixed CO_2-foam flows through the core. The pressure drop
across the core is recorded by a second Validyne differential
pressure transducer. Two fine tapered needle valves in series are
used to regulate the output flow rate of the mixture of surfactant
solution and dense CO_2. In addition to the digital readout of the
values of pressure drop across the capillary and across the core, a
two-pen recorder is used to record these simultaneous measurements.
Both of these measurements are subject to rapid short-term variations of fairly small magnitude, that seem to be indicative of the
mechanism of foam flow. All calculations were performed with
steady-state values that were averaged over these short-term
variations. In this experiment, the measured variables are $\Delta Pcap$
and $\Delta Pcore$. Knowing the $\Delta Pcap$, the flow rate of pure CO_2 into the
core is computed by using a calibration constant obtained by
measuring the flow of dense CO_2 through the capillary tube. The

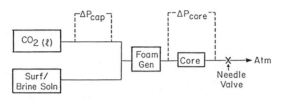

Figure 1. Basic flow system for mobility measurements.

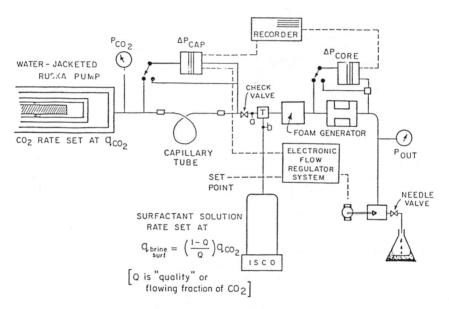

Figure 2. Schematic of the CO_2-foam mobility apparatus. (Taken from Lee, H.O. in <u>Surfactant-Based Mobility Control</u>; Smith, D.H., Ed.; ACS Symposium Series No. 373; American Chemical Society: Washington, DC, 1987; p. 378.)

setting of the Isco pump gives the flow rate of the aqueous
surfactant solution, and the total flow rate is simply the sum of
these two. From the total flow rate and the ΔPcore, the mobility λ
can be evaluated from:

$$\lambda = \frac{Q/A}{\Delta Pcore/L} \tag{1}$$

where   Q is the total flow rate in cc/sec
             A is the cross-sectional area of the rock sample in cm^2
             L is the length of the rock sample in cm
             ΔPcore is the pressure drop across the rock sample in
             standard atmospheres
        and λ is the mobility in $cm^2/(atm-sec)$ or darcy/cp.

The Effect of Surfactant Concentrations. The effect of surfactant
concentrations on CO_2-foam mobility is plotted on a log-log scale
in Figure 3. The presented data points are the average mobility
values obtained from a superficial velocity range of 2-10 ft/day,
with the CO_2-foam fraction was kept constant around 80%. With
Berea sandstone, ZS and AEGS surfactants were used. The measured
average permeability of the Berea sandstone with 1% brine was 305
md. With Baker dolomite, AEGS was used to make comparison with
Berea sandstone. The permeability of the Baker dolomite was 6.09
md measured with 1% brine solution.
     Interfacial tension measurements were used to select the
surfactant concentration for the initial experiments. The interfacial tension measurements were performed with a Rosano surface tensiometer which uses the Wilhelmy plate method. The interfacial
tension was measured at atmospheric conditions against isooctane to
simulate dense CO_2. Isooctane was selected because it is a
nonpolar compound and has low solvent power, like dense CO_2. The
initial concentration was chosen around the CMC (critical micelle
concentration) and the subsequent concentrations were increased
above CMC to well-above CMC. The mobility measurements were conducted from low to high concentrations. Each time whenever the
concentration has been increased, sufficient pore volumes of new
surfactant concentration was flowed through the rock to establish
equilibrium condition. The results in Figure 3 indicate that in
general, mobility decreases with increase of surfactant concentration. In the case of ZS, there seems to be an upper critical
concentration (0.05%), beyond which the mobility reduction is quite
negligible. With AEGS, such upper critical concentrations were not
observed for either Baker dolomite or Berea sandstone because the
measurements were not carried out beyond 0.5% AEGS. However, the
slopes of the curves of AEGS measured with both rocks show similar
behavior to that of ZS, and likely reach such plateaus. This graph
also shows that the magnitudes of mobility reached in Baker
dolomite and Berea sandstone are similar with the same surfactant
at identical concentrations. It can be noted that in Berea core,
about 30 times the CMC of the ZS was required to obtain the same
mobility magnitude as was attained by 0.05% AEGS. This suggests
that surfactant type can play indeed an important role in lowering
foam mobility.

Effect of CO_2-Foam Fraction. The effect of CO_2-foam fraction,
which is defined as the fractional flow of CO_2, on mobility was
studied with 0.1% AEGS. Four different CO_2-foam fractions were
used to study the foam fraction effect at various flow rates. The
results obtained with Baker dolomite is presented in Figure 4 and
with Berea sandstone in Figure 5. Both figures clearly demonstrate
that the mobility decreases with increase of surfactant fraction.
These experimental results indicate that the population of lamellae
present in the pore spaces is responsible for decreasing the foam
mobility as well as stabilizing the foam flow. The slopes of the
fitted lines (that is, the derivatives of the mobility with flow
velocity) in both cores suggest that more shear-thinning behavior
is evident at 90% of CO_2-foam (which indicates a surfactant
deficient environment) compared to 60% or 70% of CO_2. Figure 6
also presents the CO_2-foam fraction effect. The data presented are
the mobility measurements as a function of foam fraction at an
average velocity of 7.0 ft/day with 0.1% AEGS. These fitted curves
show the same trend noted in figures 4 and 5.

Effect of Rock Permeability. The effect of rock permeability has
been investigated by comparison of mobility measurements made with
Baker dolomite and Berea sandstone. Mobility measurements carried
out with Rock Creek sandstone (from the Big Injun formation in
Roane County, W.Va) is also reported. Rock Creek sandstone has a
permeability of 14.8 md. A direct comparison was made with Berea
sandstone and Baker dolomite measured with 0.1% AEGS. As mentioned
in an earlier section, the permeability of Baker dolomite (a
quarried carbonate rock of rather uniform texture with microscopic
vugs distributed throughout) was 6.09 md, and of Berea sandstone
was 305 md. The single phase permeabilities were measured with 1%
brine solution.
     To emphasize the difference between the results in different
kinds of rock, it is helpful to consider the relative mobility, λ_r,
which is the ratio of the measured absolute mobility to the rock
permeability. The dimension of relative mobility is reciprocal
centipoise, whereas absolute mobility is measured in (md/cp).
Thus, in the very approximate sense in which "foam" is considered
to be a single fluid that saturates the porespace (so that single-
phase permeability could be used to characterize the rock), the
"effective viscosity" of the foam could be taken to be the reciprocal of the relative mobility.
     Figure 7 is a semi-log plot which graphically presents the
relative mobility measurements made at different total flow rates
(the latter are given as Darcy velocity in the rock) in the three
kinds of rock. The CO_2-foam fraction was kept constant around 80%.
Four different surfactants from different manufacturers were used.
Concentrations of 0.1% Chembetaine BC-50 (Z) and 0.05% Alipal CD-
128 (A) were used in experiments with Rock Creek sandstone. With
Berea sandstone, 0.03% ZS and 0.05% and 0.1% AEGS were used. For
Baker dolomite, 0.1% AEGS was used to make an exact comparison with
Berea sandstone which was also tested with 0.1% AEGS. It is
evident that the effect of rock permeability overshadows that of
the surfactant type, at least in these rocks. The figure indicates
that the magnitude of the relative mobility for the Rock Creek

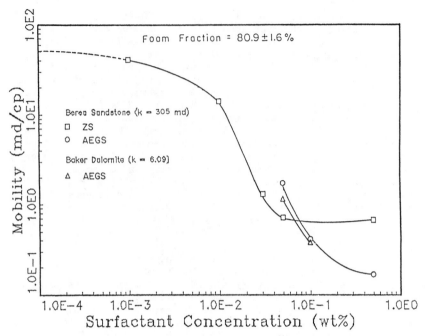

Figure 3.   Effect of surfactant concentrations.

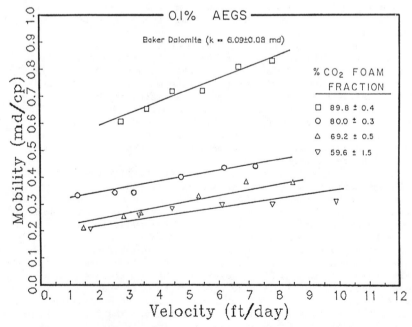

Figure 4.   Effect of CO_2-foam fraction with Baker dolomite.

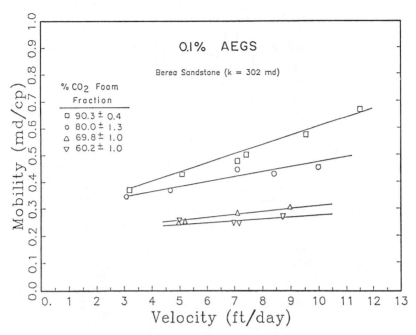

Figure 5.   Effect of CO$_2$-foam fraction with Berea sandstone.

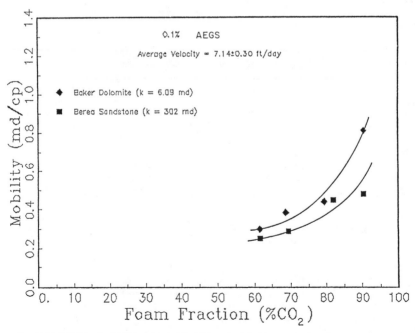

Figure 6.   Effect of CO$_2$-foam fraction at 0.1% AEGS.

sandstone is approximately 36 times higher than that of Berea
sandstone, although the inverse ratio of permeabilities is only
about 21. The similar comparison made with Berea sandstone and
Baker dolomite show that the relative mobility for the Baker
dolomite is 46 times higher than Berea sandstone. The permeability
measurements indicated that Berea was 50 times more permeable than
Baker dolomite.

   The effect of rock permeability is a significant result.
These tests show that CO_2-foam is not equally effective in all
porous media, and that the relative reduction of mobility caused by
foam is much greater in the higher permeability rock. It seems
that in more permeable sections of a heterogeneous rock, CO_2-foam
acts like a more viscous liquid than it does in the less permeable
sections. Also, we presume that the reduction of relative mobility
is caused by an increased population of lamellae in the porous
medium. The exact mechanism of the foam flow cannot be discussed
further at this point due to the limitation of the current experimental set-up. Although the quantitative exploration of this
effect cannot be considered complete on the basis of these tests
alone, they are sufficient to raise two important, practical
points. One is the hope that by this mechanism, displacement in
heterogeneous rocks can be rendered even more uniform than could be
expected by the decrease in mobility ratio alone. The second point
is that because the effect is very non-linear, the magnitude of the
ratio of relative mobility in different rocks cannot be expected to
remain the same at all conditions. Further experiments of this
type are therefore especially important in order to define the
numerical bounds of the effect.

Dynamic Adsorption Experiment

The adsorption of the surfactant on porous media is an important
variable in any enhanced oil recovery process that uses these
chemicals. It is well understood that permanent adsorption does
occur in the porous rock, which could drastically increase the cost
of the process. To maintain a specific concentration to achieve
the desired mobility reduction, it is necessary to calculate an
additional amount of surfactant that is required to satisfy its
permanent or irreversible adsorption in the flood area of the
reservoir. Furthermore, the time required for a 'front' of the
surfactant to flow through the core will be longer compared to the
brine solution. This is an indication of chromatographic delay, a
reversible adsorption of the surface active agents within the
available times determined by the flow rates in the reservoir. The
particular experiment described here will provide both the expected
chromatographic delay factor of the surfactant front in the reservoir, and a measure of the quantity of surfactant that is (or might
as well be) irreversibly adsorbed at the flow rates of the experiment. In this section, the description of the experimental
apparatus and the nature of the experiment are discussed. As
noted, dual results of the adsorption measurements are presented,
as the amount of surfactant adsorbed per gram of rock and as the
chromatographic time delay factor.

Description of the Experiment and Apparatus. The apparatus for
this experiment is shown in Figure 8. At the left side of the
figure, inside the dashed box, the procedure for core saturation is
shown. Initially, the core tested is saturated with 1% acidified
brine solution. The saturation of the core is achieved in the
following manner.

The core is put into a Hassler sleeve inside a core holder
where overburden pressure is applied outside of the sleeve by using
water. Gas is not used, since it has the potential to diffuse
through the rubber sleeve. The overburdened core holder is
connected to the Isco pump, from which 1% acidified brine is pumped
into the core. The needle valve that restricts flow from the
output end of the core is opened slightly to allow the escape of
air and air-saturated brine. The needle valve which restricts flow
from the input end of the core is opened fully. Complete saturation is attained after several pore volumes of brine have been
pumped through the core under pressure, and this fully saturated
core is used throughout the experiment.

The first of two large circles in the figure represents the
sampling valve (from Valco Instruments Corp.). This valve has six
ports which are designated alphabetically. The sample loop is
filled with a slug injection of the sample solution through port C
at the "load" position of the valve. During this operation, flow
within the valve is as shown by the solid lines. The excess of the
sample solution flows out of port D during this time. Meanwhile,
the 1% acidified brine solution is pumped continuously into port A
and out through port F of the valve.

The sample loop, V_s, has a volume of 0.25 cc. To initiate
injection at time zero of the experiment, the valve handle is
rotated 60°, and the flows within the valve are redirected along
the dashed lines as shown in the figure. The sample solution in
the loop is then displaced through port F by the 1% acidified brine
from the pump. The sample loop therefore must be long enough to
contain a sufficient "slug" of sample solution. Here, the phrase
"sample solution" refers to the surfactant solution prepared in 1%
acidified brine solution. Teflon tubing and connections are used to
minimize the uptake of any corrosion products into the 1% acidified
brine solution.

The fluid leaves the sampling valve from port F and enters port
2 of a second switching valve (Rheodyne type 7010). The Rheodyne
valve has six ports designated by Arabic numerals. The position of
this valve determines the direction of flow from port 2; either
through the core or through the capillary tube. These alternative
paths are depicted in the second circle shown in the figure, by
either dashed or solid lines. The capillary tube, V_c, has a volume
of 0.50 cc. The core, which has been saturated with the 1% acidified
brine, has a length of about 5 inches and is 1/2 inch in diameter.
The solid line shows the flow inside the valve when external flow is
through the core, and the dashed line represents the flow during an
experiment that uses the capillary tube in place of the core.

The fluid from the tube or the core leaves the valve through
port 5 and enters the inlet of the sample cell of the differential
refractometer (made by Knauer of West Germany). The residue flows
out of the sample cell to the waste. The reference cell contains

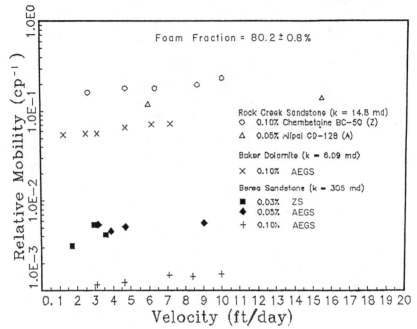

Figure 7.   Effect of rock permeabilities.

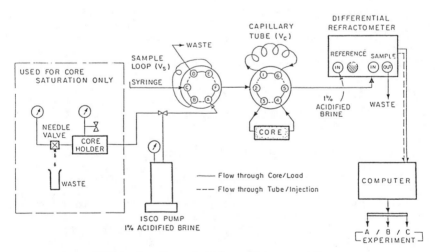

Figure 8.   Dynamic adsorption apparatus.

the 1% acidified brine, and its outlet is plugged for convenience. The differential refractometer is used to achieve high sensitivity for the surfactants under test. The refractive index detector can be interfaced with the (Terak) computer for on-line data acquisition and analyses.

Basically, three types of experiments are carried out for measurement of the adsorption parameters of a given rock sample. The response of the experiment is the measured refractive index difference, which can be readily seen on the computer screen and also recorded by the chart recorder during operation. The descriptions of the experiments are as follows.

Experiment A is a non-adsorption experiment through the core, performed to measure the time for emergence of the peak. A 1.3% (concentration higher than the standard) acidified brine is loaded into the sample loop to be used as a sample medium. This particular experiment is carried out to measure the retention time by recording the time required before the peak is observed. The retention time can also be used to compute the exact porosity of the core, under the assumption of zero adsorption of salts from the brine.

Experiment B is also a non-adsorption experiment in which flow through the capillary tube is used. The sample medium used is the surfactant solution prepared in the 1% acidified brine. Results will be combined with those from Experiment C to get the information on the permanent or irreversible adsorption on the porous medium by measuring peak areas.

Experiment C is designed to yield information on the amount of the surfactant that is actually adsorbed on the rock. This experiment measures the variation of surfactant concentration at the outlet of the core, after injection of a "slug" of surfactant. The surfactant concentration in the brine depends on the position along the core and on time. The experiment is dynamic because the changing, but near equilibrium level of the adsorbed surfactant at any point along the rock sample is a function of the concentration in the solution at that point. This is described by the adsorption isotherm from a plot of M, the mass of surfactant adsorbed per gram of rock vs. Concentration.

From type B and type C experiments, the permanent adsorption can be computed from the ratio of one peak area to the other. The mass of surfactant adsorbed, in mg per gram of rock, is computed in the following manner: By knowing the volume of the sample loop which contains the surfactant solution, the initial surfactant concentration and its density, and the ratio of one peak area to the other, the amount of surfactant adsorbed can be calculated. Similarily, the mass of the rock can be computed from the volume of the rock and the density of the solid. The density of the solid can be computed by knowing the porosity of the rock and the density of the rock material. For the computational purpose, this density was taken to be 2.7 g/cc. This result will provide necessary information to compute the additional amount of surfactant that will be required in a reservoir, over and above that required to give the necessary surfactant concentration in the aqueous part of the CO_2-foam. Furthermore, combining the data from types A and C experiments makes it possible to compute the time delay factor. The chromatographic time delay factor is simply the ratio of the time required

to observe a peak of surfactant flow to the time before the peak of
concentrated brine flow through the core is observed. The time for
the surfactant peak is longer than that measured in Experiment A
using the concentrated brine solution alone through the same core.
This indicates chromatographic delay factor or reversible adsorption. Knowing the area ratio and chromatographic delay factor, the
additional amount of surfactant that will be required, as well as
time delay to be observed in reservoir application, can be computed.

Results and Discussion on Dynamic Adsorption Measurements. Baker
dolomite was used to study the dynamic adsorption experiment. The
computed porosity of the rock was 24%. One concentration below the
CMC of AEGS, one at CMC, and two concentrations above CMC were
chosen to measure the adsorption of this surfactant with Baker
dolomite. The mass of surfactant adsorbed per gram of rock is
plotted as a function of flow rate in a semi-log plot in Figure 9.
The corresponding chromatographic time delay factors are presented
in Table I. The results of Figure 9 apparently indicate that more
adsorption of surfactant occurred at low flow rate. The shapes of
the peaks from the flow through the core are also broader at low
flow rate than at high flow rate. Also more anti-symmetry or
deviations from the bell-shaped curve was observed with increase of
concentration. Furthermore, more time was needed to reach baseline
zero at high concentration. This must be due to the fact that as
there is more adsorption, the time molecules spend in the mobile
phase decreases which results in broad peak. Compared to the broad
peak observed in core experiments, narrow and sharp peaks were
observed with the tube flow. In the concentration range of this
experiment, adsorption increased with increase of surfactant concentration. Data presented in Table I shows a chromatographic delay factor at the three flow rates. The results show that at high
concentration strong adsorption causes a longer retention time at
the rate of 10 cc/hr. However, at 0.5% with a flow rate of 20 cc/hr
and similarly at 0.1% and 0.5% with 30 cc/hr did not give an expected
trend similar to 10 cc/hr. Further tests are necessary to support
this observation.
       The slopes of the peaks in the dynamic adsorption experiment
is influenced by dispersion. The 1% acidified brine and the surfactant (dissolved in that brine) are miscible. Use of a core sample
that is much longer than its diameter is intended to minimize the
relative length of the transition zone produced by dispersion because
excessive dispersion would make it more difficult to measure peak
parameters accurately. Also, the underlying assumption of a simple
theory is that adsorption occurs instantly on contact with the rock.
The fraction that is classified as "permanent" in the above calculation depends on the flow rate of the experiment. It is the
fraction that is not desorbed in the time available. The rest of
the adsorption occurs reversibly and equilibrium is effectively
maintained with the surfactant in the solution which is in contact
with the pore walls. The inlet flow rate is the same as the outlet
rate, since the brine and the surfactant are incompressible. Therefore, it can be clearly seen that the dynamic adsorption depends on
the concentration, the flow rate, and the rock. The two parameters

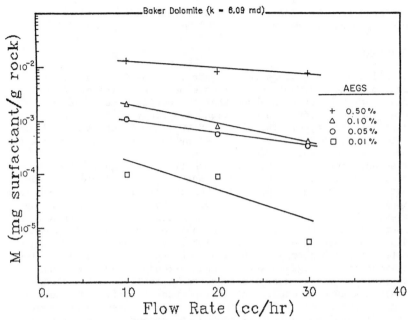

Figure 9. Mass of surfactant adsorbed per gram of rock.

Table I. Adsorption Measurements

Effects of Concentrations and Flow Rates on Chromatographic Time Delay Factors

$Q \left(\dfrac{cc}{hr} \right)$	Conc. (wt.%)	Time Delay Factor
10	0.01	1.15
	0.05	1.16
	0.1	1.21
	0.5	1.24
20	0.01	1.04
	0.05	1.05
	0.1	1.10
	0.5	1.05
30	0.01	1.10
	0.05	1.17
	0.1	1.06
	0.5	1.06

calculated from these experiments are expected to be those that would be operative in the field.

Conclusions

The following conclusions can be made from the CO_2-foam mobility measurements and dynamic adsorption experiment. The effect of surfactant concentration shows that the mobility decreases with increasing surfactant concentration regardless of the type of surfactant or rock type. The effect of CO_2-foam fraction also demonstrates that mobility can be reduced by increasing the surfactant fraction at constant surfactant concentration. The effects of surfactant concentration and CO_2-foam fraction suggest that there has to be a sufficient quantity of surfactant available, in order to stabilize the number or population of lamellae for the foam flow through the porous rock. Also, more of the shear-thinning behavior was evident at high CO_2-fraction. The effect of rock permeability illustrates that CO_2-foam is not equally effective in all porous media. The reduction of relative mobility caused by foam is much greater in the higher permeability rock. By this mechanism, displacement in heterogeneous rocks can be rendered even more uniform than could be expected by the decrease in mobility ratio alone. Hence, this type of experiment is important to define the numerical bounds of the effect. The dynamic adsorption measurements show that additional surfactant is needed to compensate for permanent adsorption, and also that, chromatographic delay cannot be ignored.

Acknowledgment

The work reported here is a part of a larger project entitled Improvement of CO_2 Flood Performance, which has been supported by the US Department of Energy, the New Mexico Research and Development Institute, and a Consortium of oil companies. We extend them our great appreciation. Acknowledgement is also made of the efforts of Mustofa Sadeq for collecting the data for adsorption measurements and James McLemore for his assistance in the experimental parts.

Literature Cited

1. Heller, J.P.; Lien, C.L.; Kuntamukkula, M.S. Soc. Pet. Eng.J. 1985, 25, 603-13.
2. Bernard, G.G.; Holm, L.W.; Harvey, C.P., SPEJ 1980, 20, 281-92.
3. Heller, J.P.; Kuntamukkula, M.S. Ind. Eng. Chem. Res. 1987, 26, 318-25.
4. Radke, C.J.; Ransohoff, T.C., SPE 15441 presented in part at the 61st Annual Technical Conference and Exhibition of the Society of Petroleum Engineers, New Orleans, LA, Oct. 1986.
5. Lee, H.O.; Heller, J.P., SPE 17363 presented at the SPE/DOE Enhanced Oil Recovery Symposium, Tulsa, OK, April 1988.
6. Casteel, J.F.; Djabbarah, N.F., SPE 14392 presented at the 60th Annual Technical Conference and Exhibition of the Society of Petroleum Engineers, Las Vegas, NV, Sept. 1985.
7. Wellington, S.L.; Vinegar, H.J., SPE 14393 presented at the

60th Annual Technical Conference and Exhibition of the Society of Petroleum Engineers, Las Vegas, NV, Sept. 1985.

8.   Heller, J.P., SPE/DOE 12644 presented at the SPE/DOE Fourth Symposium on Enhanced Oil Recovery, Tulsa, OK, April 1984.

9.   Heller, J.P.; Taber, J.J.,SPE 15001 presented at the Permian Basin Oil & Gas Recovery Conference of the Society of Petroleum Engineers, Midland, TX, March 1986.

10. Pande, P.K.; Heller, J.P., SPE 12753 presented at the California Regional Meeting, Long Beach, CA, April 1984.

11. Heller, J.P.; Boone, D.A.; Watts, R.J., SPE 14395 presented at the 60th Annual Technical Conference and Exhibition of the Society of Petroleum Engineers, Las Vegas, NV, Sept. 1985.

12. Patton, J.T.; Holbrook, S.T.; Hsu, W. Soc. Pet. Eng. J. 1983 23, 456-60.

13. Owete, O.S.; Brigham, W.E., SPERE, 1987, 2, 315-23.

14. Huh, D.G.; Cochrane, T.D.; Kovarik, F.S., SPE/DOE 17359 presented at the SPE/DOE Enhanced Oil Recovery Symposium, Tulsa, OK, April 1988.

15. Abou-Kassem, J.H.; Farouq Ali, S.M.,SPE 15954 presented at the SPE Eastern Regional Meeting, Columbus, Ohio, Nov. 1986.

16. Khatib, Z.I.; Hirasaki, G.J.; Falls, S.H., SPE 15442 presented at the 61st Annual Technical Conference and Exhibition of the Society of Petroleum Engineers, New Orleans, LA, Oct. 1986.

17. Rossen, W.R., SPE 17358 presented at the SPE/DOE Enhanced Oil Recovery Symposium, Tulsa, OK, April 1988.

18. Borchardt, J.K., Bright, D.B.; Dickson, M.K.; Wellington, S.L., SPE 14394 presented at the 60th Annual Technical Conference and Exhibition of the Society of Petroleum Enginners, Las Vegas, NV, Sept. 1985.

19. Borchardt, J.K., SPE 16279 presented at the SPE International Symposium on Oilfield Chemistry, San Antonio, TX, Feb. 1987.

20. Mannhardt, K.; Novosad, J.J., presented at the European Symposium on EOR, Hamburg, West Germany, Oct. 1987.

21. Bae, J.H.; Petrick, C.B., SPEJ 1977, 17, 353-57.

22. Grow, D.T.; Shaeiwitz, J.A., J. Colloid Inteface Sci. 1982, 86, 239-53.

23. McCutcheon's Emulsifiers & Detergents, McCutcheon Division, MC Publishing Co., Glen Rock, NJ, 1986 North American Edition, 1986, p. 276.

RECEIVED November 28, 1988

Chapter 28

Laboratory Apparatus for Study of the Flow of Foam in Porous Media Under Reservoir Conditions

C. W. Nutt,[1,2] R. W. Burley,[1,2] A. J. MacKinnon,[1] and P. V. Broadhurst[3]

[1]Heriot-Watt University, Edinburgh, United Kingdom
[2]IMOD Processes Ltd., Linlithgow, West Lothian, EH 49 7JU, United Kingdom
[3]ICI Chemicals and Polymers Ltd., Wilton, Cleveland, Great Britain

The numerous previous studies of the flow of foam in porous media and of its application for improving the displacement of oil from such media, have almost always been conducted under ambient conditions of temperature and pressure; there have been very few reports of laboratory studies under reservoir conditions. Although many interfacial properties are known to be temperature dependant, little attention has been paid to the influence of temperature upon the properties of foam. Furthermore, the rheological properties of foams, and their effectiveness for the displacement of oil are strongly dependant upon foam quality, which is in turn strongly dependant upon the pressure to which the foam is subjected.

This paper describes an apparatus for the determination of the viscous properties of foam, and for the measurement of its ability to displace oil from porous media, under reservoir conditions, and will report and discuss some of the preliminary results obtained.

It is now well established through studies in many laboratories throughout the world that foam injection shows considerable promise as an agent for the improvement of oil recovery from watered-out porous media, and for the diversion of the flow of other oil-displacing fluids from more permeable paths into less permeable paths in the medium[1]. Whilst the reasons for the effectiveness of foam for these purposes are not completely clear, the explanation is thought to lie in the behaviour of the foam lamellae

during their motion through the pore structure, and to
derive from the rheological properties of the
interfacial layers of the lamellae[2,3]. Comparison of
the behaviour of foams formed by surfactants which are
typical of the main classes of commercially available
reagents has revealed that the effectiveness of a foam
depends strongly on the chemical nature of the foaming
agent as well as on the quality of the foam[1].

Now it has been the usual practice to conduct
laboratory tests of the effectiveness of foam to
improve oil recovery from porous media at room
temperature and atmospheric pressure. Yet it is well
established that the stability of foams often
diminishes with increase of temperature, and indeed
the ability of aqueous solutions of many surfactants
to form foams disappears completely at a
characteristic temperature which is less than the
temperature within many deep oil fields. Moreover,
since the effectiveness of a foam to displace oil, or
to selectively divert flow, are functions of foam
quality and since the quality of a foam is strongly
dependant upon the pressure, the selection of a
foaming agent for EOR applications therefor requires
an understanding of the behaviour of the foam at the
elevated temperature and pressure which exists in the
oil field. Experimental study of these aspects is
therefor required, but regrettably few such
investigations have been reported[4-6]. This paper is
concerned with a description of an apparatus designed
for this purpose, and a preliminary report of some
aspects of its performance together with some
preliminary observations on the effect of pressure on
the apparent viscosity of a foam flowing in a straight
capillary tube.

APPARATUS

The central feature is a high pressure cell(1 in
Figure 1), a cylindrical vessel, internally about 0.9
m in length and 0.1 m in internal diameter mounted
with its axis horizontal. The cell is constructed of
selected stainless steel with a wall thickness and
design such that it can withstand an internal pressure
of 5250 p.s.i.g. The screwed-on end caps of this test
cell are capable of carrying various different test
units(2 in Figure 1) within the pressure cell, and are
provided with inlet and outlet lines for the test
fluid. The test units which can be incorporated
include straight, glass, capillary tubes up to 0.76 m
long and sand pack or core sample holders up to the
same length and 0.1 m in diameter. Optical glass
windows, 0.1 m long and 0.01 m wide at the top and
bottom of the test cell can permit visual observation
of the behaviour of the fluid within glass test units.

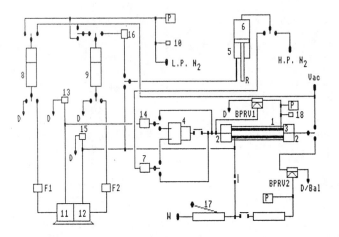

Figure 1. Schematic representation of the flow diagram of the foam rheology rig.

1	High pressure cell	2	End caps of test cell
3	Test unit	4	Foam generator
5	Rodded cell	6	Gas in rodded cell
7	Non-return valve	8	Surfactant storage vessel
9	Water storage vessel	10	Pressure relief valve
11	Surfactant pump	12	Water pump
13	pressure relief valve	14	Non-return valve
15	Pressure relief valve	16	Pressure reducing valve
17	Pressure pump	18	Pressure relief valve
	(hand operated)		

D	Discharge to drain
F1,F2	Filters
P	Pressure gauge
BPRV1	Back pressure relief valve
BPRV2	Back pressure relief valve
D/Bal	Digital balance
L.P.N2	Low pressure nitrogen blanket supply
H.P.N2	High pressure nitrogen supply
Vac	Vacuum line
W	Water supply

Another thick walled cylindrical, stainless steel, vessel(4 in Figure 1), inside diameter 0.03 m and inside length 0.075 m also capable of withstanding an internal pressure of 5250 p.s.i.g. is mounted close to the inlet feed of the high pressure test cell. Cores or sand packs can be installed within this vessel, which serves as a foam generator.

Gas can be supplied to the foam generator or to the test cell from a rodded cell(5 in Figure 1), a stainless steel cylinder, inside dimensions approximately 0.5 m long and 0.04 m diameter, mounted with its axis vertical, and designed for an internal working pressure of 6000 p.s.i.g

Electrical strip heaters, wrapped around the three stainless steel vessels enable them to be operated at any desired temperature. Thermocouples strapped to the surfaces of the vessels permit measurement of their temperatures and, coupled to controllers, maintain the vessels at constant temperature or ensure that the rate of change of the temperature of the vessels is not excessive, so avoiding unnecessary thermal stress in them.

As shown in Figure 1, for investigations at pressures up to 2000 p.s.i.g. high pressure nitrogen, from a cylinder, can be introduced into the rodded cell. The gas , 6, in the rodded cell at the desired pressure can be driven by the motion of the piston, R, via a non-return valve, 7, to the foam generator or directly to the test unit. The position of the piston can be determined by a digital position indicator gauge mounted on the end of the piston rod(R in Figure 1), which thus permits determination of the volume of gas injected and the rate of injection.

Two 4 1 cylindrical glass(QVF) vessels with stainless steel end plates, serve as reservoirs(Figure 1) for surfactant solution(8) and water(9). Facility is available to evacuate these vessels as required by means of a rotary vacuum pump with glass cold trap in line to minimise water vapour. Another pipeline permits supply of pure nitrogen, or other gas, at low pressure, to the vessels, to provide a blanket, as desired. Proper operation and safety from over pressure is ensured by a pressure relief valve(10 in Figure 1) and the pressure gauge(P in Figure 1).

The desired surfactant solution, prepared by dissolving the appropriate weight in a known volume of distilled water, can be introduced into the storage vessel(8) from which it may be discharged to a drain(D in Figure 1), or fed via a filter(F1 in Figure 1), to one side of a dual peristaltic pump(11 in Figure 1). From the pump outlet the surfactant solution flows through a line equiped with pressure relief valve(13 in Figure 1C), to a non-return valve(14 in Figure 1C), and thence to the foam generator, or to the test cell as desired. The rate of delivery of solution can be

controlled by suitable adjustment of the variable
volume setting of the pump, previously calibrated to
permit easy operation.

The other part of the dual pump(12 in Figure 1)
can similarly supply water from the reservoir(9 in
Figure 1), via a filter(F2 in Figure 1) past a
pressure relief valve(15 in Figure 1) to the underside
of the piston of the rodded cell(5 in Figure 1). This
in turn drives gas from the upper side of the piston
to the other inlet of the foam generator. To refill
the rodded cell with high pressure gas, the water from
the underside of the piston can be returned to the
reservoir(9) via a pressure reduction valve(16 in
Figure 1). The filters(F1 and F2 in Figure 1), 5
micron pore size, in the feeds to the pumps, serve to
protect these devices from damage and additional
security is provided by pressure relief valves(13 and
15 in Figure 1) on the effluent flow lines of the
pumps.

Foam created in the foam generator is fed
directly to the inlet of the test unit in the test
cell, and the pressure drop developed by its flow
through the test unit is measured by means of a
differential pressure transducer, not shown in the
figure, connected between inlet and outlet, and
provided with by-pass and isolation valves for
protection against accidental overloading.

The water pumped to the underside of the piston
of the rodded cell can also be delivered to the test
cell, to fill the jacket surrounding the test
unit(Figure 1). The contents of the jacket can be
discharged via a back-pressure relief valve(BPRV1 in
Figure 1) which is arranged to open when the water
pressure exceeds that of the foam flowing to the inlet
of the test unit.

Foam leaves the test unit via another
back-pressure relief valve(BPRV2 in Figure 1). In most
of the investigations which are planned the foam will
be collected in a suitable vessel on a continuous
weighing top loading digital balance(D/Bal in Figure
1). As shown in Figure 1, the setting of this
back-pressure relief valve is determined by the water
pressure applied to the jacket surrounding the test
unit so that flow out of the test unit occurs when the
internal pressure in the unit exceeds that in the
water jacket surrounding the test unit.

Facility is provided to bring the equipment up to
the desired absolute pressure without subjecting the
test unit to excessive pressure difference between its
interior and exterior. This is accomplished by first
filling the test cell with water by means of a hand
operated hydraulic pump(17 in Figure 1) to a suitable
value as indicated on the Bourdon dial gauges(P in
Figure 1). The pressure thus developed is used also to
control the appropriate back-pressure relief valve

which remains closed until the internal pressure in
the test unit has increased to the same value as the
external pressure. Once the necessary pressure has
been reached, the supply of water from the hand
operated pump is discontinued and the test cell water
jacket connected to the peristaltic pump used to
supply water to the rodded cell. Excessive pressure
difference between the interior of the test cell and
exterior is avoided by feeding the inlet pressure of
the test unit to the other back-pressure relief valve
which opens to permit water from the exterior of the
test unit to be discharged should the external
pressure become too high. A pressure relief valve(18
in Figure 1) provides additional protection.
Facilities are provided to fit pressure transducers at
various points on the flow lines, should it be desired
to monitor such pressures in the future.

Signals from the digital balance, the rodded
cell rod position indicator, the gauge indicating the
pressure differential over the test unit and the
output of thermocouples indicating the temperature of
foam entering and leaving the test unit are
transmitted to a data logging facility and recorded on
floppy discs on an Acorn "BBC Mode B" microcomputer,
and conveyed to an Acorn "Archimedes" computer for
processing as desired. Several hundred values for each
of the variables could be recorded during a run
lasting up to ninety minutes, in this way, for
subsequent analysis.

EXPERIMENTAL

The experiments initially conducted were designed to
test the performance of the apparatus by determination
of the effect of pressure on the viscosity/quality
spectrum of a typical foaming agent when it flows
through a straight capillary tube. The method used is
based on a technique recently developed in the
laboratories(7) for this purpose, at atmospheric
pressure.

The viscosity/foam quality spectrum at a fixed
pressure is derived from a single experiment in which
the foam generating vessel is first filled completely
with foaming agent solution, and then the gas supply
is turned on. Initially, aqueous solution is displaced
from the generator, but once gas break-through
commences a wet foam is discharged. As gas flow
continues, the foam becomes drier and drier until
ultimately only gas flows.

Determination of the rodded cell rod position, as
a function of time(Figure 2 shows a typical set of
results), permits computation of the volumetric gas
flow rate into the foam generator.

Figure 3 shows a typical set of results for the
cumulative weight of fluid discharged from the

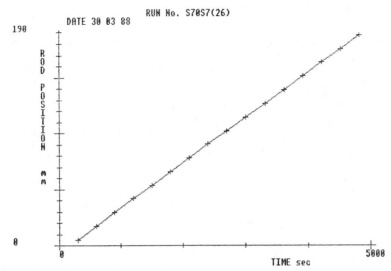

Figure 2. Typical plot of data output from the rig, showing values of the position of the rodded cell piston as a function of time, during an experiment.

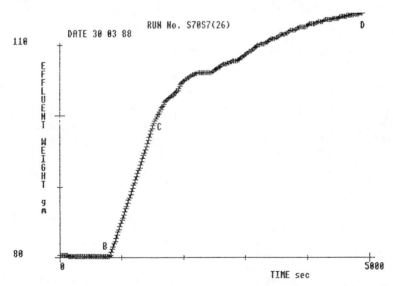

Figure 3. Typical plot of data output from the rig, showing values of the weight of effluent from the capillary test unit as a function of time, during an experiment. The portion of the curve from B - C shows the efflux of surfactant solution from the foam generator which precedes gas break-through. Portion C - D shows the efflux of foam of increasing quality. At times larger than D only "dry" gas, without liquid phase, emerges.

apparatus, during the course of an experiment. The
results demonstrate clearly how in the initial stage,
aqueous surfactant solution is eluted from the foam
generator, followed by foam once gas break-through has
commenced. The steady decline in the rate of increase
of cumulative weight of foam through the course of the
experiment shown by the results in Figure 3, clearly
demonstrates how the foam quality steadily increases
in dryness until at the end of the experiment, only
gas emerges from the foam generator.

After gas break-through. the incremental rate of
increase in effluent weight can be taken as a measure
of the incremental weight of liquid in the foam. This,
together with a knowledge of the instantaneous
volumetric gas flow rate, permits calculation of the
quality of the foam. Provided the pressure drop over
the foam generator is sufficiently small, the
volumetric flow rate for such calculations can be
taken to be constant throughout an experiment and
equal to that indicated by data such as that shown in
Figure 2. This assumption was justified in the studies
at atmospheric pressure reported previously[7], but in
the present investigations correction for changes in
pressure drop across the foam generator during the
course of an experiment may be necessary.

By conducting a series of such measurements at a
number of different gas flow rates, measurements of
the pressure drop across a capillary tube having a
known diameter and length, during each experiment
(Figure 4 shows a typical set of values of the
differential pressure gauge output as a function of
time), permit evaluation of the wall shear stress and
wall shear rate as functions of time[7], and thereby
for the range of foam qualities created during the
experiment. Previous studies[6] have shown that under
such conditions, at atmospheric pressure, foam
conforms closely to the simple power law relation for
non-Newtonian fluids:

$$\mu_{app} = \tau_w/\gamma_w = K\gamma_w^{n-1} \tag{1}$$

where K is the flow consistency,
   n is the power index,
   μ_{app} is the apparent viscosity,
   τ_w is the wall shear stress, and
   γ_w is the wall shear rate.

The results shown in Figure 2 illustrate that a
satisfactorily constant gas flow rate over a period of
time in excess of one hour could be achieved readily.

Consideration of the slope of the effluent
weight/time curve just after gas breakthrough, shown
in Figure 3, indicates that the foam first breaking
through had a quality of about 60%, whilst that
emerging near the end of the run has a quality of
approximately 92%. The foam quality at intermediate

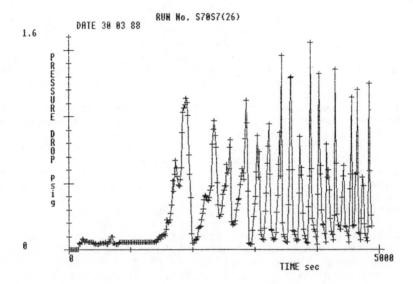

Figure 4. Differential pressure transducer output, as a function of time, during a typical experiment at atmospheric pressure.

times between these limits sometimes fluctuated irregularly, sometimes becoming drier and then showing an abrupt increase in wetness. Each such irregularity was accompanied by corresponding changes in the pressure drop observed for the flow through the capillary tube, as illustrated in Figure 4. The change in foam quality indicated by the effluent weight curve was delayed by approximately 150 secs behind that indicated by the pressure drop/time curve, at the particular gas flow rate used in that experiment. This was due to the hold-up of effluent in the back-pressure relief valve in the foam outlet line. These irregularities in foam quality were caused by the gas flow process through the sand pack in the foam generator. From time to time as the sand pack dried out, new gas flow paths developed, and associated with each such event there occurred a transient increase in foam wetness. Careful packing of the foam generator could minimise this phenomenon. Alternatively or additionally, the effect could be eliminated by inclusion of a mixing vessel of properly selected capacity, in the foam flow line between the generator and the test unit.

The rheological behaviour of a typical surfactant at atmospheric pressure, derived using this technique, has been reported and illustrated in various figures in an earlier communication[7].

FURTHER DEVELOPMENT

In the current program of investigations it is intended to study the rheological behaviour of foams formed by a range of surfactants, during flow through a capillary tube, over the quality range 50-60% to 95%, at temperatures up to 100°C and pressures up to 2000 psig. After the addition of a pressure intensifier the studies of certain selected foams will be extended to pressures up to 5,000 psig.

By the addition of other liquid reservoirs, e.g. for water and crude oils, and replacement of the capillary test unit by a sand pack, the effectiveness of the same foams for enhanced oil recovery at reservoir temperatures and pressures will be investigated.

ACKNOWLEDGMENTS

The authors wish to record their thanks and appreciation to ICI Chemicals and Polymers Ltd. and to Offshore Supplies Office for financial support and for permission to publish this paper, and to Alval Engineering Ltd for equipment manufacture and high pressure design input.

REFERENCES

(1) Ali J., Burley R.W. and Nutt C.W. "Foam Enhanced
    Oil Recovery from Sand Packs" Chem.Eng.Res.Des.
    1985 63 101.

(2) Giordano R.M and Slattery J.C.
    A.I.Chem.E. Symp.Series 1982 78 58.

(3) Nutt C.W., Burley R.W., Stavenam A. and
    Naismith S.M. "Mechanism of the Displacement of
    Oil from a Porous Medium by Foam"   (in press)

(4) Heller J.P. and Kuntamukkula M.S. "Critical Review
    of Foam Rheology Literature" Ind.Eng.Chem. 1987 26
    318.

(5) Farouq Ali S.M. and Selby R.J. "Function,
    Characteristics of EOR Foam Behaviour Covered in
    Laboratory Investigations" Technology, Oil
    & Gas J. Feb 3 1986 57-63

(6) McPhee C.A., Tehrani A.D.H. and Jolly R.P.S.
    S.P.E. 1988 Paper No.17360

(7) Assar G.R., Nutt C.W. and Burley R.W.
    "The Viscosity-Quality Spectrum of Foam Flowing in
    Straight Capillary Tubes" (in press)
    Int. J. of Eng. Fluid Mech. 1988.

RECEIVED November 28, 1988

Chapter 29

Associative Organotin Polymers

Triorganotin Fluorides in Miscible Gas Enhanced Oil Recovery

D. K. Dandge, P. K. Singh, and John P. Heller

New Mexico Petroleum Recovery Research Center, New Mexico Institute
of Mining and Technology, Socorro, NM 87801

Preparation, characterization, and properties of
silicon-containing triorganotin fluorides, both symmetrical and unsymmetrical, were investigated. It
was observed that introduction of a trimethylsilyl
group in the alkyl chain results in a considerable
enhancement of solubility in various nonpolar
solvents including dense carbon dioxide.

Enhanced oil recovery processes involving displacing fluids such
as dense CO_2 and liquified petroleum gases (LPG) are currently
being applied in different parts of the world. At moderately high
pressure and reasonable temperatures common in many reservoirs,
CO_2 is capable of extracting the light ends of the oil in the
region of contact with the oil.

The effectiveness of CO_2 in displacing oil from reservoirs
is marred, however, by its extremely low viscosity. The viscosity
of dense CO_2 remains low (in the range from 0.03 to 0.08 cp or
0.03 to 0.08 mpa) despite its relatively high density (above 0.45
g/cm^3) under reservoir conditions. This low viscosity of CO_2 as
compared to that of crude oil (1-10 cp) results in a high mobility
ratio which degrades the macroscopic efficiency of the displacement process. Therefore, some method of mobility control is
required for efficient use of CO_2, to increase greatly the
quantity of producible oil.

The desired low mobility ratio (approaching 1) could be
achieved by the viscosification of CO_2. This prompted the search
for compounds which can dissolve in and viscosify dense CO_2 and
nonpolar hydrocarbons.

Certain triorganotin fluorides are known to be effective viscosifiers for nonpolar solvents. Dunn and Oldfield (1) have
studied the solution properties of tri-n-butyltin fluoride (BUF)
in various organic solvents. They reported that, among n-alkanes,
BUF dissolves only in n-hexane and not in n-heptane. This

0097–6156/89/0396–0529$06.00/0

intriguing solubility behavior was attributed to the matching size
of solvent molecules with those of the solute. Dandge et al.
($\underline{2,3}$), however, found that increasing the alkyl chain beyond C_4 in
alkyltin fluorides (R_3SnF) substantially increases their solubility in most organic solvents. Furthermore, viscosification of
nonpolar solvents by these trialkyltin fluorides (which ranged
from triamyl to tridecyltin fluoride) was as effective as that by
BUF or even better ($\underline{2,3}$). Due to the large difference in the
electronegativity of Sn and F atoms, R_3SnF-type compounds form
polymeric chains ($\underline{1}$) by association wherein Sn assumes a penta-
coordinate structure:

```
       R   R      R   R       R   R
        \ /        \ /         \ /
    -- Sn-F ---- Sn-F ---- Sn-F --
        |          |           |
        R          R           R
```

These transient polymeric chains resulting from dipole-dipole
interactions are responsible for the increase in solvent viscosity
which is asymptotic above a certain concentration. The structure
dependence of both solubility and viscosity in triorganotin compounds is known ($\underline{1,2,3}$). For example, trimethyl and tripropyltin
fluorides are insoluble in most organic solvents while the bulky
trineophyltin fluoride ($\underline{4}$) dissolves in many solvents but does not
increase their viscosity.

Our previous work ($\underline{2,3}$) showed that while increasing the
alkyl chain in R_3SnF increases both their solubility and viscosity
in hydrocarbons such as propane (d = 0.5 g/cc) and butane (d = 0.5
g/cc) ($\underline{5}$), the solubility in CO_2 did not show a considerable
increase. Also, in a separate study ($\underline{6}$), it was observed that
silicon-containing polymers are more soluble in dense CO_2 than any
other polymer of similar molecular weight. Taking these observations into consideration, we decided to synthesize silicon substituted trialkyltin fluorides in order to increase the solubility
of these associative compounds in CO_2. The field applicability of
such viscosifiers, it is found, would depend on the price of oil.
R_3SnF compounds with long chain R groups are nontoxic and offer
little risk to the environment. This paper describes the synthesis, characterization, and solution properties of three silicon
containing organotin fluorides.

Experimental

Materials. The chemicals used in the synthesis were purchased
from Aldrich Chemical Co. and Petrach Systems.

Methods. FTIR spectra were obtained in nujol mull on a Perkin
Elmer model No. 1700GC-IR spectrophotometer. ^1H NMR spectra were
obtained at 363 MHz. ^{13}C NMR spectra were obtained at 91 MHz.
All NMR spectra were run by Spectral Data Services. No external
standard such as TMS was added to the sample as the compounds had
$(CH_3)_3Si$ groups. X-ray diffraction patterns were obtained on
Rigaku-Geiger flex x-ray diffraction machine. Melting points and

boiling points are uncorrected. Elemental analysis was carried
out by Galbraith Laboratories, Knoxville, Tennessee.

Viscosity measurements were carried out using Cannon-Fenske
viscometers of appropriate capillary diameter so as to keep the
efflux time between 200-300 seconds. Approximate shear rate at
the wall was calculated using the equation

$$\gamma_w \;=\; 4V/(\pi R^3 t)$$

where
γ_w =   Shear rate at the wall
V =   Volume of the viscometer bulb
R =   Capillary radius
t =   Efflux time

Shear rate and viscosity curves were obtained at 25°C using
Contraves low shear viscometer (Model LS-30).

General Synthesis Procedures

Tetraorganotin. A Grignard reagent was prepared in THF from
equimolar amounts of bromoalkane and metallic magnesium. A few
drops of bromoethane were used to initiate the reaction. To the
reagent was added benzene solution of stannic chloride (1/18 mole
bromo-alkane used), so as to keep the reaction mixture under
gentle reflux. Unreacted magnesium was filtered, and the filtrate
was treated with saturated ammonium chloride. The organic layer
was separated and dried. Removal of solvent and fractional
distillation yielded the corresponding tetraorganotin which was
characterized by boiling point and refractive index.

Triorganotin Bromide. Tetraorganotin compounds were brominated
with stoichiometric quantities of bromine (2 moles of bromine for
each mole of tetraorganotin). Bromination was carried out after
cooling the tetraorganotin compound in a dry ice-acetone bath. The
contents then were slowly brought to room temperature. In most
cases, the crude bromides were used as such for fluorination.

Triorganotin Fluoride. Fluorination of triorganotin bromide was
carried out by an excess of ammonium fluoride at 120°C for one
hour.

Results and Discussion

Synthesis and Characterization. Tris(trimethylsilylpropyl)tin
fluoride (PTF) and tris(trimethylsilylmethyl)tin fluoride (MTF)
were synthesized according to scheme A (Figure 1) while scheme B
(Figure 1) was followed in the synthesis of dibutyl-
(trimethylsilylpropyl)tin fluoride (BTF). MTF has been reported
in the literature (6); however, it was synthesized to compare its
solution properties with the other two novel compounds. The steps
involved in these syntheses were straight-forward and thus need no
elaboration. In the synthesis of BTF, the unsymmetrical triorganotin fluoride, bromination of tetraorganotin compound resulted

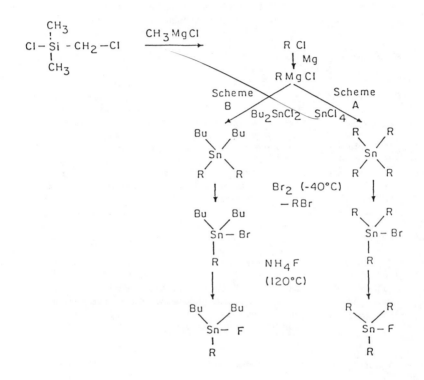

R = (CH₃)₃ Si CH₂—   or   (CH₃)₃ SiCH₂CH₂CH₂—

Figure 1. Schemes for the synthesis of silicon containing
triorganotin fluorides.

in the preferential displacement of the trimethylsilylpropyl group. Similar observations were reported by Seyferth (8).

Measured and calculated percentages of the elements in the compounds are shown in Table I. It is apparent that these values are separated by no more than the probable experimental error. Also shown in the same table are boiling points and the refractive indices of the tetraorganotin compounds, which are the precursors of MTF, PTF, and BTF.

The x-ray diffraction patterns for MTF, PTF, and BTF are given in Figure 2. The unsymmetrical BTF shows broad dispersed peaks indicating an increase in amorphousness compared to MTF, its symmetrical analogue. The peaks in the pattern for PTF, on the other hand, are broader and less sharp than those for MTF, which may be the result of an increase in the alkyl chain length by two carbon atoms.

FTIR spectra of the three fluorides are given in Figure 3. Typical frequencies of the peaks found in the region of 450 cm^{-1} to 1300 cm^{-1} region and their assignments are as follows:

525 cm^{-1}	Sn-C Stretch
615 cm^{-1}	Sn-C Stretch
690 cm^{-1}	-CH_2 Stretch
712 cm^{-1}	-CH_2 Stretch
755 cm^{-1}	Nujol
840 cm^{-1}	Si-C Stretch
860 cm^{-1}	$(CH_3)_3$ Si Stretch
1250 cm^{-1}	Si - CH_3

Due to the unavailability of appropriate cells, the Sn-F absorptions in the far infrared region cannot be observed.

The ^1H and ^{13}C NMR spectra of the three fluorides are given in Figures 4 and 5 respectively. The peak assignments for various groups in each compound are given in Tables II and III, while calculated and observed peak ratio values are given in Tables IV and V.

Solution Properties

Solubility Behavior - The solubilities of MTF, PTF and BTF in various organic solvents including CO_2 are given in Table VI. The solvents are arranged in the order of increasing solubility parameter. It has been observed, after a comparative study with tri-n-alkyltin fluorides, that replacement of a methyl group by a trimethylsilyl group enhanced the solubility in nonpolar solvents, and is well reflected in the markedly higher solubility of MTF, PTF, and BTF in CO_2. For example, tri-n-alkyltin fluorides dissolve to an extent of 1.26 gm/liter whereas compounds having trimethylsilyl group with appropriate number of CH_2- groups dissolved to an extent of 4.5 gm/liter (see Table VI).

The rate of dissolution of PTF and BTF in nonpolar solvents is considerably higher than in tri-n-alkyltin fluorides. The former dissolve instantly while the latter dissolve only by a slow swelling process. This indicated that the presence of the tri-methylsilyl group, at the end of the chain and somewhat away from

Table I

Silicon Containing Triorganotin Fluorides

Compound	m.p. °C	$25n_D$	Calc. for %				Elemental Analysis Found %			
			C	H	Si	F	C	H	Si	F
MTF	102	--	36.09	8.27	21.13	4.74	36.35	8.42	20.51	4.89
PTF	180	--	44.72	9.32	17.45	3.94	44.74	9.65	15.74	3.15
BTF	82	1.4765	45.81	8.99	7.66	5.18	46.15	8.50	7.20	4.26

MTF   Tris(trimethylsilylmethyl)tin fluoride
PTF   Tris(trimethylsilylpropyl)tin fluoride
BTF   Dibutyl(trimethylsilylpropyl)tin fluoride

Intermediates for the synthesis of the above compounds were:

Tetraorganotin Compound	b.p	$25n_D$
Tetra(trimethylsilylmethyl)tin	140°C	1.4825
Tetra(trimethylsilylpropyl)tin	205°C	1.4694
Dibutyl-di(trimethylsilylpropyl)tin	160°C	1.4674

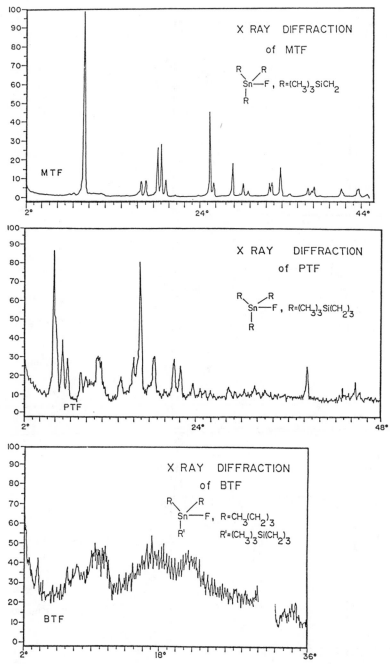

Figure 2. X-ray diffraction of silicon-containing triorganotin fluorides.

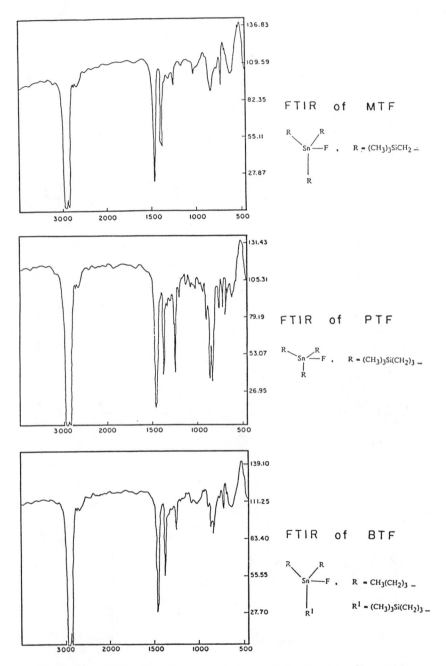

Figure 3.   FTIR of silicon-containing triorganotin fluorides.

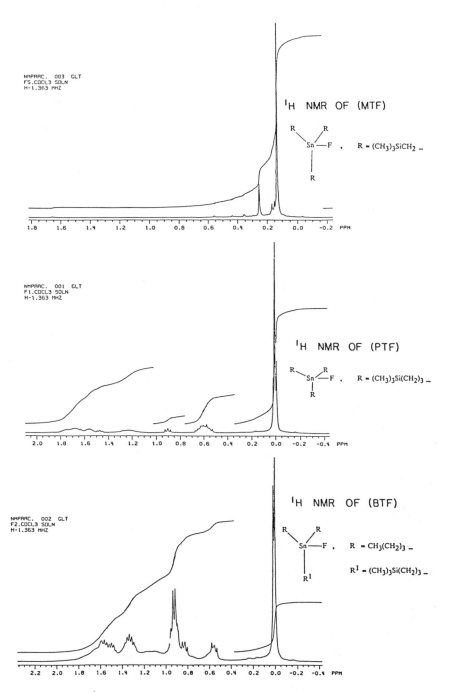

Figure 4. ¹H NMR of silicon-containing triorganotin fluorides.

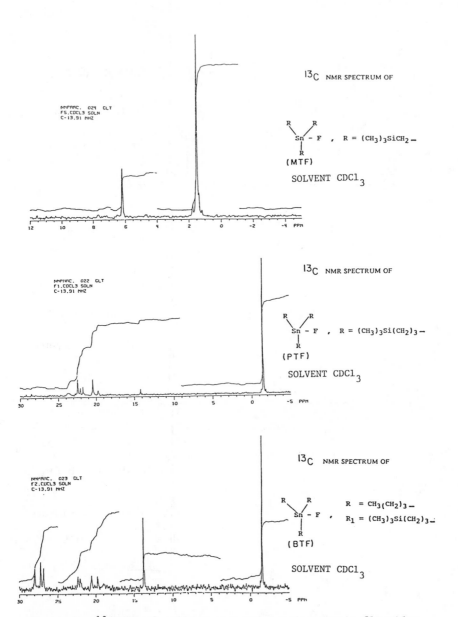

Figure 5. ^{13}C NMR of silicon-containing triorganotin fluorides.

Table II. Proton NMR, Chemical Shifts in $CDCl_3$, δppm

MTF	.190 $(CH_3)_3Si$		.25 $-CH_2$ (Near Sn)	
PTF	.013 $(CH_3)_3$ $SiCH_2-$		.6 $-CH_2$ (Near Sn)	1.6 $-CH_2$
BTF	.504 $(CH_3)_3$ $SiCH_2-$	.549/.822 CH_3-	.909 (CH_2-Sn)	1.329 $-CH_2-$	1.514 $-CH_2-$

Table III. ^{13}C NMR, Chemical Shifts, δppm

MTF	1.441 $(CH_3)_3Si-$		6.159 $-CH_2$ (Near Sn)		
PTF	-1.559 $(CH_3)_3Si-$		20.00 $-CH_2$ (Near Sn)		22.205 $-CH_2-$
BTF	-1.809 $(CH_3)_3Si$	13.880 $-CH_3$	19.80 $(-CH_2-)$ (Near Sn)	22.00 C_2-H_5	27.157 $-CH_2-$

Table IV. Calculated and Observed Ratios of Protons
MTF, PTF, and BTF

	$(CH_3)_3Si$ CH_2/CH_2		$(CH_3)_3Si$ CH_2/CH_3	
	Observed	Calculated	Observed	Calculated
MTF	5.71	5.501		
PTF	4.7 4.7	5.5 5.5		
BTF	5.5, 2.85 2.85	5.5, 3.3	4.46	4.00

the Sn-F bond, imparts higher solubility as well as higher
dissolution rates to triorganotin fluorides.

Viscosity Behavior. The polymeric nature of triorganotin fluorides dissolved in nonpolar solvents is outlined in the introduction. As a result of the transient polymer formation, these
solutions exhibit nonlinear concentration vs. viscosity curves.
The concentration at which a steep rise in this curve begins has
been termed as the critical or threshold concentration (2,3).
Figure 6 shows such typical curves for PTF and BTF in n-hexane.
Despite the fact that different shear rates are involved in
capillary viscometry, it can be qualitatively said that at a given
concentration, PTF viscosified n-hexane better than BTF. It is
clear from Figure 6 that the critical concentration for these two
compounds is above 0.7%, while analogous tri-n-alkyltin fluorides
showed a critical concentration of less than 0.4% (3). This may
be due to the presence of bulky Me_3Si-groups nearer to the Sn-F
bond, which causes some steric hindrance to auto-association.
      These results suggest that the solubility of trialkyltin
fluoride in nonpolar solvents is considerably increased by the
introduction of a trimethylsilyl group at the end of the chain but
very much so at the expense of decreasing their ability to
viscosify these solvents.
      Figure 7 shows the viscosity vs. concentration curves for PTF
in various organic solvents. As would be expected, in toluene,
the most polar among the four solvents, the viscosity was the
lowest at any given concentration. Figure 8 shows the effect of
concentration on the viscosity of CO_2 at 25°C. The observed 50%
increase in viscosity is not sufficient for enhanced oil recovery
operations.
      The viscosity vs. shear rate data for PTF and BTF (Figure 9)
were obtained in n-heptane solution at 25°C using a Contraves low
shear rate instrument. n-Heptane was used as a nonpolar solvent,
as it has a high enough boiling point to avoid losses due to
evaporation during the measurement period. The curves presented
here are reproducible in both directions of shear and are thus
time-independent.
      In conclusion, synthesis of novel silicon-containing trior-
ganotin fluorides was accomplished. It was shown that the introduction of a trimethylsilyl group on the alkyl chain considerably
enhances the solubility of trialkyltin fluorides. However, if the
bulky trimethylsilyl group is closer to the Sn-F bond, the
effectiveness of such compounds in enhancing solvent viscosity is
adversely affected. Higher solubilities in CO_2 were achieved but
the viscosification of CO_2 at the highest solubility (0.4%) was
only 50%. Additional viscosity measurements are to be made of the
other two compounds described above. Further modifications in the
molecular structure of these compounds might also increase both
their solubility and the viscosity of their CO_2 solutions.

Table V. Calculated and Observed Ratios of Carbon Atoms in MTF, PTF, and BTF

	$(CH_3)_3Si\ CH_2/CH_2$		$(CH_3)_3Si\ CH_2/CH_3$	
	Observed	Calculated	Observed	Calculated
MTF	4.4	4.5		
PTF	4.6	4.00		
	4.00	5.5		
BTF	1.11	1.00	2.05	2.00
	2.85			

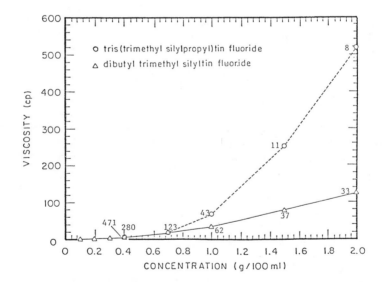

Figure 6. Viscosity as a function of concentration for PTF and BTF. Solvent n-hexane temperature 25°C. Number near each point indicates approximate shear rate at the wall.

Table VI

Solubilities of Silicon Containing Triorganotin Fluorides at 25°C

Solvent	Solvent Solubility Parameter (cal/cm^3)	Dipole Moment D†	Dielectric Constant (ε)†	Solvent†† Viscosity CP	MTF*	PTF*	BTF*
carbon dioxide**	6.0	0	1.60	0.07		4.46 ■	4.14 ■
ethane**	6.0	0		0.16			
propane**	6.4	0.084	1.61	0.22 (25°C)			
butane	6.8	0.05		0.27	0	X†	X†
pentane	7.0	0	1.84	0.4	0	X†	X†
hexane	7.3	0.08	1.89 (25°C)	0.4	X	X†	X†
heptane	7.4	0	1.92 (20°C)	0.6	0	X†	X†
decane	8.0	0	1.99	1.26			
cyclohexane	8.2	0	2.02	1.3 (17°C)	X	X†	X†
carbon tetra-chloride	8.6	0	2.24	0.5	X	X†	X†
toluene	8.9	0.36	2.38	0.6 (30°C)	X†	X†	X†
benzene	9.2	0	2.275	0.7	X	X†	X†
chloroform	9.3	1.0	4.80	0.3	X	X†	X†

†Reference 3
††Reference 4
■ In g/l

0 - insoluble
X - solubility at least 0.2 g/dl
X† - solubility >> 0.4%

* As in Table 1
**Conditions used for solubility determinations

	Pressure PSI	Temp. °C
CO$_2$:	2500	25
ethane:	1200	25
propane:	1200	25

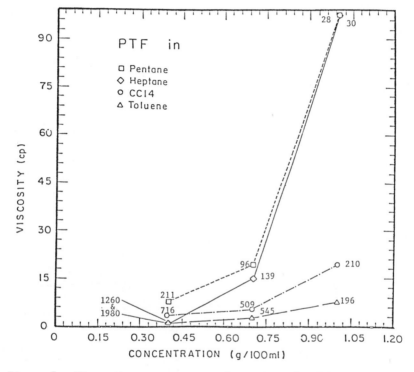

Figure 7. Viscosity vs. concentration curves for PTF in various solvents at 25°C. The number near each point indicates approximate shear rate at the wall.

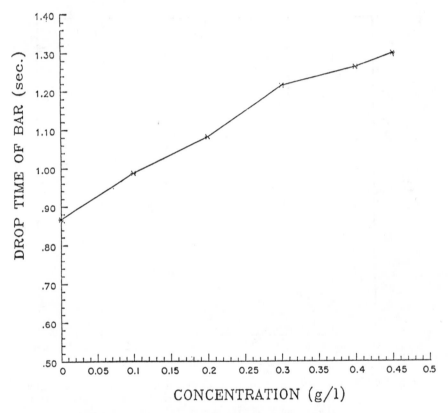

Figure 8.   Viscosity of PTF in carbon dioxide.

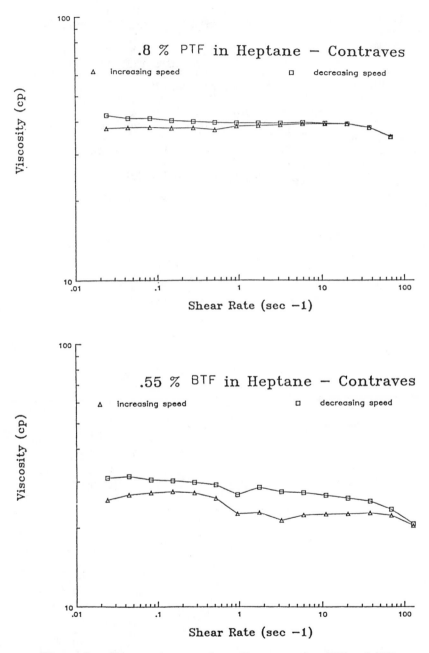

Figure 9. Shear rate vs. viscosity curve for PTF and BTF.

References

1. Dunn, P. and Oldfield, D.J.; <u>Macromol. Sci. Chem.</u>, A4, 1157
    (1970).
2. Dandge, D.K., Taylor, C. and Heller, J.P., Wilson, K.V., and
    Brumley, N.: Submitted to <u>J. Polym. Sci.-Chem.</u> (1987).
3. Taylor C., M.S. Thesis, Dept. of Chem., New Mexico Institute
    of Mining and Technology, Socorro, NM 87801.
4. Reichle, W.T.; <u>J. Inorg. Chem.</u>, 5(1), 88 (1966.)
5. Heller, J.P. and Dandge, D.K., U.S. Patent 4,607,696.
6. Heller, J.P., Dandge, D.K., Card, R. and Donamura, L.G, <u>Soc.
    Petr. Engr. J.</u>, 679 (1985).
7. Kruglaya, O.A., Kalima, G.S., Petrov, B.I., and Vyazankin,
    N.S.; <u>J. Organomet. Chem.</u>,46, 51 (1972).
8. Seyferth, D.J.; <u>J. Am. Chem. Soc.</u>, 5881 (1957).

RECEIVED November 28, 1988

FLUID–ROCK INTERACTIONS

Chapter 30

Hydrodynamic Forces Necessary To Release Non-Brownian Particles Attached to a Surface

Habib Chamoun, Robert S. Schechter, and Mukul M. Sharma

Department of Chemical and Petroleum Engineering, The University of Texas at Austin, Austin, TX 78712

The release of non-Brownian particles (diameter $\geq$ 5 μm) from surfaces has been studied. The influence of several variables such as flow rate, particle size and material, surface roughness, electrolyte composition, and particle surface charge has been considered. Experiments have been performed in a physically and chemically well-characterized system in which it has been observed that for certain particle sizes there exists a critical flow rate at which the particles are released from surfaces. This critical flow rate has been found to be a function of the particle size and composition. In addition, it has been determined that the solution pH and ionic strength has an effect on the release velocity. These observations imply that both fluid-mechanical effects and surface-force effects control the particle release.

A number of important processes depend on the permanence of particle attachment to surfaces by Van der Waal forces in the presence of flowing fluids. These include enzyme fixation, particle filtration, oil production, nuclear reaction excursions, migration of surface contaminants, etc. The release of particles attached to a surface plays an important role in these processes.

For the purpose of this study, particles are classified as Brownian or non-Brownian, where Brownian particles are defined as those for which the diameter is less than five microns and non-Brownian are those with diameter greater than five microns. The major focus of this work is on the second category. The particle release process has been studied both theoretically and experimentally, and it is found that for non-Brownian particles the surface charge and the electrolyte composition of the flowing phase are less significant factors than the hydrodynamic effects. However, Van der Waals forces are found to be important and the distortion of particles by these forces is shown to be crucial.

0097–6156/89/0396–0548$06.00/0

In this investigation we experimentally determine the factors controlling the release of non-Brownian particles. Also, we discover the initial particle release mechanism. (i.e., rolling-vs-sliding).

Literature Survey

The problem of detachment of particles under the action of a water stream has also been considered by Visser (1). He mentioned that a tangential force, Ft, due to the fluid drag, contributes to the dislodging force acting on the particle; however, he remarked that the lift force contributes negligibly to the dislodging force. The tangential force on a spherical particle of radius R in contact with a plane wall in a slow linear shear flow has been calculated theoretically by Goldman, Cox, and Brenner (2) and O'Neill (3). The tangential force that they obtained is given by the following equation:

$$FH = 1.7005(6\pi)\eta RV_{X=R} \tag{1}$$

where R is the particle radius, η is the fluid viscosity, and $V_{X=R}$ is the fluid velocity at $r = R$.

Visser uses the preceding equation for the tangential force, and makes the additional assumption of laminar flow near the wall to arrive at the following equation:

$$FH = 32 R^2\tau_0 \tag{2}$$

where R is the particle radius.

That is, the removal of spherical particles from a flat surface is determined by the magnitude of the wall shear stress, τ_0. Visser (1) also claims that since the removal mechanism is unknown, it is not possible to relate the F_H (tangential force) to the F_a (adhesive force) on theoretical grounds. Therefore, he assumes that the tangential force required for particle release is proportional to the adhesive force.

$$F_H = \alpha F_a \tag{3}$$

where F_a is the force of adhesion and α is the proportionality constant. This constant of proportionality has been determined experimentally for different systems. For example, Visser (1) found a value of 1 for α on the removal of 0.21 μ carbon-black colloidal particles from a cellophane surface. Zimon (4) experimentally found a value for α of 0.65 on the removal of submicron glass particles from a flat surface by a water stream.

For non-Brownian particles it is believed that the particle may come off by a rolling, rather than by a lifting, mechanism. Polke (5) and St. John (6) compared the normal and parallel direction forces necessary to remove gold particles from a gold surface in vacuo. They found that removing a particle by rolling requires a tangential force 10 to 50 times smaller than removal caused by a normal acting force. A general overview of the non-Brownian particle release is given by Hubb (7). He developed a theoretical model for the detachment of colloidal particles from surfaces. The final product of his analysis gives a relationship between particle radius, R, and shear stress at the wall, τ_0. This is done for different modes of incipient motion, in particular, sliding, rolling, and lifting. The case of rolling motion is presented in this paper. In order to analyze this case, a torque balance is performed between the hydrodynamic torque and the product LF_A, where L is a characteristic length and F_A is the net force of attraction acting through the particle center (see Figure 1). The characteristic length is a function of the particle material and particle size. Hubb considered the two cases: (1) surfaces with a high elastic modulus and (2) surfaces with a low elastic modulus. In this chapter we discuss only the latter one for which the characteristic length, L, is related to the contact area of deformation strictly due to attractive forces [Krupp (8), Muller et al. (9), and Johnson et al. (10)]. From the torque balance, considering that L is proportional to $R^{2/3}$ from the elasticity theory (see Hubb (7), the following equation is found:

$$\tau_{0(rolling)} = \alpha R^{-2/3} \qquad (4)$$

The above presentation due to Hubb (7) is a good initial approach which gives one physical insight into the problem, but a more detailed presentation for the release of elastic particles form rigid surfaces (specifically, cylindrical particles) may be found in Chamoun (11). In addition, extensive experimental data on the release of particles from various surfaces is reported by this author. An analysis of this experimental data is presented in Shirzadi et al. (12).

A summary of the most important experimental findings of Chamoun (11), along with a description of the experimental apparatus and procedure, is presented in this chapter. In particular, the experiments have shown which factors (such as pH, ionic strength, etc.) control the release of non-Brownian particles and also have proven that the initial particle release mechanism is rolling rather than sliding.

Experimental Methods

Two different types of studies have been carried out: flow experiments and centrifuge experiments.

Apparatus

Flow Experiments. The main components of the experimental apparatus are illustrated in Figure 2. The most important component is the glass flow cell, shown in detail in Figure 3.

The glass cell consists of a microscope slide overlying a flat glass plate, being held together with epoxy and separated by a paper gasket. The cell dimensions are 1-1/2 inches in width and 3 inches in length, with a spacing of 200 ± 10 μm between the microscope slide and the glass surface.

A flow system has been designed to pump fluid through the flow cell. A piston-cylinder arrangement is used as a "pump-like" device by

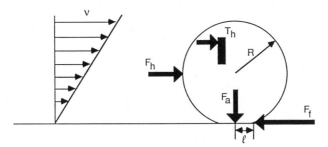

Figure 1.   Torque balance.

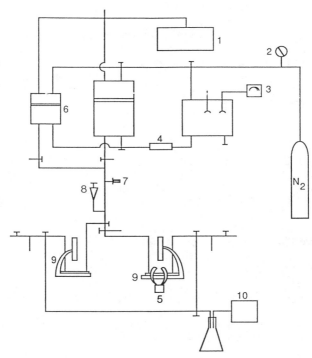

1. Fluid Reservoir
2. Pressure Gauge
3. PH Meter
4. Pump
5. Fiber Optic Illuminator

6. Cylinder Pump N2/H20
7. Flow Rate Vernier
8. Particle Syringe
9. Microscopes / Flow Cells
10. Vacuum pump

Figure 2.   Experimental apparatus.

the application of pressurized nitrogen on one side of the piston; and the flow rate of fluid is controlled by a vernier located on the other side of the piston. Other elements of the apparatus are two microscopes and a fiber-optic illuminator. These are integrated within the flow system as illustrated in Figure 2. (Note that one microscope uses transmitted light while the other uses reflected light.)

Centrifuge Experiments. The design of the centrifuge cell is virtually identical to that of the flow cell. However, the experimental procedures are different and will be explained in the procedure section of this paper.

Materials

The effect of particle size was investigated by using different sizes of microspheres. Two types of particles and several different fluid media were used on the glass surface. The particles examined were charged polystyrene spheres and glass microspheres. (All the particles used are available from Duke Scientific.) Three different sizes of polystyrene particles were used: 10, 20, and 40 μm with a standard deviation of ±1 to 2% and a specific density of 1.05 g/cc. The glass microsphere sizes were 10, 15, and 30 μm with a ±15% standard deviation and a specific density of 2.5 g/cc. The surface charge on the particle was determined experimentally for different particles through the measurement of zeta potential as a function of pH.

Procedure

Flow Experiments. The flow cell was evacuated for two hours prior to each test. Then, after having filled the cylinder with fluid from the reservoir, the cylinder was pressurized with nitrogen to 40 psia., the flow rate vernier was opened, and the entire flow system was circulated with the desired solution. After the flow system had been completely flushed, particles of the desired size were injected with a syringe. The particles were allowed to settle for a specified length of time, typically 24 hours. At each flow rate, surface concentrations of particles were measured by point counting. The point counting of particles on the surface was accomplished with the help of a grid in the microscope eyepiece. The flow was increased incrementally until the maximum allowable flow rate of the flow cell was reached. This maximum flow rate was approximately 0.6 cc/sec which corresponds to an average velocity of 8 cm/sec or a Reynolds number of about 0.8 (based on particle diameter).

Centrifuge Experiments. The particles to be studied were deposited in the cell by syringe injection. The cell was then sealed by capping the fittings and the particles were allowed to settle for the same length of time attained in the flow cell. A specific area of the cell to be studied was chosen and marked. The number of particles in this area was counted using a microscope. The cell was then placed in the centrifuge rotor perpendicular to the rotational axis (Figure 4) and spun at a constant RPM for a specified length of time, usually two to three minutes. The cell was then carefully removed from the centrifuge and the particles in the selected area were again counted. The same procedure was repeated once more at a higher RPM.

Experimental Results

Flow Experiments. A "critical velocity," v_c, is defined as that velocity at which a "measurable" release of particles is observed; above this critical velocity, release of particles is continuous. Specifically, it is taken to be the velocity at which 10% of the particles have been released. Based on this definition, v_c is a function of particle size--the larger the particle, the smaller is v_c (Figure 5).

As mentioned in the introduction, the particle composition has an effect on the release velocity. In particular, from Figure 6 it is clear that 10 µm glass particles release more easily than 10 µm polystyrene particles.

Solution pH and ionic strength also have an effect on the release velocity. Figure 7 demonstrates the moderate effect of solution pH on release velocity, with an increasing pH lowering the release velocity. Figure 8, on the other hand, shows that the effect of ionic strength is quite small, with a decreasing ionic strength causing only a marginal decrease in release velocity.

Centrifuge Experiments

For the same experimental conditions as described in the flow experiments, (i.e., 10 µm glass particles, pure water flowing solution) centrifuge tests yield a higher force to release particles. The torque required for particle release is the same in both experiments; hence, particle release in the flow cell is initiated by a rolling rather than sliding motion (Figure 9).

Discussion of Results

The DLVO theory, with the addition of hydration forces, may be used as a first approximation to explain the preceding experimental results. The potential energy of interaction between spherical particles and a plane surface may be plotted as a function of particle-surface separation distance. The total potential energy, V_t, includes contributions from Van der Waals energy of interaction, the Born repulsion, the electrostatic potential, and the hydration force potential. [Israelachvili (13)].

The Hamaker constants used in these calculations were obtained from values reported in the literature [Gregory (14)]. Zeta potentials were obtained both experimentally and from the literature [Huang and Stumm (15) and Sharma (16)].

Experimentally derived potential energy curves are shown in Figures 10 and 11. (Note that only one particle size is illustrated, namely, 10 µm.) The shape of these potential energy curves as a function of ionic strength, solution pH, particle and surface composition, etc. may be used to explain the effect of some of these variables on particle capture and

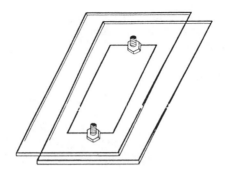

Figure 3. Glass flow cell.

Figure 4. Centrifuge illustration.

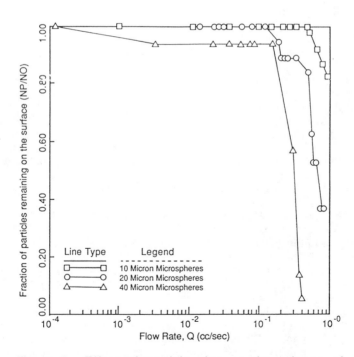

Figure 5. Effect of particle size on the release of polysytrene particles from a glass surface.

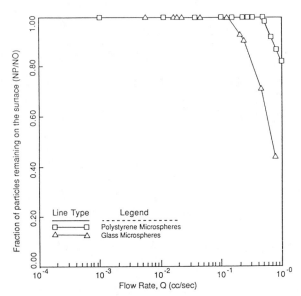

Figure 6. Effect of particle composition on the release phenomena.

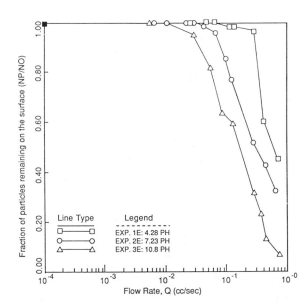

Figure 7. Effect of solution pH on the release of 10 μm glass microspheres from a glass surface.

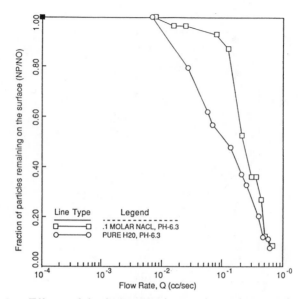

Figure 8.   Effect of ionic strength on the release of 10 μm glass particles from a glass surface.

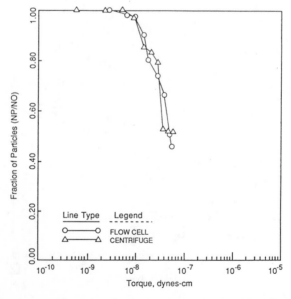

Figure 9.   Comparison of the release of 10 μm glass microspheres from a glass surface between a flow cell experiment and a centrifuge experiment.

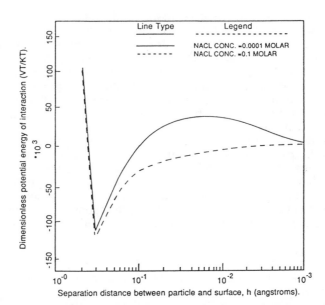

Figure 10. Potential energy of 10 μm glass particles as a function of ionic concentration, solution pH = 6.

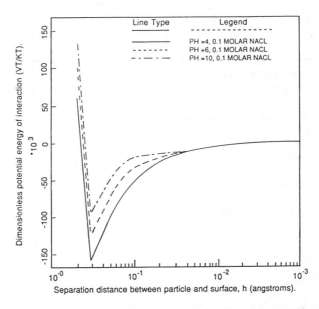

Figure 11. Effect of the solution pH on the potential energy of 10 μm glass particles interacting with glass surfaces.

release. For example, it is clear that both decreasing ionic strength and increasing pH decrease the depth of the attractive energy trough at about 3 Å. Also, a repulsive energy barrier is created at 50 Å by decreasing the solution salinity. The height of this barrier is about 24×10^3 kT. In order to see how the DLVO theory can explain these experimental results, consider the effect of pH. In the experiments an increasing pH results in the release of more particles at a given velocity. From the DLVO theory the attractive trough depth becomes smaller in this situation. This is because both the particle and the surface become more negatively charged as pH increases. As a result, a smaller hydrodynamic force is required for particle release. The relationship between potential energy and release velocity is similar for the other important variable, ionic strength, and for this variable the experiments are also consistent with the theory. On the other hand, the effects of particle composition and size are not consistent with the above theoretical predictions but can be explained using the aforementioned torque balance model [Chamoun (11)].

<u>Conclusions</u>

(1)   A critical flow rate or velocity for release of particles has been observed.
(2)   The critical velocity depends on particle radius, particle composition fluid medium, and fluid conditions such as solution pH and ionic strength.
(3)   Increasing the particle radius decreases the critical velocity required to remove it from any surface studied.
(4)   Increasing solution pH and decreasing the ionic strength decreases the critical velocity for all cases examined.
(5)   The effect of surface roughness has not yet been studied, but in all experiments the surface roughness was small on the scale of particle sizes used.
(6)   Centrifuge experiments show that the particle release mode is rolling rather than sliding.
(7)   The DLVO theory provides a qualitative explanation of the ionic strength and solution pH effects on particle release.
(8)   The effect of particle size and composition on the release problem is explained by the use of a torque-balance [Chamoun (11)].

<u>Literature Cited</u>

1.   Visser, J. <u>Surface and Colloid Science</u>; Matijeric, E., Ed.; John Wiley: New York, 1976; Vol. 8.
2.   Goldman, A. J.; Cox, R. G. ; Brenner, H. <u>Chem. Eng. Sci.</u> 1968, 23, 1293.
3.   O/Neill, M. E. <u>Chem. Eng. Sci.</u> 1968, 23, 1293.
4.   Zimon, A. D. <u>Adhesion of Dust and Powders</u>; Consultants Bureau: New York, 2nd Edition, 1982.
5.   Polke, R. <u>Bull. Soc. Chim. Fr.</u> 1970, 3241.
6.   St. John, D. F.; Montgomery, D. J. <u>J. Appl. Phys.</u> 1971, 663.
7.   Hubb, M. A. <u>Colloid Surfaces</u> 1984, 12, 151.

8. Krupp, H. Adv. Colloid Interface Sci. 1967, 1, 111.
9. Muller, V. M.; Yushchenko, V. S.; Deryagin, B. V. J. Colloid Interface Sci. 1983, 92, 92.
10. Johnson, K. L.; Kendall, K.; Roberts, A. D. Proc. R. Soc. Lond. 1971, A324, 301.
11. Chamoun, H. Ph.D. Dissertation (in progress), The University of Texas at Austin , Texas, 1988.
12. Shirzadi, S.; Sarkar, A.; Sharma, M. M. AIChE J. 1988.
13. Israelachvili, J. N. Adv. Colloid Interface Sci. 1982, 16, 31.
14. Gregory, J. Adv. Colloid Interface Sci. 1969, 2, 396.
15. Huang, C. P.; Stumm, W. J. J. Colloid Interface Sci. 1973, 43, 409.
16. Sharma, M. M. Ph.D. Dissertation, University of Southern California, Los Angeles, 1985.

RECEIVED September 26, 1988

Chapter 31

Formation Wettability Studies that Incorporate the Dynamic Wilhelmy Plate Technique

Dale Teeters,[1] Mark A. Andersen,[2] and David C. Thomas[2]

[1]Chemistry Department, University of Tulsa, Tulsa, OK 74104
[2]Amoco Production Company, Tulsa, OK 74102

The dynamic Wilhelmy plate technique, a new method for characterizing oil reservoir wettability, gave quantitative values of wetting preference and comparisons of surface energy values related to wetting properties. Water-wetting and oil-wetting systems were distinguished readily, as were hybrid-wetting systems which have both types of wetting behaviors, and interfacial properties were quantified. Oxygen contamination caused inconsistent wetting behavior for some crude oil/brine/solid systems, but operating in a newly developed anaerobic vessel gave reproducible wetting behavior. The contact angle hysteresis of dolomite, marble, glass and polytetrafluoroethylene in a series of solvents gave a qualitative evaluation of surface energies. Dolomite and marble had similar surface energies which correlated to the wetting behavior for these solids obtained with the dynamic Wilhelmy plate technique. Other advantages of the Wilhelmy technique in studying reservoir wettability are discussed.

The preferential wetting characteristics of crude oil/water/rock systems play an important role in characterizing oil reservoirs. Formation wetting preference affects the success of most conventional and enhanced recovery methods. Waterflood performance depends on the amount of imbibition which can be expected of a reservoir and the selection of enhanced oil recovery methods are affected by the formation wettability. Matching and predicting performance successfully depends on the ability to determine the degree of wetting preference of the formation. The relative permeability, capillary pressure, electrical response, and occasionally the rock mechanical response all depend on the position of the fluids in the pores. The importance of these aspects of formation wettability has been been covered in a series of thorough review papers by Anderson (1- 6).

Contact angle measurement is one method of obtaining quantitative wettability values and is usually done in the petroleum industry by the sessile drop method (7- 10) or a modification of this
technique (11, 12). The contact angle is measured at the edge of a
drop of crude oil placed between parallel crystals in a brine bath.
One crystal is displaced, creating a new contact angle when the
water advances over a portion of the crystal formerly covered by
oil, and another new angle on the other side of the drop when the
oil advances over a portion of the crystal formerly covered by
water. The displacement is repeated until the equilibrium contact
angle has been reached.
        The sessile drop method has several drawbacks. Several days
elapse between each displacement, and total test times exceeding one
month are not uncommon. It can be difficult to determine that the
interface has actually advanced across the face of the crystal.
Displacement frequency and distance are variable and dependent upon
the operator. Tests are conducted on pure mineral surfaces, usually
quartz, which does not adequately model the heterogeneous rock surfaces in reservoirs. There is a need for a simple technique that
gives reproducible data and can be used to characterize various mineral surfaces. The dynamic Wilhelmy plate technique has such a
potential. This paper discusses the dynamic Wilhelmy plate apparatus used to study wetting properties of liquid/liquid/solid systems
important to the oil industry.

The Dynamic Wilhelmy Plate Method

The Wilhelmy hanging plate method (13) has been used for many years
to measure interfacial and surface tensions, but with the advent of
computer data collection and computer control of dynamic test conditions, its utility has been greatly increased. The dynamic version
of the Wilhelmy plate device, in which the liquid phases are in
motion relative to a solid phase, has been used in several surface
chemistry studies not directly related to the oil industry (14- 16).
Fleureau and Dupeyrat (17) have used this technique to study the
effects of an electric field on the formation of surfactants at
oil/water/rock interfaces. The work presented here is concerned
with reservoir wettability.
        Figure 1 is a schematic of the apparatus used in our studies.
A Cahn Model 29 microbalance and a stepper motor were interfaced to
an IBM PC/XT through an RS-232 interface and an IEEE488 general purpose interface bus (GPIB), respectively. The microbalance rested on
top of a housing containing a flat platform on a vertical stage
moved by the stepper motor. The vessel holding the liquids rested
on the platform. The plate hanging from the balance was immersed in
and removed from the liquids in a continuous motion so that immer-
sion-emersion cycles or "wetting cycles" could be obtained. Clean
surfaces on the plates were of the utmost importance for reproducible wetting cycles. The cleaning procedures for the plates used in
this work and the preparation of the mineral plates from bulk samples have been described elsewhere (18). The initial plate position
was typically 4 mm above the liquid/vapor interface. A wetting
cycle run consisted of moving the liquid interface a certain distance up onto the plate and then the same distance down at a standard speed of 0.127 mm/sec. During this movement, 720 data values

were recorded by the microcomputer for total cycle distances of
50.4 mm and 360 values for cycles of 25.4 mm.

Figure 2 is a representation of the force balance on a Wilhelmy
plate that has gone through one phase and has been wetted by a
second phase. The three interfacial tensions are related to the
contact angle (measured through phase 2) by the familiar Young
equation

$$\gamma_{12} \cos\theta = \gamma_{S1} - \gamma_{S2} \tag{1}$$

where the subscripts 12, S1 and S2 represent the phase 1/phase 2,
solid/phase 1 and solid/phase 2 interfaces, respectively.

In the experiment described above the force, F, on a plate in
an air/liquid system which is partially submersed in the liquid is

$$F = p\, \gamma \cos\theta - B \tag{2}$$

where p is the perimeter of the plate, γ is the surface tension, θ
is the contact angle and B the buoyant force on the portion of the
plate below the general surface. When two liquids such as oil and
water are involved, the force on a plate which has passed through
the oil layer into the water layer is given by

$$F = p\, \gamma_{AO} \cos\theta_{AO} - B_O + p\, \gamma_{OW} \cos\theta_{OW} - B_W \tag{3}$$

where AO and OW indicate the air/oil and oil/water interface respectively, B_O is the buoyant force caused by the oil and B_W that caused
by the water layer.

The hexadecane/water/glass system was used as an initial model
for crude oil/brine systems. Characterization of wetting behavior
of this system from the dynamic Wilhelmy plate data shown in
Figure 3 is interpreted by using Equation 3. The mass read by the
balance was converted to a tension by multiplying by the gravita-
tional acceleration, g, and dividing by the plate perimeter, p. As
the plate was immersed, the tension increased when the solid surface
was wetted by the hexadecane ($\gamma_{AO} \cos\theta_{AO} < 90°$ in Equation 3).
Further immersion of the plate into the hexadecane resulted in a
slight decrease in the tension because of the buoyant force.
Another increase in tension was observed at the hexadecane/water
interface since the water wetted the glass surface in preference to
hexadecane and the advancing contact angle as measured through the
aqueous layer was again less than 90°. The buoyant force of the
water caused the measured tension to slightly decrease as the slide
went further into the water layer. The direction of motion was
reversed for the emersion half of the cycle and the contact angles
in Equation 3 changed from advancing to receding angles. The difference between these two angles caused the hysteresis observed in
the wetting cycle shown in Figure 3. The hexadecane/water/glass
system is a typical example of water-wetting behavior.

Oil-wetting systems can be modeled by replacing the glass plate
with a plate of polytetrafluoroethylene (PTFE), as shown in the
hexadecane/water/PTFE system in Figure 4. At the hexadecane/water
interface the water phase did not wet the PTFE surface, the contact
angle was thus greater than 90° and the tension decreased. The peak

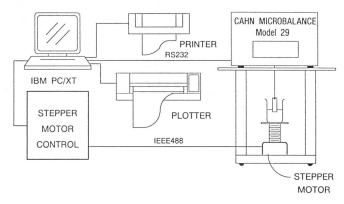

Figure 1. Schematic diagram of the dynamic Wilhelmy Plate Apparatus.

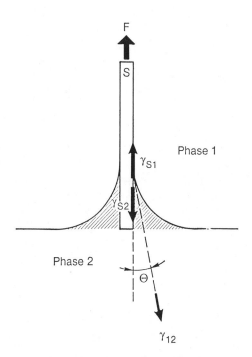

Figure 2. Forces on a thin plate with a meniscus. The surface tensions are γ_{S1} and γ_{S2} for the solid against fluids 1 and 2, respectively, and γ_{12} for the interfacial tension between the liquids at contact angle Θ. The force on the plate F is measured by a microbalance.

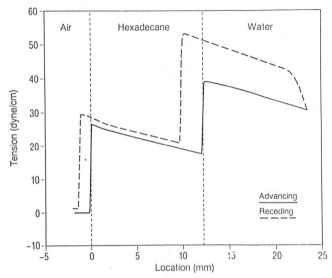

Figure 3. Hexadecane/water/glass wetting cycle exhibiting water-wetting behavior. (Reproduced with permission from Teeters, D.; Wilson, J. F.; Andersen, M. A.; Thomas, D. C. J. Colloid Interface Sci., 1988, 126 in press. Copyright 1988 Academic Press.)

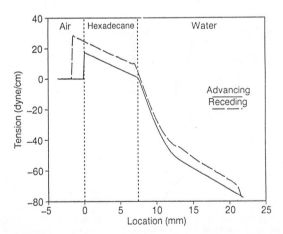

Figure 4. Hexadecane/water/PTFE wetting cycle showing oil-wetting behavior.

observed at the hexadecane/water interface on emersion was typical
of other oil-wetting systems and is believed to be caused by edge
effects at the bottom of the plates.

Crude/brine systems have been known to have large amounts of
hysteresis with advancing contact angles greater than 90° and receding angles less than 90° (8). This type of wetting behavior is
shown in Figure 5 which is a hexadecane/water/glass system in which
the hexadecane phase contains oleic acid (0.8 molar). When the
plate was immersed, the water did not displace the hexadecane-oleic
acid phase and the advancing contact angle was greater than 90°.
When the direction of motion was reversed, the line of
hexadecane/water contact on the plate was pinned and the magnitude
of the tension increased until a stable meniscus with receding contact angle less than 90° was formed. This wetting cycle demonstrated oil-wetting behavior on immersion and water-wetting on
emersion and has been termed a hybrid-wetting cycle. The large hysteresis is most likely due to adsorption of oleic acid onto the
glass surface during the immersion part of the cycle since Langmuir
(19) has observed a marked difference between contact angles of an
advancing and a receding water surface on glass covered by an oleic
acid monolayer.

The computer interface system lends itself well to the determination of interfacial tension and contact angles using Equation 3
and the technique described by Pike and Thakkar for Wilhelmy plate
type experiments (20). Contact angles for crude oil/brine systems
using the dynamic Wilhelmy plate technique have been determined by
this technique and all three of the wetting cycles described above
have been observed in various crude oil/brine systems (21) (Teeters,
D.; Wilson, J. F.; Andersen, M. A.; Thomas, D. C.; J. Colloid Interface Sci., 1988, 126, in press). The dynamic Wilhelmy plate device
also addresses other aspects of wetting behavior pertinent to petroleum reservoirs.

Advantages of Dynamic Crude Oil Wetting Measurements

The dynamic Wilhelmy plate technique directly measures the adhesion
tension, which is the product of the interfacial tension (IFT) and
the cosine of the contact angle, as illustrated in Equation 3. The
degree of wetting of a system depends on the adhesion tension rather
than the contact angle alone. As an example, consider the systems
listed in Table I. Using Craig's criteria (22) the contact angles
indicate that system A has intermediate wettability while system B
is oil-wetting. The IFT of system A is higher than the IFT of
system B, leading to a greater adhesion tension for the "intermediate" system than for the "oil-wetting" system. That is, the oil in
the "intermediate" system A adheres more strongly to the solid surface than the "oil-wetting" system B.

The Leverett J-function is used in reservoir engineering (22)
to relate the permeability k, porosity ϕ, and wetting characteristics to water saturation S_w

$$J(S_w) = \frac{P_c}{\gamma_{OW}\cos(\theta)} \sqrt{\frac{k}{\phi}} \qquad (4)$$

TABLE I. EXAMPLE OF TWO WETTING SYSTEMS

	Interfacial Tension	Contact Angle	Adhesion Tension	Wetting Character
System A	39 dyne/cm	105°	-10.1 dyne/cm	"Intermediate"
System B	12 dyne/cm	130°	- 7.7 dyne/cm	"Oil-wetting"

where P_c is the capillary pressure. Other wetting effects, such as
the pressure drop across a curved interface (Young-Laplace equation)
and the height of capillary rise also depend on both the IFT and the
contact angle (23). The adhesion tension should be used whenever
wetting forces are being compared with any other forces – it is not
sufficient to determine the contact angle alone. Since a dynamic
Wilhelmy plate measurement yields the adhesion tension it is remark-
ably well suited to petroleum industry wettability studies.

Far from a wellbore, the velocity of reservoir fluids is about
one linear foot per day. Near a wellbore, the velocity can increase
one-hundred fold. A static or quasi-static test such as the sessile
drop (contact angle) test may not represent the dynamic behavior of
the fluids in the field. The dynamic Wilhelmy device gives results
which are comparable in interface velocity to the field displacement
rate. The interface in the Wilhelmy test described here moved at a
steady rate of 0.127 mm/sec or 36 ft/day. The wetting cycle for a
hybrid-wetting crude oil system was not affected by moving at a rate
less than 1 ft/day.

Another measure of wetting character in the field is the dimensionless capillary number N_c, which is the ratio of viscous to
capillary forces. One expression for this number is (24)

$$N_c = \frac{V\,\mu}{\gamma}\tag{5}$$

where V is the macroscopic velocity of the fluid and μ is the viscosity. Residual oil saturation, which is the oil left behind after
a waterflood, is approximately constant for capillary numbers up to
1.0×10^{-5}, then begins to decrease (24). For a 2 centipoise oil
with an interfacial tension of 31 dyne/cm flowing at 1 ft/day, $N_c =
2.27 \times 10^{-5}$, and at 100 ft/day, $N_c = 2.27 \times 10^{-3}$. For such an
oil/brine system, the dynamic Wilhelmy device had an equivalent
capillary number of 8.21×10^{-4} – in contrast, the sessile drop test
is quasi-static.

Wetting preference affects the location of fluids flowing in
porous media, influencing the behavior of waterfloods and other
methods of improved oil recovery. In a waterflood, water displaces
oil from the pore space. At the leading edge of the water bank, the
capillary forces given by the water-advancing adhesion tension
influence the displacement process. An oil bank can build in front
of the water bank, so there can also be a water-receding zone ahead
of the oil bank. Both the water-advancing and the water-receding
wetting information needed to model this displacement behavior is
provided by the dynamic Wilhelmy technique. Specifically, a water-

wetting system shows hysteresis in the oil relative permeability curve, but not in the water relative permeability curve. An oil-wetting system is the opposite.

All crude oil/brine/solid systems examined for this study exhibited wetting hysteresis. In some cases the difference between advancing and receding adhesion tensions was small. These cases correspond to the water-wetting or oil-wetting cases generally studied in the laboratory which display relative permeability hysteresis only in the non-wetting phase. In several cases studied with the dynamic Wilhelmy device, the wetting hysteresis was very large and the system was characterized as hybrid, similar in form to the hexadecane/water system with oleic acid. In such a system, the solid tends to hold whichever fluid is in contact with it in preference to the displacing fluid, regardless of which fluid is in contact with the solid. Although a systematic relative permeability study has not been conducted of such systems, it is likely a rock saturated with a crude oil/brine system which displays hybrid behavior would have a relative permeability hysteresis for both phases. These systems have been observed in relative permeability tests, but a connection with hybrid wettability has not been hypothesized before.

Crude Oil Wettability Results

Dynamic Wilhelmy wettability tests on single component systems, such as hexadecane/water/glass, can be done quickly in open beakers. Some crude oils contain components which can oxidize and change wettability (12, 25) so tests on reservoir oil samples must be performed in an oxygen-free environment. Exposure to air had a marked effect on the wetting cycle of one crude oil discussed below.

The apparatus shown in Figure 6 was assembled with a cover containing oil, brine and argon/vacuum lines. Three vacuum/argon cycles purged oxygen from the vessel. These cycles also purged the lines to the brine and oil supplies. Argon bubbled through the brine storage vessel for about fifteen minutes. Brine was drained from a glass storage tank into the vessel to cover the glass plates. The vacuum pump pulled gas from the brine in the vessel for forty-five minutes. A positive pressure of argon on the brine assisted a gravity drain from the vessel until the brine level dropped below the bottom of the glass plates. The oil inlet valve was opened and the pressure in the cylinder forced oil into the vessel until the glass slides were covered with oil.

Six glass slides hung from a PTFE and glass support around the periphery of the vessel (Figure 6). The hooks holding the slides were not wetted by brine or oil. As noted above, brine contacted the plates first followed by oil to replicate the order of exposure of fluids to the rocks in an oil field. A plate could be suspended from the microbalance through the central port of the vessel for the dynamic Wilhelmy test. A beaker centered in the vessel kept the brine level under the central port higher than the general oil/brine interface. Thus, for the dynamic Wilhelmy test the plate passed through a central thin layer of oil into the brine while the suspended plates aged in a thick layer of oil around the periphery of the vessel.

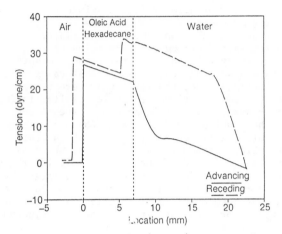

Figure 5. Hexadecane-oleic acid /water/glass wetting cycle with hybrid-wetting behavior.

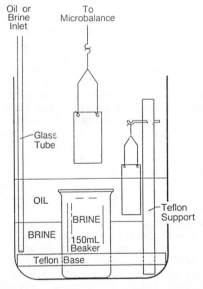

Figure 6. Schematic of interior configuration of anaerobic vessel, covering lid omitted. (Reproduced with permission from ref. 21. Copyright 1988 Society of Petroleum Engineers.)

Argon flowed through a relief valve to keep a slight positive pressure in the anaerobic vessel. To perform the test, the central stopper was removed and the plate to be tested was "fished" from its hanger and suspended from the microbalance. A small flow of argon out of the port prevented air from diffusing into the vessel during the ten minute period of the test. By testing each of the plates after extended times, the influence of the interface age was determined.

Several crude oils have been examined in both aerobic and anaerobic systems (21). Oil-wetting, water-wetting and hybrid-wetting systems have all been observed in these systems. The wetting cycles generally were performed on glass slides, although some studies using marble have been done. Quartz, limestone (or marble) and dolomite can be used to model reservoir rocks.

The behavior of each crude oil was examined quickly by testing the oil/brine system in an open beaker. Air contacted the oil; only glass covers protected the fluids from contamination by dust between tests. The brine had 50 kppm NaCl and 5 kppm $CaCl_2$ in distilled water. Most of the crude oils examined when exposed to air came to equilibrium in about one day (21). One West Texas crude oil (SS1473) tested in an open beaker (Figure 7) displayed a wetting behavior which changed in a way which was not systematic. Each wetting cycle used a new, clean glass plate but the same beaker of oil and brine. A second sample of oil from the same oil field behaved in a qualitatively similar way.

Figure 8 shows the results of the tests on SS1473 in the anaerobic vessel. The interface age noted on Figure 8 was also the aging time of the plate prior to performing the dynamic Wilhelmy test. The first plate tested immediately after preparing the vessel displayed hybrid wetting behavior in two wetting cycles. The water-advancing adhesion tension was somewhat less oil-wetting during the first cycle. Both cycles displayed strong water-wetting behavior during the water-receding portion of the cycle. After one day of soaking in crude oil and on subsequent days up to six days the behavior was oil-wetting with a small hysteresis. The water-advancing adhesion tension for all of these tests (listed in Table II) was almost constant. The water-receding adhesion tension values were generally less consistent for most tests involving crude oils. The thinning of the oil layer in the latter cycles was due to loss of volatile components from the oil.

TABLE II. ANAEROBIC CRUDE OIL ADHESION TENSION MEASUREMENTS AND CONTACT ANGLES USING IFT = 36.4 DYNE/CM

Interface	Measured Adhesion Tension		Calculated Contact Angle	
Age	Water-Advancing	Water-Receding	Advancing	Receding
15 min.	-21.72 dyne/cm	25.04 dyne/cm	127°	47°
30 min.	-30.65 dyne/cm	23.03 dyne/cm	147°	51°
22 hours	-29.76 dyne/cm	-26.54 dyne/cm	145°	137°
3 days	-32.45 dyne/cm	-25.92 dyne/cm	153°	135°
6 days	-30.80 dyne/cm	-19.53 dyne/cm	148°	122°

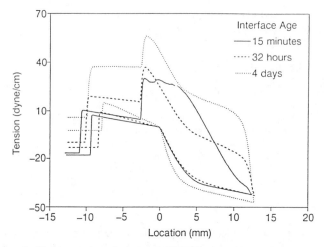

Figure 7. Wetting cycles of crude oil SS1473 tested in an open beaker. (Reproduced with permission from ref. 21. Copyright 1988 Society of Petroleum Engineers.)

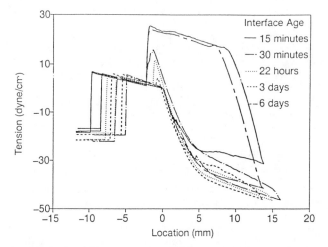

Figure 8. Wetting cycles of crude oil SS1473/brine/glass tested in an anaerobic vessel. (Reproduced with permission from ref. 21. Copyright 1988 Society of Petroleum Engineers.)

The IFT of this crude oil and brine was 36.4 (+/- 0.5) dyne/cm, measured using the maximum bubble pressure method. This value was used to calculate the contact angles shown in Table II. The equilibrium value of the water-advancing contact angle was 149°, and the water-receding contact angle was 131°. The value for sessile drop water-advancing contact angle measurements on six other samples of oil from the same reservoir ranges from 112° to 145°, with an average of 131° (12).

In many oil-wetting systems a film of oil adhered to the plate below the oil/brine meniscus. In some cases, the film was not persistent; during the minute or so the bottom of the plate was submerged in brine, it was swept clear of the oil film and the receding cycle switched from oil-wetting to water-wetting as the cycle progressed. After about one day of equilibration between the oil and the brine, this short-time behavior was no longer seen.

For water-wetting oil/brine systems, the receding portion of the wetting cycle was more consistent (smoother) than the advancing portion, which is opposite to the effect noted with the oil-wetting crude oils discussed above. This has been seen with other water-wetting materials in our laboratory. In both the oil-wetting and water-wetting cases, the smoother curves were noted when the relative motion of the plate pulled the wetting fluid into the non-wetting fluid. The more ragged features were seen when the non-wetting fluid advanced over the thin edge of the wetting fluid. This raggedness appeared to be a result of irregular detachment of the wetting fluid along the contact line, a localized "stick-slip" behavior.

This study demonstrated two aspects of measurement of wettability of crude oils. Exposure to air can cause changes in the wetting cycle. This was not true of normal paraffins such as hexadecane, which yielded stable wetting cycles for days and weeks when exposed to air. Equilibration of the crude oil/brine/solid system also caused changes in the wetting behavior. From this study it is not clear whether the changes were due to equilibration of the oil and brine phases or the aging of the solid in the oil phase. It is likely that both affect the measurement.

The four-to-six day duration of the dynamic Wilhelmy tests (wherein equilibrium actually occurred after one day) were much shorter than the times generally required for the sessile drop test. The conventional contact angle measurements on oil from the fields mentioned above required up to 48 days (12).

Surface Energy Study Related to Wettability

The wetting behavior of liquid/liquid/solid systems is not only dependent on the two liquid phases, but upon the interaction of the solid surface with these liquids (see Equation 1). An example is in the wetting cycles for glass and PTFE in a hexadecane/water system. A wetting cycle for a glass slide in a hexadecane/water system has the typical water-wetting cycle shown previously in Figure 3. Figure 4 shows the data for PTFE used as the solid phase with the same liquid/liquid system where an oil-wetting cycle is observed. When wetting cycles for plates of the minerals dolomite and marble were obtained for the hexadecane/water system, hybrid wetting cycles such as those shown in Figure 5 were seen.

The surface free energy of the solid is important in understanding the interaction of a solid with liquids. Solids with low
surface energies, such as PTFE, tend to be wetted by only those liquids that have low surface tensions; in an oil/water system such
solids tend to have oil-wetting behavior as shown in Figure 4 for
hexadecane/water/PTFE. Solids with higher surface energies, such as
glass, are wetted by water with its high surface tension in preference to hydrocarbons of low surface tension such as hexadecane.
Surface energy values can thus be very important in understanding
the wetting phenomenon. Unfortunately, surface energies of single
crystalline solids are inaccessible to direct measurement and postulating values for heterogeneous reservoir rock complicates matters
even more. However, approximate values of surface energies can be
deduced by contact angle studies with probe liquids (26).
     Surface free energies calculated by this technique use only
advancing contact angles or "equilibrium" angles which are somewhere
between the advancing and receding angles; the surface energy is
represented with a single value. This practice is probably not adequate for samples with heterogeneous surfaces. In light of these
difficulties, Penn and Bowler (27) used both the advancing and
receding contact angles for a series of liquids to compare the surface energies of solids. They suggested that advancing and receding
contact angle data for a series of probe liquids against a solid be
presented in bar graph form, describing a "fingerprint" for that
solid. The bar graphs for one solid are compared to those for
another solid; if the graphs are similar, the solids have similar
surface energies and good adhesive performance (27).
     This same technique should be helpful in understanding wetting
properties important in the oil industry since wetting is very
dependent on mineral surface energies. The use of contact angle
hysteresis information may allow a better understanding of the
effects of surface heterogeneities of natural mineral samples. The
dynamic Wilhelmy plate technique is ideally suited for such experiments.
     Plates of glass, PTFE, dolomite and marble were used. Dolom-
ite and marble were chosen because they represent minerals found in
oil reservoirs. Glass and PTFE were investigated because they represent high and low surface energy solids respectively and are good
model systems for data comparisons. Liquid/solid wetting cycles
were obtained for each of the solids in the liquids listed in
Table III.
The contact angles were calculated by using Equation 2 and the surface tension values listed in Table III for the respective liquids
(determined with the Wilhelmy maximum pull method). All liquids
used were of the highest purity and these surface tension values and
accepted literature values are in close agreement. Figure 9 shows
the bar graph presentation of these data. The abscissa of the graph
is an ordered listing of the surface tensions of the probe liquids.
Several runs of each liquid/solid system were averaged. The bottom
of each bar is the cosine of the advancing contact angle and the top
is the cosine of the receding angle of the liquid/solid system. The
data for those liquid/solid systems having little or no hysteresis
between advancing and receding contact angles were very reproducible. The data for the PTFE, dolomite and marble for the liquids
showing the largest hysteresis had scatter of 0.06 cosine units or

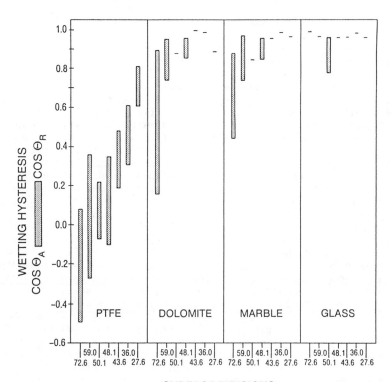

Figure 9. Bar graphs giving contact angle information for a series of probe liquids on PTFE, dolomite, marble, and glass.

TABLE III. SURFACE TENSIONS OF LIQUIDS USED FOR
SOLID SURFACE CHARACTERIZATION

Liquid	Surface Tension (dyne/cm)
Water	72.6
Formamide	59.0
Methylene iodide	50.1
Ethylene glycol	48.1
1-Bromonaphthalene	43.6
N,N-Dimethylacetamide	36.0
Hexadecane	27.6

less. Advancing contact angles calculated in this study for
water/PTFE and methylene iodide/PTFE of 119° and 93°, respectively,
compare well with reported values of 112° (28) and 88° (29).
    Figure 9 shows that the dolomite and marble had very similar
hysteresis response while the glass and PTFE were much different.
This corresponds with the hybrid wetting seen for dolomite and
marble in a hexadecane/water system as compared with the water-wetting behavior of glass and the oil-wetting nature of PTFE in the
same liquid/liquid system. Calculated values for the surface energy
of glass, marble and PTFE are approximately 300 (30), 200 (31) and
19 ergs/cm² (32) respectively. No literature values for dolomite
were found. Glass and marble have surface energies that are very
similar; however the Wilhelmy technique is still able to distinguish
between the wetting behavior of the two systems. The importance of
this type of surface characterization has been shown in wetting studies using glass and marble in a crude oil/brine system under anaerobic conditions (Andersen, M. A.; Thomas, D. C.; Teeters, D. C.
The Log Analyst, Society of Core Analysts paper 8801, to be published). The two solids exhibited different wetting cycles as would
be predicted from the work presented here. From the similarity of
the bar graphs for marble and dolomite, we believe that dolomite
must have a surface energy near that of marble. The very oil-wetting PTFE surface is easily distinguished from the hybrid-wetting
marble and dolomite and the water-wetting glass. Studies such as
these make it possible to more fully characterize the wetting behavior of reservoir rock.

Other Applications of the Dynamic Wilhelmy Plate Technique

In addition to the work described in this paper, the dynamic Wilhelmy plate technique lends itself to other surface studies important to the oil industry. This technique has been used to
investigate drilling fluid emulsions by characterizing the
liquid/liquid/solid interactions of weighing agents such as barite
and hematite with drilling fluid oils containing emulsifiers (Cline,
J. T.; Teeters, D.; Andersen, M. A., SPE Preprint 18476, Society of
Petroleum Engineers, in press). Because tensions as low as
1 dyne/cm can be observed, it can be used to study the change in
adhesion tension and contact angles with the adsorption of surfac-

tants on rock surfaces (17) and used to study micellar properties (33).

Conclusions

The dynamic Wilhelmy plate technique provides a novel way to study wetting behavior. It can be used to easily and quickly distinguish water-wetting, oil-wetting and hybrid-wetting systems, under anaerobic conditions if necessary. Quantitative values of contact angles can also be obtained. This technique can be used to compare surface energies of natural mineral systems making the prediction of reservoir rock wettability more reliable. The dynamic Wilhelmy plate technique has much potential for future studies of importance to the oil industry.

Acknowledgments

We would like to thank Amoco Production Company for funding this work and for allowing its publication. We would also would like to acknowledge Dr. Jeffrey T. Cline of Amoco for providing the marble samples.

Legend of Symbols

B	- buoyant force
F	- force on Wilhelmy plate
g	- gravitational acceleration
h	- capillary rise
k	- permeability
N_c	- capillary number
p	- perimeter of plate in Wilhelmy test
ΔP	- pressure drop across a curved interface
P_c	- capillary pressure ($P_o - P_w$)
r	- radius of a narrow capillary
S_w	- water or brine saturation
v	- fluid velocity
γ	- surface tension or interfacial tension
μ	- viscosity
θ	- contact angle
ϕ	- porosity
$\Delta\rho$	- density difference between two fluid phases

Subscripts

A	- air	S	- solid
O	- oil	1	- Phase 1
W	- water or brine	2	- Phase 2

Literature Cited

1.   Anderson, W. G., J. Pet. Technol. 1986, 38(10), 1125.
2.   Anderson, W. G., J. Pet. Technol. 1986, 38(11), 1246.
3.   Anderson, W. G., J. Pet. Technol. 1986, 38(12), 1371.
4.   Anderson, W. G., J. Pet. Technol. 1987, 39(10), 1283.
5.   Anderson, W. G., J. Pet. Technol. 1987, 39(11), 1453.

6. Anderson, W. G., J. Pet. Technol. 1987, 39(12), 1605.
7. Neumann, A. W.; Good, R. J. In Surface and Colloid Science; Good, R. J.; Stronberg, R. R., Eds.; Plenum Press: New York, 1977; Vol. 11, pp 31-91.
8. Hjelmeland, O. S.; Larrondo, L. E. SPE Reservoir Eng., 1986, 1(4), 321.
9. McCaffery, F. G.; Mungan, N. J. Can. Petrol. Technol. 1970, 9, 185.
10. McCaffery, F. G. J. Can. Petrol. Technol. 1972, 11, 26.
11. Leach, R. O.; Wagner, O. R.; Wood, H. W.; Harpke, C. F. J. Petrol. Technol. 1962, 14, 206.
12. Treiber, L. E.; Archer, D. L.; Owens, W. W. Soc. Petrol. Engrs. J. 1972, 12, 531.
13. Wilhelmy, L. Ann. Physik. 1863, 119, 117.
14. Dognon, A.; Abribat, M. Bull. Soc. Chim. Biol. 1941, 23, 62.
15. Bendure, R. L. J. Colloid Interface Sci. 1973, 42, 137.
16. Johnson, R. E.; Detter, R. H.; Brandreth, D. A. J. Colloid Interface Sci. 1977, 62, 205.
17. Fleureau, J.-M.; Dupeyrat, M. J. Colloid Interface Sci., 1988, 123, 249.
18. Teeters, D.; Smith, B.; Andersen, M. A.; Thomas, D. C. In Symposium on Advances in Oil Field Chemistry, Toronto 1988; American Chemical Society Preprints, Division of Petroleum Chemistry, Inc.; Washington, DC, 1988, p 146.
19. Langmuir, I. Science 1938, 87, 493.
20. Pike, F. P.; Thakkar, C. R. In Colloid and Interface Science; Kerker, M., Ed.; Academic Press: New York, 1976, Vol. 3, p. 375.
21. Andersen, M. A.; Thomas, D. C.; Teeters, D. 1988 SPE/DOE Symposium on Enhanced Oil Recovery; SPE Preprint 17368; Society of Petroleum Engineers: Richardson, Texas, 1988, pp 529-537.
22. Craig, F. F. The Reservoir Engineering Aspects of Waterflooding; Monograph Series 3, Society of Petroleum Engineers: Richardson, TX 1971.
23. Adamson, A. W. Physical Chemistry of Surfaces; 4th ed.; Wiley: New York 1982.
24. Stalkup Jr., F. I. Miscible Displacement; Monograph Series 8, Society of Petroleum Engineers: New York 1983.
25. Bartell, F. E., Niederhauser, D. O. Fundamental Research on Occurrence and Recovery of Petroleum, American Petroleum Institute: New York 1946-1947, pp 57-80.
26. Kaelble, D. H. J. Adhes. 1970, 2, 66.
27. Penn, L. S.; Bowler, E. R. Surf. Interface Anal. 1981, 3, 161.
28. Damm, J. R. J. Colloid Interface Sci. 1970, 32, 302.
29. Fowkes, F. M.; McCarthy, D. C.; Mostafa, M. A. J. Colloid Interface Sci. 1980, 78, 200.
30. Shartsis, L.; Smock, A. W. J. Amer. Ceram. Soc. 1947, 30, 130.
31. Janczuk, B.; Chibowski, E.; Staszczuk, P. J. Colloid Interface Sci. 1983, 96, 1.
32. Zisman, W. A. In Contact Angle, Wettability, and Adhesion; Fowkes, F. M., Ed.; Advances in Chemistry No. 43; American Chemical Society: Washington, DC, 1964.
33. Thomas, D. C.; Christian, S. D. J. Colloid Interface Sci. 1980, 78, 466.

RECEIVED November 28, 1988

Chapter 32

Enhanced Oil Recovery by Wettability Alteration

Laboratory and Field Pilot Waterflood Studies

H. H. Downs[1] and P. D. Hoover[2,3]

[1]Baker Performance Chemicals, Inc., Houston, TX 77227–7714
[2]Santa Fe Energy Company, Torrance, CA 90503

Thin Film Spreading Agents are alkoxylated nonylphenol resins which displace asphaltene molecules from oil-water interfaces and mineral surfaces. Laboratory studies on demulsification, wettability alteration and oil recovery efficiency indicate that TFSA molecules recover incremental oil by coalescing near wellbore emulsions, making reservoir rock surfaces water-wet, and improving areal sweep efficiency. In a 36 acre waterflood pilot study, a 0.1 pore volume bank containing 239 mg/kg of TFSA was injected into an irregular pattern of 1 injector surrounded by 9 producing wells. An interwell chemical tracer study established fluid flow patterns within the pilot. Decline curve analysis showed that TFSA injection recovered more than 8150 ± 850 bbl of incremental oil, and provided a 54 % DCF rate of return for the 18 month pilot project.

Clean mineral surfaces are strongly water-wet and when in contact with an aqueous phase are positively, neutrally or negatively charged depending on the zeta potential of the surface. When crude oil is added to the system, surface active species in the crude oil (eg., asphaltenes) diffuse through the intervening water film and adsorb on the charged mineral surfaces(1). Polar portions of the asphaltene molecules are oriented towards the charged surface, while nonpolar portions are directed away from the interface and thereby render the surface oil-wet. The chemical identity of the mineral surface, physiochemical properties of the asphaltenes, salinity, temperature, pressure and history of the system all influence asphaltene adsorption and desorption, and thereby determine the wetting state of the surface. Thus, water-wet sedimentary rocks can become oil-wet in the presence of crude oil. Authoritative studies(2,3) have indicated that as many as 50 % of all silicate reservoirs and 80 % of all carbonate reservoirs are oil-wet.

[3]Current address: Petroleum Underground Pump Specialists, Inc., Torrance, CA 90501

It is very difficult to determine the wetting state of
reservoirs. whenever the composition or distribution of fluids in
reservoir rock changes, the equilibrium between surface-adsorbed
and solubilized surfactants is disturbed. As a result, the
wettability of the rock surface can also change. During core
sampling and testing, the wettability of native state rock can be
significantly altered by the flushing action of drilling fluids,
the presence of surfactants in drilling fluids, and changes in pH,
pressure and temperature[1].

Wetting conditions control the location, distribution and
flow properties of oil and brine through porous media[4]. Due to
differences in capillary forces and pore pressures for wetting and
non-wetting fluids, only the wetting fluid is located in small
pores while both the wetting and non-wetting fluids occupy the
larger pores. When a uniformly oil-wet medium is waterflooded,
the non-wetting water phase channels through the larger pores and
bypasses much of the oil located in the smaller pores[5,6]. Water
breakthrough occurs early in the flood[7] and most of the economically produced oil is recovered after water breakthrough[6].
Compared to a waterflood in a water-wet medium, the flowing
fraction of oil in a waterflood of the corresponding oil-wet
medium is lower[4,5], substantially more water must be injected
into and produced from the oil-wet medium in order to recover a
given amount of oil[6,8-10], and the oil saturation at the
economic limit of the waterflood is higher than it would have been
had the flood been conducted in a water-wet medium[5,7,9-12].
Furthermore, because the relative brine permeability is higher and
the relative oil permeability is lower in an oil-wet medium than
in the corresponding water-wet medium, the water-oil mobility
ratio for a waterflood in an oil-wet medium is higher and the
areal sweep efficiency is lower than in a waterflood of a water-
wet medium[5]. Thus, it is widely recognized that the efficiency
of a waterflood increases as the wettability of the uniformly
wetted porous medium is varied from oil-wet to water-wet[4,6-
8,10,11,13-15].

In general, reservoir rock surfaces are not wetted uniformly.
Different mineral types and crystalline faces having different
surface properties are exposed to the reservoir fluids of region-
ally varying chemical composition. Surface active components in
the crude oil can be strongly adsorbed at certain locations within
the porous media and the wettability of the rock surface can vary
throughout the reservoir[1]. In a fractionally wetted medium[16],
the oil-wet and water-wet regions have sizes on the order of a
single pore. In a mixed wettability medium[17], the oil-wet
regions form continuous paths for oil to flow through several of
the larger pores. It is likely that all petroleum reservoirs,
even those which have been waterflooded, contain regions of oil-,
water-, intermediate-, fractional-, and mixed-wettability over
various length scales. Furthermore, during the lifetime of a
waterflood, these regions of wettability can change from water-wet
to oil-wet as well as from oil-wet to water-wet. Oil is displaced
from fractionally wetted porous media in a similar manner to its
displacement from uniformly wetted media[4]. As the fractional
water-wet surface area increases, the flowing fraction of oil

increases and substantially less water must be injected into and
produced from the reservoir in order to recover a given amount of
oil. In addition, the oil saturation at the economic limit of the
waterflood decreases, and oil recovery efficiency increases as the
reservoir becomes more widely water-wet(4,18).

Increasing the water-wet surface area of a petroleum reservoir is one mechanism by which alkaline floods recover incremental
oil(19). Under basic pH conditions, organic acids in acidic crudes
produce natural surfactants which can alter the wettability of
pore surfaces. Recovery of incremental oil by alkaline flooding
is dependent on the pH and salinity of the brine(20), the acidity
of the crude and the wettability of the porous medium(1,19,21,22).
Thus, alkaline flooding is an oil and reservoir specific recovery
process which can not be used in all reservoirs. The usefulness
of alkaline flooding is also limited by the large volumes of
caustic required to satisfy rock reactions(23).

It appears that many of the limitations of alkaline flooding
can be overcome by injecting a synthetic surfactant into the oil-
bearing formation instead of generating the surfactant in situ
with caustic. Thin Film Spreading Agents (designated TFSA) are a
class of surface active, alkoxylated resins which are designed to
generate high spreading pressures and displace asphaltene
molecules from rock surfaces(24). As a result of TFSA treatment,
the wettability of reservoir rock is changed from oil-wet and
fractionally-wet to strongly water-wet. Recovery of incremental
oil by wettability alteration with TFSA molecules depends on the
wettability of the rock surface, the efficacy of the TFSA in
making the rock surface more strongly water-wet, and the oil
saturation at the beginning of the TFSA flood(25). In addition,
members of this class of resin may function as demulsifiers.
Thus, TFSA molecules can also promote the recovery of oil by
coalescing emulsions in the near-wellbore region of production
wells(26).

TFSA molecules have been extensively and successfully used as
steam additives in cyclic steam operations(27-32). Recently,
results of a TFSA-waterflood which was conducted in West Texas
were reported(33). The purpose of the work described in this
paper was to further evaluate the feasibility of recovering
incremental oil in a mature waterflood by injection of surfactants
which change the wettability of reservoir rock surfaces. In this
paper, we present the results of laboratory studies with Thin Film
Spreading Agents and the results of a carefully conducted TFSA-
waterflood pilot in the Torrance Field located in the Los Angeles
Basin of California.

EXPERIMENTAL SECTION

THIN FILM SPREADING AGENTS. The Thin Film Spreading Agent used in
this study was an alkoxylated substituted phenol formaldehyde
resin of relatively high molecular weight. The surfactant was
prepared by a previously described procedure(34). Magnaflood 907
is a commercial product of Baker Performance Chemicals, Inc. and
contains the Thin Film Spreading Agent formulated with anionic
surfactants.

SAND DEOILING. Sand deoiling tests provide information about the
relative performance of TFSAs in recovering crude oil from oil-wet
sand and formation brine. To conduct the tests, sand/oil mixtures
were prepared by saturating clean, dry sand (99.8% silica, 60-120
mesh) with crude oil in an 8/1 weight ratio. 9.0 g samples of the
oil-sand mixture were then transferred to citrate bottles,
blanketed with nitrogen, capped and aged overnight in a 85 °C
oven. After cooling to room temperature, the oil-sand samples
were contacted with 0.030 dm³ of the formation brine containing
the specified concentrations of TFSA and placed in a 60 °C water
bath for a total of 20 minutes. After 10 min in the water bath,
all tubes (including the blank) were gently inverted to insure
that the TFSA contacted the entire oil-sand mixture. At the end
of 20 minutes, the samples were removed from the bath and cooled
to room temperature. Ten milliliters (0.010 dm³) of xylene were
then slowly added and gently mixed with the oil phase at the top
of the citrate bottles. The amount of oil recovered from the sand
was determined spectrophotometrically from a calibration curve of
the crude oil in xylene using a Bausch and Lomb Spectronic 88 set
at a wavelength of 430 nm. Each TFSA was tested in duplicate to
obtain an average oil recovery and the percent deviation from the
mean.

CONTACT ANGLE MEASUREMENTS. Asphaltene molecules were precipitated from crude oil onto borosilicate glass surfaces so that the
effect of TFSA molecules on the wettability of asphaltene-modified
surfaces could be quantified. Initially, the optically smooth
glass surfaces were cleaned with chromic acid, then thoroughly
rinsed with deionized water and dried in a vacuum oven. When the
borosilicate surface was rigorously cleaned by this procedure,
water droplets spread on the uniformly water-wet surface. A
number of procedures for depositing a film of asphaltene molecules
on the borosilicate surface were investigated; one procedure was
particularly convenient and gave reproducible results. The clean,
dry borosilicate glass was aged for 3 days in crude oil. During
the aging process, successively larger volumes of pentane were
gradually added to the crude oil. At the end of the aging-
precipitation process, the borosilicate surface was coated with a
heterogeneous asphaltene layer. The contact angles of water
droplets on the asphaltene-modified surface were measured with a
Rame-Hart contact angle goniometer. While the magnitude of the
contact angle varied with the details of the surface treatment,
all contact angles for surfaces prepared by the above procedure
were consistently between 85° and 95°·

DEMULSIFICATION TESTS. Demulsification tests were conducted using
standard bottle test procedures to evaluate the relative performance of Thin Film Spreading Agents in coalescing emulsions of
formation brine in crude oil under reservoir conditions.
Specified concentrations of TFSA were injected into sample bottles
containing 0.1 dm³ of the untreated crude oil emulsion. The
sample bottles were then capped, vigorously shaken and heated to
the appropriate reservoir temperature. The volume percentage of
the total water phase which separates from the oil, the clarity of

the separated oil phase, oil-water interface characteristics and
the residual level of water remaining in the crude oil phase were
measured as functions of time and TFSA concentration.

RESERVOIR DESCRIPTION. The TFSA-waterflood pilot study was
conducted in Santa Fe Energy Company's Torrance Field. The field
was discovered in 1922 and produces from Miocene and Pliocene
sands located at depths of 3100 ft to 4400 ft (945 m to 1340 m)
subsea. Within the pilot area, the net pay thickness of the Main
Zone averages 96 ft (29 m) and varies from less than 90 ft (27 m)
in the center of the pattern to more than 110 ft (34 m) in the
northwest and southeast sections of the pilot.

Wells which are completed to the Main Zone inject brine and
produce fluids from four sands, termed the C-, D-, E- and F-sands.
The C-sand is well isolated from the D-, E-, and F-sands which
were not discrete. Spinner surveys obtained before and after
injection of TFSA indicate that the TFSA did not adversely alter
the injectivity of the formation and that the formation was not
damaged by TFSA injection.

Low salinity brine is produced from and injected into the
Main Zone of the Torrance Field. Brine produced from Well TU-101,
located within the pilot pattern, contains 16,000 mg/kg of total
dissolved solids (TDS) while brine injected into the pilot
injector, Well TU-120, contains 15010 mg/kg of TDS. The brine
contains approximately 560 mg/kg of Ca^{2+} and 250 mg/kg of Mg^{2+}.

The crude oil produced from the Main Zone of the Torrance
Field has an API gravity of 18° and contains 5.3 weight percent
asphaltenes. The solubility of the asphaltene molecules in Main
Zone oil was measured by the Oliensis Test(35). In this test, the
solubility parameter of the oil was lowered by adding to the oil
successively larger volumes of hexadecane, a poor solvent for
asphaltene molecules. The minimum volume (in milliliters) of
hexadecane, which when added to 5 g of crude oil, will cause the
chromatographic separation of the asphaltene fraction is termed
the Oliensis Number. The Oliensis Number for the Main Zone crude
oil is 3, indicating that the asphaltene molecules are not well-
solubilized in the oil. Small changes in the solubility parameter
of the Main Zone oil can cause the asphaltenes to precipitate.
Such changes can occur during production.

PILOT DESCRIPTION. The site of the TFSA-waterflood pilot was
chosen on the basis of four criteria: the site was representative
of reservoir conditions and production operations in the Main
Zone, extensive historical production data was available for each
of the production wells in the pattern, production wells
completely surrounded the injection well, and the pattern appeared
to be well isolated from adjacent patterns and injection wells.
The production wells surrounding Well TU-120, an injection well,
appeared to meet most of the site selection criteria.

The pilot, shown in Figure 1, was an irregular pattern with
nine production wells surrounding a single injector, Well TU-120.
Based on results of the tracer study, all nine producing wells are
in communication with the injector. With the exception of Well
TU-127 which came on production on February 12, 1986, all produc-

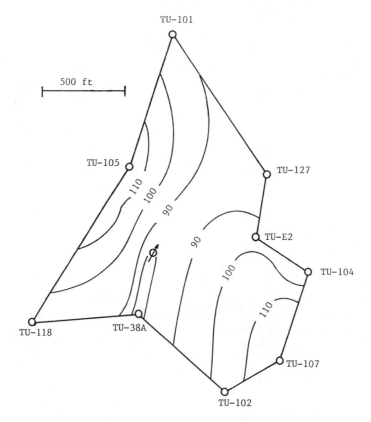

Figure 1.   36 acre TFSA-waterflood pilot showing bottom-hole
            well locations and net pay thickness in feet.

tion wells have extensive well-documented production histories. The pilot appears to be bounded on the southeastern flank by a fault and on the northeast by the water-oil contact. The pilot area was estimated to encompass 35.6 acres (1.44 x 10^5 m^2) with a net pay thickness of 3450 acre-ft (4.26 x 10^6 m^3). At the start of the TFSA-waterflood, the oil saturation was estimated to be 36 % and the oil cut in fluids produced from the pilot was 3.1 %.

To establish the well drainage boundaries and fluid flow patterns within the TFSA-waterflood pilot, an interwell chemical tracer study was conducted. Sodium thiocyanate was selected as the tracer on the basis of its low adsorption characteristics on reservoir rocks(36-38), its low and constant background concentration (0.9 mg/kg) in produced fluids and its ease and accuracy of analysis(39). On July 8, 1986, 500 lb (227 kg) of sodium thiocyanate dissolved in 500 gal (1.89 m^3) of injection brine (76700 mg/kg of thiocyanate ion) were injected into Well TU-120. For the next five months, samples of produced fluids were obtained three times per week from each production well. The thiocyanate concentration in the produced brine samples were analyzed in duplicate by the standard ferric nitrate method(39) and in all cases, the precision of the thiocyanate determinations were within 0.3 mg/kg. The concentration of the ion in the produced brine returned to background levels when the sampling and analysis was concluded.

RESULTS AND DISCUSSION

RECOVERY MECHANISMS. Being surface active, TFSAs lower oil-water interfacial tension, but not by the three orders of magnitude needed to increase the capillary number sufficiently to recover a substantial amount of incremental oil. Instead, TFSAs enhance the recovery of oil by changing the wettability of reservoir rock surfaces from oil-wet and intermediate wettability to strongly water-wet, and by coalescing emulsions in the near-wellbore region of the production wells.

A number of screening tests were conducted to select the best TFSA surfactant for use in the Main Zone of the Torrance Field and results of these screening tests illustrate the recovery mechanisms. Sand deoiling tests were conducted to evaluate the relative performance of TFSA molecules in recovering oil from the surface of oil-wet sand grains. Results of the sand deoiling experiments are shown for three different crude oils in Table I. For each crude oil studied, oil recovery increased as the concentration of TFSA increased from 0 to 500 mg/kg. The biggest percentage increase was observed for the California crude; oil recovery from oil-wet sand increased from 2.7 % to 33.5 % of the oil in place as the concentration of TFSA in the formation brine was increased from 0 mg/kg to 500 mg/kg. Compared to microemulsion floods which typically require surfactant concentrations in excess of 2 % to generate ultralow interfacial tensions, a substantial amount of incremental oil can be recovered by the wettability alteration mechanism using very low concentrations of TFSA surfactant. That high oil recoveries can be achieved with low concentrations of surfactants may be due to the low reservoir rock adsorption

Table I. Recovery of Oil from Oil-Wet Sand Increases As
Concentration of Thin Film Spreading Agent Increases

| | Percent Oil Recovery[1] With Increasing TFSA Concentrations | | | |
	0 mg/kg	100 mg/kg	250 mg/kg	500 mg/kg
Crude A West Texas	5.6	7.4	11.7	18.6
Crude B Alberta	4.8	--	7.5	20.3
Crude C California	2.7	--	13.2	33.5

[1] Oil recoveries after the oil-wet sand was contacted with equal
volumes of brine containing the reported concentrations of
Thin Film Spreading Agents.

characteristics reported for anionic-nonionic surfactant mixtures
of certain members of the TFSA family(36).

Contact angle measurements for a water droplet on an
asphaltene modified borosilicate surface confirmed that low
concentrations of TFSA molecules change the wettability of the
surface from fractionally-wet to water-wet. Table II shows the
results of the contact angle measurements; all reported results
are the average of 10 separate measurements, none of which varied
from the mean by more than 5°. As the concentration of the TFSA

Table II. Thin Film Spreading Agents Make Asphaltene-Modified
Surfaces More Strongly Water-Wet

| | Contact Angle, θ, At Time T | | |
	Initial	5 min	10 min
Brine Alone	91	90	90
200 mg/kg TFSA	88	74	43
400 mg/kg TFSA	80	66	45
TFSA Alone	15	0	0

in the brine droplet increased, the contact angle of the brine droplet on the asphaltene modified surface decreased from approximately 90° to 0°·   Although the numerical values of the contact angles reported in Table II are dependent on the details of the surface treatment, the data show that surfaces become more strongly water-wet in the presence of TFSA. Recovery efficiency by the wettability alteration mechanism depends on the wetting state of the reservoir rock, the efficacy of the TFSA in making the rock surface more strongly water-wet, and the oil saturation at the beginning of the TFSA flood(25).

In addition to adsorbing at mineral-oil interfaces, asphaltene molecules also adsorb at oil-water interfaces. Strong intermolecular dipole-dipole, hydrogen bonding, electron donor-acceptor and acid-base interactions cause the surface-adsorbed asphaltene molecules to form rigid "skins" at oil-water interfaces(41-43). When water droplets are dispersed in an oil which contains asphaltene molecules, molecularly thick, viscous asphaltene films form around the water droplets, inhibit the drainage of intervening oil and sterically stabilize the water-in-oil emulsion.

Flow properties of macroemulsions are different from those of non-emulsified phases(19,44). When water droplets are dispersed in a non-wetting oil phase, the relative permeability of the formation to the non-wetting phase decreases. Viscous energy must be expended to deform the emulsified water droplets so that they will pass through pore throats. If viscous forces are insufficient to overcome the capillary forces which hold the water droplet within the pore body, flow channels will become blocked with persistent, non-draining water droplets. As a result, the flow of oil to the wellbore will also be blocked.

Alkoxylated phenol formaldehyde resins are a well-known class of demulsifier, and the emulsion coalescence data in Table III confirm that Thin Film Spreading Agents, which belong to this class, can also function as chemical demulsifiers. When water in

Table III.   Thin Film Spreading Agents Are a Well-Known Class of Chemical Demulsifier

| | Percent Coalescence with 60 mg/kg of TFSA | | | |
	15 min	30 min	60 min	120 min
Crude A West Texas	43	68	73	77
Crude B Alberta	--	18	54	72
Crude C California	16	80	89	90

crude oil emulsions are treated with low concentrations of TFSA
molecules, a clear water phase separates from the oil and the
volume percent of dispersed water in the crude oil decreases to
very low levels.   While the mechanisms by which demulsifiers
coalesce water-in-oil emulsions are poorly understood, it is clear
that TFSA molecules generate high spreading pressures(31,41) and
displace asphaltene molecules from oil-water interfaces(41).   When
adsorbed at oil-water interfaces, the TFSA molecules present an
energy barrier to coalescence of water droplets which is
significantly lower than the energy barrier presented by thick,
viscous asphaltene films.   Thus, Thin Film Spreading Agents can
promote recovery of oil from porous media by coalescing "emulsion
blocks"(26) and perhaps by lowering oil-water interfacial
viscosity(45).

TFSA-WATERFLOOD PILOT.   A 36 acre (1.44 x 10^5 m^2) TFSA-waterflood
pilot was recently conducted in the Torrance Field in the Los
Angeles Basin of Southern California.   To characterize the fluid
flow patterns within the pilot, an interwell chemical tracer study
was conducted with sodium thiocyanate.   Results of the tracer
study are shown in Table IV.   Only 61.6 % of the injected tracer
was recovered in the produced fluids, indicating that as much as
38.4 % of the injected fluids were flowing out of the pattern.
Furthermore, since only 1604 bbl/d (255 m^3/d) of brine was
injected into the pattern, as much as 75.9 % of the total fluids
produced by pilot wells were from outside the pattern.
        Starting on May 19, 1986, Magnaflood 907 was continuously
metered into the injection brine.   The TFSA concentration averaged
239 mg/kg and the total TFSA bank size was 0.09 pore volumes.
During the TFSA injection stage, the brine injection rate averaged
1604 bbl/d (255 m^3/d) and varied from 1226 bbl/d (195 m^3/d) in
February 1987 to 1838 bbl/d (292 m^3/d) during the first week in
November 1987.   The injection pressure which averaged 1380 psig
(9.51 MPa) was maintained between 1250 psig (8.62 MPa) in February
1987 and 1450 psig (10.0 MPa) during the third week in October
1986.   Therefore, the injectivity during the TFSA-waterflood
varied by less than 16 % from the average of 1.16 bbl/d/psig
(26.8 m^3/d/MPa).   These results, along with results of spinner
surveys obtained before and after TFSA injection, clearly indicate
that injection of TFSA did not adversely affect the injectivity of
the formation.
        The effects of injecting TFSA into the pilot are shown for
Well TU-E2 in graphs of water cut versus cumulative oil (Figure 2)
and oil cut versus date (Figure 3).   After TFSA injection began,
the oil cut in the fluids produced from Well TU-E2 increased by 96
%, from an oil cut of 2.9 % to an oil cut of approximately 5.7 %.
This increase in oil cut upon injection of TFSA represented a
significant 180 % increase in oil production rate from 15 bbl/d
(2.4 m^3/d) just prior to TFSA injection to 42 bbl/d (6.7 m^3/d)
after TFSA injection.   The tracer study indicated that 40.5 % of
Well TU-E2's gross production is from the pilot pattern (see Table
IV).   Other wells in the pilot have lower percentages of their
gross production from the pilot.   Consequently, the response in
oil production rate due to TFSA injection was proportionate to the

Table IV. Tracer Study Established Fluid Flow Patterns Within the Pilot Area

Well Number	Production Rate, Bbl/d		NaSCN Tracer Recovered			Predicted Tracer Recovery[1]
	Total Fluids	Pilot Fluids	Weight, lb.	% of Injected	% of Produced	% of Produced
TU-38A	1075	423	84.7	16.9	27.5	25.1
TU-107	1096	248	49.6	9.9	16.1	15.7
TU-104	1240	218	39.4	7.9	12.9	13.8
TU-E2	727	290	34.5	6.9	11.2	18.4
TU-127	797	168	29.4	5.9	9.5	10.6
TU-102	608	111	21.9	4.4	7.1	7.0
TU-118	424	37	20.4	4.1	6.6	2.3
TU-101	483	43	16.9	3.4	5.5	2.7
TU-105	210	66	11.0	2.2	3.6	4.2
TOTALS	6660	1604	307.8	61.6	100.0	cc = 0.91

(1) Results of simulation model.

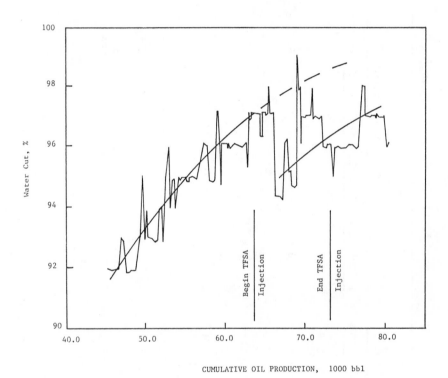

CUMULATIVE OIL PRODUCTION,   1000 bbl

Figure 2.   The efficiency of the waterflood increased upon
            injection of TFSA.

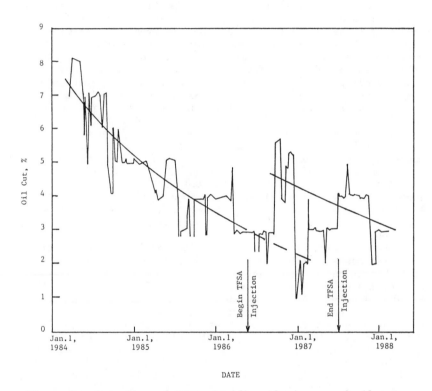

Figure 3.   Injection of TFSA significantly increased oil cut
in fluids produced from pilot wells.

percentage of each well's gross production which was from the
pilot pattern.

To establish the amount of incremental oil which was produced
by TFSA injection, the historical production decline rate for
wells in the pattern was determined from the three years of
production history prior to TFSA injection. The site of the TFSA-
waterflood was specifically chosen because all wells in the pilot
had extensive, well-documented production histories. Semiloga-
rithmic graphs of oil production rate versus date, shown in Figure
4 for one representative well (TU-E2), are not linear; the
production decline rate prior to TFSA injection was best fit by a
hyperbolic decline model. Hyperbolic decline models are widely
used to fit the oil production decline observed in water-drive
reservoirs having good pressure maintenance. The use of a hyperbolic decline curve to model the production decline rate in this
pilot is supported by the fact that Santa Fe Energy Company
produces oil from the Main Zone with good pressure maintenance.
Compared to an exponential decline model, the hyperbolic model
gives a conservative estimate of the incremental oil produced by
TFSA.

Incremental oil production for each of the pilot wells was
calculated by subtracting the extrapolated production decline
curve which was established prior to TFSA injection from the
actual production after TFSA injection. Results of this analysis
indicate that a total of 8150 $\pm$ 850 bbl (1295 $\pm$ 135 m^3) of
incremental oil were obtained due to injection of TFSA.
Incremental oil production was assumed to have ceased by October
10, 1987 when the two high brine producing Wells TU-107 and TU-104
were shut in.

The economics of the TFSA-waterflood project were evaluated
for three cases - each case was based on different assumptions.
In the first case, the minimum values for the economic yardsticks
were evaluated assuming that a conservative 7300 bbl (= 8150 bbl -
850 bbl; 1160 m^3 = 1295 m^3 - 135 m^3) of incremental oil had been
produced by the end of the project. Maximum values for the
economic data were calculated by assuming that 9000 bbl (= 8150
bbl + 850 bbl; 1430 m^3 = 1295 m^3 + 135 m^3) of incremental oil were
produced by only 61.6 % of the TFSA which had been injected into
Well TU-120; this assumption is based on the results of the tracer
study which showed that as much as 38.4 % of the injected fluids
flowed out of the pilot pattern. In the final case, the most
probable values for the economic yardsticks were calculated
assuming the 8150 bbl (1295 m^3) of incremental oil were produced
by 90 % of the TFSA.

Results of the economic analysis are summarized in Table V.
At a sales price of $13.00 per barrel, the value of the
incremental oil produced by the TFSA was between $94,900.00 and
$117,000.00. This revenue was generated at a chemical cost of
between $1.93 and $3.87 per incremental barrel. Cumulative
incremental oil production, shown in Figure 5, indicates that the
volume of incremental oil produced reached a constant and maximum
value 18 months after the pilot was started. Of the total
incremental oil recovered, 37.5 % was produced in the first six
months of the pilot and 81.25 % was produced by the end of the

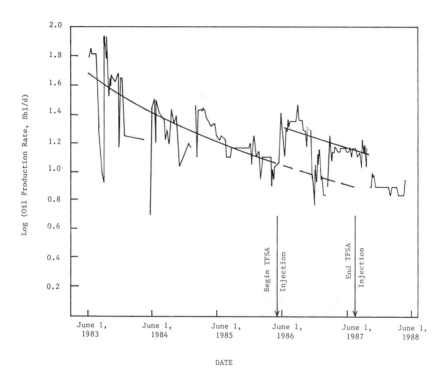

Figure 4. Production decline rate for wells in the pilot
was best fit by a hyperbolic decline model.

Table V. Economic Analysis of TFSA-Waterflood Production Results

	Minimum Value [1]	Most Probable Value [2]	Maximum Value [3]
Incremental Oil Produced, bbl	7300	8150	9000
Value of Incremental Oil at $13.00/bbl	$94,900.00	$105,950.00	$117,000.00
Cost of TFSA Utilized	$28,253.40	$25,428.06	$17,404.09
Chemical Cost per Bbl Incremental Oil	$3.87	$3.12	$1.93
Discounted-Cash-Flow Rate of Return[4]	35 %	54 %	108 %

[1] Assumes minimum amount of incremental oil production and
     that none of the TFSA flowed out of the pilot.
[2] Assumes mean value for incremental oil production and that
     10% of the TFSA flowed out of the pilot.
[3] Assumes maximum amount of incremental oil production and that
     38.4% of the TFSA flowed out of the pilot.
[4] Assumes the project life is 18 months, semi-annual
     compounding, chemical cost is the only expense, and cost
     is paid in full before the start of the project.

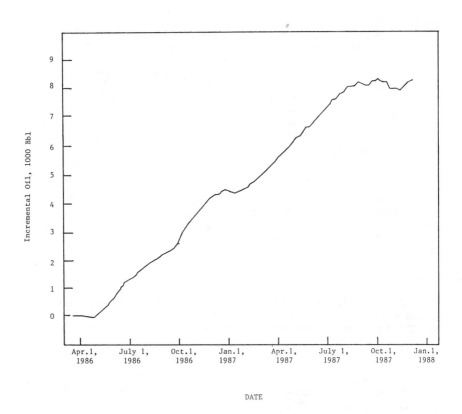

Figure 5.   Cumulative incremental oil production indicates that
            8150 bbl of incremental oil were recovered by TFSA.

first year. Based on semi-annual compounding, the discounted cash
flow rate of return for the project was approximately 54 %.

SUMMARY AND CONCLUSIONS

1) Changing the wettability of reservoir rock surfaces from
oil-wet to water-wet, increases the permeability of the formation
to oil, decreases the permeability to water, decreases mobility
ratio, increases sweep efficiency, increases the flowing fraction
of oil at every saturation, and increases oil recovery at the
economic limit of the waterflood.
2) Thin Film Spreading Agents are synthetic surfactants
which change the wettability of reservoir rock surfaces from oil-
wet and intermediate wettability to water-wet.
3) A 36 acre TFSA-waterflood pilot study was conducted in
the Main Zone of Santa Fe Energy Company's Torrance Field.
4) Fluid flow patterns within the pilot pattern were
characterized by an interwell chemical tracer study which showed
that as much as 38.4 % of the fluids injected into the pilot
flowed out of the unconfined pattern and 75.9 % of the produced
fluids are from outside the pattern.
5) Beginning on May 19, 1986, a 0.09 pore volume bank
containing an average TFSA concentration of 239 mg/kg was injected
into the reservoir.
6) Analysis of the hyperbolic decline curves indicated that
8150 $\pm$ 850 bbl (1295 $\pm$ 135 m³) of incremental 18° API oil were
recovered during the 18 month pilot project.
7) The discounted cash flow rate of return for the pilot
TFSA-waterflood project was 54 %.

ACKNOWLEDGMENT

The authors gratefully acknowledge the helpful discussions,
guidance and technical support given by Dr. Charles M. Blair,
Mr. Richard E. Scribner and Mr. John E. Frederiksen.

LITERATURE CITED

1. Anderson, W. G. J. Petrol. Techn., 38, 1125 (1986).
2. Treiber, L. E.; Archer, D. L.; Owens, W. W. Soc. Petrol.
   Eng. J., 531 (1972).
3. Chilinger, G. V.; Yen, T. F. Energy Sources, 7, 67 (1983).
4. Anderson, W. G. J. Petrol. Techn., 39, 1453, 1605 (1987).
5. Craig, F. F. The Reservoir Engineering Aspects of
   Waterflooding; Society of Petroleum Engineers of AIME:
   Dallas, 1971.
6. Raza, S. H.; Treiber, L. E.; Archer, D. L. Producers
   Monthly, 32, 2 (1968).
7. Owens, W. W.; Archer, D. L. J. Petrol. Techn., 873 (1971).
8. Kyte, J. R.; Naumann, V. O.; Mattax, C. C. J. Petrol.
   Techn., 579 (1961).
9. Donaldson E. C.; Thomas, R. D. SPE 3555 (1971).
10. Mungan, N. Soc. Petrol. Eng. J., 247 (1966).
11. Mungan, N. World Oil, 192, 149 (1981).

12. Jennings, H. Y. J. Petrol. Techn., 116 (1966).
13. Kinney, P. T.; Nielsen, R. F. Producers Monthly, 14, 29 (1950).
14. Kinney, P. T.; Nielsen, R. F. World Oil, 132, 145 (1951).
15. Bobek, J. E.; Bail, P. T. J. Petrol. Techn., 950 (1961).
16. Brown, R. J. S.; Fatt, I. Trans. AIME, 207, 262 (1956).
17. Salathiel, R. A. J. Petrol. Techn., 1216 (1973).
18. Fatt, I.; Klikoff, W. A. Trans. AIME, 216, 426 (1959).
19. Johnson, C. E. J. Petrol. Techn., 85 (1976).
20. Cooke, C. E.; Williams, R. E.; Kolodzie, P. A. J. Petrol. Techn., 1365 (1974).
21. Wagner, O. R.; Leach, R. O. Trans. AIME, 216, 65 (1959).
22. Castor, T. P.; Somerton, W. H.; Kelly, J. F. In Surface Phenomena in Enhanced Oil Recovery; Shah, D. O., Ed.; Plenum Press: New York, 1981.
23. Breit, V. S.; Mayer, E. H.; Carmichael, J. D. In Enhanced Oil Recovery: Proceedings of the Third European Symposium on Enhanced Oil Recovery; Fayers, J. F., Ed.; Elsevier: New York, 1981; Chapter 13.
24. Blair, C. M. U.S. Patent 4 341 265, 1982.
25. Since Thin Film Spreading Agents do not produce ultralow interfacial tensions, capillary forces can trap oil in pore bodies even though the oil has been displaced from the surface of the porous medium. Therefore, recovery of incremental oil is dependent on the formation of an oil bank.
26. Muggee, F. D. U.S. Patent 3 396 792, 1968.
27. Blair, C. M. SPE 14906 (1986).
28. Stout, C. A.; Blair, C. M.; Scribner, R. E. J. Can. Petrol. Techn., 24, 37 (1984).
29. Adkins, J. D. SPE 12007 (1983).
30. Blair, C. M.; Scribner, R. E.; Stout, C. A. SPE 11739 (1983).
31. Blair, C. M.; Scribner, R. E.; Stout, C. A. J. Petrol. Techn., 34, 2757 (1982).
32. Blair, C. M.; Scribner, R. E.; Stout, C. A. SPE 10700 (1982).
33. Blair, C. M.; Stout, C. A. Oil and Gas J., 83, 55 (1985).
34. Blair, C. M. U.S. Patent 4 337 828, 1982.
35. ASTM D 1370-58.
36. Scamehorn, J. F.; Schechter, R. S.; Wade, W. H. J. Colloid Interface Sci., 85, 494 (1982).
37. Greenkorn, R. A. J. Petrol. Techn., 97 (1962).
38. Wagner, O. R. J. Petrol. Techn., 1410 (1977).
39. Brigham, W. E.; Smith, D. H. SPE 1130 (1965).
40. Standard Methods for the Examination of Water and Waste Water; Rand, M. C.; Greenberg, A. E.; Taras, M. J.; Franson, M. A., Eds.; American Public Health Association: Washington, DC, 1976; p 383.
41. Blair, C. M. Chem. and Ind., 538 (1960).
42. Neumann, H. J. Erdol V. Kohle, Erdgas Petrochenue, 18, 776 (1965).
43. Strassner, J. E. J. Petrol. Techn., 303 (1968).
44. McAuliffe, C. D. J. Petrol. Techn., 727 (1973).
45. Kimbler, O. K.; Reed, R. L.; Silberberg, I. H. Soc. Petrol. Eng. J., 153 (1966).

RECEIVED November 28, 1988

Chapter 33

Bound Water in Shaly Sand

Its Determination and Mobility

Ying-Chech Chiu

Department of Chemistry, Chung Yuan Christian University, Chung-Li, Taiwan 32023, Republic of China

Specific ion electrodes were used for anion-free water determination of clay minerals at equilibrium with electrolyte solutions. A new equation was developed for determining anion-free water. Mobility of the anion-free water was determined by compaction experiments with pressure up to 10,000 psi. At NaCl concentrations of 0.2 M or higher, the anion-free water is immobile. At lower concentrations, it is movable under high pressure. Under ordinary flowing conditions, the anion-free water gives a good indication of the immobile water. At high pressure, the amount of the immobile water seems to be related to the porosity of the rock. The actual amount of the immobile water can be found by the method described in this paper. Through compaction of the clay-water slurry, bound water can be concentrated in the sample to facilitate NMR measurements and other studies.

Recent interest in clay hydration water and its effect on various petrophysical properties of shaly sand (1-13) has prompted the author to reinvestigate the subject of "bound water". It has long been suggested that water associated with the clay mineral surfaces causes deviations from the normal petrophysical measurements (14-16). Since bound water exists in the interfacial region between liquid and solid, it is quite difficult to obtain accurate and meaningful results concerning the nature and amount of water in this region. In the past, information concerning bound water has mainly been extracted from vapor phase adsorption of water (8,9,17,18). In recent years, NMR studies have provided much insight into the problem. It is generally recognized now by NMR study that some water molecules are preferentially oriented at the clay surfaces (10,13,19-21). The mobility of water in the interfacial region depends on the type of clay and the amount of sorbed water (22-24). Many authors believe that the tightly bound water occupies only a small fraction of a monolayer (25-27).

Little direct, quantitative measurement has been done on bound water existing in equilibrium with clay mineral substances in

0097–6156/89/0396–0596$06.00/0

aqueous solution. Dmitrenko (28) was the first to devise a method
for determining bound water in natural sediments. The method is
based on the determination of "nonsolvent water". By assuming the
bound water a nonsolvent and assuming chloride ion adsorption negligible, Dmitrenko (28) calculated the amount of bound water after
chemical analysis and material balance of water and chloride in the
sample. Hill et al. (2), using the Dmitrenko method in conjunction
with the "anion exclusion (29-31) technique" based on electrical
double layer theory, determined the amount of bound water in twenty-
eight samples from six fields. They called the water determined by
this method "anion-free water" and assumed this water to be the clay
hydration water (2). By using the amount of anion-free water and
assuming that the anion-free water is immobile for normal flow processes, Hill et al. (2) developed procedures for (1) obtaining an
estimate of brine permeability, (2) correcting mercury injection
curves to estimate oil-water or gas-water capillary curves, (3) obtaining a more realistic estimate of formation water salinity from core
water salinity and (4) calculating oil or water saturations for predicting whether oil or water will be produced.

This paper discusses the use of specific ion electrodes for determining the anion-free water. This method is simpler and more
accurate at low electrolyte concentration than ordinary chemical
methods. It is potentially useful for oilfield application and
laboratory automation. The mobility of this water is also examined
under forced conditions with pressure gradients. It is expected
that by using the methods developed in this paper, one may obtain a
better understanding of the clay properties.

Experimental

Bound Water Determination. A modified method of Hill et al. (2) was
used in determining the bound water (anion-free water). No separation of solid and liquid was made during this determination. The ion
concentrations were measured by specific ion electrodes. The major
equipment consisted of an Orion Model 801 digital pH/mv meter, a
Beckman 39278 sodium electrode, an Orion 94-17 chloride electrode and
an Orion 90-01 reference electrode. Glen Rose Shale (Baroid Division
of NL Industries supplied this sample in a fine powder) was used as
the sample. Samples in bottles were weighed and dried at $110^{\circ}C$
overnight. A known quantity of distilled water or NaCl solution was
added to the bottle. After shaking with a floor shaker for 2 hours,
the clay and sand were allowed to settle overnight. The ion potential was measured by placing the electrode in the clear liquid on
top. All experiments were carried out at $24\pm1^{\circ}C$. The electrodes
were calibrated with NaCl solutions of known concentration. The
concentration of the unknown solution was determined from the calibration curves.

The amount of anion-free water was calculated by a material
balance of chloride and water in the system. The calculation can be
simplified by using volume concentrations. Details of the calculation are illustrated in Appendix I.

Compaction Experiments and Other Related Measurements. After the

anion-free water was determined, the shale-electrolyte slurry was
poured into a filtration or a compaction cell. Gas or hydraulic
pressure was applied to force the water out of the system. A Milli-
pore filter (No. 4004700) or a Fann Filter press was used when the
gas pressure was below 100 psi. Hydraulic pressure of 400 to 10,000
psi was applied through a compaction cell having a configuration as
described by Darley (32). The solution forced out of the cell was
collected, weighed, and its chloride concentration was determined.
When no more liquid could be forced out, the mineral cake was weighed
and dried at 110°C to constant weight. The amount of water in the
cake is taken as the residual water in clay. The volume of the cake
was also measured.

The pore size distribution of the dried sample was measured by a
Aminco 60,000 psi Mercury-Intrusion Porosimeter.

The NMR measurement on water retained in the compacted sample
was performed by using a Bruker CXP-200 NMR Spectrometer.

Some properties of the rock used in this study were measured :
The cation exchange capacity (cec) was determined by the barium
sulfate method as described by Mortland and Mellor (33). Surface
area was measured by using a Digisorb Meter (Micromeritics Instrument
Corporation) through nitrogen adsorption. Estimation of mineral
composition and indentification of the rock were performed by X-ray
diffraction.

Results and Discussion

Bound Water Determination at Low Electrolyte Concentration. From the
theoretical and experimental investigation of the anion-free water in
the literature, it is most reasonable to conclude that the anion-free
water is the water inside the electrical double layer. This concept
has been discussed in more details using the "dual water model" (1)
in which the anion-free water is referred to as the "clay water".
Figure 1 compares the experimentally determined anion-free water
contents. The three circles in Figure 1 are data taken from Ref. 2
determined in the NaCl concentration region 0.228 -5.41 M. The other
data are taken from this investigation in the low concentration
region. Ws is the anion-free water in g/100 g rock. The cec value
and other properties of the sample, Glen Rose Shale, are given in
Table I. In order to chech the agreement between the values obtained
in Ref. 2 and in this study, four points have been gathered around
0.2 M NaCl ($C_O^{-\frac{1}{2}}$=2.4). All the points coincide well and are shown as
one point in Figure 1. The agreement between the two studies is good.
A straight line passing through the origin correlates all the points
in Figure 1. At extremely high electrolyte concentrations, flocculation of clay should occur. Therefore, when $C_O^{-\frac{1}{2}}$=0, Ws=0.

Figure 2 compares data from this investigation with those from
the literature. Most literature data in Figure 2 are from Figure 3
of Ref. 2. The equations mentioned in Figure 2 have the following
forms : Equation 1 comes from Schofield (29) and Bolt et al. (30-31),

$$\frac{\lambda^-}{SC_O} = \frac{2}{(BC_O)^{\frac{1}{2}}} - \frac{4S}{Bcec} \tag{1}$$

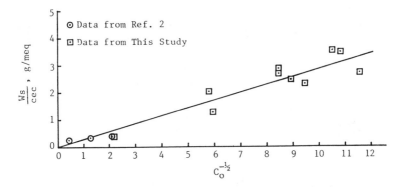

Figure 1.   Anion-free Water As A Function of NaCl Concentration.

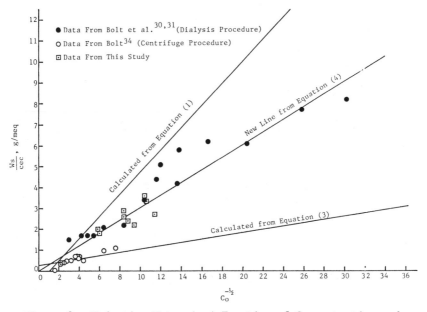

Figure 2.   Hydration Water As A Function of Concentration and
Method of Determination.

Table I. Properties of the Rock Sample

Sample	Surface Area m²/g	cec meq/100g	Surface Charge Density (meq/m²) x 10³	Composition, % from X-Ray Diffraction									
				Quartz	Calcite	Dolomite	Pyrite	Feld-spar	Clay	Clay			
										Illite	Chlorite	Kaolinite	Mixed layer Illite
Grundite Illite (Magco-bar)	42.7	13.33	3.11	48			2		50			30	70
Glen Rose Shale	16.8	10.83	6.45	45	25	3	2	5	20	60	40		
Berea Sand-stone	0.97	0.45	4.65	80		3		5	15	10		90	
Fithian Illite (Ward's)	51.5	13.73	2.66	40	5	2	3		50		20		80
Rochester Illite (Ward's)	19.0	5.0	2.63	45	5	5			50	90	10		

where λ^- is the negative anion adsorption, C_0 has the same meaning as the equilibrium solution concentration, S is the area of the charged surface, and B is a constant. By assuming cec/S to be constant and λ^-/C_0 to be equivalent to Ws, Hill et al. (2) defined equation 2,

$$\frac{Ws}{cec} = AC_0^{-\frac{1}{2}} - B \tag{2}$$

where A and B are coefficients affected by the ratio, cec/S. By fitting their data into equation 2, Hill et al. (2) obtained equation 3 in this form :

$$\frac{Ws}{cec} = 0.084 \ C_0^{-\frac{1}{2}} + 0.22 \tag{3}$$

Data from this study seem to coincide better with the data of Bolt et al.. A straight line passing through the origin best fits all the points in Figure 2. This line has the following form :

$$\frac{Ws}{cec} = 0.31 \ C_0^{-\frac{1}{2}} \tag{4}$$

Surface Charge Density. In the previous section, it was assumed that the surface charge density, cec/S, is a constant. Table I gives the surface charge density for five samples. It varies from 2.63 x 10^{-3} to 6.45 x 10^{-3} meq/m^2. Figure 3 shows a plot of these points. Four of the points can be fitted closely by a straight line with a slope of 2.9 x 10^{-3} meq/m^2. One point deviates from the line. Patchett (35) has plotted some cec vs.S for nine API standard clays on a log-log paper. The best line representing the 9 points has a slope of 2.4 x 10^{-3} meq/m^2. When the data in Figure 3 were plotted in Patchett's diagram, the five points lie around his line with about the same degree of scattering as his own data. It appears that the assumption of cec/S being a constant is a reasonable assumption for many rocks.

Mobility of The Anion-Free Water. It is well known that water in the electrical double layer is under a field strength of 10^6-10^7 V/cm and that the water has low dielectric constants (36). Since anion-free water is thought to be the water in the electrical double layer between the clay and the bulk solution, at high electrolyte concentrations, the double layer is compressed; therefore, the water inside is likely quite immobile. At low electrolyte concentrations, the electrical double layer is more diffuse, the anion-free water is expected to be less immobile. Since the evaluation of the shaly formation properties requires the knowledge of the immobile water, experiments were conducted to find out the conditions for the anion-free water to become mobile.

By definition, the anion-free water is free of salt. When pressure is applied to a clay-brine slurry to force out water (as that described in the experimental section), the solution that flows out of the cell should maintain the same chloride concentration as the brine's if the anion-free water is immobile. Otherwise, the concentration of the chloride decreases. Pressure forces water to flow through the pores with a certain velocity; meanwhile, the pore size

is reduced. By accounting material balances during the experiment,
useful information was deduced.

Table II shows the result of compaction experiments with Glen
Rose Shale. Column 2 gives the equilibrium NaCl concentration of the
solution before the compaction experiment. Column 3 gives the anion-
free water calculated as shown in Appendix I. Column 4 gives the
amount of the bulk solution which has the NaCl concentration given in
Column 2. Column 5 gives the total amount of fluid flowing out of
the cell. Column 6 indicates the pressure applied and Column 7, the
initial flow rate. The flow rate decreased with time and could be
measured or estimated from equations (37). Column 8 gives the water
retained in the clay sample after compaction by determining the
weight loss after heating at $110^{\circ}C$. Porosity of the dried clay
sample was determined by comparing the volume of residual water and
total volume of the sample and is given in Column 9. Water density
was assumed to be $1g/cm^3$. Column 10 gives the average pore diameter
of the dried clay sample.

Experiments No. 1,2 and 3 were performed at gas pressure beginning at 15 psi and stepping up to 77 psi. The total fluid collected
was less than the bulk solution in the system. The concentration of
chloride in the fluid collected in these three runs was about the
same as the values given in Column 2. It was concluded that under
these conditions, the anion-free water was immobile. It was observed
that under the same applied pressure, the higher the NaCl concentration, the faster the flow rate -- consistent with observations reported by Engelhardt and Gaida (38).

In order to increase the flow rate without too much pressure,
Experiment 4 was performed with a Fann filter press which has a
wider cross sectional area. A constant air pressure of 100 psi was
applied, the flow rate was 26 times that of Experiment 1 while the
NaCl concentration was only slightly higher than that of Experiment 1.
Although the flow rate was much increased in Experiment 4, the result
was similar to Experiment 1. The water retained in the clay (Column
8) determined by drying was found to be close to the amount of anion-
free water. The porosity of the sediment was 0.4 and the average
pore diameter was 4466 $\overset{\circ}{A}$. It was concluded from this experiment,
that the anion-free water was immobile even at 100 psi and 7.4 ft/day.
The pore size distribution of the sample showed 90% of the pores to
have a diameter above 350 $\overset{\circ}{A}$ and less than 3% of the pores to have a
diameter below 100 $\overset{\circ}{A}$ (Figure 4).

It was decided to increase the pressure in subsequent experiments to push the anion-free water out. Experiments 5 and 6 were
performed at 400 psi at a NaCl concentration around 0.01 M. Experiments 5-10 were performed in the compaction cell as described in the
experimental section. This apparatus was rated for 10,000 psi. The
pressure regulation at 400 psi region was about $\pm$ 100 psi. Some
evaporation occurred that made the total fluid collection less than
expected from total material balance. NaCl concentration of the
collected fluid could not be measured accurately. However, the
amount of fluid collected and the amount of water retained in the
sediments in Experiments 5 and 6 clearly indicated some anion-free
water was mobilized.

Experiment 7 was the first run under 400 psi hydraulic pressure
followed by 10,000 psi. Most of the fluid was collected at the low

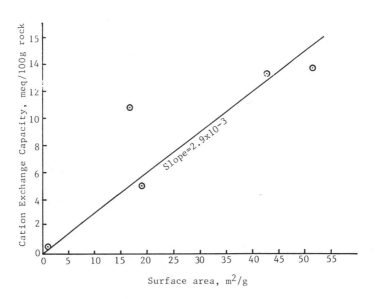

Figure 3. Cation Exchange Capacity As A Function of Surface Area.

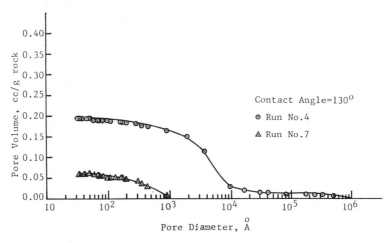

Figure 4. Pore Size Distribution By Mercury Intrusion.

Table II. Compaction Experiments of Glen Rose Shale

Run No.	NaCl Concn. M	Anion-free H_2O, g	Bulk Soln. g	Fluid Collected g	Pressure psi	Initial cm/sec	Flow Rate ft/day	Residual H_2O in clay, g	Porosity	Av. Pore Dia., Å
1	0.0086	18.5	31.3	29.6	15-77 (N2 or air)	0.00010	0.31			
2	0.0300	10.9	38.9	32.2	same as above	0.00013	0.36			
3	0.212	2.0	48.2	32.9	same as above	0.00026	0.74			
4	0.0126	12.7	35.7	34.8	100 (air)	0.0026	7.4	12.0	0.40	4466
5	0.0091	18.8	53.2	54.8	400 (hydraulic)	0.0015	4.2	9.8	0.42	2728
6	0.0140	15.6	37.8	39.1	same as above	0.00067	1.9	9.6	0.34	3337
7	0.216	1.8	46.2	32.8	400 (hydraulic)	0.0024	6.7			
				6.0	10,000 (hydraulic)			3.2	0.15	440
8	0.0096	16.6	27.4	29.0	400 (hydraulic)		0.7			
				3.3	10,000 (hydraulic)			3.6		
9	0.033	9.1	32.5	15.6	10,000 (hydraulic)		>10	3.1	0.16	491
10	0.235	1.5	37.0	13.6	same as above		>10	3.1	0.17	487

pressure. NaCl concentration of the collected fluid was the same as the bulk concentration, 0.216 M. The high pressure of 10,000 psi was applied for three days. Only 0.1 g liquid was collected during the last two days because of evaporation. NaCl concentration of the fluid was not measured. In this experiment, the amount of water retained in the sediment was less than that of anion-free water.

Experiments 8,9 and 10 were all performed up to 10,000 psi hydraulic pressure. Evaporation occurred during Experiments 9 and 10, although high flow rate was attained. It appears that At this high pressure, the porosities of the dried sediment decreased to 0.15-0.17 and the residual water contents in the clay decreased to 3.1-3.6 g, regardless the original NaCl concentration. At 0.2M NaCl concentration, the residual water content was higher than that of anion-free water; whereas at lower NaCl concentrations, the residual water content is lower. Even at such a high pressure (10,000 psi), the anion-free water was not mobile at the NaCl concentrations above 0.2 M. The porosity wad decreased by pressure close to those of some Berea rocks. Figure 4 shows the pore size distribution curve for the dried sediment of Run No. 7. The curve is quite typical for samples compressed at 10,000 psi such as Run No. 8 and 9. The average pore diameter of the dried sediments for these three runs was in the range Of 440-491 Å.

The experiments listed in Table II suggested that the anion-free water was not pressed out until the bulk water had been squeezed and sufficient pressure was applied. Data obtained by Von Engelhardt and Gaida (38) also follow the same trend. They measured the change in NaCl concentrations for pore solutions in compressed montmorillonite clay under the pressure range of 440 psi to 470,000 psi. The concentration of NaCl in the solution decreased very slowly until the porosity of the clay was decreased to 0.5. Then the concentration decreased very quickly with further decrease of the porosity. At extremely high pressure, the NaCl concentration showed a slight tendancy of reversing the trend. When kaolinitic clays were compressed, the concentration change in pore solution was not observed. Von Engelhardt and Gaida (38) did not measure the anion-free water. The decrease of pore solution concentration would be expected, only if the bulk solution was pressed out first, followed by the anion-free water. Since the cec value of kaolinitic clay is low, the amount of anion-free water should also be low; thus the concentration change is not expected.

NMR Measurement of The Residual Water. The residual water obtained in Experiments No. 5 and No. 6 was measured with a broadline NMR spectrometer, in which, proton signal of the water gives a very broad line. The full width at half height of the NMR signal is 15750 Hz and 18000 Hz for samples obtained in Experiments No. 5 and No. 6, respectively. The line width for liquid water is normally less than 5 Hz while the line width for polycrystalline ice is 56000 Hz (39). Therefore, the residual water is expected to have a mobility closer to ice than to liquid water. The wider line given by sample obtained in Experiment No. 6 seems to agree with the expectation that the water is more immobile at higher electrolyte concentration.

This study suggests a method for sample preparation to increase the concentration of bound water. By compressing the clay-water

slurry, most of the mobile water can be removed. The resulting
sample should show more of the bound water characteristics. The use
of NMR in studying this kind of sample could provide much insight
into the subject.

Conclusions

Under the experimental conditions of this study, several conclusions
can be drawn :
(1) Anion-free water determined by using a chloride ion electrode
agrees well with data given in the literature. (2) A new equation
has been proposed for the bound water calculation. (3) The mobility
of the anion-free water was found to be affected by pressure, porosity and electrolyte concentration. (4) Compaction experiments indicated that the anion-free water will not move until all the bulk
water has been removed. (5) It is possible to increase the ratio of
bound water to bulk water in a sample through compaction experiment.

Acknowledgment

The author wishes to thank J.F. Huang and L.J. Chen for their assistance in this work. This research was supported by the National
Science Council, R.O.C., grant number NSC 77-0402-E033-01. Special
thanks are due to Mrs. Yuh-Shinq Chang Wu for typing the manuscript.

Appendix I. Calculation of Anion-Free Water

	Bottle A	Bottle B
wt. of Glen Rose Shale, g.	33.1	42.6
wt. of added 0.200 M NaCl, g.	0	50.0
wt. of added distilled water, g.	49.8	0
Volume of added NaCl solution, ml	0	49.9
Volume of added distilled water, ml	49.9	0
Final concentration, Na^+, M	0.0158	0.208
Cl^-, M	0.0023	0.210

Let V_f = final volume of the solution, ml
material balance of Cl^- gives

$$0.210\ V_f = 0.200 \times 49.9 + 0.0023 \times 49.9 \times \frac{42.6}{33.1}$$

$$V_f = 48.2\ ml.$$

volume of anion-free water = 49.9 - 48.2 = 1.7 ml.

$$\text{wt. of anion-free water} = 1.7 \times 0.998 \times \frac{100}{42.6}$$

$$= 4.0\ g/100\ g\ rock.$$

Carrying out the six step calculation listed in Ref. 2 gives the same
answer.

Literature Cited

1. Clavier, C.; Coates, G.; Dumanoir, J. 52nd Annual Fall Tech. Conf.
    Soc. Pet. Eng. AIME, 1977, SPE paper No. 6859.
2. Hill, H.J.; Shirley, O.J.; Klein, G.E. The Log Analyst 1979, 3.
3. Juhasz, I. SPWLA 20th Annual Logging Sym., 1979.
4. Holditch, S.A. J. Pet. Tech. 1979, 1515.
5. Jones, F.O.; Owens, W.W. J. Pet. Tech. 1980, 1631.

6. Fertl, W.H.; Frost, E. Jr. J. Pet. Tech. 1980, 1641.
7. Low, P.E. Soil Sci. Soc. Am. J. 1980, 44, 667.
8. Branson, K.; Newman, A.C.D. Clay Minerals 1983, 18, 277.
9. Odom, J.W.; Low, P.F. Soil Sci. Am. J. 1983, 47, 1039.
10. Fripiat, J.; Cases, J.; Francois, M; Letellier, M. J. Coll. Interface Sci. 1982, 89, 378.
11. Fripiat, J.; Letellier, M; Levitz, P. Philos. Trans. R. Soc. London 1984, Ser. A., 311 (1517), 287.
12. Lipsicas, M.; Straley, C; Costanzo, P.M.; Giese, R.F. Jr. J. Coll. Interface Sci. 1985, 107, 221.
13. Giese, R.F. Jr.; Costanzo, P.M. ACS Sym. Ser. 1986, 323, 37.
14. Klinkenberg, L.J. in Drilling and Production Practice; Am.Pet. Inst., 1951, P.200.
15. Baptist, O.C.; Sweeney, S.A. Bur. Mines Report Inv. 5180, 1955.
16. Von Engelhardt, W.; Tunn, W.L.M. State Geological Survey Circular 194, 1955, Urbana, Illinois.
17. Grim, R.E. Clay Mineralogy; McGraw-Hill Co., 1953; pp. 161-183.
18. Martin, R.T. 9th National Conf. on Clays and Clay Minerals, 1962, p.28.
19. Woessner, D.E.; Snowden, B.S. Jr. J. Coll. Interface Sci. 1969, 30, 54.
20. Woessner, D.E. Proc. NMR Spectrosc. Pestic. Chem. Symposium, 1974, 279.
21. Woessner, D.E. Mol. Phys. 1977, 34, 899.
22. Slonimskaya, M.V.; Raitburd, Ts. M. Dokl. Akad. Nauk SSSR 1965, 162, 176.
23. Kitagawa, Y. Am. Mineral 1972, 57, 751.
24. Laughlin, L.J. U.S. Gov. Report Announce. Index 1979, 79, 111.
25. Matyash, I.V.; Litovchenko, A.S.; Vasilev, N.G. Kolloid Zhurnal 1974, 36, 531.
26. Kvlividze, V.I.; Krasnushkin, A.V. Dokl. Phys. Chem. 1975, 222, 480.
27. Ananyan, A.A.; Golovanova, G.F.; Volkova, E.V. Merzlotnye Issled. 1976, 15, 182.
28. Dmitrenko, O.I. Kolloid Zhurnal 1957, 20, 157.
29. Schofield, R.K. Nature, 1947, 160, 408.
30. Bolt, G.H.; Warkentin, B.P. Kolloid Z. 1958, 156, 41.
31. Bolt, G.H. VIe Congres International De La Science Du Sol 1956, Paris.
32. Darley, H.C.H. J. Pet.Tech. 1969, 883.
33. Mortland, M.M.; Mellor, J.L. Proc. Soil Sci. Soc. Am. 1954, 18, 363.
34. Bolt, G.H. Ph.D. Thesis, Cornell University, 1954.
35. Patchett, J.G. Trans. SPWLA 16th Annual Logging Symposium 1975, Paper V.
36. Kavanau, J.L. Water and Solute-Water Interactions; Holden-Day Inc., 1964, pp. 28-78.
37. Von Engelhardt, W.; Schindewolf, E. Kolloid Z. 1952, 127, 150.
38. Von Engelhardt, W.; Gaida, K.H. J. Sed. Petrology 1963, 33, 919.
39. Barnall, D.E.; Lowe, I.J. J. Chem. Phys. 1967, 46, 4808.

RECEIVED December 12, 1988

Chapter 34

Acid Wormholing in Carbonate Reservoirs

Validation of Experimental Growth Laws
Through Field Data Interpretation

Olivier Liétard and Gérard Daccord

Dowell Schlumberger, BP 90, 42003 Saint-Etienne Cedex, France

Actual responses of two carbonate petroleum reservoirs to
matrix injection of hydrochloric acid are compared with a
recently proposed experimental model for wormholing. This
model is shown to be applicable in undamaged primary porosity reservoirs, and should be useable in damaged double
porosity ones. Formations of no primary porosity are shown
to respond very differently.
        The dissolution channels (wormholes), obtained under
certain conditions of attack of carbonate rocks by hydrochloric acid, have been recently proven to have a fractal
geometry. An equation was proposed, relating the increase
of the equivalent wellbore radius (i.e. the decrease of the
skin) to the amount of acid injected, in wellbore geometry
and in undamaged primary porosity rocks. This equation is
herein extended to damaged double porosity formations
through minor modifications.
        Two cases of initially undamaged reservoirs have been
selected for proper validation of modeling equations. From
pressure and rate data recorded all along their treatment,
the skin variations during acid attack are derived according to a recently published methodology. Their analysis
validates the proposed model. This would mean that wormholes in reservoirs do scale up with laboratory ones
according to the proposed law.

One third of the world production of hydrocarbons originates from carbonate reservoirs, which, in addition, are supposed to contain half of
the reserves of these compounds (1). The economical importance of such
reservoirs is therefore large enough to justify the research of new
methods aiming at better producing oil from these rocks.
        Most of these reservoirs have natural permeabilities below 10 mD,
and the stimulation of their production is achieved through acid
fracturing operations. Viscosified hydrochloric acid is pumped into
wells at pressures larger than formation parting pressure. Irregular
etching of the fracture walls by the acid is expected to create highly

0097–6156/89/0396–0608$06.00/0

conductive channels which remain open to oil flow when the injection
pressure is released and the fracture heals up.

Some carbonate reservoirs are naturally fairly permeable (more than
10 mD) and are not candidates for acid fracturing. Such a treatment
would neither be technically feasible (large fluid loss rates would
prevent fracture propagation) nor boost originally large productivities.
However, wells drilled through such rocks usually demonstrate produc-
tions lower than one could expect according to their permeability. This
is the consequence of the presence, around the wellbore, of a zone of
reduced conductivity, refered to as damage. Many causes of damage have
been recognized (2). In carbonate reservoirs, the most common ones are
invasion by drilling muds and precipitation of scales or petroleum heavy
ends due to the production pressure drop at wellbore.

The damaged zone has variable extension and severity, the latter
being defined as the ratio of undamaged to damaged permeabilities, k_u
and k_d respectively. This zone is responsible for an additional, large
resistance to oil flow, dramatically impairing production. When initiating production by imposing a given pressure drawdown ΔP, part of the
latter is lost through the damaged zone. This results in lower production rate Q and productivity index $Q/\Delta P$. The differential pressure
through the damaged zone is given by :

$$\Delta P_d \;=\; \frac{\mu Q}{2\pi k_u h}\; S \tag{1}$$

where μ is the viscosity of the flowing fluid and h the height of the
producing interval. The skin S is a coefficient representative of the
characteristics of the damaged zone :

$$S \;=\; \int_{r_w}^{r_d} \left[\frac{k_u}{k(r)} - 1\right] \frac{dr}{r} \tag{2}$$

where r_w is the wellbore radius, r_d the radial extension of the damaged
zone, and $k(r)$ the permeability profile throughout it. When the severity is constant from r_w to r_d, Equation 2 reduces to (3) :

$$S \;=\; \left(\frac{k_u}{k_d} - 1\right)\; \log\left(\frac{r_d}{r_w}\right) \tag{3}$$

The aim of a matrix acidizing job, which, in carbonate reservoirs,
generally consists in injecting plain hydrochloric acid at pressures
below the formation parting pressure, is to decrease S down to zero.
This operation is more often refered to as damage removal. It is not,
strictly speaking, a reservoir stimulation, since it merely recovers the
natural productivity index of the formation. In addition, the wording
''damage removal'', originating from similar operations in sandstone
reservoirs, is not fitted to carbonate ones. It would be better to talk
about ''damage by-pass'', since in carbonates the damaging materials
are seldom soluble in hydrochloric acid, whereas the rock is.

Matrix acidizing jobs are inexpensive treatments, but their effect
on well productivity is often spectacular, with common 5 to 20-fold
increases of the productivity index. This one fact would justify devo-
ting lots of effort in the understanding and the modeling of phenomena
occuring while acidizing a carbonate reservoir. However, the particularly puzzling behavior of the acid in such rocks prevented progress to
be made in this area despite numerous efforts and studies.

Acid Wormholing in Carbonate Rocks

When submitting a carbonate rock to the flow of an acidic solution, the attack, most of the time, leads to the preferential growth of large pores. The dissolution of the rock is not uniform, and the final pore size distribution is much broader than the original one. This fact has been recognized and described in detail more than fifteen years ago (4-5). Macroscopic pores, which are the end result of such an unstable attack, have received the name of wormholes.

It has been shown that the development of wormholes in carbonate rocks is a consequence of diffusion-limited (mass-transfer-limited) kinetics of attack (6). Such kinetics prevail in most of these rocks, i.e. limestones and dolomites, providing that, for the latter, the temperature is larger than about 200°F (90°C) (7-8).

The determination of the evolution of the permeability of these rocks during acidizing is necessary when attempting to predict the evolution of the skin (Equation 2). Previous studies (6) have tried to model the shift of the pore size distribution due to acid attack. Then, permeability profiles were computed by integrating the contributions to the overall flow of each of the rock pores, all over the considered volume of rock. The main limitation of this method lies in the disregarding of the spatial correlation between rock pores.

Another approach of the problem has been proposed recently (9). Based on a comprehensive experimental study, it considers that wormholes have an almost infinite conductivity in comparison with the original pores of the rock. Consequently, a fair approximation of the permeability profile of an acidized piece of rock of length L is a step function. From its inlet up to almost the tip of the wormholes, the permeability is infinite. In the rest of the rock, it is equal to the original one, i.e. k_0, and the overall permeability of this core becomes :

$$k = k_0 \left[\frac{L}{L - L_e} \right] \qquad (4)$$

where L_e, strictly speaking, is not the length of the largest wormhole, but an equivalent distance over which the permeability can reasonably be assumed to be infinite as compared to k_0 (10).

Similarly, in 3D-radial geometries of interest for petroleum engineers, an equivalent wellbore radius r_e is defined. The near-wellbore region, including radially distributed wormholes from r_w up to r_e, is infinitely permeable and therefore becomes a mere radial extension of the wellbore itself. Equation 2 can be used to calculate the pseudo-decrease of the skin when an undamaged primary porosity formation of permeability k_0 includes wormholes as described hereabove :

$$S = \int_{r_w}^{r_e} \left[\frac{k_0}{k(r)} - 1 \right] \frac{dr}{r} = -\log \left(\frac{r_e}{r_w} \right) \qquad (5)$$

since $k(r) \to \infty \quad \forall\, r$ with $r_w \leq r \leq r_e$, and $k(r) = k_0 \quad \forall\, r > r_e$.

Therefore, the problem reduces to the definition of the relation linking r_e with the volume V of injected acid. In (9), the following equation was proposed for 3D-radial acid injection into an undamaged primary porosity reservoir :

$$\frac{r_e}{r_w} = \left[1 + A(N_{ac}, D, Q...) \times \left(\frac{V}{h}\right) \right]^{1/df}, \quad A(...) = \frac{bN_{ac}}{\pi\phi r_w{}^{df}}\left(\frac{Q}{Dh}\right)^{-1/3} \tag{6}$$

where A is a slowly varying function. Q, h, r_w, r_e and V are defined as before, ϕ is the initial rock porosity, and D is the diffusivity constant. df is the fractal dimension of the wormholing structure, experimentally determined as being equal to 1.6 ± 0.1 (11).

b is a constant that reflects the relative conductivity of wormholes compared with the rock permeability. It is equivalent to the constant B in (10), and was given a value of 1.7×10^4 S.I. units in (9). N_{ac}, the acid capacity number, is defined as in (12):

$$N_{ac} = \frac{\phi C M}{\beta(1-\phi)\rho} \tag{7}$$

where C is the acid concentration, M the molar weight of the chemical species constituting the rock, ρ the density of the latter, and β the stoichiometric coefficient of the reaction (2 in the case of carbonate and hydrochloric acid).

One difficulty in Equation 6 lies in the determination of the diffusivity constant for highly concentrated acids over a broad range of temperature. However, available data (13), combined with viscosity values of hydrochloric acid, lead to estimates shown in Tables I and II for the field cases that will be described later on.

This wormholing model was shown to agree with previously published experimental data (14).

Extension to Damaged Double Porosity Reservoirs

Most of the carbonate reservoirs consist in thick layers of double porosity rocks. There are few limestones, such as chalks, that show primary porosity only (voids being intergranular spaces). On the other hand, there are also some deep dolomites of secondary porosity only: intergranular spaces have disappeared due to large overburden pressures and diagenetic recrystallization. Their secondary porosity consists in cracks, fissures, fractures, faults and vugs. The vast majority of carbonate reservoirs lie in between these two extremes. The main part of their permeability (50-500 mD) originates from a pattern of diagenetic and/or tectonic fissures (secondary porosity), whereas the permeability of a piece of rock containing no cracks is much lower, in the order of several milliDarcies.

In double porosity carbonate reservoirs, whatever the nature of the damaging materials, the damage is preferentially located in the secondary porosity, and its distribution seems fairly regular. Considering a thin slice of reservoir at a given depth, a good approximation of its permeability profile along any direction perpendicular to the wellbore axis is the following (Figure 1):

- from r_w to r_d, a crown of damaged rock wherein the original double-porosity-related permeability k_{DP} has been reduced to its sole primary porosity contribution k_{PP}, natural fissures being almost completely filled with and sealed by damaging materials (such as drilling mud cakes, for instance);

- from r_d up to the reservoir boundary, the permeability is constant and equal to k_{DP}.

It is likely that, in natural reservoirs, the extension r_d of the damaged zone is not even all over the height of the producing interval. This is a consequence of variable conditions of hydrostatic head during drilling, or of variable pressure drawdown during production. For the sake of simplicity, however, we will here consider that r_d is constant.

Injecting acid in such an interval produces wormholes in the near-wellbore region, which is damaged, so that the acid flow only takes place in the clean primary porosity of the rock. This superposes to the original step function of the permeability profile a second, similar function, stating that the acidized profile now includes in the vicinity of the wellbore, from r_w up to r_e, a zone of infinite conductivity (Figure 1). Using Equation 2 again, it comes :

$$S = -\log\left(\frac{r_e}{r_w}\right) + \left(\frac{k_{DP}}{k_{PP}} - 1\right)\log\left(\frac{r_d}{r_e}\right) \qquad \forall\ r_e \leq r_d \qquad (8)$$

The initial skin S_0 of this reservoir was (Equation 3) :

$$S_0 = \left(\frac{k_{DP}}{k_{PP}} - 1\right)\log\left(\frac{r_d}{r_w}\right) \qquad (9)$$

so that Equation 8 becomes :

$$S = S_0 - \frac{k_{DP}}{k_{PP}}\ \log\left(\frac{r_e}{r_w}\right) \qquad \forall\ r_e \leq r_d \qquad (10)$$

On the other hand, Equation 6 can be written :

$$\log\left(\frac{r_e}{r_w}\right) = \frac{1}{df}\ \log(1 + A'V)\ , \qquad A' = \frac{A}{h} \qquad (11)$$

so that the skin value of this reservoir is related to the cumulative volume of acid V. It comes :

$$S(V) = S_0 - \frac{1}{df}\frac{k_{DP}}{k_{PP}}\ \log(1 + A'V) \qquad \forall\ r_e \leq r_d \qquad (12)$$

keeping in mind that the proportionality factor A' includes a porosity term which is the one related to the sole primary porosity (corresponding to k_{PP}).

When r_e becomes larger than r_d, a complete modification of the flow and attack conditions happens. Etching of the secondary-porosity fissures walls becomes the predominant phenomenon. Modeling equations established for wormholing propagation in primary porosity rocks most likely do not hold true anymore. However, at the time the wormholes tips reach the undamaged part of the reservoir, the treatment is virtually finished, its objective being already fulfilled. There is no need, consequently, to develop modeling equations for the rest of the treatment : a correct job design methodology would just aim at allowing the volume of acid necessary to achieve breakthrough, i.e. damage by-pass.

For undamaged primary porosity reservoirs, Equation 12 reduces to :

$$S(V) = -\frac{1}{df}\ \log(1 + A'V) \qquad \forall\ V \qquad (13)$$

Comparison with Actual Skin Histories during Acid Jobs

In damaged double porosity reservoirs, the influence of the fractal dimension df is masked by the severity ratio k_{DP}/k_{PP} (Equation 12).

Therefore, it has been chosen here to study very particular and rare treatments performed in undamaged reservoirs. Equation 13 is then applicable, and a direct access to the values of both parameters df and A' becomes possible. Two cases have been considered. First, a clean reservoir of secondary porosity only, where Equation 13 should not fit with actual data. Second, an undamaged, primary porosity formation, where it should.

Matrix acidizing treatments are more often performed, nowadays, with sensors and data acquisition systems continuously recording the surface pressure and rate histories. According to a recently proposed methodology (15), these records can be used to compute downhole rate and pressure evolutions. The bottomhole pressure history is then compared to the theoretical response of an equivalent reservoir wherein a non-reactive fluid would have been injected according to an identical rate schedule. Following this method, the difference between both theoretical and actual pressure responses originates from the evolution of the skin of the true reservoir under the influence of the acid attack. Equation 1 is then used to derive the skin decrease from this pressure difference.

Well A.   This oil producer was to be converted into a water injector. Though its production period had been relatively smooth, with no sign of production damage build-up, the operator wanted to be sure to obtain a maximum injectivity index before initiating water injection. The well data and acid job main parameters are summarized in Table I.

Analyses of three rock samples from the pay zone show almost pure calcium carbonate (no magnesium) associated with traces of quartz, kaolinite and pyrite. Solubilities in hydrochloric acid range in between 94.6 and 97.5 %. Nitrogen permeabilities are below 1 mD, except for one sample containing a visible crack for which a permeability of about 10 mD was measured when submitting the sample to a minimum confining pressure. These laboratory data were obtained after the acid job was performed on this well. They show a formation having secondary porosity only, with an overall permeability that should amount several milliDarcies. Consequently, the skin history should not be interpretable according to the hereabove proposed modeling equations.

An injection/fall-off test was performed in this well, one hour and a half prior to the acid treatment. Two cubic meters (12.6 barrels) of water were injected during 14 minutes at an average rate of 1300 bpd. Then the pressure was recorded during 45 minutes. The fall-off data were interpreted on the basis of an homogeneous primary porosity reservoir, in the absence of suitable interpretation tools for naturally fissured formations. The derived values of the permeability, skin and wellbore storage coefficient are given in Table I. The negative pseudo-skin value, together with the large wellbore storage coefficient, are indicators of an undamaged fissured reservoir. The overall permeability is in line with laboratory data.

Then, 11 m^3 (70 barrels) of 15 % HCl were pumped downhole through a coil tubing. The bottomhole pressure history is in this case precisely derived from surface data, since the latter are collected in the open annular space between the coil tubing and the casing. No computation of friction pressure drops all along the injection string is needed, which removes a major source of errors in the derivation of bottomhole data. The bottomhole pressure is just equal to the sum of the surface

Table I. Characteristics of Well A

Nature :	oil producer, to be converted to water injector
Deviated depth, mid-perforations :	6534 feet (1992 m)
Completion :	cased hole, perforation density 4 shots/foot
Wellbore radius :	3 inches (0.0762 m)
Formation height :	88.6 feet (27 m)
Bottom Hole Static Temperature (BHST) :	ca. 150°F (65°C)
Bottom Hole Static Pressure (BHSP) :	3000 psi
Formation fluid :	oil (viscosity not communicated)
Formation nature :	secondary porosity limestone
Reservoir porosity :	not communicated
Reservoir permeability :	14.1 mD
Initial skin value :	-2.6
Wellbore storage coefficient :	3.05×10^{-3} bbl/psi
Bottom Hole Injection Temperature :	ca. 115°F (45°C)
Average injection rate :	1400 bpd (2.57×10^{-3} $m^3.s^{-1}$)
Acid capacity number :	0.0121
Acid diffusivity :	2.3×10^{-9} $m^2.s^{-1}$

one and of the hydrostatic head. Surface rate and bottomhole pressure histories are depicted in Figure 2.

The relative skin $(S - S_0)$ evolution with respect to the cumulative acid volume is given in Figure 3 (solid line). Peaks at about 0.5 and 3.5 m^3 are artifacts due to hammering effects following rapid variations of the injection rate. According to Equations 6 and 13, the skin evolution, if the formation were a primary porosity one, should be :

$$\Delta S = -0.625 \ \log(1 + 3.17 \times V) \tag{14}$$

with V expressed in cubic meters (dashed curve in Figure 3). We used $\phi = 0.13$, as if the permeability of this formation originated from primary porosity, and according to common relations between these two characteristics in carbonate rocks (16). Actually, the best fit to the skin curve of Well A is :

$$\Delta S_{AC} = -0.165 \ \log(1 + 21.1 \times V) \tag{15}$$

(dotted curve in Figure 3). There is a large discrepancy between these two equations, as evident in Figure 3. Not only the curvatures are different; the amplitudes are visibly not comparable.

The actual response of Well A to the injection of acid cannot fit with any kind of theoretical equation for undamaged primary porosity or damaged double porosity reservoir. Therefore, the acid attack in secondary porosity formations proceeds very differently, as expected.

Well B.   Well B is an exploratory well in an oil bearing formation. Its characteristics are summarized in Table II. As common for exploratory wells, whose purpose is not economical, its perforation density is low. Only 163 perforations had been opened, the perforated intervals being regularly spaced all over the entire formation height. Most of the initial skin value originates from this low density.

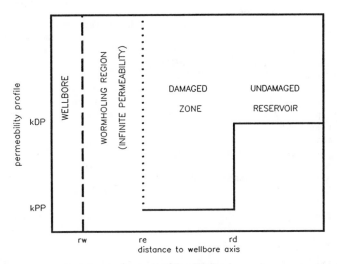

Figure 1 : Permeability profile in a damaged double porosity
reservoir during acidizing. r_w: wellbore radius, r_e: wormholes
penetration, r_d: damage radius, k_{DP}: undamaged reservoir permeability (total contribution of both primary and secondary porosities),
k_{PP}: damaged permeability (primary porosity contribution only).

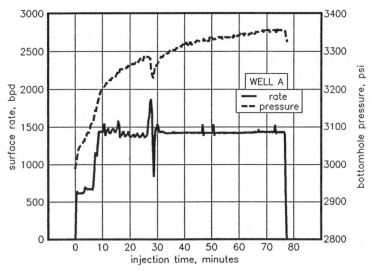

Figure 2 : Surface rate and bottomhole pressure evolutions during
treatment of Well A.

Table II. Characteristics of Well B

Nature :	exploratory well
Deviated depth, mid-perforations :	13130 feet (4002 m)
Completion :	cased hole, perforation density 1 shot/foot
Wellbore radius :	3 inches (0.0762 m)
Formation height :	561 feet (171 m)
Bottom Hole Static Temperature (BHST) :	240°F (115°C)
Bottom Hole Static Pressure (BHSP) :	6100 psi
Formation fluid :	oil, viscosity 6.85 cps. at BHST
Formation nature :	mainly primary porosity, faulted
Reservoir porosity :	0.13 (estimate, according to (16))
Reservoir permeability :	16.4 mD
Initial skin value :	+3.3
Wellbore storage coefficient :	2.5×10^{-4} bbl/psi
Bottom Hole Injection Temperature :	ca. 170°F (75°C)
Average injection rate :	5200 bpd (9.56×10^{-3} $m^3.s^{-1}$)
Acid capacity number :	0.0121
Acid diffusivity :	3.2×10^{-9} $m^2.s^{-1}$

Other wells in the same field had shown a limestone reservoir of fair permeability, with major tectonic faults and some associated fissures. Well data were provided by the operator. The low value of the wellbore storage demonstrates no coupling of the well with secondary porosity. However, the operator suspected the presence of some kind of fault not far from Well B, whose production test was unusual. He decided to acidize in an attempt to establish communication.

The original wellbore fluid was a sodium chloride brine. A 3500 meter-long coil tubing was lowered into the well, then most of the brine was displaced by 9.7 m^3 (61 bbl) of 15 % HCl, in circulating while pulling out the coil. There remained 3.7 m^3 (23 bbl) of brine at the bottom of the well. Then the wellbore fluids were squeezed into the formation by directly pumping 5.7 m^3 of water then 12 m^3 of 22 % HCl through the tubing. The surface rate and bottomhole pressure histories corresponding to this squeeze are given in Figure 4. The rest of the job is not provided, since ball sealers were then dropped into the well to divert the acid flow to low intake intervals. From this moment the skin curve is not interpretable anymore.

The skin history, given in Figure 5, shows six steps :
        - Brine injection (negative values of the volume): the relative skin is constant (no attack) and nil, which confirms the formation data provided by the operator.
        - 15 % HCl injection (0 - 5 m^3): the arrival of the acid is clearly visible and the onset of the attack fairly rapid.
        - The rest of the 15 % HCl injection (5 - 9.7 m^3) shows a much more rapid skin decrease; this is interpreted as being the sign of the establishment of communication between the wellbore and some adjacent fissures by propagating wormholes.
        - Water injection (9.7 - 15.4 m^3), demonstrating a very erratic response at a more or less constant relative skin of -2.2.
        - 22 % HCl stage (15.4 - 27.4 m^3), resuming a very sharp decrease of the skin, whose absolute final value is equal to -3.2,

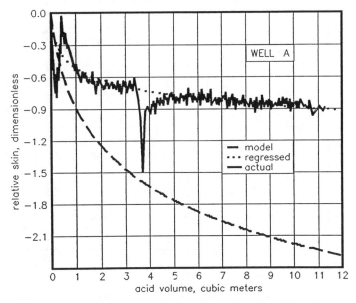

Figure 3 : Comparison of actual, regressed and model skin curves during treatment of Well A.

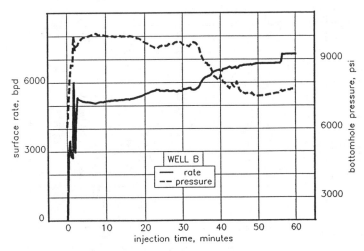

Figure 4 : Surface rate and bottomhole pressure evolutions during treatment of Well B.

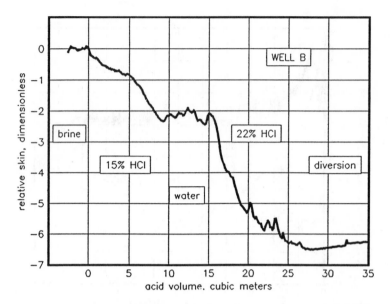

Figure 5 :   Actual skin evolution during treatment of Well B.

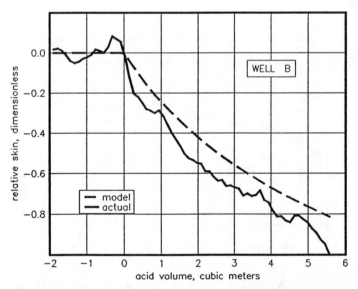

Figure 6 :   Comparison of actual and model skin curves during
wormholing of 15 % HCl (first acid stage of Well B treatment).

showing that the objective of the treatment was already apparently fulfilled. However, the flow equation used to derive these data are merely adapted to primary porosity reservoirs. Because communication has been established with some kind of secondary porosity, they are not applicable anymore, so that such negative values of the absolute skin have to be considered with extreme care.

- Onset of the diversion stage (27.4 m^3), with the skin now increasing due to the closure of some perforations by ball sealers (the rest of the curve would have shown that the relative skin value reached -2.5 at the end of the diversion stage).

Figure 6 is a scale-up of the first two steps of this skin history (solid curve), relevant to the present study. It corresponds to the injection of brine and 15 % HCl in the primary porosity zone surrounding the wellbore, before communication was established with adjacent fissures and/or faults.

Again, according to Equations 6 and 13, the theoretical response during the acid stage, if wormholing were to take place, would be :

$$\Delta S_{TH} \;=\; -0.625 \; \log(1 + 0.48 \times V) \tag{16}$$

The curve corresponding to this expression is given in Figure 6 (dashed). Though not perfect, the agreement between the two curves validates Equation 6. The best fit with actual data, when fixing $1/df$ equal to 0.625, would give $A' \simeq 0.65$. Such a discrepancy is acceptable, keeping in mind the imprecisions attached, on one hand, to flow equations in heterogeneous reservoirs, computations of friction pressure drops in tubings and, on the other hand, to the determination of treatment and reservoir data (mainly ϕ and D).

Conclusion

Because of the highly unstable nature of the acid attack in most of the carbonate reservoirs (propagation of wormholes), the development of a descriptive model of the skin evolution was not possible until the recent advent of the theory of fractals. In addition, the characteristics of the damaged zone greatly affect the behavior of the skin during acid injection in any type of reservoir, but particularly in carbonate ones. A correct validation of descriptive models stems from either the perfect description of the damage, or from its absence : hence the choice of the two described cases.

Two examples are certainly not sufficient for validating a model. However, some preliminary conclusions can already be drawn :

- the orders of magnitude are respected in the case of primary porosity reservoirs ; therefore the model could be used to describe unstable acid attacks in damaged primary and double porosity formations ;

- the fractal nature of wormholes, assessed by laboratory studies, seems to scale up at formation size according to the present model ;

- secondary porosity formations do not obey the proposed modeling equations, as was expected.

Literature Cited

1. Pettijohn F.J. Sedimentary Rocks; Harper and Row Publishers; 3rd Edition, 1975; chapter 10, 317.

2. Piot B.M.; Liétard O.M. In Reservoir Stimulation; Economides M.J.
    and Nolte K.G. Ed.; Schlumberger Educational Services, Houston,
    1987; chapter 12, ''Nature of Formation Damage''.
3. Dake L.P. Fundamentals of Reservoir Engineering; Elsevier, 1978;
    chapter 5, 149.
4. Horton H.L.; Hendrickson A.R.; Crowe C.W. API Division of Produc-
    tion 1965, paper number 906-10-F.
5. Nierode D.E.; Williams B.B. SPE 1973, paper number 3101.
6. Schechter R.S.; Gidley J.L. AIChE J. 1969, 15, 3, 339-50.
7. Daccord G.; Lemanczyck R.Z. In Reservoir Stimulation; Economides
    M.J. and Nolte K.G. Ed.; Schlumberger Educational Services,
    Houston, 1987; chapter 13, ''Acidizing Physics''.
8. Lund K.; Fogler H.S. Chem. Eng. Sci. 1973, 28, 691-700.
9. Daccord G.; Touboul E.; Lenormand R. SPE 1987, paper number 16887,
    to appear in SPE Prod. Eng., February 1989.
10. Daccord G. Phys. Rev. Lett. 1987, 58, 479-82.
11. Daccord G.; Lenormand R. Nature 1987, 325, 41-3.
12. Lund K.; Fogler H.S. Chem. Eng. Sci. 1976, 31, 381-92.
13. Lund K.; Fogler H.S. Chem. Eng. Sci. 1975, 30, 825-35.
14. Hoefner M.L.; Fogler H.S. AIChE J. 1988, 34, 45-54.
15. Prouvost L.; Economides M.J. Pet. Sci. Eng. 1987, 1, 145-54.
16. Timmerman E.H. Practical Reservoir Engineering; PennWell Books,
    1982; Vol. 1, 85.

RECEIVED December 12, 1988

Chapter 35

A Unique Source of Potassium for Drilling and Other Well Fluids

John T. Patton[1] and W. T. Corley[2]

[1]Department of Chemical Engineering, New Mexico State University, Las Cruces, NM 88003
[2]Mayco Wellchem, Inc., 1525 North Post Oak Road, Houston, TX 77055

Laboratory and field tests of a clear well fluid, formulated with tetrapotassium pyrophosphate, TKPP, indicate increased productivity from inhibition of clay swelling, sloughing and dispersion. Adding small amounts of multivalent cations to TKPP fluids further suppresses clay-water interactions. TKPP completion fluids can be converted to superior drilling fluids by viscosifing with xanthan gum. Handling characteristics of TKPP solutions are superior to most other well fluids. They are not acidic, nontoxic and non-corrosive, a definite benefit for both rig personnel and the environment. They are compatible with most formation fluids, but will precipitate when mixed with spent acid. Unlike well fluids containing zinc and calcium, pH can be adjusted. TKPP is available to petroleum operators worldwide as free-flowing powder or a 60% wt. solution.

Since the invention of rotary rigs, used for drilling oil and gas wells, scientists have been searching for fluids that will both facilitate the drilling process and provide maximum productivity once the well is completed. There are two prime requisites for all well fluids. These are:

(1) They must perform their required functions in a superior manner; e.g., completion fluids must balance the formation pressure and, thus, avoid the possible loss of the well due to blow-out. Drilling fluids should minimize the cost of making hole.

(2) Filtrate lost to the formation or particulates deposited in oil flow passages must not impair a well's flow potential when the well is returned to operation.

The most sensitive test for identifying damage characteristic of a well fluid involves flowing the well fluid through a reservoir core. For fluids containing particulates, it is necessary to test both the whole fluid and the filtrate

0097–6156/89/0396–0621$06.00/0

in order to assess the total damage potential and identify the mechanisms responsible.

Clay minerals, inevitably present in a petroleum formation, are sensitive to the type and concentration of ions contained in the well fluid filtrate lost to the reservoir. This sensitivity is demonstrated by a reduction in the permeability caused by the well fluid filtrate flowing through the core under investigation.

Clays have the capacity to exchange ions with well fluids. When calcium ions, incorporated in the clay matrix, are exchanged for sodium or even potassium ions, clays can swell and/or become dispersed in the fluid. This mechanism significantly reduces permeability, i.e., causes large productivity damage[1,2]. The equilibrium concentration of the various ions strongly favors the adsorption of multivalent ions by the clay minerals. For this reason, it is possible to counteract the swelling effect of well fluids on clays by the addition of small amounts of polyvalent ions. Calcium is by far the most effective, but the least soluble in the presence of other salts. Magnesium has been found to be advantageous and also synergistic when used in conjunction with calcium ions. Fluids containing divalent ions usually cause less permeability damage in petroleum reservoir cores.

A second cause of reduced productivity is mechanical blockage. When the well fluid contains particulates, damage occurs when the solids become tightly wedged in the tiny interstices through which oil flows. This damage, which can be so severe as to render a well uneconomic, has prompted many investigators to regard particulate blocking as the prime cause of productivity impairment[3,4]. The now popular "clear" well fluids were introduced to avoid particulate damage. These fluids are often viscosified to minimize leak-off[5].

Both field and laboratory results indicate that formation damage can only be minimized through the selection of well fluids whose chemical composition is formulated to avoid both fluid and particulate damage. This objective has been difficult if not impossible to achieve with current well fluid additives. Recognition of the chemistry required to formulate such a nondamaging well fluid is not in itself sufficient to prevent damage. Kruger has observed that strict quality control of a fluid's physical and chemical properties, especially during the completion process, is essential. In addition, damage can be further minimized through the use of clean pipe and continuous circulation and filtration of the fluid being used in the well operation[6].

What has long been sought and badly needed is a soluble weighting agent less corrosive than, and equally non-damaging as, the popular calcium and zinc halides. Preliminary laboratory and field data indicate TKPP may be the solution.

Work by Jones identified the desirability of incorporating TKPP as a scale inhibitor for packer fluids[7]. Subsequent work by Siskorski identified TKPP's superior properties in a blend for preparing high density, clear well fluids[8]. TKPP has long been recognized as a builder for liquid detergents. The fact that detergency is not a requisite property of a superior well fluid probably obscured excellent petroleum applications until now. This paper addresses the opportunities that exist for preparing improved well fluids, especially for wide application in drilling, based on TKPP.

GENERAL:

TKPP ($K_4P_2O_7$) is a merchant chemical produced by many major companies throughout the world. It is available commercially as a free flowing, white, granular solid or as a 60% wt. solution. In addition to its widespread use in the liquid detergent formulations, TKPP is also utilized in the food industry. It has been certified by the U. S. Department of Transportation as nontoxic. The solubility of TKPP in water at 60°F is shown in Table I. It is possible to achieve increased density at temperatures above 60°F because of increased solubility.

Table I. Concentration - Density of TKPP Solutions

Density lb/gal @ 16°C	TKPP Wt %	Concentration lb/bbl
9.0	8.7	33
9.6	16	64
10.7	28	126
12.0	41	206
14.0	57	335
15.2 (Solubility limit)	66	422

DAMAGE ASSESSMENT; CORE TESTS:

Tests were run to compare conventional completion fluids, i.e., those containing calcium, zinc, and potassium salts. Each well fluid was passed through a five micron filter and then flowed through a fresh Berea sandstone core having a brine permeability of approximately 800 md. Prior to testing a well fluid, the core's initial permeability had been established by flowing a filtered reservoir brine through the core. The damage imparted by a filtered well fluid was determined by noting the decrease effected in the core's permeability compared to its initial permeability.

The results, shown in Table II, quantify the adverse effect of common well fluid ions on permeability. Especially relevant are the data indicating that even calcium chloride/calcium bromide fluids cause some permeability reduction, although the cause of the adverse interaction is obscure.

With the exception of saturated zinc bromide, all of the fluids tested minimized permeability damage relative to fresh water which reduced permeability an average of 54% for three cores.

Table II. Formation Damage Effected by Clear Well Fluids

| Fluid | | Ion Conc., wt. % | | | Core Perm., md | | Damage |
Type	K	Ca	Mg	Zn	Init.	Final	%
TKPP [+]	3.8	.04	.41	–	874	844	3
TKPP	3.8	–	–	–	837	718	14
TKPP+	11.3	.04	.41	–	874	738	16
CaCl$_2$/CaBr$_2$	–	.16	–	–	711	583	18
KCl	1.6	–	–	–	794	625	22
TKPP	11.3	–	–	–	837	580	32
CaCl$_2$/CaBr$_2$	–	.10	–	–	855	581	32
KCl	5.2	–	–	–	794	531	33
KCl	11.3	–	–	–	794	381	52
ZnBr$_2$	–	–	–	.22	673	34	95

Although potassium chloride fluids performed better than the calcium and zinc halides, damage was still measurable. These results, confirmed in triplicate, were unexpected since it is well accepted that even 0.5 weight % KCl should protect Berea cores against permeability damage. The most plausible explanation lies in variation between the test procedures.

The formation brine used to establish the core's initial permeability contained 2.7% total dissolved solids, TDS, with a monovalent-divalent (calcium) ratio of 30. Once a core is equilibrated with this brine, any increase in the ratio or drastic decrease in TDS has the potential for decreasing permeability. Obviously fresh water represents a significant decrease in TDS and, hence, the 54% permeability damage. Adding KCl helps overcome the decreased salinity but, in so doing, increases the ionic ratio resulting in still measureable but usually reduced permeability damage.

Even very small amounts of calcium provide a desirable decrease in the Na/Ca ratio. Prior studies indicating potassium chloride totally negates permeability reduction may have utilized water that contained some small amount of calcium ion to measure KCl solution permeability. A second factor, which might explain the lack of KCl damage reported in prior studies is a low ionic concentration, especially calcium, in the water used to equilibrate the cores prior to the KCl tests.

Increasing KCl concentration lowers inhibition as shown in Table II. The fact that damage increased with KCl concentration is consistant with the ionic ratio hypothesis and suggests a base exchange mechanism whereby calcium ions are more easily extracted from the clay and replaced by potassium ions as the potassium concentration increases.

Similar tests were performed with 10% and 20% by weight aqueous solutions of TKPP having approximately the same potassium ion concentration as the KCl tests. Damage, although measurable, was reduced, indicating for the first time that an anion, in this case pyrophosphate, can contribute to reducing permeability damage.

The damage potential of TKPP fluids was further reduced by the addition of multivalent cations. When these salts were added at approximately their solubility limit in TKPP solutions, damage decreased, lending additional credence to the observation that the addition potassium ions alone is not sufficient to completely eliminate formation damage.

The severe interaction of the zinc bromide fluid, 19.2 ppg (2.32 g/cc), was unexpected. Severe plugging of the core occurred, caused by precipitation of zinc hydroxide, as the injected solution mixed with and was neutralized by formation brine. Tests in which the zinc bromide fluid was simply titrated with distilled water also produced a precipitate, 0.0036 g/cc.   Titration in the presence of the common reservoir clay, montmorillonite, increased both the rate of precipitation, and total quantity to 0.03 g/cc.

At the conclusion of a few selected damage tests, an additional experiment was performed in which solutions containing 10% TKPP, 0.1% calcium chloride, and 0.4% magnesium chloride were pumped through the damaged cores. In each instance, the permeability of the core recovered dramatically, as shown in Figure 1.

DAMAGE ASSESSMENT; CLAY SWELLING:

Tests were conducted to determine the effect of TKPP solutions on the hydration and dispersion tendencies of clay particles. The test involved immersing a standard bentonite Volclay pellet in the solutions of TKPP and observing the degree of disintegration as a function of time. Comparable tests were also conducted using fresh water, halide completion fluids, and diesel oil for comparative purposes. TKPP solutions having densities ranging from ten to fifteen pounds per gallon, ppg, were tested to define the effect of TKPP concentration on clay inhibition. These tests indicate that TKPP, when used at higher solution densities, can inhibit the hydration and disintegration of water-sensitive clays almost completely. When used in conjunction with polyvalent cations, inhibition is evident at concentrations as low as 75 ppb (0.21 g/cc). Even at this low concentration the inhibitive character of the fluid is much superior to that of the calcium bromide and zinc bromide clear completion fluids. This can be seen from the data shown in Table III.

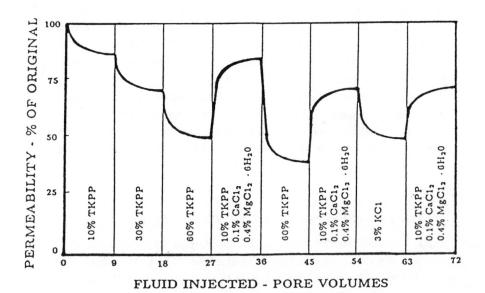

Figure 1.   Composition and Concentration Effects of Potassium
Based Clear Well Fluids on Core Permeabilities.

Table III. Clay Disintegration Due to Fluid Interaction

Time, Minutes	Well Fluid Properties	Disintegration State of Volclay Tablet					
		Conventional			Tetrapotassium Pyrophosphate		
	Salt Type	CaBr$_2$	ZnBr$_2$	KCl	TKPP*	TKPP*	TKPP*
	K$^+$, ppm.	0	0	127,000	81,000	128,400	176,000
	Density, ppg.	14.2	19.2	9.7	9.7	10.6	11.6
0.5		2	2	2	1	1	1
1.0		3	2	2	1	1	1
5.0		3	3	3	2	2	1
10.0		3	3	3	2	2	1
14.0		3	3	3	3	2	1
18.0		3	3	3	3	3	1
30.0		3	3	3	3	3	2

1 - No Visible Effect   2 - Cracks/Sloughing   3 - Complete Disintegration
*TKPP solutions contain 0.21% CaCl$_2$ and 0.65% MgCl$_2$

At solution densities of 13 ppg and above, the TKPP virtually elimi-
nates any hydration and disintegration effects an aqueous fluid might have on
water-sensitive clays. In this concentration range, the results of TKPP plus
polyvalent cations are comparable to those obtained with a non-aqueous fluid
such as diesel oil. However, as evidenced by the permeability regain effects
noted in damaged sandstone cores, TKPP solutions invading the producing
formation adjacent to the well bore might have the capacity to actually remove
damage whereas the diesel oil has no ability to restore permeability in a similar manner. If so, the responsible mechanism would be the dehydration and
subsequent shrinking of clays by the highly structured, concentrated TKPP
solution.

TKPP IN DRILLING FLUIDS:

The long-recognized ability of potassium ions to minimize the swelling and
disintegration of water-sensitive clays would infer that TKPP should be a de-
sireable additive for drilling fluids[9]. Current practice is to formulate drilling
fluids with KCl or KOH to supply the required potassium ion concentration.
However, both chemicals generate problems that are often expensive to overcome. Potassium hydroxide is little used because of the high pH (above 12.0)
and loss of desirable mud properties when enough is added to supply the requisite amount of potassium ion. It is good practice to avoid pH's above 11.5

because of danger to rig crews, increased problems in controlling shale slough-ing or clay swelling and difficulty in maintaining acceptable mud rheology. Therefore, the use of KOH is generally limited to pH control for KCl muds.

There are two major problems associated with adding KCl to a drilling fluid. First, the corrosivity of the fluid is increased, often several fold[10]. The increased corrosion of drill pipe, drill collars, bits, and pumps can be a significant expense.

Of even more economic importance is the rheological impact of the ad-dition of KCl to conventional water-base drilling fluids. KCl causes undesirable increases in both yield point and gel strength that can only be eliminated by chemical disperants or by dilution with fresh water. Dilution in turn requires more KCl for clay inhibition, and the cycle continues with mud costs escalating exponentially.

Tests were conducted to compare the effect of KCl and TKPP on the rheology of an unweighted (nondispersed) and a weighted (dispersed) drilling fluid. All fluids were mixed in an industry standard Hamilton Beach mixer. The rheology of each fluid was then measured in a Fann viscometer at 120°F at 300 rpm and 600 rpm.

The results of these tests and the fluid formulations are shown in Table IV. The adverse effect on fluid rheology of adding potassium, KCl, Test B, and TKPP, Test C, is evident. Both salts cause increases in yield point, YP, and initial gel strength. TKPP produced only about half as much increase as did KCl and can be added in much higher concentrations before the rheological properties of the drilling mud become unacceptable.

Similar results were obtained with a barite-weighted, dispersed drilling mud. The base fluid, Test D, has acceptable rheology; however, the addition of only 10 ppb of KCl, Test E, elevated the yield point and initial gel to un-acceptable levels. Although the addition of TKPP, Test F, caused increases in both YP and gel, they were only about 75% as great, as increases caused by KCl. This reduced adverse effect translates into a significant reduction in mud costs, as detailed in the results of actual drilling in experimental wells. The decrease in plastic viscosity for the weighted system, Test F, is benefi-cial. It indicates an increased pseudoplasticity that, in practice, will provide better hole cleaning and more horsepower available at the bit during drilling operations.

When suitably viscosified, TKPP solutions become superior low solids drilling fluids. Many water-soluble polymers were tested to identify satis-factory viscosifiers. Most commercially available polymers were found to be insoluble in TKPP solutions at densities above 11 ppg. Only xanthan gum

Table IV. Effect of Potassium Salts on Conventional Drilling Muds

Composition/Property	A	B	C	D	E	F
Concentration, ppb						
Gel	15	15	15	15	15	15
Drill Solids	30	30	30	30	30	30
Chrome lignosulfonates	0	0	0	6	6	6
K Lignite	0	0	0	3	3	3
Barite	0	0	0	216	216	216
KCl	0	11	0	0	10	0
TKPP	0	0	11	0	0	10
Weight, ppg	9.0	9.0	9.0	12.0	12.0	12.0
Plastic Vis., cp	7	4	5	17	17	15
YP, lbs/100 ft^2	6	23	11	4	54	39
Initial Gel lbs/100 ft^2	2	10	6	2	42	30

exhibited enough solubility to be useful in viscosifying TKPP solutions up to 14.5 ppg. This common drilling polymer is available in both powder and liquid form in the oil field. Its use is well known and does not require any special techniques or mixing equipment.

A series of TKPP solutions were prepared; varying in weight from 10 to 12 ppg, these were viscosified with 2 ppb of Kelzan XC, a commercial xanthan drilling product. The rheological properties were measured at 120°F using a standard Fann Viscometer. Table V summarizes the rheological properties of the solutions so prepared. It was found that hydration of xanthan is retarded by high concentrations of TKPP. When preparing solutions having a density above 12 ppg, it is advisable to hydrate the xanthan in relatively fresh water first. After the xanthan is completed hydrated, granular TKPP can be added to achieve the desired solution density. The presence of the xanthan does not retard the rapid dissolution of the TKPP.

Table V. Rheology of TKPP - Xanthan Drilling Muds

TKPP Conc.		Density		Apparent	Plastic	Yield Point
ppb	Wt%	ppg	g/cc	Visc., cp	Visc., cp	lb/100 Ft.^{2}
0	0	8.33	1.00	15.5	6	19
86	20	10	1.20	26.5	7	23
145	31	11	1.32	31	10	26
206	41	12	1.44	42.5	19	33

There may be instances where a high concentration of TKPP is required to inhibit clay hydration, but a lower density or higher lubricity is desired. This can be achieved by emulsifying diesel or mineral oil in a TKPP solution viscosified with xanthan. Emulsion of oil in TKPP solutions can be achieved with a viscosifier such as xanthan alone although it is usually desirabale to add a surfactant to improve stability. Many different surfactant structures were tested for suitability as an emulsifying agent for oil in TKPP solutions. The best additive identified is a phosphate ester surfactant, Pluradyne OF-90, manufactured by BASF Wyandotte. This material has been tested in the past for improving producing well productivity by reversing clay swelling and mitigating paraffin deposition problems.

A TKPP solution having a density of 12 ppg was formulated by dissolving 1.5g of Kelco's KOD-85, a finely ground commercial xanthan gum product, in 299 cc tap water by mixing at high speed for 5 minutes in a standard Hamilton Beach mud mixer. 206 g of TKPP, followed by 18 ml of diesel oil, was then added and the viscosified solution blended for an additional 5 minutes with a Hamilton Beach mixer. A smooth, water-continuos emulsion formed immediately that contained a small amount of entrained air. One percent of BASF Pluradyne OF-90 was added to the emulsion and blended for an additional 5 minutes. The resulting fluid was a smooth, creamy, drilling fluid containing finely dispersed air bubbles which broke out slowly. After standing overnight, the fluid was a homogeneous, water-containing emulsion containing no entrained air. Fluid rheology measured at room temperature in a Fann Viscometer. produced the following results:

Apparent vis.=18.5 cp,   Plastic vis.=13.0 cp,   Yield point=11 lb/100 ft.^{2}
These values are characteristic of a superior, low-solids, drilling fluid that promotes a high drilling rate and good solids removal. In addition, maximum inhibition of hydrating clays is provided.

DRILLING FIELD TEST

Two wells were drilled in Frio County, Texas, using TKPP for clay/shale inhibition. They were compared to three wells that had just been finished in the same field, to the same depths, using the same rig and crews; all three using a KCl/polymer drilling fluid. The KCl polymer wells experienced numerous hole problems including balling, stuck pipe, poor hole cleaning (gel sweeps were necessary), high torque, drag, and washout. The average mud cost for these KCl polymer wells was $45,276. Two wells, drilled using TKPP and polymer, experienced no hole problems. The average mud cost for the two wells was $13,952, a 69% cost reduction in mud alone. Dollar savings on rig time were even more significant. The new additive was used on six wells in

Lavaca County, Texas. Once again, the wells were free of the hole problems experienced in preceeding offset wells.

COMPATIBILITY OF TKPP SOLUTIONS:

Solutions of TKPP were mixed with aqueous fluids commonly encountered in drilling or completion of wells. Unlike saturated zinc bromide, concentrated TKPP solutions can be mixed in any proportion with fresh water with the only result being a decrease in solution density. Similar results were obtained with conventional oil field brines containing as much as 400 parts per million polyvalent cations, mostly calcium. Saturated solutions of calcium hydroxide also can be added to TKPP in any proportion without promoting precipitation as can concentrated hydrochloric acid solutions, conventionally used for well stimulation. The acid tends to generate a slight haze as the pH is reduced from 11.5 to approximately 8; however, this haze rapidly disappears as the pH is lowered by further addition of acid.

The only fluid, common to oil field operations, that has a significant interaction with TKPP solutions was concentrated calcium chloride. Solutions of calcium chloride, spent acid, are generated during the acidization of a limestone or a dolomite formation. When solutions containing 10% calcium chloride were mixed in equal proportions with 14.5 ppg TKPP solutions, massive precipitation occurred. Similar precipitation was observed with oil field brines having calcium concentrations above 400 ppm.

In normal operations, there is little chance for spent acid to contact the completion fluid as the well will usually be produced after perforation, effecting the removal of completion fluid prior to acidization. The fact that a calcium precipitation reaction can occur should be recognized by those using TKPP solutions as a clear completion fluid in well operations. A KCl spacer is recommended to avoid completion problems in formations having high calcium brines.

Not only should well fluids be compatible with reservoir fluids and minerals, but they must also be stable at surface conditions. One disadvantage of conventional halide fluids is their tendency to crystallize at ambient temperatures. Solutions containing varing amounts of TKPP were exposed to constant temperatures as low as -76°F (-60°C) for a period of 30 hours. No crystallization was observed at any temperature above 36°F (2°C) even at concentrations as high as 60% by weight. At low fluid densities, less than 11.7 ppg, there is no evidence of TKPP crystallization even at temperatures as low as 30°F(-1°C), the freezing point of most fresh water drilling fluids.

The addition of TKPP lowers the freezing point of the fluid significantly. This offers the opportunity of using a subfreezing TKPP mud to drill through strata, i.e., the permafrost in Alaska and Canada, without thawing of the formation. Problems in stability of the drilled hole as well as the rig foundation could thus be minimized.

WELL LOGGING:

Another advantage, unique to using TKPP in drilling fluids, would be its contribution to more accurate log interpretation. The popular neutron lifetime logs respond to the presence of chloride ions in the formation pore space. The addition of TKPP does not add any chloride ions, and the drilling fluid reacts to neutron logs much like conventional fresh water muds. The contrast in log response between TKPP and chloride fluids is also useful in the potential log-inject log procedure for determining porosity and oil saturations. As experience is gained in logging wells drilled with TKPP drilling fluids, additional benefits may become apparent.

CORROSION:

One property of drilling fluids that has received only minimal attention in the past is corrosiveness imparted by various additives. Additives improve the requisite properties of well fluid substantially; and, hence, increase in corrosiveness is usually ignored.

This is particularly true with respect to drilling fluids containing potassium chloride to inhibit shale hydration. In 1974 Bodine, et al., recognized the potential of the phosphate ion to combat the increased corrosion rate produced by potassium chloride[11]. Accelerated corrosion is especially severe at the elevated temperatures now being encountered in deep wells and geothermal reservoirs. In most systems, phosphate corrosion inhibition is enhanced in solutions containing oxygen[12]. All drilling fluids are constantly aerated; and, hence, it should be no surprise to find TKPP-based fluids with reduced corrosion rates one or more orders of magnitude lower than fluids containing the same amount of potassium ion introduced by the addition of potassium chloride.

Radenti et al. reported the corrosion rate of a typical potassium chloride fluid of 247 mils/year at 212°F. In contrast, they found by substituting potassium carbonate for potassium chloride, the corrosion rate was reduced to 3 mils/year[10]. Unfortunately, potassium carbonate is not optimum as a drilling fluid additive because it can produce massive amounts of calcium precipitation, may elevate the pH to undesirable levels, and in all cases reduces the calcium ion concentration to such a low level as to promote destabililzing cation exchange with clay minerals.

Corrosion rates for TKPP solutions measured at 194°F were low, 2 mils/year. Data at 400°F are equally impressive, 14 mils/year for 1020 steel

coupons exposed to 14.5 ppg TKPP solutions for 60 days. Of special significance is the fact that TKPP reacts with steel to form a protective coating
of potassium iron pyrophosphate. This coating protects the steel surface, and
corrosion rates after 60 days were less than 2 mils/year even at 400°F.

Although the importance of considering corrosion in the formulation
of a drilling fluid is of secondary importance for shallow wells, the effect is
still unmistakable. As drilling continues to deeper, hotter horizons, the more
expensive rigs and tubular goods, such as those utilized in offshore operations,
magnify the importance of considering corrosion in formulating an optimum
drilling mud.

Paralleling the corrosion problem is one involving compatibility of any
well fluid with nonmetallic materials used in well completion apparatus. All
injection wells and many producing wells are equipped with packers to isolate
the casing annulus from the high temperature, pressure, and salinity characteristic of the petroleum reservoir environment. Conventional packers, as well
as other well tools, utilize elastomeric materials to mechanically seal appropriate locations.

At the request of an international petroleum company, a major manufacturer and supplier of down-hole equipment performed tests of the various
elastomers commonly used in the construction of packers and other oil field
tools. Seven of the nine most commonly used thermoplastic materials were
found to be completely inert to TKPP solutions. The test included continual
immersion in saturated TKPP for 21 days at 280°F. Only two elastomers, Vi-
ton and Fluorel, showed any adverse reaction. O-rings made from these two
elastomers showed minor cracking at the termination of the test. A listing of
the elastomers that tested inert to TKPP solutions include nitrile, saturated
nitrile (HNBR), Aflas, Kalrez, PEEK, Glass-filled Teflon, and Ryton. Several
of these elastomers are attacked or degraded by conventional clear completion fluids containing calcium and zinc halides. The inertness of commonly
employed elastomers to TKPP is an important advantage for TKPP fluids in
normal operations.

ENVIRONMENTAL CONSIDERATIONS:

Governmental regulations relative to acceptable environmental practices have
multiplied many fold in recent years. The impetus for these new regulations
comes from an increased awareness and public concern for the value of maintaining the existing quality of our environment. There are many additives
for well fluids, now being phased out, which were adopted without adequate
regard for the effect accidental spills and disposal operations might have on
fish and other forms of wildlife.

An example of some of the newer regulations is the restriction against the use of the popular mud dispersant, chrome lignosulfonate. It is expected that this regulation is merely the initial step toward ruling out the use of all heavy metal salts commonly employed in the formulation of well fluids because of their toxicity to aquatic life and humans. This means that the use of zinc and lead, in addition to chromium, may not be allowed in the future. At least one major oil company has already taken steps in this direction by ruling out the use of heavy metal salts in any well fluid in their worldwide operations.

Although at high concentration, TKPP is toxic to fish and shrimp. When diluted, 500-1000 PPM, it is non-toxic and expected to be both environmentally acceptable and of low risk to humans. The widespread use of TKPP in detergent formulations supports this expectation. It is listed by the U. S. Department of Transportation as a nontoxic substance relative to shipping regulations.

CONCLUSIONS:

Solutions of TKPP have been shown to have unique and advantageous properties for use in formulating a wide variety of well fluids. Its reasonable cost, worldwide availability, and nontoxic properties make it a preferred additive for use in many petroleum applications. It has been shown to be a most effective salt with respect to inhibiting hydration and swelling of clay minerals commonly encountered in drilling operations and/or reservoirs. Avoiding clay problems is the major impetus for the incorporation of potassium ions in well fluids, and the use of TKPP provides advantages over and above those available from other potassium salts.

The use of TKPP solutions as drilling fluids for clay inhibition and increased density appears to be the prime application, at least initially. Excellent rheology and fluid loss is obtained by viscosifing with xanthan. Higher density, above 15 ppg, can be achieved by incorporating finely ground weighting agents, such as barite or hematite, in the drilling fluid. As a source of potassium ions, the addition of TKPP to a conventional drilling fluid for clay control has been shown to produce less of an adverse effect on rheology than the addition of a comparable amount of potassium chloride.

Perforating fluids are conventionally prepared with soluble salts for weighting and usually filtered to avoid formation damage. Many operators perforate with the hydrostatic pressure slightly below the expected formation pressure (under balanced) to eliminate fluid loss and subsequent formation damage after the well is perforated. The nondamaging character of TKPP perforating fluids reduces the need for under balance and provides a higher degree of safety in perforation operations.

Gravel pack completions involve placing an annular layer of gravel between the liner and the producing formation. The gravel is carried into the annular space, suspended in a clear fluid, usually viscosified with hydroxyethylcellulose (HEC). Much of the carrying fluid is lost, often purposely, to the formation and, hence, should be formulated to be nondamaging. Solutions of TKPP viscosified with xanthan are outstanding in this respect. HEC cannot be used with high concentrations of TKPP.

Packer fluids are used to provide hydrostatic balance to partially offset the reservoir pressure and eliminate a large pressure differential across the packer element. TKPP solutions more than meet the requirements of being noncorrosive, nondestructive to elastomers, variable in density, and stable for many years. It is desirable to perforate with a well bore filled with packer fluid so that the packer can be set and the well produced as soon as the perforation operation is complete. Since TKPP solutions make superior perforation fluids, their dual use as a packer fluid is further enhanced.

Work-over fluids are used routinely to kill wells for remedial operations, wash-out fill, or provide a safe environment for special logging or other well diagnostic procedures. In such operations it is often necessary to store the work-over fluid in tanks at the well site. TKPP solutions are stable even at sub-freezing temperatures, which provides a distinct advantage over solutions of halide salts that sometimes crystallize at ambient conditions encountered in rig operation. The avoidance of a crystallization problem coupled with the noncorrosive nature of TKPP work-over fluids makes them attractive with respect to other clear work-over fluids now popular in the industry.

LITERATURE CITED:

1. Jones, F.O., Jr. J. Pet.Tech. April 1964, 441-446.
2. Mungan, N. "Permeability Reduction through Changes in pH and Salinity," J. Pet. Tech. Dec. 1965, 1449-1453.
3. Glenn, E.E.; Slusser, M.L. J. Pet. Tech. May 1957, 132-39.
4. Tuttle, R.N.; Barkman, J.H. J. Pet. Tech. Nov. 1974, 1221-1226.
5. Chatterji, J.; Borchardt, J.K. J. Pet. Tech. Nov. 1981, 2042-2056.
6. Krueger, R.F. J. Pet. Tech., Feb. 1986, 131-152.
7. Jones, L. W. U. S. Patent 3 481 869, 1969.
8. Sikorski, C. F. U. S. Patent 4 521 316, 1985.
9. Steiger, Ronald P. J. Pet. Tech. Aug. 1982, 1661-1670.
10. Radenti, G., Palumbo, S.; Zucca, G. Pet. Eng. Intr. Sept. 1987, 32-40.
11. Boding, O. K.; Sauber, C. A. U. S. Patent 4 000 076, 1976.
12. Uhlig, H. H. I & E Chem., 44, 1952.

RECEIVED January 27, 1989

Chapter 36

Impedance Spectroscopy

A Dynamic Tool for the Design of Corrosion Inhibitors

F. B. Growcock[1] and R. J. Jasinski

Dowell Schlumberger, P.O. Box 2710, Tulsa, OK 74101

The corrosion of steel in HCl, with and without inhibitors, shows relatively straightforward impedance spectroscopy (IS) phenomenology and can be represented by simple equivalent circuits of primarily passive electrical elements. Inductive effects are, apparently, a consequence of the measurement altering the surface being measured. IS reveals that during steel corrosion in hot concentrated HCl, the heterogeneity of the surface is established rapidly and can be simulated with a single type of equivalent electrical circuit. Chemisorbing and electrostatic inhibitors in all cases reduce surface heterogeneity. At the same time, all of the inhibitors increase the charge-transfer resistance without producing a concomitant decrease in the interfacial capacitance. Time constant analysis suggests this arises from specific adsorption of the inhibitor to form insulating "islands", rather than uniform adsorption to give an electronically conductive monolayer. In some cases, polymerization of the inhibitor occurs subsequent to adsorption; this can lead to formation of a protective multilayer film atop the adsorbed monolayer, but, at the same time, result in significant loss of inhibitor in the solution.

Corrosion of steel during oil well acidizing or acid pickling treatments can be controlled effectively and economically with organic corrosion inhibitors. These additives interact with the steel surface to form an adherent barrier, the nature of which depends on the additives' physicochemical properties. Work to date has established that acetylenic alcohols chemisorb and subsequently polymerize on steel surfaces (1-5). α,β-Unsaturated aldehydes and α-alkenylphenones appear to behave in a similar manner (6,7). The nature of

[1]Current address: Amoco Production Company, Tulsa, OK

0097–6156/89/0396–0636$06.00/0

the chemisorption process is, however, not very well understood. In addition, although the polymer films appear to contribute to the long-term protection of the underlying steel, it is not at all clear how important polymerization is initially.

Electrochemical techniques have been utilized for many years to study metal corrosion. Two of these techniques, linear polarization (LP) and cyclic voltammetry (CV), complement each other, LP providing corrosion rates under conditions where the surface is minimally altered and CV furnishing information about the corrosion mechanism. With the advent of impedance spectroscopy (IS), both kinds of information can be gleaned simultaneously and more rapidly, while leaving the surface almost intact. In this paper, we discuss the application of IS to the study of rapid steel corrosion and describe a study we undertook to elucidate the roles played by adsorption and film formation in the inhibition mechanisms of the above-named compounds. For comparison, we also investigated two quaternary nitrogen salts, which appear to adsorb electrostatically and presumably do not form macroscopic films (8).

Experimental Approach

Materials. 1-Octyn-3-ol (98%, from Aldrich), trans-cinnamaldehyde (TCA - 99%, from Cofinil, Milan, Italy) and benzoyl allyl alcohol (BAA - 80% solution in isopropanol, prepared by D. G. Hill) were chosen to represent acetylenic alcohols, α,β-unsaturated aldehydes and α-alkenylphenones, respectively. A quinoline quat (QQ - 97%) and n-dodecylpyridinium bromide (DDPB - 80% solution in isopropanol), which were prepared as described in Reference (8), served as model quaternary nitrogen salts. The HCl solutions were prepared by dilution of reagent-grade 37% HCl (Mallinckrodt). API N80 steel, obtained from U.S.X. Corp., had a mixed structure of ferrite (α-Fe, grain size ASTM No. 12) and a fine aggregate of ferrite and cementite (Fe_3C); impurity levels were 0.34% C, 0.015% S, 0.010% P, 1.31% Mn, 0.02% Mo and 0.06% Si, with Cr, Ni and Cu below detectable levels.

Procedure. All IS experiments were conducted under a blanket of N_2 in a thermostatted IBM Instruments cell, using 20 mL of deaerated solution and a 3-electrode configuration. The Pt foil counter electrode measured 4.0 cm^2. The potential between the N80 steel rotating disk working electrode (WE - 0.20 cm^2, 1000 rpm) and the Ag/AgCl reference electrode (with Luggin capillary) was maintained at the open-circuit value with a Solartron 1286 ECI. The AC frequency was varied from 60 kHz to 0.01 Hz (10 mV peak to peak) with a 1250 FRA, 5 or 10 steps/decade using a 1-cycle minimum integration time (6 min total time). Some experiments were run only to 0.1 or 1.0 Hz (1 min total time) in order to minimize the scan period. The WE was ground to a 600-grit finish before most tests and wiped with a methanol-saturated soft cloth. Prior to immersion in the test solution, the WE was exposed to a stream of N_2 maintained at the solution temperature, so that the test temperature was reached and deaeration was complete within seconds after immersion.

Equivalent Circuit Analysis. IS measurements yield values of Z' and Z'', the real and imaginary components of the impedance, as a function of f, the AC frequency. The data are usually displayed as Nyquist plots (Z'' vs. Z') or Bode plots (impedance modulus, $|Z| = \sqrt{(Z'')^2 + (Z')^2}$, and phase angle shift, Φ, vs. f). The electrochemical system is then simulated with an electrical circuit that gives the same impedance response. Ideally this electrical circuit is composed of linear passive elements, e.g. resistors and capacitors, each of which represents individual physicochemical steps in the electrochemical reaction.

The impedance data were fitted to candidate electrical circuits using the non-linear weighted least-squares fitting program "EQIVCT" developed by Boukamp (9). Graphical analysis was utilized to furnish reasonable first guesses of the circuit parameters for input to EQIVCT.

Characterization of the Corroding Surface. Mechanistic studies of steel corrosion in aggressive environments, such as hot concentrated HCl, have to deal with a rapidly changing and, hence, poorly characterized surface. For example, exposure of a mild steel to uninhibited 15% (4.4 N) HCl at 65°C gives a typical corrosion rate of 1 $g/cm^2/day$, which corresponds to a surface removal rate of 150 A/sec. Exacerbating this problem is surface area development, which proceeds in a manner dictated by physical and chemical heterogeneities of the surface. Characterization of an inhibited steel surface also has some fundamental complications. Although the steel surface is changing less rapidly, the chemical and physical nature of the film, as well as the rate of film formation, depends on acid concentration, temperature and exposure time (6). Thus, the film itself is often a poorly characterized complex mixture of polymeric materials.

Nevertheless, we will show that all of the systems studied exhibited relatively straightforward electrochemical phenomenology and could be represented by simple equivalent circuits involving primarily passive electrical elements.

Corrosion of Steel in Uninhibited HCl

Consider first the corrosion of low alloy steel in HCl per se, i.e. before the addition of organic inhibitors. As shown in Figures 1 and 2 for N80 steel in 15% and 28% HCl at 65°C, Nyquist plots for steel in concentrated HCl typically have only one distinct feature: a single capacitance loop (a loop above the Z' axis) with a hint of a second capacitance loop at lower frequencies. The low-frequency loop is more fully developed in 28% HCl than in 15% HCl. Mass transport limitations are not evident except under extreme conditions, e.g. above 28% HCl and 65°C.

This impedance response, in general, is similar to that elicited from an Armstrong electrical circuit, shown in Figure 3, which we represent by $R_\Omega + C_d/(R_t + C_a/R_a)$. R_Ω is identified with the ohmic resistance of the solution, leads, etc.; C_d with the double-layer capacitance of the solution/metal interface; R_t with its resistance to charge transfer; and C_a and R_a with the capacitance and resistance

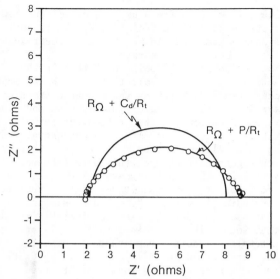

Figure 1. Nyquist plot of N80 steel in 15% HCl at 65°C.

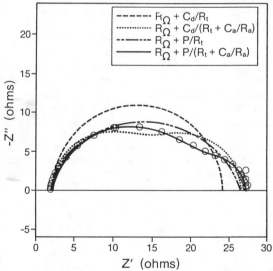

Figure 2. Nyquist plot of N80 steel in 28% HCl at 65°C.

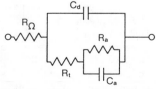

Figure 3. Armstrong electrical circuit.

associated with a slow (low frequency) intermediate process ($\underline{10},\underline{11}$). The total resistance at the limit of $f \to 0$, $R_t + R_a$, is equivalent to the corrosion resistance measured in a DC experiment and is inversely proportional to the net corrosion rate.

Harrington and Conway ($\underline{12}$) point out that in the case of the H_2 evolution reaction, C_a is equivalent to the H^+ adsorption pseudo-capacitance only when H^+ adsorption and electron transfer (the Volmer reaction) occur in an almost reversible fashion and electrochemical desorption or recombination is the rate-determining step. Although we believe such is the case for steel in concentrated HCl, the low-frequency loop is usually so poorly developed that neglect of the R_a/C_a sub-circuit generally changes the data fits very little. This is especially true for the 15% HCl data. The calculated value of C_a is usually 1 to 2 orders of magnitude higher than that of C_d, in agreement with studies in weaker HCl and consistent with an adsorption pseudocapacitance. But R_a is considerably smaller than R_t, opposite of that found in the other work ($\underline{10}$), and so contributes little to the total corrosion resistance. Thus, the data in many cases can be represented equally well by $R_\Omega + C_d/R_t$ and by the Armstrong circuit. Fits of the data in Figure 2 to these two circuits are shown in that figure for comparison.

The C_d-values (per unit surface area) are, in most cases, considerably higher than C_d-values expected for a perfectly smooth surface; for example, on mercury, $C_d \sim 20$-30 $\mu F/cm^2$ ($\underline{13}$), whereas our values range from 30 to 500 $\mu F/cm^2$. C_d is high because the steel surface is rough on a microscale, even with a 600-grit finish, so that the "true" surface area is actually greater than the geometric surface area used here. Corrosion develops surface roughness to an even greater extent; in fact, the true surface area increases continuously with exposure time. This is apparent from visual inspection of the surface. Thus, it is not uncommon to be working with a steel surface whose true surface area is several-fold greater than its apparent surface area ($\underline{14},\underline{15}$).

Surface roughness is also expected to result in depression of the capacitance semi-circle. This phenomenon, which is indeed apparent in both Figures 1 and 2, is, however, unrelated to surface area. Rather, it is attributable to surface heterogeneity, i.e. the surface is characterized by a distribution of properties. Macdonald ($\underline{16}$) recently reviewed techniques for representing distributed processes. A transmission line model containing an array of parallel R/C units with a distribution of values is physically attractive, but not practical. An alternative solution is introduction of an element which by its very nature is distributed. The Constant Phase Element (CPE) meets such a requirement. It has the form

$$P = Y_o \cdot w^n$$

where Y_o is a combination of properties related to both the surface and the electroactive species, and $n=1$ for a pure capacitance. In our case, we can substitute P for C_d and the value of n can be used as a gauge of the heterogeneity of the surface ($\underline{17}$). Thus, fitting the data in Figure 1 via EQIVCT to the CPE-containing circuit $R_\Omega + P/R$ generates a well-matched calculated curve (see Figure 1) with n = 0.84. Rammelt and Reinhard ($\underline{18}$) have treated the case of roughened polycrystalline iron in H_2SO_4 using a CPE in parallel with C_d and R_t; however, fits of our data to that circuit are very poor and generate n-values less than 0.5, which is not theoretically possible.

Although the true surface area increases with increasing exposure time, Figure 4 shows that the impact of heterogeneity on electrochemical measurements is established quickly and remains constant with time.

Utilizing the $R_\Omega+P/R_t$ circuit, we find that n reaches a relatively constant value almost immediately and changes very little over the next two hours. The n-values for different steels range from 0.76 to 0.92 at 1-min exposure, depending on the metal and the environment. The relative surface areas displayed in Figure 4 for comparison with n were determined from capacitance measurements of the steel in 4M LiCl at pH=1, immediately after the corrosion experiments. These results are consistent with Mulder and Sluyters' argument that the CPE is related to the fractal surface properties of the roughened steel (17). Furthermore, the near-constancy of the n-values implies that, at least in the case of steel in HCl, the fractal properties of the surface are relatively constant throughout the exposure period.

The CPE appears to arise solely from roughening of the surface by the corrosion process. This was verified with IS experiments on iron and several steels in 15% HCl at 25°C. The electrodes were polished with alumina and maintained at 150 mV cathodic of the rest potential. Complex plane plots of the impedance responses were near-perfect semi-circles centered on the Z' axis. Analyses via EQIVCT using the $R_\Omega+P/R_t$ circuit, gave rise to n-values of the CPE in excess of 0.93 in all cases and remained constant throughout the tests.

Thus, it appears that in most cases we can treat steel corrosion in concentrated HCl adequately with the circuit $R_\Omega+P/R_t$. In cases where the two capacitance loops are sufficiently distinguishable, we must resort to the full circuit $R_\Omega+P/(R_t+R_a/P_a)$, where P_a is the CPE counterpart of C_a; when the full circuit is prescribed, the fitted n-value of P_a is always less than the n-value of P.

Consider next the effect of adding corrosion inhibitors.

Effect of Corrosion Inhibitors

Adsorption and Film Formation.   Inhibition of HCl corrosion by organic compounds is a complicated multi-step process. Nevertheless, the effect of an inhibitor on corrosion of a metal is often treated mathematically with an equilibrium adsorption model for displacement of water (19,20):

(a) $M_S\{H_2O\}_m$ + Inhibitor $\rightleftharpoons$ $M_S\{Inhibitor\}$ + mH_2O

where M_S represents an active surface site(s) and the brackets denote sorbed species. Equilibrium (a) competes with adsorption of the corrosive species on M_S and is assumed to be fast, so that the corrosion rate rapidly reaches a "steady state" value. However, chemical analyses of products on the steel surface have revealed that in many cases a macroscopic film is formed via polymerization reactions and that the film thickens with increasing time (3).

To account for the effect of polymerization on the corrosion rate, it has been proposed (4) that film formation be considered as a series of iterative elementary steps, the sum of which may be expressed as

(b) $M_s\{Inhibitor\} + M_s\{H\} + n$ Inhibitor $\longrightarrow M_s\{Inhibitor\}\{Film\}$
$$+ M_s\{Film\}$$

where the polymer film is draped over uncovered active surface sites, i.e. $M_s\{Film\}$, and anchored by sorbed inhibitor, i.e. $M_s\{Inhibitor\}\{Film\}$. Coverage of inactive surface sites also occurs, but at a slower rate (21).

Equivalent Electrical Circuit. In spite of the complex nature of the inhibition process, the inhibited systems actually display simple impedance responses.

Tests were run with N80 steel in 15% and 28% HCl at 25°C with and without octynol for periods extending up to 2 hours. Immediately after injection of octynol into the acid, two phenomena were observed. First, near the low-frequency limit of the tests, a prominent inductive loop (below the Z' axis) appeared which then vanished within a few minutes. Secondly, fits of the data above 1 Hz to the $R_\Omega + P/R_t$ circuit, i.e. ignoring the inductive loop, gave rise to a higher CPE n-value, which then remained relatively constant for the duration of each experiment. This result is shown in Figure 5.

Other inhibitors, namely TCA and BAA, gave a similar response to octynol. Furthermore, the octynol and BAA systems were tested with and without rotation at 1000 rpm of the steel disk, giving virtually the same result, which indicates that the sorbed inhibitor and polymer film are strongly adherent. In all cases, the standard deviations of the fitted parameters, σ_i, fell considerably on addition of inhibitor, e.g. in the case of 15% HCl with octynol, σ_n fell from 1 to 3% (no inhibitor) down to 0.2 to 0.6% (with inhibitor). Generally, the inhibited systems gave less scatter in the data, a result we attribute to slower evolution of H_2 and lower variability in the size of attached bubbles, i.e. lower variability in instant average contact area of specimen with solution.

Inductive loops, such as that described above, are sometimes associated with adsorption of an oxidizable or reducible intermediate. Indeed, the polymer film and perhaps the adsorbed form of octynol are created by surface-mediated reductions. However, when we re-ran the octynol tests above using a peak-to-peak voltage of 4 mV instead of 10 mV, the size of the inductive loop fell by an order of magnitude, which suggests it is an artifact of the IS technique whereby the measurement alters the surface being measured. More detailed investigation of this inductive behavior may yield clues about the nature of the polymerization process. For our purposes, it suffices to know that its effect is small at the rest potential (open-circuit potential) and we shall ignore it in the discussion that follows.

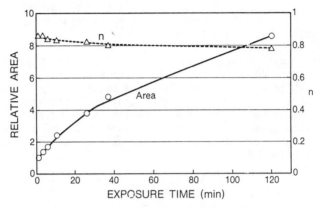

Figure 4.   Effect of exposure time on CPE n-value and relative
surface area.

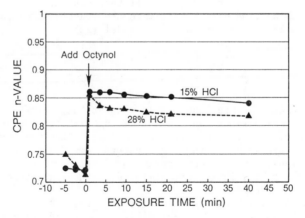

Figure 5.   Effect of octynol on CPE n-value.

An interesting incidental effect observed when an inhibitor is present in such a great excess that it forms a separate phase is the appearance of a large low-frequency capacitance loop which we attribute to precipitation (physical adsorption). This effect, however, plays no role in the experiments discussed above, since they all deal with aqueous single-phase solutions.

In conclusion, though the inhibitor chemistry is complex, IS sees only a slightly roughened surface covered with an adsorbed film. Since this is representable by a single type of equivalent circuit, analysis of the inhibition process per se is relatively straightforward, as will be discussed below.

<u>Adsorption versus Polymerization</u>. It is instructive to examine further the time dependence of the corrosion inhibition. In acid corrosion inhibition tests, steady state is customarily assumed to be reached within 10 to 20 min after initial exposure of the metal specimen. Since the inhibitors function by reducing the available active surface area, we expect an increase in R_t and a corresponding decrease in P. The degree of corrosion protection the inhibitor provides is given by

$$\Psi = \frac{R_{to}^{-1} - R_t^{-1}}{R_{to}^{-1}} \tag{1}$$

where R_{to} and R_t are the resistances in the $R_\Omega + P/R_t$ circuit without and with inhibitor, respectively. If the sorbed inhibitor is impermeable (blocking), we can approximate the fractional surface coverage by the parallel-plate two-capacitor model (<u>22</u>):

$$\theta = \frac{P_o - P}{P_o - P_\infty} \tag{2}$$

where P_o, P and P_∞ are the CPE's without the inhibitor (coverage = 0), with the inhibitor and with excess inhibitor (coverage = 1), respectively, and are evaluated at $\omega_z{''}_{max}$ (the frequency at which Z'' is a maximum). Using C_d instead of P, we obtain identical results. Thus, θ may be considered the fraction of the <u>total</u> surface area covered by the inhibitor, while Ψ is the fraction of the <u>active</u> surface that is covered.

When step (a) reaches equilibrium, both Ψ and θ should have steady state values. To test this hypothesis, we determined the effect of exposure time on Ψ and θ for N80 steel in 15% HCl at 25°C and in 28% HCl at 65°C, as shown in Figures 6 and 7, respectively. Here [octynol] = 3.5 x 10^{-3}M in 15% HCl and 2.1 x 10^{-2}M in 28% HCl. It is immediately apparent from these plots that **neither Ψ nor θ reaches a steady state** value during the course of the experiments. Although Ψ <u>appears</u> to be relatively constant throughout the 15% HCl test and during the first stage of the 28% HCl test, it is continually increasing in both cases.

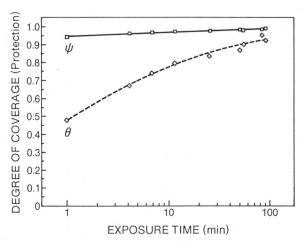

Figure 6. Effect of exposure time on θ and Ψ: 15% HCl with 3.5 x 10^{-3} M octynol, 25°C.

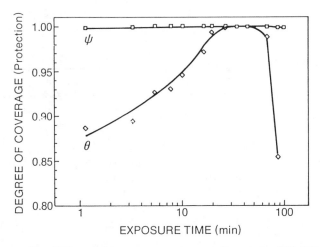

Figure 7. Effect of exposure time on θ and Ψ: 28% HCl with 2.1 x 10^{-2} M octynol, 65°C.

Adsorption reactions on nonporous surfaces are generally quite rapid (unless there is a large activation energy barrier). By contrast, surface polymerization reactions are usually much slower. Thus it is likely that the initial high level of Ψ arises from adsorption, while the subsequent small, but continuous, increase in Ψ is caused by the thickening polymer film.

The precipitous drop in Ψ observed in Figure 7 for 28% HCl is also attributable to film formation. Previous work has shown that octynol itself is stable under these conditions in the absence of a steel surface (3). In the presence of the steel, however, polymerization occurs, which consumes octynol. These reactions slow with increasing exposure time, but they do not stop. We surmise, therefore, that the sudden decrease in inhibition after 50 min arises from the concentration of octynol in the solution reaching too low a level to maintain the sorbed monolayer; the inhibitor adsorption equilibrium shifts so that octynol desorbs and the corrosion rate returns to the level attained in uninhibited acid. This result was verified by spectrophotometric measurements of [octynol], which showed a large decrease in [octynol] with exposure time. The decrease in inhibition was also corroborated with companion weight-loss tests, which showed that catastrophic failure generally occurred at sufficiently long exposure times. In the long run, then, although the polymer film contributes to the protection of the steel, the polymerization reactions prove to be deleterious.

Adsorption Mechanism. If P_∞ is negligibly small and the surface is homogeneous (all surface sites are equally active), the surface coverage, θ, and the degree of protection, Ψ, will be identical. It is clear, however, from Figures 6 and 7 that, although the two parameters track the same way, θ does not go hand-in-hand with Ψ. Significantly, the discrepancy between θ and Ψ is greatest at short exposure times. Ψ reaches a very high value — near its "steady state" value — almost immediately, whereas θ initially is quite low and increases slowly with increasing exposure time. If adsorption of octynol were entirely responsible for the inhibition, the initial difference between θ and Ψ would imply that sorbed octynol can provide a high degree of protection to the steel while only covering a small fraction of the surface. This arises from the assumption that the P_∞ value corresponds to a surface covered with a monolayer of sorbed octynol when, in fact, it actually corresponds to the film-covered surface. Indeed, the initial discrepancy between θ and Ψ is consistent with the following limiting adsorption mechanisms: (I) octynol forms an insulating barrier but adsorbs non-uniformly (a permeable, or porous, barrier), covering active sites to a greater extent than inactive — or less active — sites; and (II) octynol adsorbs uniformly (an impermeable barrier), covering the entire surface with a monolayer that is electronically and ionically conductive. While the monolayer may or may not be a good electrical insulator, the aged polymer film, by virtue of being a saturated hydrocarbon, is a good electrical insulator. The slow but large rise of θ with exposure time results from the decrease in P as the interface becomes more non-polar.

In the end, analysis of Ψ vs θ (the "classical" approach) is not quantitative, a problem we associate with surface area variability and ambiguity arising from the interpretation of P_∞. To help determine which adsorption mechanism is operative, we turn to an alternative parameter, the IS impedance "time constant", τ, which does not suffer from these drawbacks.

Time Constant Analysis. τ is the relaxation time of the corrosion process and is dependent on the dielectric properties of the interface. τ is given by $\tau = R_t P$, but can be measured independently: $\tau = \omega_z{}''_{max}{}^{-1}$. Since R_t and P vary with surface area in exactly opposite fashion, τ (or $\omega_z{}''_{max}$) should be independent of surface area. To verify that this is indeed the case, we examined the corrosion of N80 steel in uninhibited 15% HCl at 65°C. With increasing exposure time, we observed a continuous decrease in R_t (hence an increase in corrosion rate) and a concomitant increase in P. And, as expected, $\omega_z{}''_{max}$ did not vary at all (see Figure 8).

All of the inhibitors we examined reduce $\omega_z{}''_{max}$ substantially, although they vary considerably in mode of inhibition, e.g. QQ and DDPB are electrostatic adsorbers, whereas TCA forms a thick film with carbonyl character and octynol forms a thin hydrocarbon film. Figures 8 and 9 show this effect in plots of $\omega_z{}''_{max}$ versus exposure time and inhibitor concentration, respectively, for several inhibitors and inhibitor mixtures. Because of variability in reactivity among the specimens of N80, the $\omega_z{}''_{max}$ values were normalized to the same initial value. Initial inhibitor concentrations ranged from 0.005 to 0.01 M. Figure 8 shows that all of the inhibitors produce a large, almost instantaneous, initial drop in $\omega_z{}''_{max}$. Thus, all of the inhibitors increase τ, i.e. they slow the kinetics of the corrosion charge-transfer reaction.

According to Figure 8, QQ and DDPB give a relatively constant value of $\omega_z{}''_{max}$ within a few minutes; interestingly, though, the response with QQ is considerably slower than with DDPB, suggesting that QQ undergoes a secondary reaction at the surface, e.g. reorientation or electrochemical reduction. Octynol, BAA and the mixtures of DDPB with BAA or TCA do not give a constant value of $\omega_z{}''_{max}$; rather, $\omega_z{}''_{max}$ continues to fall, albeit slowly, even after 30 min exposure. Considering that octynol, BAA and TCA are all film-forming inhibitors, we conclude that the large magnitude of the initial decrease in $\omega_z{}''_{max}$ corresponds to adsorption and the slow fall-off in $\omega_z{}''_{max}$ to film-forming reactions. Furthermore, the large magnitude of the initial decrease in $\omega_z{}''_{max}$ indicates that adsorption is the dominant mode of inhibition.

The experiments on which Figure 9 ($\omega_z{}''_{max}$ vs [Inhibitor]) is based were all run with repolished electrodes pre-exposed for 5 minutes; this is sufficient time to achieve a constant temperature, yet too short a period for extensive polymerization. It is evident from Figure 9 that, at a high enough concentration, every one of the inhibitors reduces $\omega_z{}''_{max}$ by at least an order of magnitude.

Indeed, the relative values of $\omega_z{}''_{max}$ for all of the inhibitors are almost the same, falling in the range 10 to 20, e.g. $\omega_z{}''_{max}$ decreases by factors of 16-22 for DDPB, 12-26 for TCA, and 17 for

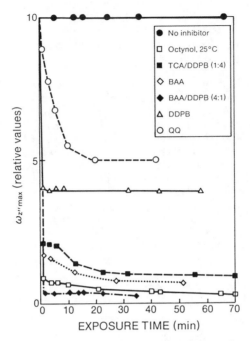

Figure 8. Effect of exposure time on $\omega_z{}''_{max}$ for various inhibitors in 15% HCl at 65°C.

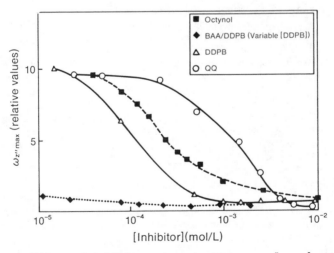

Figure 9. Effect of inhibitor concentration on $\omega_z{}''_{max}$ for various inhibitors in 15% HCl at 65°C.

BAA. That $\omega_z"_{max}$ is so nearly independent of the nature of the inhibitor lends additional credence to mechanism (I). By contrast, if mechanism (II) were operative, we would expect $\omega_z"_{max}$ to vary greatly with the electronic conductivity (and polarity) of the inhibitor barrier.

Conclusions

The impedance spectroscopy of steel corrosion in concentrated HCl, with and without inhibitors, exhibit relatively straightforward electrochemical phenomenology and can be represented by simple equivalent circuits involving primarily passive electrical elements. Analysis of these circuits for steel corroding in HCl per se reveals that the heterogeneity of the surface is established rapidly and can be simulated with a simple electrical circuit model.

Classical IS measurements indicate that corrosion inhibitors reduce surface heterogeneity and function primarily by adsorption. Furthermore, the sorbed monolayer is either (I) permeable and insulating or (II) impermeable and conductive. Analysis of the time-constant, τ, for the corrosion process, suggests that mechanism (I) is operative.

Non-steady state corrosion rate behavior appears to be a general phenomenon and is associated with polymerization reactions. The latter, which results in formation of a film on the sorbed monolayer, provides a smaller increment of protection than does adsorption and occurs at the expense of inhibitor loss from the solution. In some cases, however, the increased protection provided by the film is substantial and merits further investigation.

Acknowledgments

We thank Dowell Schlumberger for permission to publish this work.

Literature Cited

1. Poling, G. W. J. Electrochem. Soc., 1967, 114, 2109.
2. Duwell, E. J.; Todd, J. W.; Butzke, H. C. Corros. Sci., 1964, 4, 435.
3. Growcock, F. B.; Lopp, V. R.; Jasinski, R. J. J. Electrochem. Soc., 1988, 135, 823.
4. Growcock, F. B.; Lopp, V. R. Corros. Sci., 1988, 28, 397.
5. Growcock, F. B. Corrosion '88, 1988, Paper No. 338.
6. Growcock, F. B.; Lopp, V. R. Corrosion, 1988, 44, 248.
7. Frenier, W. W.; Lopp, V. R.; Growcock, F. B. Corrosion, in press.
8. Growcock, F. B.; Frenier, W. W. Corrosion '84, 1984, Paper No. 121.
9. Boukamp, B. A. Solid State Ionics, 1986, 20, 31.
10. Armstrong, R. D.; Race, W. P.; Thirsk, H. R. J. Electroanal. Chem., 1968, 16, 517.
11. Mansfeld, F. Corrosion, 1981, 36, 301.

12. Harrington, D. A.; Conway, B. E. Electrochim. Acta, 1987, 32, 1703.
13. Hurt, R. L.; Macdonald, J. R. Solid State Ionics, 1986, 20, 111.
14. de Levie, R. Electrochim. Acta, 1965, 10, 113.
15. Armstrong, R. D.; Burnham, R. A. J. Electroanal. Chem., 1976, 72, 257.
16. Macdonald, J. R. J. Electroanal. Chem., 1987, 223, 25.
17. Mulder, W. H.; Sluyters, J. H. Electrochim. Acta, 1988, 33, 303.
18. Rammelt, U.; Reinhard, G. Corrosion Sci., 1987, 27, 373.
19. Uhlig, H. K. Corrosion and Corrosion Control, 2nd Ed.; John Wiley: New York, 1971; pp. 265-271.
20. Ateya, B. G. J. Electroanal. Chem., 1977, 76, 191.
21. Growcock, F. B.; Frenier, W. W.; Lopp, V. R. Proc. 6th European Conf. on Corrosion Inhibition, 1985.
22. Schuhmann, D. Electrochimica Acta, 1987, 32, 1331.

RECEIVED December 7, 1988

Chapter 37

Use of Starved Bacteria To Increase Oil Recovery

Hilary M. Lappin-Scott, Francene Cusack, F. Alex MacLeod, and J. William Costerton

Department of Biological Sciences, University of Calgary, Calgary, Alberta T2N 1N4, Canada

Limitations in present oil recovery methods leave the majority of oil as unobtainable in the reservoir. Therefore, any technique that increases or enhances the recovery rate of this resource would have great potential for field applications. Our interest lies in utilizing microorganisms to assist in enhanced recovery by virtue of their growth properties. We report on our recent laboratory data in model rock strata. Our laboratory data demonstrate a new method of blocking the high permeability formations using starved forms of bacteria. It also has applications to control oilwell coning.

   Estimates suggest that a maximum range of between 8-30% of the total oil is presently recovered from petroleum reservoirs leaving vast quantities underground as the focus for developing new techniques to increase recovery rates. One recovery method now in use is the waterflooding of differing permeability rock strata (Fig. 1). As the waterflood commences the high and low permeability zones are oil saturated (Fig. 1A). The waterflood follows the routes of least resistance, that is the high permeability channels, and acts as an energy force to push out any oil along its path. However, once the oil has been displaced from these zones the water continues to follow the same course leaving the lower permeability zones unswept and therefore full of oil (Fig. 1B). The process of blocking off the higher permeability strata and diverting the waterflood to other unswept zones is called selective plugging. In order to be effective, the plug must penetrate throughout the high permeability zone or the water will return preferentially from the lower back to the higher permeability strata. If a shallow plug is established (Fig. 1C) the waterflood is initially blocked from the higher permeability strata and pushes out some oil. However, beyond the plug the waterflood returns to the high permeability zone leaving much oil in the low permeability strata. With a deeper plug in the high permeability strata the waterflooding is forced to stay in the lower permeability zone and push the oil out (Fig. 1D) until most of the oil is drained (Fig. 1E).

0097-6156/89/0396-0651$06.00/0

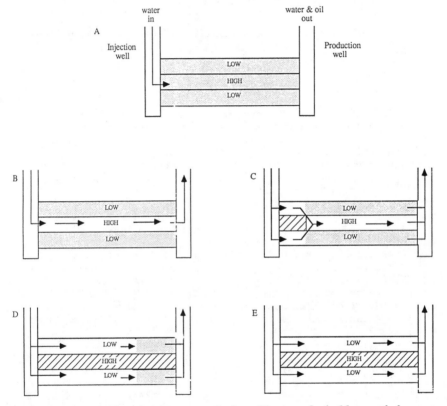

Figure 1. Schematic diagram of the effects of shallow and deep
plugs on enhanced oil recovery.

Several agents are currently used for plugging high permeability strata. These include small fibers that are carried in the
waterflood and deposited in the high permeability zones[1] and
chemical reactions forming insoluble precipitations[2]. Some of the
current methods available, for example polymers or foams, are
subject to deterioration and are costly. This gives them limited
application as they are not able to penetrate deep into the strata.
    In some reservoirs another problem develops that reduces oil
recovery, called coning. This occurs most commonly when water lies
below heavy oil in the reservoir. During primary recovery instead
of the oil being pumped out, water, being of lower viscosity, is
preferentially pulled to the surface. Continued pumping only
succeeds in pulling up more water, thus reducing or halting further
oil recovery[3]
    Crawford[4,5] reported that bacteria could be used as selective
plugging agents and have the ability to penetrate the high permeability areas. Another advantage is that bacteria can grow in the
rock and produce plugging both by their growth and their growth
products[6]. Research at the University of Oklahoma[7,8] demonstrated
that when bacteria were injected into two sandstone cores of differing permeabilities the bacteria preferentially grew and plugged up
the higher permeability core first. However, when full-sized
bacteria, 2.0 μm or more in length, are injected into solid matrices
the bacteria collect at the inlet in a sticky mass containing microbial growth and growth products, called biofilms[9,10] and the plug is
referred to as a skinplug. Reducing the concentration of bacteria
to approximately 10[6]/ml reduced the opportunity for skinplug
formation[11] and this could be further reduced by using dormant,
smaller forms of bacteria. The smaller size of dormant bacteria,
that is spores or starved bacteria, together with their absence of
sticky growth products may allow them to travel further into rock
strata.
    Microorganisms have been reported to decrease substantially in
size as a response to low nutrient conditions[12]. Then, after a
period of starvation when the organisms exist in a dormant state,
they commence growth again and return to full size when given
nutrients[13]. Our work at the University of Calgary attempted to
harness these changes in cell size during starvation-resuscitation
as a method for enhanced oil recovery. We considered that the
smaller cell size may enable them to penetrate deeper into rock
strata than full-sized bacteria. By injecting organisms in a
starved state into rock then giving them growth nutrients would
allow them to grow and plug the rock. Further waterflooding would
bypass these plugged regions and sweep the areas containing oil.
Starved bacteria may also be used to control coning. They may be
injected at the oil:water interface then resuscitated with
nutrients. Bacterial growth will forma a deep pancake of sticky
slime to physically separate the water from the oil and prevent any
more water being sucked to the surface.
    <u>Klebsiella</u> <u>pneumoniae</u> was isolated as a representative
microorganism from produced water[14]. The bacterium was starved in
phosphate buffer salts solutions at concentrations of either 10[4]/ml
or 10[8]/ml. During starvation periods of up to 24 days the bacterial
cells changed in size and shape from rod-shaped, up to 2.2 μm long,

producing sticky slimes of polysaccharide-containing biofilms to spherical or small rods 0.5 µm by 0.25 µm with little or no biofilm[13] (Fig. 2). Such microorganisms are termed ultramicrobacteria (UMB). We investigated the ability of the UMB to grow and resuscitate using different nutrients. One contained a rich mixture of growth substances, called Brain Heart Infusion or BHI. The BHI was added at half of the manufacturers recommended concentration as this was sufficient to support rapid growth. Another nutrient contained only one carbon source, called sodium citrate medium or SCM. The SCM was added at concentrations of 7.36 g/l. BHI was a fast acting nutrient and supported rapid resuscitation in 4 hours, SCM was a slow acting nutrient and supported resuscitation in 8 hours[14].

After establishing that microorganisms from oilwells could decrease in size and form UMB then return to full size when given nutrients in laboratory growth cultures, we investigated whether the UMB were able to penetrate deeper when injected into porous matrices than their full sized counterparts. The experimental core flood apparatus consisted of a constant pressure, variable flow rate injection system[9]. The pressure was maintained at 3.5 psi. In a comparative study using sintered glass bead cores with permeabilities of between 6 and 7 Darcys, one set of cores were injected with 10^8/ml K. pneumoniae starved for 4 weeks, the other with 10^8/ml of the full sized bacteria grown on SCM[10]. The full sized cultures blocked the cores quickly and reduced the permeability to less than 1% of the original value with the addition of 500 pore volumes. The starved bacteria only reduced the permeability to 82% despite the addition of in excess of 800 pore volumes[10]. After the injections were completed the glass bead cores were cut up into equal size sections and examined by electron microscopy to establish the position of the bacteria within the cores. The electron microscopy of different sections of cores treated with full sized cells showed that a mass of large rod-shaped bacteria was located at the core inlet. The bacteria had produced polysaccharides and the biofilm which plugged the core inlet. With the starved cultures the UMB were evenly distributed throughout each of the core sections with little or no biofilm apparent. From this work, we were able to conclude that starved K. pneumoniae was able to penetrate deeper into solid matrices than full sized cells as a result of their smaller size, less sticky glycocalyx and reduced biofilm production.

Another series of experiments used sandstone cores previously injected with starved bacteria to investigate the ability of the bacteria to grow within rock cores when given a suitable nutrient[15]. Berea sandstone cores of 200 and 400 millidarcy (md) permeabilities were used as they were considered to be more representative of reservoir conditions than the glass bead cores. The sandstone cores were injected with 300 to 450 pore volumes of 10^5/ml starved bacteria until the cores contained an even distribution of bacteria (Fig. 3A & B) and the core permeabilities were between 13% and 18%. SCM nutrient was injected through the cores (Fig. 3C) until the core permeability fell to 0.1%, this required 360 pore volumes of SCM. The starved bacteria resuscitated by utilizing the SCM and grew within the sandstone forming a deep bacterial plug composed of cells

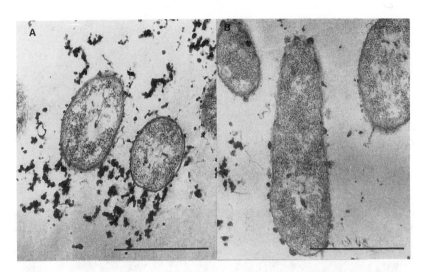

Figure 2. Starved (A) and full-sized (B) <u>Klebsiella</u> <u>pneumoniae</u> in laboratory cultures viewed through an electron microscope. The sizes and shapes of the cells differ markedly. The bar represents 1 μm.

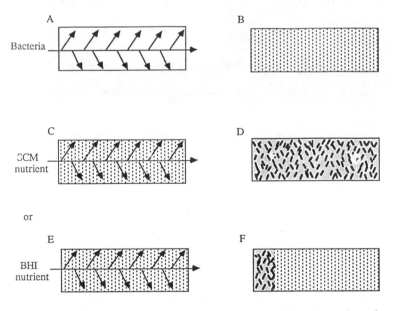

Figure 3. Diagrammatic representation of the plugging of rock cores with resuscitated starved bacteria. See text for details.

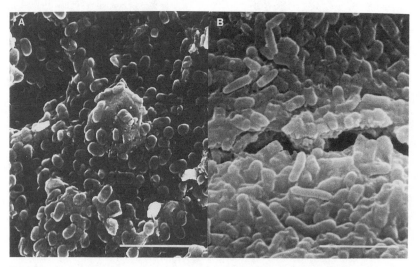

Figure 4. Scanning electron microscopy of starved (A) and nutrient stimulated (B) Klebsiella pneumoniae in sandstone rock cores. The bar represents 5 μm.

and glycocalyx (Fig. 3D). The resuscitation followed a similar
pattern to that reported in laboratory batch growth systems[15], that
is, a difference in resuscitation rates was observed with SCM and
BHI. With SCM the slower resuscitation permitted a nutrient flow to
all of the starved bacteria in the sandstone and the subsequent
growth produced a deep plug throughout the entire core (Fig. 3C &
3D). With BHI, a 95 pore volume injection resulted in a drop in
permeability to 0.5%. Resuscitation was so rapid that the growth of
bacteria by the core inlet blocked off the supply of nutrients to
lower down the core resulting in a shallow bacterial plug only at
the top of the core (Fig. 3E & 3F). Sectioning and examination of
the sandstone cores using scanning electron microscopy showed large
differences in the bacteria before and after nutrient additions
(Fig. 4A & B). The nutrient fed bacteria were observed to have
increased noticeably in size and a mass of biofilm was produced.
The starved bacteria were still tiny and singular with little or no
biofilm.

  We also investigated several cost reduction exercies, such as,
a) giving the starved bacteria short bursts of nutrient (less than
50 pore volumes) instead of a continuous flow and b) injecting fewer
starved bacteria into the core (150 pore volumes) before nutrient
injection. Both still resulted in deep bacterial plugs when SCM was
used as a nutrient.

  Other experiments are planned to study the location, distribution and resuscitation of ultramicrobacteria in large three-
dimensional sandpacks. Such studies will allow a more realistic
approximation of reservoir conditions than the unidirectional core
studies. We do not consider that the ultramicrobacteria will reach
or grow in areas where residual oil is located. Selective plugging
involves blocking the high permeability zones already drained of
oil. We consider that the injection of ultramicrobacteria will be
carried, like waterflood operations, to the areas of the strata
already drained of oil and permit them to disperse through pore
spaces and resuscitate in these areas.

  Conditions differ in each reservoir with respect to
temperature, pressure and salinity. No one microorganism will be
expected to survive, grow and plug all these different reservoirs.
We suggest that a bacterium is chosen from laboratory collections
that cover a range of environmental conditions best suited to the
particular well.

  In summary, care must be taken to inject nutrients that do not
encourage rapid growth as undesirable shallow bacterial plugs form
(Fig. 3F). With the correct nutrient package, such as SCM in this
instance, a deep plug will form throughout the strata (Fig. 3D). In
conclusion, our laboratory based studies demonstrate that starved
bacteria may be used to physically block rock strata already drained
of oil. Further recovery operations can then deal with strata still
containing oil and thus enhance recovery rates.

Literature Cited

1. Moses, V. Microbiol. Sci. 1987, 4, 306-309.
2. Breston, J. N. J. Petrol. Technol. March 1957, 26-31.

3. Hower, W. F.; Ramos, J. J. Petrol. Technol. January 1957, 137–140.
4. Crawford, P. B. Producers Monthly 1961, 25, 10–11.
5. Crawford, P. B. Producers Monthly 1962, 26, 12.
6. Jenneman, G. E.; Knapp, R. M.; McInerney, M. J.; Menzie, D. E.; Revus, D. E. Soc. Petrol. Eng. J. Feb. 1984, 33–37.
7. Raiders, R. A.; Freeman, D. C.; Jenneman, G. E.; Knapp, R. M.; McInerney, M. J.; Menzie, D. E. Paper SPE 14336 presented at 60th Annual Technical Conf. & Exhibition of Soc. Petrol. Eng., Las Vegas, NV, 1985.
8. Raiders, R. A.; McInerney, M. J.; Revus, D. E.; Torbati, H. M.; Knapp, R. M.; Jenneman, G.E. J. Industr. Microb., 1986, 1, 195–203.
9. Shaw, J. C.; Bramhill, B.; Wardlaw, N. C.; Costerton, J. W. Appl. & Envir. Microbiol., 1985, 50, 693–701.
10. MacLeod, F. A.; Lappin-Scott, H. M.; Costerton, J. W. Appl. & Envir. Microbiol., 1988, 54, 1365–1372.
11. Jang, L.-K.; Chang, P. W.; Findley, J. E.; Yen, T. F. Appl. & Envir. Microbiol., 1983, 46, 1066–1072.
12. Novitsky, J. A.; Morita, R. Y. Appl. & Envir. Microbiol., 1976, 33, 635–641.
13. Torella, F.; Morita, R. Y. Appl. & Envir. Microbiol., 1981, 41, 518–527.
14. Lappin-Scott, H. M.; Cusack, F.; MacLeod, F. A.; Costerton, J. W. J. Appl. Bacteriol., 1988, 64, 541–549.
15. Lappin-Scott, H. M.; Cusack, F.; Costerton, J. W. Appl. & Envir. Microbiol., 1988, 54, 1373–1382.

RECEIVED January 13, 1989

Appendix

Weldon M. Harms

Halliburton Services, P.O. Drawer 1431, Duncan, OK 73536—0426

Reviews of Oil & Gas Fracturing(Ref. 1)

- **Clark**, J.B., Fast, C.R., and Howard, G.C.: "A Multiple Fracturing Process for Increasing the Productivity of Wells," 1952 Spring Meeting of the Midcontinent District API, Wichita, KS, March 19-21.
- **Prusick**, J.H. and Morgan, Z.V: "The Use of Emulsions and Related Techniques in the Treatment of Oil and Gas Wells," Pet. Eng. (May 1954) B54-B56.
- **Hurst**, R.E., Moore, J.M., Ramsey, D.E.: "Development and Application of 'Frac' Treatments in the Permian Basin," T.P. 4032, Soc. Pet. Eng. J. (1955) Trans., AIME, 204.
- **Alderman**, E.N., Mack, D.J., and Ousterhout, R.S.: "Evaluation of Fracturing Materials for Optimum Well Stimulation," API paper 801-37J, 1961 Spring Meeting of Pacific Coast District, Los Angeles, May 11-12.
- **Hassebroek**, W.E. and Saunders, C.D.: "Hydraulic Fracturing," Modern Well Completion Series, Part 13, Pet. Eng. (July, 1961), B55,B63,B69,B72,B74,B75,B77,B78,B81.
- **Grubb**, W.E.; Martin, F.G. "A Guide to Chemical Well Treatments," Part 1, Pet. Eng., May 1963, 94-110; Part 2, ibid June 1963, 100-104; Part 3, ibid, July 1963; Part 4, ibid August 1963, 100-110; Part 5, ibid September 1963, 94-108; Part 6, ibid October 1963, 122-126; Part 6, ibid October 1963, 118-128; and Part 7, ibid, November 1963, 122,126.
- **Hassebroek**, W.E. and Waters, A.B.: "Advancements Through 15 Years of Fracturing," J. Pet. Technol. (July 1964) 760-764,787.
- **Hill**, K.E. and Coqueron, F.G.: "The Petroleum Industry A Look Backward...and Forward," SPE paper 1622, 1966 SPE Annual Meeting, Dallas, October 2-56
- **Howard**, G.C. and Fast, C.R.: Hydraulic Fracturing, Monograph Series, SPE(1970)2.
- **Alderman**, E.N.: "Super Thick Fluids Provide New Answers to Old Fracturing Problems," SPE paper 2852, 1970 Spring Symposium SPE, Fort Worth, March 8-10.
- **Dysart**, G.R., Richardson, D.W., and Kannenberg, B.G.: "Second Generation Fracturing Fluids," API paper 906-15-H, 1970 Spring Meeting of the Southwestern District Division of Production, Odessa, March 18-20.
- **Coulter**, G.R. and Wells, R.D.: "The Advantages of High Proppant Concentration in Fracture Stimulation," SPE paper 3298, 1971 SPE Annual Meeting, New Orleans, October 3-6.
- **Rosene**, R.B. and Shumaker, E.F.: "Viscous Fluids Provide Improved Results from Hydraulic Fracturing Treatments," SPE paper 3347, 1971 SPE Rocky Mountain Regional Meeting, Billings, June 2-4.
- **Weiland**, D.R., "Recent Trends in Hydraulic Fracturing," SPE paper 3659, 1971 SPE Eastern Regional Meeting, Charleston, November 4-5.
- **Coulter**, G.R.: "Hydraulic Fracturing - New Developments," Transactions Gulf Coast Association of Geological Societies, XXIII(1973), 47-53.
- **White**, J.L., Rosene, R.B., and Hendrickson, A.R.: "New Generation of Frac Fluids," 1973 SPE Annual Meeting, Edmonton, May 8-12.
- **Buechley**, T.C. and Lord, D.L.: "Hydraulic Fracturing Fluid Mechanics - State of the Art," Fluid Mechanics of Oil and Gas Production Symposium, 1973 National Meeting, A.I.Ch.E., New Orleans, March 11-15.
- **Krueger**, R.F.: "Advances in Well Completion and Stimulation During JPT's First Quarter Century," J. Pet. Technol., December, 1973, 1447-1462.
- **Anderson**, R.W. and Baker, J.R.: "Use of Guar Gum and Synthetic Cellulose in Oilfield Stimulation Fluids," SPE paper 5005, 1974 SPE Annual Meeting, Houston, October 6-9.
- **White**, J.L. and Free, D.L.: "Properties of Various Frac Fluids as Compared to the Ideal Fluid," Proc. Symp. Stimul. Low Perm. Reserv. Colo. School Mines.(1976), 1-14.
- **Kohlhaas**, C.A.: "Fracturing," Subsurf. Geology (4th Ed)J, edited by LeRoy, et al. (1977) 400-406.
- **Waters**, A.B.: "Stimulation of Hydrocarbon Reservoirs," 1980 Industrial Mineral Meeting, Houston, September 11-12.
- **Waters**, A.B.: "Hydraulic Fracturing Deep Gas Wells," 1981 International Gas Research Conference.
- **Holman**, G.B.: "State-of-the-Art Well Stimulation," J. Pet. Technol., February 1982, 239-241.
- **Waters**, A.B.: "History of Hydraulic Fracturing," 1982 SPE Hydraulic Fracturing Symposium, Lubbock, 1-20.

0097—6156/89/0396—0659$06.00/0
© 1989 American Chemical Society

- **Veatch**, R.W., Jr.: "Current Hydraulic Fracturing Treatment and Design Technology," SPE paper 10039, 1982 SPE International Petroleum Exhibition and Technical Symposium, Bejing, China, March 28-26.
- **Pai**, V.J. and Garbis, S.J.: "Review of the Completion Practices in the Morrow Formation in Eddy, Chaves, and Lea Counties of Southeast New Mexico," SPE paper 11335, 1982 SPE Production Technology Symposium, Hobbs, Nov.
- **Veatch**, R.W., Jr.: "Overview of Current Hydraulic Fracturing Design and Treatment Technology - Part 1," J. Pet. Technol., April 1983, 677-864.; ibid. Part 2, J. Pet. Technol., May 1983, 853-864.
- **Matson**, B.G. and Baker, J.R.: "Advanced Stimulation Technology," Drilling-DCW April 20, 1977, 64-65,
- **Menjivar**, J.A.: "The Use of Water-Soluble Polymers in Oil Field Applciations: Hydraulic Fracturing," 1984 Proc. Annu. M.I.T. Seagrant Colloq. Program Lecture Seminar (3rd), Cambridge, Mass.
- **Ely**, J.: "Fracturing Fluid Systems State of the Art," 1981 Proc. Annu. Southwest Pet. Short Course, Lubbock, April 23-24, 1981.
- **Almond**, S.W. and Garvin, T.R.: "High Efficiency Fracturing Fluids for Low Temperature Reservoirs," 1984 Proc. Annu. Southwest Pet. Short Course, Lubbock, May 1984, 76-88.
- **Krueger**, R.F.: "An Overview of Formation Damage and Well Productivity in Oilfield Operations," J. Pet. Technol., February 1986, 131-152.
- **Ely**, J.W.: "Stimulation Treatment Handbook," Penwell Publications: Tulsa, OK, 1985.
- **Veatch**, R.W., Jr. and Moschovidis, Z.A.: "An Overview of Recent Advances in Hydraulic Fracturing Technology," SPE paper 14085, 1986 International Meeting on Petroleum Engineering, Bejing, China, March 17-20.
- **Allen**, T.O. and Roberts, A.P. "Production Operations," Vol. 2, Penwell Publications, 2nd ed.:Tulsa,OK, 1982.
- **Bleakely**, W.B.: "How Chemicals Improve Ultimate Recovery," PET. ENG. Internat., May 1986, pages 53-58.
- **Gray**, G.R., Darley, H.C.H., Rogers, W.F.: "Compositions and Properties of Oil Well Drilling Fluids" 4th Ed, Gulf Publishing:Houston, TX 1980.
- **Economides**, M.J.; Nolte, K.G.: "Reservoir Stimulation," Schlumberger Educational Services:Houston,TX. 1987.
- **Harris**, P.C.: "Fracturing-Fluid Additives," J. Pet. Technol., Oct. 1988, 1277-79.
- **Gidley**, J. and Holditch, S.A., eds, "Recent Advances in Hydraulic Fracturing," Ely, J.W., in Chapter 7, "Fracturing Fluids and Additives," Society of Petroleum Engineers, Dallas, TX, 1989.

Fracturing Fluid Leakoff(Ref. 4.)

- **Settari**, A.: "A New General Model of Fluid Loss in Hydraulic Fracturing," SPE/DOE paper 11625, 1983 SPE/DOE Symposium on Low Permeability, Denver, March 14-16.
- **Pye**, D.S. and Smith, W.A.: "Fluid Loss Additive Seriously Reduces Fracture Proppatn Conductivity and Formation Permeability," 1973 SPE Annual Meeting, Las Vegas, Sept. 30-Oct. 3.
- **Zweigle**, M.L. and Lamphere, J.C.: "Crosslinked, Water-Swellable Polymer Microgels," US Patent 4,172,066(1979).
- **Ely**, J.W.: "Methods of Using Aqueous Gels," US Patent 4,321,968(1982).
- **McDaniel**, R.R., Deysarkar, A.K., Callanan, M.J., and Kohlhaas, C.A.: "An Improved Method for Measuring Fluid Loss at Simulated Fracture Conditions," Soc. Pet. Eng. J., August 1985, 482-490.
- **Block**, J.: "Viscosifier and Fluid Loss Control System," US Patent 4,349,443(1982).
- **Penny**, G.S., Conway, M.W., and Lee, W.S.: "The Control and Modelling of Fluid Leak-Off During Hydraulic Fracturing," SPE paper 12486, 1984 Formation Damage Symposium, Bakersfield, January.
- **Green**, P.C. and Black, J.C.: "High Temperature Stable Viscosifier and Fluid Loss Control System," US Patent 4,486,318(1984).
- **Bellis**, H.E and McBride, E.F.: "Composition and Method for Temporarily Reducing Permeability of Subterranean Formations," US Patent 4,715,967(1987).
- **McGowen**, J.M. and McDaniel, B.W.: "The Effects of Fluid Preconditioning and Test Cell Design on the Measurement of Dynamic Fluid Loss Data," SPE paper 18212, 1988 SPE Annual Technical Conference and Exhibition, Houston, October 2-5.
- **Woo**, G.T. and Cramer, D.D.: "Laboratory and Field Evaluation of Fluid-Loss Additive Systems Used in the Williston Basin," SPE paper 12899, 1984 Rocky Mountain Regional Meeting, Casper, May 21-23.
- **Hall**, B.E. and Houk, S.G.: "Fluid-Loss Control in the Naturally Fractured Buda Formation," SPE paper 12152, 1983 SPE Annual Technical Conference and Exhibition, San Francisco, October 5-8.
- **Gulbis**, J.: "Dynamic Fluid Loss of Fracturing Fluids," SPE paper 12154, 1983 SPE Annual Technical Conference and Exhibition, San Francisco, October 5-8.
- **Penny**, G.S.: "Nondamaging Fluid Loss Additives for Use in Hydraulic Fracturing of Gas Wells," SPE paper 10659, 1982 SPE Formation Damage Control Symposium, Lafayette, March 24-25.

- Harris, P.C. and Penny, G.S.: "Influence of Temperature and Shear History on Fracturing Fluid Efficiency," SPE paper 14258, 1985 SPE Annual Technical Conference and Exhibition, Las Vegas, September 22-25.
- Cantu, L.A. and Boyd, P.A.: "Laboratory and Field Evaluation of a Combined Fluid-Loss Control Additive and Gel Breaker for Fracturing Fluids," SPE paper 18211, 1988 SPE Annual Technical Conference and Exhibition, Houston, October 2-5.
- Ford, W.G.F and Penny, G.S.: "Influence of Downhole Conditions on the Leakoff Properties of Fracturing Fluids," SPE Prod. Eng., 1988, 3(1), 43-51.
- Harris, P.C.: "Dynamic Fluid-Loss Characteristics of CO_2-Foam Fracturing Fluids," SPE Prod. Eng., 1987, 2(2), 89
- Roodhart, L.P.: "Fracturing Fluids: Fluid-Loss Measurements Under Dynamic Conditions," Soc. Pet. Eng. J., 1985, 25(5), 629-36.
- Zigrye, J.L., Whitfill, D.L., and Sievert, J.A.: "Fluid-Loss Control Differences of Crosslinked and Linear Fracturing Fluids," J. Pet. Technol., 1985, 37(2), 315-20.
- Crowell, R.F.: "Formation Fracturing Method," US Patent 4,442,897(1984).
- Pacholke, G., Rettig, D., Eins, I., Shul'ts, G, Foerster, M., Ballschuh, D., Rusche, J., Ohme, R., and Markert, H.: "Stimulation of Reservoir Rocks," German Patent DD 159,657(1983).
- Ely, J.W.: "Water Flooding and Fracturing Using Clean, Nondamaging Fracturing Fluids," US Patent 4,265,311(1981).
- King, G.E.: "Low Fluid Loss Foam," US Patent 4,217,231(1980).
- Pacholke, G., Rettig, D., Foerster, M., Stark, H.J., and Tide, R.: "Solution for Hydrofracturing an Oil- and Gas-Saturated Stratum," SU Patent 683,640(1979).
- Dill, W.R. and Elphingstone, E.A.: "Preparing and Using Acidizing and Fracturing Compositions, and Fluid Loss Additives for Use Therein," US Patent 4,107,057(1978).
- Hill, O.F., Ward, A.J., and Clement, C.C.: "Austin Chalk Fracturing Design Using a Crosslinked Natural Polymer as a Diverting Agent," J. Pet. Technol., 1978, 30(12), 1795-804.
- Fisher, H.B.: "Sealing a Permeable Stratum with Resin," US Patent 3,525,398(1970).
- McClaflin, G.G. and Jacocks, C.L.: "Polyolefin-Encapsulated Silica: Low-Fluid-Loss Additive for Fracturing Fluids," US Patent 3,466,242(1969).
- Harper, B.G. and Smith, C.F.: "Fluid-Loss Control in Oil Well Treatment," US Patent 3,409,548(1968).
- Kuhn, D.A. and Brown, J.L.: "Loss Control Additive for Subterranean Fracturing Fluids," US Patent 3,408,296(1968).
- Kuhn, D.A.: "Low Liquid Loss Composition," US Patent 3,405,062(1968).
- Gibson, D.L.: "Low-Fluid-Loss Additive," US Patent 3,351,079(1967).

Modelling of Fracture Behavior(Ref. 12)

- Barby, G.B. and Barbee, W.C.: "Ultra-High Conductivity Fracture Stimulations: A Case History," SPE paper 16222, 1987 SPE Production Operations Symposium, Oklahoma City, March 8-10.
- Odeh, A.S. and Yang, H.T.: "Flow of Non-Newtonian Power Law Fluids Through Porous Media," SPE paper 7150, 1977 SPE of AIME Annual Fall Technical Conference and Exhibition, Denver, October 9-12.
- Warpinski, N.R., Clark, J.A., Schmidt, R.A., and Huddle, C.W.: "Laboratory Investigation on the Effect of In-Situ Stresses on Hydraulic Fracture Containment," Soc. Pet. Eng. J., June 1982, 333-339.
- Hanson, M.E., Anderson, G.D., Shaffer, R.J., and Thorson, L.D.: "Some Effects of Stress, Friction, and Fluid Flow on Hydraulic Fracturing," Soc. Pet. Eng. J., June 1982, 321-332.
- Settari, A.: "A New General Model of Fluid Loss in Hydraulic Fracturing," SPE/DOE paper 11625, 1983 SPE/DOE Symposium on Low Permeability, Denver, March 14-16.
- Dickey, P.A. and Andresen, K.H., "The Behavior of Water-Input Wells," API, Secondary Recovery of Oil in the United States, Second Edition, 1950, 332-40.
- Scott, P.P., Jr., Bearden, W.G., and Howard, G.C.: "Rock Rupture as Affected by Fluid Properties," Soc. Pet. Eng. J.(1953) 111-124; Trans., AIME, 198.
- Hubbert, M.K. and Willis, D.G.: "Mechanics of Hydraulic Fracturing," Soc. Pet. Eng. J.(1957) 153,166,167; Trans., AIME, 210.
- Khristianovic, S.A. and Zheltov, Y.P.: "Formation of Vertical Fractures by Means of Highly Viscous Liquid," Proc., Fourth World Pet. Cong. (1954) Section II/T.O.P., 579-586.
- Perkins, T.K. and Kern, L.R.: "Widths of Hydraulic Fractures," J. Pet. Technol.(Septemeber 1961) 937-949.
- Geertsma, J. and de Klerk, F.: "A Rapid Method of Predicting Width and Extent of Hydraulically Enduced Fractures," J. Pet. Technol.(1969) 1571-1581.
- Whitsitt, N.F. and Dysart, G.R.: "The Effect of Temperature on Stimulation Design," J. Pet. Technol.(April 1970) 493,495-502.
- Daneshy, A.A.: "Opening of a Pressurized Fracture in an Elastic Medium," SPE paper 7616, 1971 Annual Technical Meeting of SPE of CIM, Banff, June 2-4.
- Nordgren, R.P.: "Propagation of a Vertical Hydraulic Fracture," Soc. Pet. Eng. J.(August 1972), 306-314.

■ **Harrington**, L.J., Hannah, R.R., and Williams, D.: "Dynamic Experiments on Proppant Settling in Crosslinked Fracturing Fluids," SPE paper 8342, 1979 SPE Annual Technical Conference and Exhibition, Las Vegas, September 23-26.

■ **Hannah**, R.R., Harrington, L.J. and Potter, J.S. Jr.: "Post Fracturing Behavior of Fracturing Gels and Its Influence on Fracture Closure and Proppant Distribution," SPE paper 9331, 1980 Annual Fall Technical Conference and Exhibition of SPE of AIME, Dallas, September 21-24.

■ **Cloud**, J.E. and Clark, P.E.: "Stimulation Fluid Rheology III. Alternatives to the Power Law Fluid Model for Crosslinked Fluids," SPE paper 9332, 1980 SPE Annual Technical Conference and Exhibition, Dallas, September 21-24.

■ **Rogers**, R.E., Veatch, R.W., Jr. and Nolte, K.G.: "Pipe Viscometer Study of Fracturing Fluid Rheology," SPE paper 10258, 1981 SPE Annual Technical Conference and Exhibition, San Antonio, October 5-7.

■ **White**, J.L, and Daniel, E.F.: "Key Factors in MHF Design," J. Pet. Technol., August 1981, 1501-1512.

■ **Clark**, P.E. and Quadir, J.A.: "Prop Transport in Hydraulic Fractures: A Critical Review of Particle Settling Velocity Equations," SPE/DOE paper 9866, 1981 SPE/DOE Low Permeability Symposium, Denver, May 27-29.

Clark, P.E. and Guler, N.: "Prop Transport in Vertical Fractures: Settling Velocity Correlations," 1983 SPE/DOE paper 11636, SPE/DOE Symposium on Low Permeability, Denver, March 14-16.

■ **Torr**, R.S.: "Particle Settling in Viscous Non-Newtonian Hydroxyethyl Cellulose Polymer Solutions," AlChE Journal (Vol. 29, No. 3) May, 1983, pages 506-508.

■ **Meyer**, B.R.: "Generalized Drag Coefficient Applicable for All Flow Regimes," Oil Gas J., May 26, 1986, 71-77.

■ **Lord**, D.L.: "Turbulent Flow of Stimulation Fluids: An Evaluation of Friction Loss Scale-Up Methods," SPE paper 16889, 1987 SPE Annual Technical Conference and Exhibition, Dallas, September 27-30.

■ **Lee**, W.S. and Daneshy, A.A.: "Fracture Geometry and Proppant-Transport Computation for Multiple-Fluid Treatment," SPE Prod. Eng. J., (November 1987) 257-266.

■ **McDaniel**, B.W. and Hoch, O.F.: "Realistic Proppant Conductivity Data Improve Hydraulic Fracturing Treatment Design," paper No. 87-38-73, 1987 Annual Technical Meeting of the Petroleum Society of CIM, Calgary, June 7-10.

■ **Daneshy**, A.A.: "Numerical Solution of Sand Transport in Hydraulic Fracturing," J. Pet. Technol., January, 1978, 132-140.

■ **Clark**, P.E., Harkin, M.W., Wahl, H.A., and Sievert, J.A.: "Design of a Large Vertical Prop Transport Model," SPE paper 6814, 1977 Annual Fall Technical Conference and Exhibition of the SPEngineers of AIME, Denver, October 9-12.

■ **Shah**, S.N.: "Proppant Settling Correlations for Non-Newtonian Fluids Under Static and Dynamic Conditions," SPE paper 9330, 1980 Annual Fall Technical Conference and Exhibition of the SPE of AIME, Dallas, September 21-24.

■ **Torrest**, R.S.: "Aspects of Slurry and Particle Settling and Placement for Viscous Gravel Packing (AQUAPAC)," SPE paper 11009, 1982 Annual Fall Technical Conference and Exhibition of the SPE of AIME, New Orleans, September 26-29.

■ **Torrest**, R.S.: "The Flow of Viscous Polymer Solutions for Gravel Packing Through Porous Media," SPE paper 11010, 1962 Annual Fall Technical Conference and Exhibition of the SPE of AIME, New Orleans, September 26-29.

■ **Roodhart**, L.P.: "Proppant Settling in Non-Newtonian Fracturing Fluid," SPE paper/DOE 13905, 1985 Low Permeability Gas Reservoirs, Denver, May 19-22.

■ **Kirkby**, L.L.: "Proppant Settling Velocities in Nonflowing Slurries," SPE paper/DOE 13906, 1985 Low Permability Gas Reservoirs, Denver, May 19-22.

■ **Acharya**, A.: "Particle Transport in Viscous and Viscoelastic Fracturing Fluids," SPE Prod. Eng., March 1986, 104-110.

■ **Peden**, J.M. and Luo, Y.: "Settling Velocity of Variously Shaped Particles in Drilling and Fracturing Fluids," SPE Drilling Engineering, December 1987, 337-343.

■ **Wahlmeier**, M.A. and Andrews, P.W.: "Mechanics of Gravel Placement and Packing: A Design and Evaluation Approach," SPE Prod. Eng., February 1988, pages 69-82.

■ **Baumgartner**, S.A. and Mack, D.J.: "On-Site Computer Monitoring of Foamed Stimulation Fluids," SPE paper 17531, 1988 SPE Rocky Mountain Regional Meeting, Casper, May 11-13.

■ **Novotny**, E.J.: "Proppant Transport," SPE paper 6813, 1977 Annual Fall Technical Conference and Exhibition of the SPE of AIME, Denver, October 9-12.

■ **Misak**, M.D., Atteberry, R.D., Venditto, J.J. and Fredrickson, S.E.: "A Fracturing Technique to Minimize Water Production," SPE paper 7563, 1978 Annual Fall Technical Conference and Exhibition of the SPE of AIME, Houston, October 1-3.

■ **Teeuw**, D. and Hesselink, F.T.: "Power-Law Flow and Hydrodynamic Behavior of Biopolymer Solutions In Porous Media," SPE paper 8982, 1980 SPE Fifth International Symposium on Oilfield and Geothermal Chemistry, Stanford, May 28-30.

■ **Babcock**, R.E., Prokop, C.L. and Kehle, R.O.: "Distribution of Propping Agents in Vertical Fractures," paper No. 851-41-A, 1967 API Spring Meeting of the Mid-Continent District Division of Production, Oklahoma City, March 29-31.

- **Bedeaux**, D.: "The Effective Viscosity for a Suspension of Spheres," J. Coll. Interface Sci., 1987, 118(1), 80-86.
- **Morita**, N., Whitfill, D.L., and Wahl, H.A.: "Stress-Intensity Factor and Fracture Cross-Sectional Shape Preclictions From a 3-D Model for Hydraulically Induced Fractures," SPE paper 14262, 1985 SPE Annual Technical Conference and Exhibition, Las Vegas, September 22-25.
- **Cunningham**, R.D. and Nelson, R.G.: "A New Method for Determining a Well's In-Place Hydrocarbons From a Pressure Buildup Test," J. Pet. Technol.(July 1967) 859-866.
- **Acharya**, A. and Deysarkar, A.K.: "Rheology of Fracturing Fluids at Low-Shear Conditions," SPE paper 16917, 1987 Annual Technical Conference and Exhibition, Dallas, September 27-30.
- **Wilson**, Ed, B. and Hsu, T.R.: "A Technique for the Measurement of Fracture Toughness of Oil Sands," Proc. Tar Sands Symp. 1986, 150-5.
- **Jefri**, M.A., Nichols, K.L., and Jayaraman, K.: "Sedimentation of Two Contacting Spheres in Dilute Polymer Solutions," Proc. Symp. Recent Dev. Struct. Continua., 1985, 21-5.
- **Al-Atter**, H.H.: "The Combined Effect of Oil Viscosity, Initial Water Saturation and Water Injection Rate on the Performance of Fractured Oil Reservoirs," J. Pet. Res., 1984, 3(2), 1-16.
- **Daneshy**, A.A.: "On the Design of Vertical Hydraulic Fractures," J. Pet. Technol.,(1973) 83-97.
- **van Domselaar**, H.R. and Visser, W.: "Proppant Concentration In and Final Shape of Fractures Generated by Viscous Gels,"(1974) 531-536.
- **Harrington**, L. and Hannah, R.R.: "Fracturing Design Using Perfect Support Fluids for Selected Fracture Proppant Concentrations in Vertical Fractures," 1975 SPE Annual Meeting, Dallas, September 28-October 1.
- **Smith**, J.E.: "Effect of Incomplete Fracture Fill-Up at the Wellbore on Productivity Ratio," 1975 Proc. Annu. Southwest Pet. Short Course, 135-144.
- **Nolte**, K.G. and Smith, M.B.: "Interpretation of Fracturing Pressures," SPE paper 8297, 1979 SPE Annual Technical Conference and Exhibition, Las Vegas, September 23-26.
- **Dobkins**, T.A.: "Methods to Better Determine Hydraulic Fracture Height," SPE paper 8403, 1979 SPE Annual Fall Technical Conference and Exhibition, Las Vegas, September 23-26.
- **Cleary**, M.P.: "Comprehensive Design Formulae for Hydraulic Fracturing," SPE paper 9259, 1980 SPE Annual Technical Conference and Exhibition, Dallas, September 21-24.
- **Cleary**, M.P.: "Mechanisms and Procedures for the Production of Desirable Fracture Shapes in Representative Reservoir Formations," SPE paper 9260, 1980 SPE Annual Technical Conference, Dallas, September 21-24.
- **van Eekelen**, H.A.M.: "Hydraulic Fracture Geometry: Fracture Containment in Layered Formations," Soc. Pet. Eng. J., June 1982, 341-349.
- **Nolte**, K.G.: "Fracture Design Considerations Based on Pressure Analysis," SPE paper 10911, 1982 SPE Cotton Valley Symposium, Tyler, May 20. See also Nolte, K.G.: "Principles for Fracture Design Based on Pressure Analysis," SPE Prod. Eng., February 1988, 22-30; Nolte, K.G.: "Application of Fracture Design Based on Pressure Analysis," SPE Prod. Eng., February 1988, 31-42.
- **McLennan**, J.D. and Roegiers, J.C.: "How Instantaneous are Instantaneous Shut-In Pressures?," SPE paper 11064, 1982 SPE Annual Technical Conference and Exhibition, New Orleans, September 26-29.
- **Smith**, M.B., Rosenberg, R.J., and Bowen, J.F.: "Fracture Width-Design vs. Measurement," SPE paper 10965, 1982 SPE Annual Technical Conference and Exhibition, New Orleans, September 26-29.
- **Soliman**, M.Y.: "Modifications to Production Increase Calculations for a Hydraulically Fractured Well," J. Pet. Technol.(1983) 170-172.
- **McLeod**, H.O., Jr.: "A Simplified Approach to Design of Fracturing Treatments Using High Viscosity Crosslinked Fluids," SPE/DOE paper 11614, 1983 SPE/DOE Symposium on Low Permeability, Denver, March 14-16.
- **Ahmed**, U., et al.: "Effect of Stress Distribution on Hydraulic Fracture Geometry: A Laboratory Simulation Study in One Meter Cubic Blocks," SPE/DOE paper 11637, 1983 SPE/DOE Symposium on Low Permeability, Denver, March 14-16.
- **Nierode**, D.E.: "Comparison of Hydraulic Fracture Design Methods to Observed Field Results," SPE paper 12059, 1983 SPE Annual Technical Conference and Exhibition, San Francisco, October 5-8.
- **Lee**, W.S.: "Pressure Decline Analysis with the Christianovich and Zheltov and Penny-Shaped Geometry Model of Fracturing," SPE/DOE paper 13872, 1985 SPE Low Permeability Gas Reservoirs, Denver, May 19-22.
- **Soliman**, M.: "Fracture Conductivity Distribution Studied," Oil Gas J., February 10, 1986, 89-93.
- **Medlin**, W.L. and Masse, L.: "Plasticity Effects in Hydraulic Fracturing," J. Pet. Technol.,(1986) 995-1006.
- **Conway**, M.W., McGowen, J.M., Gunderson, D.W., and King, D.G.: "Prediction of Formation Response from Fracture Pressure Behavior," SPE paper 14263, 1985 SPE Annual Technical Conference and Exhibition, Las Vegas, September 22-25.

- Settari, A. and Cleary, M.P.: "Development and Testing of a Pseudo-Three-Dimensional Model of Hydraulic Fracture Geometry," SPE Prod. Eng., July 1986, 449-466.
- Biot, M.A., Masse, L., and Medlin, W.L.: "A Two-Dimensional Theory of Fracture Propagation," SPE Prod. Eng., January 1986, 17-30.
- Thiercelin, M.J. and Lemanczyk, Z.R.: "Stress Gradient Affects the Height of Vertical Hydraulic Fractures," SPE Prod. Eng., July 1986, 245-254.
- Nolte, K.G.: "Determination of Proppant and Fluid Schedules from Fracturing-Pressure Decline," SPE Prod. Eng., July 1986, 255-265.
- Shelley, R.F. and McGowen, J.M.: "Pump-In Test Correlation Predicts Proppant Placement," SPE paper 15151, 1986 SPE Rocky Mountain Regional Meeting, Billings, May 19-21.
- Sookprasong, P.A.: "Plot Procedure Finds Closure Pressure," Oil Gas J., September 8, 1986, 110-112.
- Lord, D.L. and McGowen, J.M.: "Real-Time Treating Pressure Analysis Aided by New Correlation," SPE paper 15367, 1986 SPE Annual Technical Conference and Exhibition, New Orleans, October 5-8.
- Poulsen, D.K. and Soliman, M.Y.: "A Procedure for Optimal Hydraulic Fracturing Treatment Design," SPE paper 15940, 1986 SPE Eastern Regional Meeting, Columbus, November 12-14.
- Acharya, A. and Kim, C.M.: "Hydraulic Fracture Treatment Design Simulation for the Rotliegendes Formation," SPE/DOE paper 16414, 1987 SPE/DOE Low Permeability Reservoirs Symposium, Denver, May 18-19.
- Branagan, P., Cipolla, C., and Lee, S.J.: "Designing and Evaluating Hydraulic Fracture Treatments in Naturally Fractured Reservoirs," SPE/DOE paper 16434, 1987 SPE/DOE Low Permeability Reservoirs Symposium, Denver, May 18-19.
- Lee, W.S.: "Fracture Propagation Theory and Pressure Decline Analysis with Langrangian Formulation for Penny-Shaped and Perkins-Kern Geometry Models," SPE paper 17151, SPE Formation Damage Control Symposium, Bakersfield, February 8-9.
- Cooper, G.D., Nelson, S.G., and Schopper, M.D.: "Improving Fracturing Design Through the Use of an On-Site Computer System," SPE paper 12063, 1983 SPE Annual Technical Conference and Exhibition, San Francisco, October 5-8.
- Roegiers, J.-C. and Ishijima, Y.: "A Coupled Fracturing Model and Its Application to Hydraulic Fracturing," SPE paper 12311, Eastern Regional Meeting, Champion, November 9-11.
- Branagan, P.T., et al.: "Comprehensive Well Testing and Modeling of Pre- and Post-Fracture Well Performance of the MWX Lenticular Tight Gas Sands," SPE paper/DOE 13867, 1985 SPE/DOE Low Permeability Gas Reservoirs, Denver, May 19-22.
- Read, D.A. and Wells, G.L.: "Measurement While Fracturing for Comparing and Optimizing the Performance of Well Stimulation Treatments," 1985 Proc. Annu. Southwest Pet. Short Course, 171-176.
- Warpinski, N.R., et al.: "Fracturing and Testing Case Study of Paludal, Tight, Lenticular Gas Sands," SPE/DOE paper 13876, 1985 SPE/DOE Low Permeability Gas Reservoirs, Denver, May 19-22.
- Warembourg, P.A., et al.: "Fracture Stimulation Design and Evaluation," SPE paper 14379, 1986 SPE Proc. Rocky Mountain Regional Meeting, Billings, May 18-21, 359-370.
- Meyer, B.R.: "Design Formulae for 2-D and 3-D Vertical Hydraulic Fractures: Model Comparison and Parametric Studies," SPE paper 15240, 1986 SPE Unconventional Gas Technology Symposium, Louisville, May 18-21.
- Lam, K.Y., Cleary, M.P., and Barr, D.T.: "A Complete Three-Dimensional Simulator for Analysis and Design of Hydraulic Fracturing," 1986 spe Unconventional Gas Technology Symposium, Louisville, May 18-21.
- Crockett, A.R., Willis, R.M. Jr., and Cleary, M.P.: "Improvement of Hydraulic Fracture Predictions by Real-Time History Matching on Observed Pressures," SPE paper 15264, 1986 SPE Unconventional Gas Technology Symposium, Louisville, May 18-21.
- Vandamme, L., Jeffrey, R.G., and Curran, J.H.: "Pressure Distribution in Three-Dimensional Hydraulic Fractures," SPE paper 15265, 1986 SPE Unconventional Gas Technology Symposium, Louisville, May 18-21.
- Elbel, J.L.: "Designing Hydraulic Fractures for Efficient Reserve Recovery," SPE paper 15231, 1986 SPE Unconventional Gas Technology Symposium, Louisville, May 18-21.
- Nolte, K.G.: "Fluid Flow Considerations in Hydraulic Fracturing," SPE Paper 18537, 1988 SPE Eastern Regional Meeting, Charleston, November 1-4.
- McLeod, H.O., Jr.: "The Effect of Perforating Conditions on Well Performance," SPE paper 10649, 1982 SPE Formation Damage Control Symposium, Lafayette, March 24-25.
- Jopling, M.W. and Ketcher, N.W.: "Real Time Decision-Making Through Use of a Log/Log Plot During a Frac Job," SPE paper 16223, 1987 SPE Production Operations Symposium, Oklahoma City, March 8-10.
- Cipolla, C.L. and Lee, S.J.: "The Effect of Excess Propped Fracture Height on Well Productivity," SPE paper 16219, 1987 SPE Production Operations Symposium, Oklahoma City, March 8-10.
- Lee, W.S.: "Study of the Effects of Fluid Rheology on Minifrac Analysis," SPE paper 16916, 1987 SPE Annual Technical Conference and Exhibition, Dallas, September 27-30.

■ Ben Naceur, K., Touband, E., and Roegiers, J.-C.: "Numerical Investigation of the Effects of Fluid Rheological Properties on 3-D Fracture Propagation,"CIM paper 87-37-26, 1986 Annual Technical Meeting of CIM, 8620268, Calgary, June 8-11.

■ Vandamme, L., Jeffrey, R.G. and Curran, J.H.: "Effects of Three-Dimensionalization on a Hydraulic Fracture Pressure Profile," 1986 Proc. U.S. Sumposium on Rock Mechanics; Key to Energy Production: 580-590, Tuscaloosa, Alabama, June 23-25.

■ Ehlig-Economides, C.: "Use of the Pressure Derivative for Diagnosing Pressure-Transient Behavior," J. Petr. Technol., October 1988, 1280-1282.

■ McDaniel, B.W.: "Realistic Fracture Conductivities of Proppants as a Function of Reservoir Temperature," SPE/DOE paper 16453, 1987 SPE/DOE Low Permeability Reservoirs Symposium, Denver, May 18-19.

■ McDaniel, B.W.: "Use of Wet Gas Flow for Long-Term Fracture Conductivity Measurements in the Presence of Gel Filter Cakes," SPE paper 17543, 1988 SPE Rocky Mountain Regional Meeting, Casper, May 11-13.

■ McDaniel, B.W. and Parker, M.A.: "Accurate Design of Fracturing Treatment Requires Conductivity Measurements at Simulated Reservoir Conditions," SPE paper 17541, 1988 SPE Rocky Mountain Regional Meeting, Casper, May 11-13.

■ Cunningham, R.D. and Nelson, R.G.: "A New Method for Determining a Well's In-Place Hydrocarbons From a Pressure Buildup Test," J. Petr. Technol. July 1967, 859-866.

■ Parker, M.A. and McDaniel, B.W.: "Fracturing Treatment Design Improved by Conductivity Measurements Under In-Situ Conditions," SPE paper 16901, 1987 SPE Annual Technical Conference and Exhibition, Dallas, September 27-30.

■ Cunningham, R.D. and Nelson, R.G.: "A New Method for Determining a Well's In-Place Hydrocarbons From a Pressure Buildup Test," J. Pet. Technol.(July 1967), 859-866.

■ Warpinski, N.R.: "In Situ Measurements of Hydraulic Fracture Behavior," Sandia Report SAND83-1826, July 1985.

■ Aguilera, R.: "Detection and Evaluation of Naturally Fractured Reservoirs from Logs," SPE paper 4398, 1973 SPE Rocky Mountain Regional Meeting, Casper, May 15-16.

■ Nelson, R.A.: "An Approach to Evaluating Fractured Reservoirs," J. Pet. Technol., September 1982, 2167-2170.

■ Kiel, O.M.: "The Kiel Process - Reservoir Stimulation by Dendritic Fracturing," SPE paper 6984.

■ Celik, M.S., Ananthapadmanabhan, K.P., and Somasundaran, P.: "Precipitation/Redissolution Phenomena in Sulfonate/AlCl, Solutions," SPE paper 11796, 1983 SPE International Symposium on Oilfield and Geothermal Chemistry, Denver, June 1-3.

■ Settari, A. and Price H.S.: "Simulation of Hydraulic Fracturing in Low Permeability Reservoirs," SPE/DOE paper 8939, Soc. Pet. Eng. J., April 1984, 141-52.

Chemistry of Gelled Oils(Ref. 42)

■ Minich, A. and Nowak, M.: "Oil Soluble Gelling Agent," US Patent 2,618,596, 1952.

■ Lundberg, R.D., Peiffer, D.G., Sedillo, L.P. and Newlove, J.C.: "Hydrocarbon Soluble Polymer Complexes Useful as Viscosifiers in Fracturing Operations," US Patent 4,579,671, 1986.

■ Hochwalt, C.A., Lum, J.H., Malowan, J.E., and Dyer, C.P.: "Alkyl Esters of Phosphoric Acid," Ind. Eng. Chem., 1942, 34(1), 20-25.

■ Self, E.S.: "Oil Base Drilling Fluid," US Patent 2,461,483(1949).

■ Pellegrini, J.P. Jr. and Strange, H.O.: "Mineral Oil Containing A Rare Earth Metal Diester Phosphate," US Patent 2,983,679(1961).

■ Revukas, A.J.: "Hydrocarbon Fuel Composition," US Patent 3,334,978(1967).

■ Kerschner, P.M. and Hess, F.G.: "Amine Salts of Metal Organo Orthophosphates," US Patent 3,338,935(1967).

■ Kiel, O.M.: "A New Hydraulic Fracturing Process," J. Pet. Technol.(January 1970), 89-96.

■ Poklacki, E.S.: "Polyamine Salts of Aluminum Alkyl Orthophosphates," US Patent 4,007,128(1977).

■ Griffin, T.J. Jr.: "Phosphate Ester-Type Reaction Product and Method of Preparing Same," US Patent 4,153,649(1979).

■ Griffin, T.J. Jr.: "Gelling of organic Liquids," US Patent 4,152,289(1979).

■ Griffin, T.J. Jr.: "Method of Fracturing with Gelled Organic Liquids," US Patent 4,174,283(1979).

■ Burnham, J.W. and Tiner, R.L.: "Fracturing Compositions and Method of Preparing and Using the Same," US Patent 4,200,539(1980).

■ Burnham, J.W.: "Method for Fracturing Subterranean Formations," US Patent 4,200,540(1980).

■ Burnham, J.W.: "Gelled Oil Base Compositions and Methods of Preparation and Use of Same," US Patent 4,316,810(1982).

■ Daccord, G., Lemanczyk, R., and Vercaemer, C.: "Method for Obtaining Gelled Hydrocarbon Compositions, the Compositions According to Said Method and Their Application in the Hydraulic Fracturing of Underground Formations," US Patent 4,507,213(1985).

■ Harris, L.E., Holtmyer, M.D. and Pauls, R.W.: "Method for Fracturing Subterranean Formations," US Patent 4,622,155(1986).

- **Hassen**, B.R., Porter, K.E., and McCorriston, L.L.: "Improving Oilwell Stimulations with Compatible Oils," <u>J. Can. Pet. Technol.</u>,(November-December, 1986) Montreal, 32-38.
- **Gross**, J.M.: "Gelling Organic Liquids," European Patent Application 0,225,661(1986).
- **Gordon**, D.A.: "Gels Increase Well Fracturing Efficiency," <u>Drilling Contractor</u>(November 1985) 54-55.
- **Kucera**, C.H., Smith, C.F. and Braunlich, F.H.: "New Oil Gelling Systems Prevent Damage in Water-Sensitive Sands," SPE paper 3503, 1971 SPE Annual Meeting, New Orleans, October 3-6.
- **Chatterji**, J., Holtmyer, M.D. and Tiner, R.L.: "Gelling Liquid Hydrocarbons," US Patent 3,900,070(1975).
- **Clark**, H.B.: "Use of Fluorochemical Surfactants in Nonaqueous Stimulation Fluids," <u>J. Pet. Technol.</u>(October 1980) 1695-1697.
- **Swanson**, B.L.: "Oil Displacement Method Using Shear-Thickening Compositions," US Patent 4,289,203(1981).
- **Gregorcyk**, W, Pauls, R., Venditto, J., Chisholm, P., and Holtmyer, M.: "Successful Stimulation of the Olmos Formation Using Oil-Base Fluids and High Proppant Concentrations in the A.W.P. Field of McMullen County, Texas," SPE 1984 Spring Symposium, Corpus Christi, March.
- **Rutinton**, R.J., Jr.: "Organic Gels," US Patent 4,537,700(1985).
- **Pellegrini**, J.P., Jr. and Strange, H.O.: "Synthetic Oil Containing a Rare Earth Metal Diester Phosphate," US Patent 2,983,678(1961).
- **Copes**, J.P., Mayhew, R.L., and Williams, E.P.: "Petroleum Hydrocarbon Compositions," US Patent 3,012,966(1961).
- **Matt**, J.W.: "Conductivity Additive for Liquid Hydrocarbons," US Patent 3,758,283(1973).
- **Jaggard**, W. and Scales, A.A.: "Gel-Like Composition for Use as a Pig in a Pipeline," US Patent 4,003,393(1977).
- **Horn**, J.M., Johnston, B.E., Napier, R.P. and Williams, T.N.: "Process for the Preparation of Dialkyl Phosphoric Acids," US Patent 4,288,392(1981).
- **Rueggeberg**, W.H.C. and Chernack, J.: "Alcoholysis of Ethyl Phosphate. The Preparation of Mixed Ethyl Butyl Phosphates," <u>J. Amer. Chem. Soc.</u>, 1948, 70, 1802-1804.
- **Gay**, R.L., Schlott, R.J., and Burroughs, J.E.: "Gelling Agents for Hydrocarbon Compounds," US Patent 4,104,173(1978).
- **Turner**, S.R., Walker, T.O. and Thaler, W.A.: "Sodium Styrene Sulfonate-Co-Sodium-N-(4-Sulfo-Phenyl)-Maleimide- An Improved Viscosity Control Additive," US Patent 4,478,727(1984).
- **Agarwal**, P.K. and Lundberg, R.D.: "Viscoelastic Behavior of Concentrated Oil Solutions of Sulfo Polymers. 2. EPDM and Zinc Sulfo-EPDMs" <u>Macromolecules</u> 1984, 17, 1918-1928.
- **Thaler**, W.A., Brois, S.J. and Ferrara, F.W.: "Sulfomaleation of EPDM Polymers," <u>Macromolecules</u>, 1987, 20, 254-258.
- **Duvdevani**, I., Wagensommer, J., and Agarwal, P.K.: "Novel Interpolymer Complexes of Sulfonated octene-1 Copolymer," US Patent 4,634,542(1987).
- **Zulauf**, M and Eicke, H.F.: "Inverted Micelles and Microemulsions in the Ternary System H_2O/Aerosol-OT/Isooctane as Studied by Photon Correlation Spectroscopy," <u>J. Phys. Chem.</u>, 1979, 82(4), 480-486.
- **Patel**, A.D. and Salandanan, C.S.: "Thermally Stable Polymeric Gellant for Oil-Base Drilling Fluids," SPE paper 13560, 1985 International Symposium on Oilfield and Geothermal Chemistry, Phoenix, April 9-11.
- **Deguchi**, K. and Mino, J.: "Solution Properties of Long-Chain Dialkyldimethylammonium Salts," <u>J. Coll. Interfac. Sci.</u>, 1978, 65(1), 155-161.
- **Kunitake**, T.: "Chemistry of Synthetic Bilayer Membranes," <u>J. Macromol. Sci. - Chem.</u>, 1979, Al3(5), 587-602.
- **McNeil**, R. and Thomas, J.K.: "On the Nature of Surfactant Vesicle and Micelle Systems," <u>J. Coll. Interfac. Sci.</u>, 1980, 73(2), 522-529.
- **Grantham**, C.K. and McLaurine, H.C.: "Thixotropy Without Viscosity: A New Approach to Rheology Control of Oil Muds," SPE paper 15415, 1986 SPE Annual Technical Conference and Exhibition of AIME, New Orleans, October 5-8.
- **MacDonald**, R.G. and Frank , J.: "Sand Fracturing the Slave Point Carbonate," <u>J. Can. Pet. Technol.</u>, November-December 1986, Montreal, 39-47.
- **Portnoy**, R.C., Lundberg, R.D., and Werlein, E.R.: "Novel Polymeric Oil Mud Viscosifier for High-Temperature Drilling," IADC/SPE paper 14795, 1986 IADC/SPE Drilling Conference, Dallas, February 10-12.
- **Culter**, J.D. and McClaflin, G.G.: "Method of Friction Loss Reduction in Oleaginous Fluids Flowing Through Conduits," US Patent 3,692,676(1972).
- **Driscoll**, P.L., Bowen, J.G., and Roberts, M.A.: "Oil Base Foam Fracturing Applied to the Niobrara Shale Formation," 1980 SPE Annual Fall Technical Conference and Exhibition of AIME, Dallas, September 21-24.
- **Sedilo**, L.P., Newlove, J.C., Peiffer, D.G., and Lundberg, R.D.: "Hydrocarbon Soluble Polymer Complexes Useful as Viscosifiers in Fracturing Operations," US Patent 4,615,393(1986).
- **Morgenthaler**, L.N. and Mikols, W.J.: "Oil-Base Well-Treating Fluids Thickened by Block Copolymers," US Patent 4,595,513(1986).

■ **Goncalves**, A.A. and Saiago, C.A.: "Composition for Use in Fracturing Petroleum-Containing Formations," Brazil Patent 84/214(1985), <u>Chemical Abstracts</u> 104(10):71519c(1986).
■ **Daccord**, G., Lemanczyk, R., and Vercaemer, C.: "Gelled Hydrocarbon Mixtures and Their Use in Hydraulic Crack Formation in Underground Formations," German Patent 3,247,758(1983), <u>Chemical Abstracts</u> 99(12):90859h(1983).
■ **Burnham**, J.W.: "Fracturing Underground Formations," German Patent 2,915,455(1979), <u>Chemical Abstracts</u> 92(16):131842y(1980).
■ **Griffin**, T.J., Jr.: "Fracturing with Gelled Organic Liquids," US Patent 4,174,283(1979).
■ **Hill**, D.G.: "Reducing the Viscoisty of an Organic Liquid, and Fracturing an Underground Formation," Brazil Patent 74/10408(1975), <u>Chemical Abstracts</u> 87(12):87655k(1977).
■ **Gay**, R.L., Schlott, R.J. and Burroughs, J.E.: "Gelling Agents for Hydrocarbon Compounds," CA Patent 974,539(1975), <u>Chemical Abstracts</u> 90(10):74182f(1979).
■ **Crawford**, D.L., Earl, R.B., and Monroe, R.F.: "Oil Well Fracturing Fluid Additive," GB Patent 1,355,080(1974), <u>Chemical Abstracts</u> 81(18):108337b(1974).
■ **Ely**, J.W. and Tiner, R.L.: "Hydraulic Fracturing Method Using Benzoic Acid to Further Increase the Viscosity of Liquid Hydrocarbon," US Patent 3,799,267(1974).
■ **Slagel**, R.C. and Bloomquist, A.E.: "Friction Reduction in Flowing Hydrocarbon Fluids," US Patent 3,779,969(1973).
■ **Shearn**, G.P. and Kiel, O.M.: "Compositions for Use in Hydraulic Fracturing Operations Involving Wells Penetrating Subterranean Formations," US Patent 3,501,198(1971).
■ **Malone**, W.T., Holtmyer, M.D., Tinsley, J.M., and Chattopadhyay, J.: "Additive for Reducing Friction Pressure Loss of Liquid Hydrocarbons Flowing Through Pipes," German Patent 2,056,700(1971), <u>Chemical Abstracts</u> 75(12):78860y(1971).
■ **Weltmann**, R.N. and Green, H.: "Rheological Properties of Colloidal Solutions, Pigment Suspensions, and Oil Mixtures," <u>J. Applied Physics</u>, 1943, 14, 569-576.
■ **Lundberg**, R.D. and Peiffer, D.G.: "Viscosification of Hydrocarbon Fluid," US Patent 4,448,926(1984).

Acidizing, Fracture Acidizing(Ref. 54)
■ **Hessert**, J.E. and Bertus, B.J.: "Method for Acidizing Subterranean Formations," US Patent 4,068,714(1978).
  Tate, J.F.: "Secondary Recovery Process," US Patent 4,200,151(1980).
■ **Dill**, W.R. and Elphingstone, E.A.: "Method of Preparing and Using Acidizing and Fracturing Compositions, and Fluid Loss Additives for Use Therein," US Patent 4,107,057(1978).
■ **Peiffer**, D.G., Lundberg, R.D., and Turner, S.R.: "Intramolecular Polymeric Complexes-viscosifiers for Acid, Base and Salt (Aqueous) Solutions," US Patent 4,461,884(1984).
■ **Abrams**, A., Schuerman, R.F., Templeton, C.C., and Richardson, E.A.: "Higher-pH Acid Stimulation Systems," <u>J. Pet. Technol.</u>, December 1983, 2175-2184.
■ **Crowe**, C.W.: "Guidelines for Selecting Iron Stabilizers for Use in Acidizing Treatments," 1980 Proc. Annu. Southwest. Pet. Short Course, Lubbock, April 17-18, 39-45.
■ **Pabley**, A.S.: "Method for Stimulating Siliceous Subterranean Formations," US Patent 4,428,432(1985).
■ **Pabley**, A.S. and Holcomb, D.L.: "A New Method of Acidizing or Acid Fracturing: Crosslinked Acid Gels," 1980 Proc. Annu. Southwest. Pet. Short Course, Lubbock, April 17-18, 31-38.
■ **Engelhardt**, F., Kuhlein, K., Riegel, U., von Halasz, S., Dawson, J.C., and Reed, A.R.: "Water Soluble Copolymers for Use in Fracture-Acidizing of Wells," US Patent 4,500,437(1985).
■ **Holcomb**, D.L., Jr.: "A New Concept in Hydrochloric-Hydrofluoric Acid Mixtures for Acidizing Low Permeability Sandstone Formations," 1974 Proc. Annu. Southwest. Pet. Short Course, Lubbock, 41-8.
■ **Clampitt**, R.L.: "Gelled Prod® Fluid for High Temperature Fracturing," 1975 Proc. Annu. Southwest. Pet. Short Course, Lubbock.
■ **Hall**, B.E.: "A New Technique for Generating In-Situ Hydrfluoric Acid for Deep Clay Damage Removal," SPE paper 6512, 1977 SPE Annual California Regional Meeting of AIME, Bakersfield, April 13-15.
■ **Watkins**, D.R. and Roberts, G.E.: "On-Site Acidizing Fluid Analysis Shows HCl and HF Contents Often Varied Substantially From Specified Amounts," <u>J. Pet. Technol.</u>, May 1983, 865-871.
■ **Hartley**, R. and Bosma, M.G.R.: "Fracturing in Chalk Completions," <u>J. Pet. Technol.</u>, January 1985, 73-79.
■ **Crowe**, C.W.: "New Treating Technique to Remove Bacterial Residues from Water-Injection Wells," <u>J. Pet. Technol.</u>, May 1968, 475-478.
■ **McLeod**, H.O., Jr., Ledlow, L.B., and Till, M.V.: "The Planning, Execution, and Evaluation of Acid Treatments in Sandstone Formations," SPE paper 11931, 1983 SPE Annual Technical Conference and Exhibition, San Francisco, October 5-8.
■ **Clementz**, D.M., Patterson, D.E., Aseltine, R.J., and Young, R.E.: "Stimulation of Water Injection Wells in the Los Angeles Basin Using Sodium Hypochlorite and Mineral Acids," SPE paper 10624, 1982 SPE International Symposium on Oilfield and Geothermal Chemistry, Dallas, January 25-27.

- **Harris**, L.E.: "High Viscosity Acidic Treating Fluids and Methods of Forming and Using the Same," US Patent 4,324,668(1982).
- **Crowe**, C.W.: "Status Report: Acid Fracturing," PET. ENG., International, June 1988, 39,41.
- **Sutton**, G.D. and Lasater, R.M.: "Aspects of Acid Additive Selection in Sandstone Acidizing," SPE paper 4114, 1972 SPE Annual Fall Meeting of AIME, San Antonio, October 8-11.
- **Norman**, L.R., Conway, M.W. and Wilson J.M.: "Temperature Stable Acid Gelling Polymers. Laboratory Evaluation and Field Results," SPE paper 10260, 1981 SPE Annual Fall Technical Conference and Exhibition of AIME, San Antonio, October 5-7.
- **Broaddus**, G.C.: "Preflush Concepts in Fracture Acidizing," 1975 Proc. Annu. Southwest Pet. Short Course, 91-97.
- **van Domselaar**, H.R., Schols, R.S. and Visser, W.: "An Analysis of the Acidizing Process in Acid Fracturing," Soc. Pet. Eng. J., August, 1973, 239-250.
- **Staudt**, J.G.: "Acidizing," Pet. Eng.(July 1954) B55-B58.
- **Knox**, J.A.: "Acidizing -- Past, Present, and Future," 1973 Annual Technical Meeting of the Petroleum Society of CIM, Edmonton, May 8-12.
- **Goins**, W.C., Jr. and McGlothlin, B.B.: "Two-Stage Injection of Acid-Producing Chemicals for Stimulating Wells," US Patent 3,707,192(1972).
- **Burns**, L.D.: Europ. Pat. App. 0 281 131 A2(1988).
- **Snow**, S.W.; Hough, E.V. "Field and Laboratory Experience in Stimulating Ekofisk Area North Sea Chalk Reservoirs," SPE paper 18225; 1988 SPE Annual Technical Conference and Exhibition, Houston.
- **Coulter**, A.W., Alderman, E.N., Cloud, J.E., and Crowe, C.W. "Mathematical Model Simulates Actual Well Conditions in Fracture Acidizing Treatment Design," SPE paper 5004, 1974 SPE Annual Fall Meeting of AIME, Houston.
- **Anderson**, M.S.; Fredrickson, S.E. "Dynamic Etching Tests Aid Fracture Acidizing Treatment Design," SPE/DOE paper 16452, 1987 SPE/DOE Low Permeability Reservoirs Symposium, Denver.
- **Swanson**, B.L. and Roper. L.E.: "Well Treating Compositions," US Patent 4,205,724(1980).
- **Walker**, M.L., Fredrickson, S., Norman, L., and Hoch, O.: "Heated Acids for Improved Stimulation," J. Can. Pet. Technol., 1987, 26(5), 57-9.
- **Nazarov**, S.N.: "Hydrochemical Fracture of Low-Permeability Materials," Neft. Khoz., 1987, (5), 59-63.
- **Baumgartner**, S.A. and Harrington, L.J.: "A Novel Approach to Acid Fracturing Treatment Design," 1987 Proc. Annu. Southwest. Pet. Short Course, Lubbock, 50-8.
- **Dawson**, J.C., McDaniel, R.R., Sedillo, L.P.: "Aqueous Acid Gels and Their Use," US Patent 4,624,795(1986).
- **Gdanski**, R.D. and Norman, L.R.: "Using the Hollow-Core Test to Determine Acid Reaction Rates," SPE Prod. Eng., 1986, 1(2), 111-16.
- **Dawson**, J.C.: "Aqueous Acid Gels and Their Use," US Patent 4,604,218(1986).
- **Katona**, J., Udovecz, G, and Vas, A.: "Acidic Fracturing Fluid for Geological Strata," Chemical Abstracts, 105(18):155989q(1986).
- **Perlman**, W.: "Fracturing Coal Formations," Chemical Abstracts, 103(24):198382n(1985).
- **Graham**, J.W. and Sinclair, A.R.: "Acidizing Propped Fractures," US Patent 4,527,627(1985).
- **Hitzman**, D.O.: "Gelled Acid Composition," US Patent 4,515,700(1985).
- **Khalil**, C.N. and Dacier, C.S.D.: "Gelled Liquid Acid Composition," Chemical Abstracts, 102(18):151945n(1985).
- **Ford**, W.G.F. and Roberts, L.D.: "The Effect of Foam on Surface Kinetics in Fracture Acidizing," J. Pet. Technol., 1985, 37(1), 89-97.
- **Conway**, M.W. and Norman, L.R.: "Treating Subterranean Formations Utilizing Foamed Viscous Fluids," US Patent 4,453,596(1984).
- **Katona**, J., Udovecz, G., and Vas, A.: "Acidic Fracturing Liquid," Chemical Abstracts, 101(8):57519j(1984).
- **Josephson**, C.B.: "Method and Compositions for Acidizing and Fracturing Wells," Chemical Abstracts, 101(6):40812b(1984).
- **Deysarkar**, A.K., Dawson, J.C., Sedillo, L.P. and Knoll-Davis, S.: "Crosslinked Acid Gel," J. Can. Pet. Technol., 1984, 23(1), 26-32.
- **Smith**, M.A., Dawson, J., and Scoggins, D.: "High Temperature, Crosslinked High-Strength Acid System," 1983 Proc. Annu. Southwest Pet. Short Course,Lubbock, 163-73.
- **Nazarov**, S.N. and Kholbaev, T.Kh.: "Hydraulic-Acid Fracturing of Formations," SU Patent 953,190(1982).
- **Chauhan**, K.J., Malhotra, B.D., and Goyal, K.L.: " Development of Acid Gels as Fracture Fluids," ONGC, Bull. 1981, 18(1), 53-76.
- **Church**, D.C., Quisenberry, J.L., and Fox, K.B.: "Field Evaluation of Gelled Acid for Carbonate Formations," J. Pet. Technol., 1981, 33(12), 2471-4.
- **Crowe**, C.W., Martin, R.C., and Michaelis, A.M.: "Evaluation of Acid-Gelling Agents for Use in Well Stimulation," Soc. Pet. Eng. J., 1981, 21(4), 415-24.
- **Pabley**, A.S. and Holcomb, D.L.: "Crosslinked Acid Gels Offer Advantages," Oil Gas J., 1981, 79(39), 286, 288, 291-2.

- **Petryk,** R.P. and Goruk, B.W.: "Fracture Acidizing with Foamed Acid: Success in the Crossfield D-1 Carbonate," J. Can. Pet. Technol., 1980, 19(3), 57-74.
- **Needham,** R.B., Thomas, C.P. and Wier, D.R.: "Well Treatment Method," US Patent 4,231,428(1980).
- **Roper,** L.E. and Swanson, B.L.: "Well Treating Compositions," US Patent 4,205,724(1980).
- **Pabley,** A.S. and Holcomb, D.L.: "A New Method of Acidizing or Acid Fracturing: Crosslinked Acid Gels," 1980 Proc. Annu. Southwest Pet. Short Course, Lubbock, 31-8.
- **Knox,** J.A. and Ripley, H.E.: " Fracture Acidizing in Carbonate Rock," J. Can. Pet. Technol., 1979, 18(4), 77-90.
- **Jahnke,** R.W.: "Thickened Aqueous Compositions for Well Treatment," US Patent 4,061,580(1977).
- **Aldrich,** R.G.: "Spontaneous Fracture of Coal," Fuel, 1977, 56(3), 345.
- **Swanson,** B.L.: "Composition for Acidifying Underground Formations," German Patent 2,657,443(1977).
- **Ugolev,** V.S. and Konyushendo, N.V.: "Reaction of Acid Foams with a Carbonaceous Porous Media under Dynamic Conditions," Chemical Abstracts, 86(2):6940d(1977).
- **Broaddus,** G.C. and Fredrickson, S.E.: "Fracture Acidizing Method," US Patent 3,918,524(1975).
- **Williams,** B.B., Gidley, J.L., Schechter, R.S.: "Acidizing Fundamentals," Monograph Series SPE, Dallas(1979).
- **Wieland,** D.R. and Vinson, M.E.: "Engineered HCl-HF Treatments Provide Successful Stimulation in Cook Inlet," SPE paper 4120, 1972 SPE Annual Fall Meeting of AIME, San Antonio, October 8-11.
- **Bailey,** D.E. and Wickham, J.F.: "Sand Fracturing vs Fracture Acidizing," SPE paper 12898, 1984 Rocky Mountain Regional Meeting, Casper, May 21-23.
- **Misak,** M.D., et al.: "Edwards Limestone Completions Improved by New Stimulation Technique," SPE paper 7115, 1978 Southwest Texas Regional SPE Meeting, Corpus Christi, April 14.
- **Fredrickson,** S.E. and Broaddus, G.C.: "Selective Placement of Fluids in a Fracture by Controlling Density and Viscosity," J. Pet. Technol.(May 1976) 597-602.
- **Sutton,** G.D. and Lasater, R.M.: "Aspects of Acid Additive Selection in Sandstone Acidizing," SPE paper 4114, 1972 SPE Annual Fall Meeting of AIME, Dallas, October 2-5.
- **Holcomb,** D.L.: "Low Concentration Hydrochloric-Hydrofluoric Acid Mixtures for Stimulation in Low Porosity, Low Permeability Sandstone Formations," Proc. Symp. Stimul. Low Permeability Reservoirs,1976, 72-86, Colo. Sch. Mines, Golden.
- **Elphingston,** E.A. and Norman, L.R.: "Treating Subterranean Well Formations," US Patent 4,231,882(1980).
- **Williams,** B.B. and Nierode, D.E.: "Design of Acid Fracturing Treatments," J. Pet. Technol., 1972, 24(July), 849-59.

Friction and Drag Reducers(Ref. 110)
- **Zakin,** et al.: "Variables Affecting Drag Reduction by Nonionic Surfactant Additives," Chem. Eng. Commun., 1983, 23, 77-88.
- **Ousterhout,** R.S.: "Reduction of Friction Loss in Fracturing Operations," J. Pet. Technol.(March, 1961) 217-222.
- **Root,** R.L.: "Method for Decreasing Friction Loss in a Well Fracturing Process," US patent No. 3,254,719(1966).
- **Sarem,** A.M.: "Method of Decreasing Friction Loss in a Well Fracturing Process," US Patent 3,357,525(1970).
- **Phillips,** K.G.: "Method for Reducing Friction Loss in a Well Fracturing Process," US Patent 4,152,274(1979).
- **Sylvester,** N.D. and Tyler, J.S.: "Dilute Solution Properties of Drag-Reducing Polymers," Ind.Eng.Chem.Prod.Res.Develop., 1970, 9(4), 548-553.
- **Kim,** O.K., Little, R.C., and Ting, R.Y.: "The Correlation of Drag-Reduction Effects with Polymer Intrinsic Viscosity," J. Colloid Interface Sci., 1974, 47(2).
- **Savins,** J.G.: "A Stress-Controlled Drag-Reduction Phenomenon," SPE paper 1724, 1966 SPE Symposium on Mechanics of Rheologically Complex Fluids, Houston, December 15-16.
- **McCormick,** C.L., et al.: "Development of Laboratory Screening Tests to Predict Polymer Performance in Enhanced Oil Recovery (I). Shear Degradation, Viscosity, and Electrolyte Studies," prepared for DOE under contract No. EF-77-S-05-5603, 1977.
- **Thorn,** D.J. and Burnham, J.W.: "Dissolving Polymeric Materials in Hydrocarbon Liquids," US Patent 4,068,676(1978).
- **Culter,** J.D. and McClaflin, G.G.: "Method of Friction Loss Reduction in Oleaginous Fluids Flowing Through Conduits," US Patent 3,692,676(1972).
- **Chang,** H.D., Darby, R.: "Effect of Shear Degradation on the Rheological Properties of Dilute Drag-Reducing Polymer Solutions," J. Rheol., 1983, 27(1), 77-88.

Xanthan Gum(Ref. 138)
- **Abdo,** M.K., et al.: "Field Experience with Floodwater Diversion by Complexed Biopolymers," SPE/DOE paper 12642, 1984 SPE/DOE Symposium on Enhanced Oil Recovery, Tulsa, April 15-18.

- **Crowe**, "Method of Inhibiting Crosslinking of Aqueous Xanthan Gums in the Presence of Ferric Acid Ions," US Patent 4,317,735(1982).
- **Shu** and Szolek, W.: "Zirconium Crosslinked Gel Compositioning Methods of Preparation and Application In Enhanced Oil Recovery," US Patent 4,676,930(1987).
- **Hannah**, R.R.: "New Fracturing Technique Leads to Improved Performance in the Mississippian Trend," J. Pet. Technol.(August 1976) 859-864.
- **Hill**, O.F., Ward, A.J., and Clement, C.C.: "Austin Chalk Fracturing Design Using a Crosslinked Natural Polymer as a Diverting Agent," J. Pet. Technol.(December 1978) 1795-1804.
- **Norton**, C.J., Falk, D.O., and Luetzelschwab, W.E.: "Xanthan Biopolymer Semi-Pilot Fermentation," SPE paper 8420, 1979 SPE Annual Technical Conference and Exhibition, Las Vegas, Septemeber 23-26.
- **Chauveteau**, G. and Kohler, N., "Influence of Microgels in Xanthan Polysaccharide Solutions on Their Flow Through Various Porous Media," SPE paper 9295, 1980 SPE Annual Fall Technical Conference and Exhibition of AIME, Dallas, September 21-24.
- **Bragg**, J.R.: "Injectivities of Biopolymer Solutions," Chemical Abstracts, 99:P197764a(1983).
- **Jamieson**, A.M., et al.: "Dynamical Behavior of Xanthan Polysaccharide in Solution," Polymer Physics(1982) Edt., 20, 1513-1524.
- **Southwick**, J.G., et al.: "Self Association of Xanthan in Aqueous Solvent Systems," Carbohydrate Research(1980) 84, 287-295.
- **Narayan**, K.S. and Ramasubramanian, V.: "Rheological Properties of Polysaccharide Gums," Indian J. Technol., 1982, 20(9), 333-338.
- **Patton**, J.T.: "Modified Heteropolysaccharides," US Patent 3,729,460(1970).
- **Kohler**, N. and Chauveteau, G.: "Xanthan Polysaccharide Plugging Behavior in Porous Media - Preferential Use of Fermentation Broth," J. Pet. Technol. 1981, 33(9), 349-357.
- **Salamone**, J.C., Clough, S.B., Salamone, A.B., Reid, K.I.G., and Jamison, D.E.: "Xanthan Gum - A Lyotropic, Liquid Crystalline Polymer and its Properties as a Suspending Agent," Soc. Pet. Eng. J. August 1982, 555,556.
- **Sato**, T., Norisuye, T., and Fujita, H.: "Double-Stranded Helix of Xanthan: Dimensional and Hydrodynamic Properties in 0.1 M Aqueous Sodium Chloride," Macromolecules 1984, 17, 2696-2700.
- **Burkholder**, L.: "Xanthan Gel System Effective for Profile Modification," Oil Gas J. 1985, 68, 69.
- **Chow**, A.W. and Fuller, G.G.: "Response of Moderately Concentrated Xanthan Gum Solutions to Time-Dependant Flows Using Two-Color Flow Birefringence," J. Rheol. 1984, 28(1), 23-43.
- **McNeely**, W.H.: "Process for Producing a Polysaccharide," Canadian Patent 821 534(1969).
- **Hartfiel**, A.H.: "Clay-Free, Thixotropic Wellbore Fluid," US Patent 4,247,402(1981).
- **Schuppner**, H.R., Jr.: "Heat Reversible Gel and Method for Preparing Same," US Patent 3,557,016(1971).
- **Jordan**, W.A.: "Thickening Compositions Containing Xanthomonas Gum and Hydroxyalkyl Ether of Guar Gum," US Patent 3,748,201(1973).
- **Jordan**, W.A. and Carter, W.H.: "Blends of Xanthomonas and Guar Gum," US Patent 3,765,918(1973).
- **Norton**, C.J. and Falk, D.D.: "Polyalkeneoxide and Polysaccharide Gum Derivatives Mobility Control Agent and Process," US Patent 3,919,092(1975).
- **Karl**, C.L.: "Hydroxyalkyl Locust Bean/Xanthomonas Hydrophilic Colloid Blends," US Patent 4,038,206(1977).
- **Patton**, J. T.: "Chemical Treatment Enhances Xanflood Polymer," SPE paper 4670, 1973 SPE Annual Meeting, Las Vegas, Sept. 30-Oct. 3.
- **Abdo**, M.K.: "Waterflood Oil Recovery Process Employing Stabilized Biopolymers," US Patent 4,141,842(1979).
- **Clark**, P.E., Halvaci, M., and Ghaeli, H.: "Proppant Transport by Xanthan and Xanthan-Hydroxypropyl Guar Solutions: Alternatives to Crosslinked Fluids," SPE/DOE 13907, 1985 Low Permeability Gas Reservoirs, Denver, May 19-22.
- **Clifford**, P.J. and Sorbie, K.S.: "The Effects of Chemical Degradation on Polymer Flooding," SPE paper 13586, 1985 SPE International Symposium on Oilfield and Geothermal Chemistry, Phoenix, April 9-11.

Frac Gel Residue and Formation Damage(Ref. 142)
- **van Poollen**, H.K., Tinsley, J.M. and Saunders, C.D.: "Hydraulic Fracturing: Fracture Flow Capacity vs. Well Productivity," SPE paper 890-G, 1957 SPE Annual Fall Meeting of AIME, Dallas, October 6-9.
- **Methven**, N.E.: "Effects of Drilling Fluids on Gas Well Productivity," SPE paper 3504, 1971 SPE Annual Meeting, New Orleans, October 3-6.
- **Carney**, M.J. and Wieland, D.R.: "Stimulation of Low Permeability Gas Wells in the Rocky Mountain Area," SPE paper 4396, 1973 Rocky Mountain Regional Meeting of AIME, Casper, May 15-16.
- **Smith**, C.F.: "Gas Well Fracturing Using Gelled Non-Aqueous Fluids," SPE paper 4678, 1973 SPE Annual Meeting of AIME, Las Vegas, September 30-October 3.

- **Pye**, D.S. and Smith, W.A.: "Fluid Loss Additive Seriously Reduces Fracture Proppant Conductivity and Formation Permeability," 1973 SPE Annual Meeting, Las Vegas, Sept. 30-Oct. 3.
- **Tindell**, W.A., Misak, M.D., and Gras, E.H.: "The Use of Alcohol-Water Mixtures in Fracture Stimulation of Gas Wells," 1974 Proc. Annu. Southwest. Pet. Short Course, Lubbock, 61-65.
- **Tiner**, R.L., Stahl, E.J. Jr., and Malone, W.T.: "Developments in Fluids to Reduce Potential Damage from Fracturing Treatments," SPE paper 4790, 1974 SPE Symposium on Formation Damage Control, New Orleans, February 7-8.
- **Tuttle**, R.N. and Barkman, J.H.: "The Need for Nondamaging Drilling and Completion Fluids," SPE paper 4791, 1974 SPE Symposium on Formation Damage Control, New Orleans, February 7-8.
- **Christian**, W.W. and Ayres, H.J.: "Formation Damage Control in Sand Control and Stimulation Work," SPE paper 4775, 1974 SPE Symposium on Formation Damage Control, New Orleans, February 7-8.
- **Tannich**, J.D.: "Liquid Removal From Hydraulically Fractured Gas Wells," J. Pet. Technol., November, 1975, 1309-1317.
- **Cooke**, C.E., Jr.: "Effect of Fracturing Fluids on Fracture Conductivity," J. Pet. Technol., October 1975, 1273.
- **Pence**, S.A.: "Evaluating Formation Damage in Low Permeability Sandstone," 1975 SPE Annual Meeting, Dallas, September 28-October 1.
- **Barth**, H.G. and Smith, D.A.: "High-Performance Size-Exclusion Chromatography of Guar Gum," J. Chromatog.,(1981) 206, 410-415.
- **Gall**, B.L. and Raible, C.J.: "The Use of Size Exclusion Chromatography to Study the Degradation of Water-Soluble Polymers Used in Hydraulic Fracturing Fluids," Polym. Mater. Sci. Eng. 1986, 55, 572-75, Anaheim.
- **Southwick**, J.G. and Manke, C.W.: "Molecular Degradation, Injectivity, and Elastic Properties of Polymer Solutions," SPE paper 15652, 1986 SPE Annual Technical Conference and Exhibition, New Orleans, October 5-8.
- **Clifford**, P.J. and Sorbie, K.S.: "The Effects of Chemical Degradation on Polymer Flooding," SPE paper 13586, 1985 International Symposium on Oilfield and Geothermal Chemistry, Phoenix, April 9-11.
- **Barth**, H.G. and Regnier, F.E.: "High-Performance Gel Permeation Chromatography of Water-Soluble Cellulosics," J. Chromatog.(1980) 192, 275-293.
- **Raible**, C.J.: "Formation Damage Due to Hydraulic Fracturing Fluids," National Institute for Petroleum and Energy Research, Cooperative Agreement DE-FC01-83FE60149, October 1, 1983-September 30, 1988.
- **Volk**, L.J., et al.: "A Method for Evaluation of Formation Damage Due to Fracturing Fluids," SPE paper/DOE 11638, 1983 SPE/DOE Symposium on Low Permeability, Denver, March 14-16.
- **Gall**, B.L. and Raible, C.J.: "A Method to Study Fracture Fluid Polymer Degradation Using Size Exclusion Chromatography," United States Department of Energy, DOE/BETC/RI-83/10(DE84008316), February 1984.
- **Sattler**, A.R., Raible, C.J., and Gall, B.R.: "Integration of Laboratory and Field Data for Insight on the Multiwell Experiment Paludal Stimulation," SPE paper/DOE 13891, 1985 SPE/DOE Low Permeability Gas Reservoirs, Denver, May 19-22.
- **Raible**, C.J. and Gall, B.L.: "Laboratory Formation Damage Studies of Wester Tight Gas Sands," SPE paper/DOE 13903, 1985 SPE/DOE Low Permeability Gas Reservoirs, Denver, May 19-22.
- **Sattler**, A.R., et al.: "Laboratory Studies for the Design and Analysis of Hydraulic Fractured Stimulations in Lenticular, Tight Gas Reservoirs," SPE paper 15245, 1986 SPE Unconventional Gas Technology Symposium, Louisville, May 18-21.
- **Gall**, B.L., et al.: "Permeability Damage to Natural Fractures Caused by Fracturing Fluid Polymers," SPE paper 17542, 1988 SPE Rocky Mountain Regional Meeting, Casper, May 11-13.
- **Sattler**, et al.: "Stimulation-Fluid Systems for Naturally Fractured Tight Gas Sandstones: A General Case Study from Multiwell Experiment Stimulations," SPE paper 17717, 1988 SPE Gas Technology Symposium, Dallas, June 13-15.
- **Tuttle**, R.N. and Barkman, J.H.: "The Need for Nondamaging Drilling and Completion Fluids," paper 4791, 1974 SPE Symposium on Formation Damage Control, New Orleans, February 7-8.
- **van Poollen**, H.K.: "Do Fracture Fluids Damage Productivity?," Oil Gas J.(1957) May 27, 120-124.
- **Leon**, L.: "The Role of the Service Company in Minimizing and Reducing Formation Contamination," SPE paper 4660, 1973 SPE Annual Meeting, Las Vegas, September 30-October 3.
- **Roodhart**, L., Kuiper, T.O. and Davies, D.R.: "Proppant Rock Impairment During Hydraulic Fracturing," SPE paper 15629, 1986 SPE Annual Technical Conference and Exhibition, New Orleans, October 5-8.
- **Cooke**, C.E. Jr.: "Effect of Fracturing Fluids on Fracture Conductivity," J. Pet. Technol.(October 1975) 1273-1282.
- **Almond**, S.W.: "Factors Affecting Gelling Agent Residue Under Low Temperature Conditions," SPE paper 10658, 1982 SPE Formation Damage Control Symposium, Lafayette, March 24-25.

- Almond, S.W. and Bland, W.E.: "The Effect of Break Mechanism on Gelling Agent Residue and Flow Impairment in 20/40 Mesh Sand," SPE paper 12485, 1984 Formation Damage Control Symposium, Bakersfield, February 13-14.
- Kim, C.M. and Losacano, J.A.: "Fracture Conductivity Damage Due to Crosslinked Gel Residue and Closure Stress on Propped 20/40 Mesh Sand," 1985 SPE Annual Technical Conference and Exhibition, Las Vegas, September 22-25.
- Abrams, A. and Vinegar, H.J.: "Impairment Mechanisms in Vicksburg Tight Gas Sands," SPE paper/DOE 13883, 1985 SPE Low Permeability Gas Reservoirs, Denver, May 19-22.
- Tinsley, J.M. and Williams, J.R. Jr.: "A New Method of Providing Increased Fracture Condutivity and Improving Stimulation Results," SPE paper 4676, 1973 SPE Annual Meeting, Las Vegas, September 30-October 3.
- Hough, E.W. and Allen T.O.: "Laboratory Techniques and Results of Tests to Determine Formation Damage from Fracturing Fluids," 1957 Spring Meeting of the Southern District Division of Production, Shreveport, March 20-22.
- Kotb, A.K. and Kasraie, B.: "Laboratory Investigation of Damage from Guar Gum Base Gels," SPE paper 3660, 1971 SPE Appalachian Section, Charleston, November 4-5.
- Pober, K.W., Hoff, M.H., and Darlington, R.K.: "Water-Insoluble Residue Following Acid Hydrolysis of Water-Soluble Polysaccharides," J. Pet. Technol.(December 1983) 2185-2191.
- Ely, J.W.: "Methods of Water Flooding and Fracturing Using clean, Non-Damaging Fracturing Fluids," US Patent 4,265,311(1981).
- Gall, B.L. and Raible, C.J.: "Formation Damage from Exposure to Hydraulic Fracturing Fluids," NIPER-63, 1985 National Institute for Petroleum and Energy Research Topical Report, March 7.
- Gall, B.: "Degradation of Fracturing Fluid Polymers," NIPER-132, 1986 National Institute for Petroleum and Energy Research Final Report, February.
- Gall, B.L. and Raible, C.J.: "Characterization of Degraded Polymer Fracturing Fluids," NIPER-48, National Institute for Petroleum and Energy Research Topical Report, February, 1985.
- Raibel, C.J.: "Formation Damage Studies of Low-Permeability Sands," NIPER-130, National Institute for Petroleum and Energy Research Final Report, February 1986.
- Woodroof, R.A. Jr. and Anderson, R.W.: "Synthetic Polymer Friction Reducers Can Cause Formation Damage, SPE paper 6812, 1977 SPE Annual Technical Conference and Exhibition, Denver, October 9-12.
- Hawkins, G.W.: "Molecular Weight Reduction and Physical Consequences of Chemical Degradation of Hydroxypropylguar in Aqueous Brine Solutions," Polym. Mat. Sci. Eng.(1986) 55, 588-93.
- Hawkins, G.W.: "Laboratory Study of Proppant-Pack Permeability Reduction Caused by Fracturing Fluids Concentrated During Closure," SPE paper 18261, 1988 Annual Technical Conference and Exhibition, Houston, Oct. 2-5.
- Branagan, P.T., et al: "Case History of Hydraulic Fracture Performance in the Naturally Fractured Paludal Zone: The Transitory Effects of Damage," SPE paper/DOE 16397, 1987 SPE Low Permeability Reservoirs Symposium, Denver, May 18-19.
- Harris, J.F.: "Acid Hydrolysis and Dehydration Reactions for Utilizing Plant Carbohydrates," Applied Polymer Symposium, 1975, 28, 131-144.
- Barth, H.G. and Smith D.A.: "High-Performance Size-Exclusion Chromatography of Guar Gum," J. Chromatog. (1981) 206, 410-415.
- Gall, B.L. and Raible, C.J.: "A Method to Study Fracture Fluid Polymer Degradation Using Size Exclusion Chromatography," United States Department of Energy, No. DOE/BETC/RI-83/10 (DE84008316), February 1984.
- Gall, B.L. and Raible, C.J.: "Molecular Size Studies of Degraded Fracturing Fluid Polymers," SPE paper 13566, 1985 International Symposium on Oilfield and Geothermal Chemistry, Phoenix, April 9-11.
- Wojtanowicz, A.K., Krilov, Z., Langlinais, J.P.: "Study on the Effect of Pore Blocking Mechanisms on Formation Damage," SPE paper 16233, SPE Production Operations Symposium, Oklahoma City, OK, March 8-10, 1987.
- Gall, B.L.: "Permeability Damage to Cracked Core by Fracturing Fluids," National Institute for Petroleum and Energy Research, proposal No. NIPER 85-68B, (ECR) M-Report, January 23, 1986.
- Gall, B.L., Maloney, D.R., and Raible, C.J.: "Permeability Damage to Artificially Fractured Cores," Final Report, National Institute for Petroleum and Energy Research, May 1988.
- McDaniel, B.W.: "Use of Wet Gas Flow for Long-Term Fracture Conductivity Measurements in the Presence of Gel Filter Cakes," SPE paper 17543, 1988 SPE Rocky Mountain Regional Meeting, Casper, May 11-13.
- McDaniel, B.W. and Parker, M.A.: "Accurate Design of Fracturing Treatment Requires Conductivity Measurements at Simulated Reservoir Conditions," SPE paper 17541, 1988 SPE Rocky Mountain Regional Meeting, Casper, May 11-13.
- McLeod, H.O. Jr.: "The Effect of Perforating Conditions on Well Performance," SPE paper 10649, 1982 SPE Formation Damage Control Symposium, Lafayette, March 24-25.
- McDaniel, B.W.: "Use of Wet Gas Flow for Long Term Fracture Conductivity Measurements in the Presence of Gel Filter Cakes," SPE paper 17543, 1988 Rocky Mountain Regional SPE Meeting, May 11-13.

- Gray, D.H. and Rex, R.W.: "Formation Damage in Sandstones Caused by Clay Dispersion and Migration," 1966 Fourteenth National Conference on Clays and Clay Minerals," 355-366.
- Parker, M.A. and McDaniel, B.W.: "Fracturing Treatment Design Improved by Conductivity Measurements Under In-Situ Conditions," SPE paper 16901, 1987 SPE Annual Technical Conference and Exhibition, Dallas, September 27-30.

Crosslinked Gels from Other Industries(Ref. 204)
- Lagally, P. and Lagally H.: "Atomistic Approach to the Crosslinking of Cellulose Fibers and Their Reaction with Fillers," TAPPI, Vol. 39, No. 11, November 1956, 747-754.
- Corben, L.D.: "Micro-Inclusions and Method of Making Same," US Patent 3,201,353(1965).
- Roy, G.L., Laferriere, A.L. and Edwards, J.O.: "A Comparative Study of Polyol Complexes of Arsenite, Borate, and Tellurate Ions," J. Inorg. Nucl. Chem.(1957) Vol. 4, 106-114.
- Schultz, R.K. and Myers, R.R.: "The Chemorheology of Poly(vinyl Alcohol)-Borate Gels," Macromolecules(1969) Vol. 2.
- Gorin, P.A.J. and Mazurek, M.: "Carbon-13 Resonance Spectroscopic Studies on the Formation of Borate and Diphenylborinate Complexes of Polyhydroxy Compounds," Can. J. Chem. Vol. 51, 1973, pages 3277-3286.
- Gorin, P.A.J. and Mazurek, M.: "C Magnetic Resonance Spectroscopic Evidence for Formation of Borate Complexes of Polyhydroxy Compounds," Carbohydrate Research, 27(1973), 325-339.
- Haug, A.J.: "Guar Mannogalactan Studies: II. Effect of Certain Variables, Including Borax, on the Rate of Oxidation of the Purified Mucilage," TAPPI, Vol. 36, No. 1, January 1953, 53-62.
- Lagally, P.: "Preparation and Stabilization of Water Soluble Reactive Titanium Derivatives," US Patent 2,950,174(1960).
- Gash, V.W.: "Metal Chelates and Process of Preparing Same," US Patent 2,976,285(1961).
- Nossal, R.: "Network Formation in Polyacrylamide Gels," Macromolecules, (1985) 18, 49-54.
- Rondestvedt, C.S., Jr.: "Titanium Compounds, Organic" in Kirk Othmer Encyclopedia of Chemical Technology, 3rd Ed, V.23, 176-244, Wiley, New York, 1983.
- Barnhart, R.E. and Sawyer, F.C.: "Gelled Ammonium Nitrate Blasting Explosive and Process," US Patent 3,072,509(1963).
- Moe, W.A., Miller, S.E. and Iwen, M.H.: "Investigation of the Reserve Carbohydrates of Leguminous Seeds. I. Periodate Oxidation," J. Amer. Chem. Soc., 1947, 69, 2621-2625.
- McIrvine, J.D.: "TNT-Tetraborate Gelled Aqueous Explosive Slurry," US Patent 3,108,917(1963).
- Kiefer, J.E. and Touey, G.P.: "Cellulose Acetate Spinning Solutions and Process of Spinning Fine Denier Filaments," US Patent 3,033,698(1962).
- Chrisp, J.D.: "Method for Gelling Water-Bearing Explosive Compositions Containing Galactomannan Gums," US Patent 3,202,556(1965).
- Jordan, W.A.: "Organo-Metallic Gel-Producing Compositions and Processes for Preparing Organo-Metallic Gels," US Patent 3,251,781(1966).
- Chrisp, J.D.: "Gelled Compositions Containing Galactomannan Gums," US Patent 3,301,723(1967).
- Revukas, A.J.: "Hydrocarbon Fuel Composition," US Patent 3,334,978(1967).
- Needles, H.L. and Whitfield, R.E.: "Crosslinking of Collagens Employing a Redox System Comprising Persulphate and a Reducing Agent," US Patent 3,427,301(1969).
- Angstadt, R.L. and Tyree, S.Y.: "The Nature of Zirconyl Chloride in Strong Hydrochloric Acid: Light Scattering," J. Inorganic Chem., 1962, 24, 913-917.
- Kraitzer, I., McTaggart, K., and Winter, G.: "Esters of Titanium," J. Oil Colour Chem. Assoc., 1948, 405-417.
- Balthis, J.: "Polymeric Hydroxyl-Containing Titanium Carboxylates and Methods for Preparing Same," US Patent 2,621,194(1950).
- Davidson, D.: "Explosive Compositions," US Patent 2,768,073(1956).
- Hock, A.L.: "Zirconium Compounds: The Industrial Importance of Their Aqueous Chemistry," Chemistry and Industry, November 2, 1974.
- Vladimirova, Z.A., Prozorovskaya, A.N., and Komissarova, L.N.: "Investigation of Complex Formation by Zirocnium and Hafnium with Formic Acid by a Kinetic Method," Russian Journal of Organic Chemistry, 1975, 20(10), 1477-1480.
- Konunova, Ts.B., Popov, M.S., and Venichenko, A.S.: "Study of the Interaction of Zirconium with Certain Monocarboxylic Acids by the Metal-Indicator Method," Russian Journal of Inorganic Chemistry, 1975, 20(6), 861-865.
- Russell, C.A.: "Process for the Preparation of Stabilized Organotitanium Compounds," US Patent 2,894,966(1959).
- Shokal, E.C.: "Sulfur-Containing Resinous Products from Polyepoxides," US Patent 2,633,458(1953).
- Mills, J.A.: "Association of Polyhydroxy Compounds with Cations in Solution," Biochem. Biophys. Res. Commun., 1961/62, 6(6), 418-21.

Quality Control of Fracturing Fluids(Ref. 335)
- **Huebinger**, T., Webster, D., Chisholm, P., Venditto, J., and Hunt, J.: "Optimizing Program Increases Field's Profits," Oil Gas J., Aug 29, 1988, 35-39.
- **Osborne**, M.W., McLeod, H.O., Jr. and Schroeder, H.D.: "The Analysis and Control of Hydraulic Fracturing Problems," SPE/DOE paper 9868, 1981 SPE/DOE Low Permeability Symposium, Denver, May 27-29.
- **Schlottman**, B.W., Miller, W.K., II and Lueders, R.K.: "Massive Hydraulic Fracture Design for the East Texas Cotton Valley Sands," SPE paper 10133, 1981 SPE Annual Technical Conference and Exhibition of AIME, San Antonio, October 5-7.
- **Pai**, V.J.: "Predicting, Preventing and Remedying Hydraulic Fracturing Screen Outs," 1984 Proc. Annu. Southwest. Pet. Short Course, Lubbock, 122-135.
- **Conway**, M.W. and Harris, L.E.: "A Laboratory and Field Evaluation of a Technique for Hydraulic Fracturing Stimulation of Deep Wells," SPE paper 10964, 1982 SPE Annual Technical Conference and Exhibition of AIME, New Orleans, September 26-29.
- **Payne**, K.L., and Harms, S.D.: "Recent Developments in Polymer Fracture Fluid Technology," 1984 AIChE National Meeting, Anaheim, May 20-24.
- **Freck**, J. and Gottschling, J.: "A Field and Laboratory Study of Polysaccharides in Fracturing Treatments," 1984 Proc. Annu. Southwest. Pet. Short Course, Lubbock, 141-156.
- **Evans**, D.W. and Willis, K.: "Fracture Execution - An Essential Part of Every Fracture Design," paper No. 86-37-70, 1986 Annual Technical Meeting of the Petroleum Society of CIM, Calgary, June 8-11.
- **Zigrye**, J.L., Osborne, M.W., and Westbrook, G.H.: "Field Analysis and Flow Loop Tests Diagnose Problems with Delayed Action Crosslinked Fracturing Fluid Systems," SPE paper 15633, 1986 Annual Technical Conference and Exhibition of the SPE, New Orleans, October 5-8.
- **Hodge**, R. M. and Baranet, S.E.: "Eavluation of Field Methods to Determine Crosslink Times of Fracturing Fluids," SPE paper 16249, 1987 International Symposium on Oilfield Chemistry, San Antonio, February 4-6.
- **Tyssee**, D.A. and Vetter, O.J.: "Chemical Characterization Problems of Water-Soluble Polymers," Soc. Pet. Eng. J., December 1981, 721-730.
- **Lescarboura**, J.A., Sifferman, T.R. and Wahl, H.A.: "Evaluation of Frac Fluid Stability Using a Heated, Pressurized Flow Loop," SPE paper 10962, 1982 SPE Annual Technical Conference and Exhibition, New Orleans, September 26-29.
- **Gardner**, D.C. and Eikerts, J.V.: "Effects of Shear and Proppant on the Viscosity of Crosslinked Fracturing Fluids," SPE paper 11066, 1982 SPE Annual Fall Technical Conference and Exhibition, New Orleans, September 26-29.
- **McLeod**, H.O., Jr.: "A Simplified Approach to Design of Fracturing Treatments Using High Viscosity Crosslinked Fluids," SPE/DOE paper 11614, 1983 SPE/DOE Symposium on Low Permeability, Denver, March 14-16.
- **Gardner**, D.C. and Eikerts, J.V.: "Rheological Characterization of Crosslinked and Delayed Crosslinked Fracturing Fluids Using a Closed-Loop Pipe Viscometer," SPE paper 12028, 1983 SPE Annual Technical Conference and Exhibition, San Francisco, October 5-8.
- **Shah**, S.N. and Watters, L.T.: "Time and Shear Effects on Rheological Properties of Crosslinked Fluids - Evaluation Method," SPE Prod. Eng.(January 1986) 55-61.
- **Constien**, V.G., et al.: "Automated Rheology Laboratory. 1." American Chemical Society Symposium Series 313 (1986): 105-113.
- **Graves**, G. and Fellin, E.: "Automated Rheology Equipment Facilitates Acid Fracturing Fluid Characterization," Presented at the SPE Computer Technology Symposium, Texas Tech University, Lubbock, March 6, 1986.
- **King**, M.T., Constien, V.G., and Fellin, E.L.: "Automated Rheology Laboratory. 2," American Chemical Society Symposium Series, 313 (1986): 114-122.
- **Saucier**, R.J.: "A Field Development Program in Hydraulic Fracture Stimulation," SPE paper 18257, 1988 SPE Annual Technical Conference and Exhibition, Houston, October 2-5.
- **Shah**, S.N., Lee, Y.N., and Jensen, D.G.: "Frac Treatment Quality Improved with Field Rheology Unit," Oil Gas J. (February 4, 1985), 47-51.
- **Ely**, J.W., Haskett, S.E., Holditch, S.A.: "Field Measurement of Fracturing Fluid Viscosity at In-Situ Conditions of Temperature and Shear," SPE paper 16915, 1987 SPE Annual Technical Conference and Exhibition, Dallas, September 27-30.
- **England**, A.A. and Davis, N., II: "The Characterization of a Drilling Fluid Additive," IADC/SPE paper 17199, 1988 IADC/SPE Drilling Conference, Dallas, February 28-March 2.
- **Craigie**, L.J.: "A New Method for Determining the Rheology of Crosslinked Fracturing Fluids using Shear History Simulation," SPE/DOE paper 11635, 1983 SPE/DOE Symposium on Low Permeability, Denver, March 14-16.

Emulsion Frac Fluids(Ref. 342)
- **Matthews**, T.M.: "Field Use of 'Superfrac' - A New Hydraulic Fracturing Technique," SPE paper 2625, 1969 SPE Annual Meeting, Denver, September 28-October 1.
- **Sinclair**, A.R.: "Rheology of Viscous Fracturing Fluids," J. Pet. Technol.(June 1970) 711-719.

- Kiel, O.M. and Weaver, R.H.: "Emulsion Fracturing System," Oil Gas J.(February 21, 1972) 72-73.
- Dauben, D.L. and Froning, H.R.: "Micellar Solutions Stimulate Wells," Oil Gas J.(September 18, 1972) 83, 72-78.
- Kiel, O.M.: "Method of Fracturing Subterranean Formations Using Oil-In-Water Emulsions," US Patent 3,710,865(1973).
- Gogarty, W.B. and Olson, R.W.: "Use of Microemulsions in Miscible-Type Oil Recovery Procedure," US Patent 3,254,714(1966).
- Kremesec, V.J. and Treiber, L.E.: "Effect of System Wettability on Oil Displacement by Micellar Flooding," SPE paper 6001, 1976 SPE Annual Technical Conference and Exhibition, New Orleans, October 3-6.
- Sinclair, A.R., Terry, W.M. and Kiel, O.M.: "Polymer Emulsion Fracturing," J. Pet. Technol.(July 1974), 731-738.
- Fast, C.R., Holman, G.B. and Covlin, R.J.: "The Application of Massive Hydraulic Fracturing to the Tight Muddy "J" Formation, Wattenberg Field, Colorado," J. Pet. Technol.(January 1977), 10-16.
- Graham, J.W., Gruesbeck, C., and Salathiel, W.M.: "Method of Fracturing Subterranean Formations Using Oil-in-Water Emulsions," US Patent 3,977,472(1976).
- Gidley, J.L., et al.: "Stimulation of Low-Permeability Gas Formations by Massive Hydraulic Fracturing - A Study of Well Performance," J. Pet. Technol.(April 1979), 525-531.
- McClaflin, G.G.: "Method of Transporting Viscous Hydrocarbons," US Patent 4,249,554(1981).
- Sifferman, T.R.: "Method of Transporting Viscous Hydrocarbons," US Patent 4,265,264(1981).
- Kiel, O.M.: "A New Hydraulic Fracturing Process," J. Pet. Technol.(January 1970), 89-96.
- Salathiel, W.M.; Muecke, T.W.; Cooke, C.E.; Li, N.N.: "Well Treatment with Emulsion Dispersions," US Patent 4,233,165(1980).
- Blauer, R.E., Phillips, A.M., and Craig, D.P.: "Rheological and Physical Differences Between Foam and Emulsion Fracturing Fluids," SPE paper 18214, 1988 SPE Annual Technical Conference and Exhibition, Houston, October 2-5.
- Miller, K.A.: "Pumping and Loading Emulsion Slurry Blasting Composition," European Patent Appl. 182,661A2(1986), Chemical Abstracts 105(16):136821n(1986).
- Krishan, K., Kapoor, S., and Goyal, K.L.: "Development of Polymer Emulsion as Fracturing Fluid," Chemical Abstracts 103(20):162915z(1985).
- Matveev, D.F.: "Hydrophobic Thermally Stable Emulsions for Treatment of Productive Strata," Chemical Abstracts 102(14):116185t(1985).
- Matveev, D.F., Starikova, T.V., Fuki, B.I., Makarenko, P.P., Stetsenko, G.I., Sergeev, V.P.: "Hydrophobic Emulsion Composition for Hydraulic Fracturing of a Formation," Chemical Abstracts 98(22):182419x(1983).
- Salathiel, W.M., Muecke, T.W., Cooke, C.F., Jr., and Li, N.N.: "Well Treatment with Emulsion Dispersions," US Patent 4,359,391(1982).
- Matveev, D.F., Reznikov, I.G., Kulikova, A.V., Belolapotnikov, G.G., Zhadanova, K.M., Grebennikova, A.I., and Perel, Z.P.: "Composition for Hydraulic Fracturing of Oil-Producing Formation," SU Patent 729,334(1980), Chemical Abstracts 93(14):134836f(1980).
- Matveev, D.F., Reznikov, I.G., Fuki, B., Kulikova, A.V., Belolapotkov, G.G., Zhadanova, K.M., Grebennikova, A.I., and Starikova, T.V.: "Hydrophobic Emulsion for Hydraulic Breakage of a Formation," SU Patent 731,998(1980), Chemical Abstracts 93(14):134835e(1980).
- Kuvandykov, I.Sh.: "Use of an Emulsion in the Repair of Wells," Neftepromysl. Delo. (9), 36-9, Chemical Abstracts 92(12):96410g(1980).
- Thorn, D.J. and Burnham, J.W.: "Dissolving Polymeric Materials in Hydrocarbon Liquids," US Patent 4,068,676(1978).
- Perkins, L.F.: "Propping Subterranean Formation Fractures," US Patent 4,029,149(1977).
- Graham, J.W., Gruesbeck, C., and Salathiel, W.M.: "Fracturing Subterranean Formations Using Oil-in-Water Emulsions," US Patent 3,977,472(1976).
- Gogarty, W.B.: "Fracturing Fluids," GB Patent 1,388,909(1975), Chemical Abstracts 83(22):181880u(1975).
- Christopher, C.A. and Allen, J.C.: "Hydraulic Fracturing," US Patent 3,865,190(1975).
- Sinclair, A.R., Terry, W.M., and Kiel, O.M.: "Polymer Emulsion Fracturing," J. Pet. Technol., 1974, 26(July), 731-8.
- Kiel, O.M.: "Hydraulic Fracturing," CA Patent 943,860(1974), Chemical Abstracts 81(18):108342z(1974).
- Crawford, D.L., Earl, R.B., and Monroe, R.F.: "Oil Well Fracturing Fluid Additive," GB Patent 1,355,808(1974).
- Kiel, O.M.: "Fracturing Using Acid External Emulsions," US Patent 3,799,266(1974).
- Kiel, O.M.: "Treating a Well Using a Volatile Hydrocarbon Liquid," US Patent 3,766,986(1973).
- Kiel, O.M.: "Treatment of Wells with Fluids Containing Complexes," US Patent 3,760,881(1973).
- Hummel, J.L. and Svaldi, M.A.: "Use of Micellar Solutions to Improve the Perforation of Oil-Bearing Subterranean Formations," US Patent 3,754,599(1973).

■ **Gogarty, W.B.:** "Hydraulic Fracturing Subterranean Formations Using a Temperature-Inverted Fracturing Fluid," US Patent 3,734,189(1973).
■ **Bernard, P.J.:** "Neutral Cationic Emulsion Containing a Slow-Fracturing Hydrocarbon Binder," Chemical Abstracts, 78(12):74596f (1973).
■ **Son, M.O., Jr.:** "Micellar Dispersion for Use in Multiple Fracturing of Subterranean Formations," US Patent 3,613,789(1971).
■ **Son, M.O., Jr.:** "Hydraulic Fracturing of Subterranean Formations Using Hydrocarbon Micellar Dispersions," US Patent 3,603,400(1971).
■ **Kiel, O.M.:** "Hydraulic Fracturing of a Subterranean Formation with a Viscous Oil-In-Water Emulsion," US Patent 3,552,494(1971).
■ **Kiel, O.M.:** "New Hydraulic Fracturing Process," J. Pet. Technol., 22(Jan), 89-96.
■ **Siegele, F.H.:** "Oil Well Fracturing Wax Emulsion Fluid Loss Additive," US Patent 3,477,512(1969).
■ **Watkins, D.R. and O'Donnell, E.J.:** US Patent 4,614,236(1986).

Non-Emulsifiers, Water Blockage Additives(Ref. 361)
■ **Cross, B., Scher, H.B., Eds.:** "Pesticide Formulations" Chapter 13. Berger, P., et al.: "Dynamic Surface Tensions of Spray Tank Adjuvants," ACS 1988, Washington, DC.
■ **Berger, P.D., Hsu, C., and Arendell, J.P.:** "Designing and Selecting Demulsifiers for Optimum Field Performance on the Basis of Production Fluid Characteristics," SPE Prod. Eng., November 1988, 522-526.
■ **Robb, D.J.M. and Alexander, A.E.:** "Adsorption and Wetting of a Polymer Latex by Cationic Surfactants," Soc. Chem. Ind. Monograph No. 25, 1967, 292-299.
■ **Hsu, C. and Berger, P.D.:** "Surface Chemistry of Surfactants Used to Prevent Surface Coatings Defects," 1988 Symposium on Coatings Defects; Federation of Soc. for Coatings Technol., Orlando, May 17-18.
■ **Mungan, N.:** "Permeability Reduction Through Changes in pH and Salinity," J. Pet. Technol., December 1965, 1449-53.
■ **Hurst, R.E.:** "Using Chemicals in Well Completion and Stimulation: What Kind and How Much," Oil Gas J., April 13, 1970, 80-85.
■ **Kellerhals, G.E. and Chiou, C.S.:** "Use of Perspective Plots to Aid in Determining Factors Affecting Interfacial Tensions Between Surfactant Solutions and Crude Oil," Soc. Pet. Eng. J., June 1982, 350-352.
■ **Shaughnessy, C.M. and Kline, W.E.:** "EDTA Removes Formation Damage at Prudhoe Bay," SPE paper 11188, 1982 SPE Annual Fall Technical Conference and Exhibition of AIME, New Orleans, September 26-29.
■ **Meyers, K.O., Skillman, H.L., and Herring, G.D.:** "Control of Formation Damage at Prudhoe Bay, Alaska, by Inhibitor Squeeze Treatment," J. Pet. Technol., June 1985, 1019-1034.
■ **Schulman, J.H. and Cockrain, E.G.:** "Molecular Interactions at Oil/Water Interfaces. Part I. Molecular Complex Formation and the Stability of Oil in Water Emulsions," Trans. Faraday Soc., 1940, 36, 24.
■ **Bonnet, C.F.:** " Treatment of Oil Sands with Surface-Active Chemicals," 1941 Mid-Year Meeting of the American Petro-Institute, Division of Production, Tulsa, May 22. See also: Division of Production Transactions, Vol. 22M[IV] (1941).
■ **Clason, C.E. and Hower, W.F.:** "Removal of Flow Restrictions in Well Completions," API paper 851-27-C, 1953 API Spring Meeting of the Mid-Continent District, Division of Production, Tulsa, March 18-20.
■ **Ribe, K.H.:** "Production Behavior of a Water-Blocked Oil Well," SPE paper 1295-G, 1959 SPE Annual Fall Meeting of AIME, Dallas, October 4-7.
■ **Kusakov, M.M. and Nekrasov, D.N.:** "Hysteresis in Capillary Rise," Russian Journal of Physical Chemistry, 1960, 37(7),764-767.
■ **Jarvis, N.L. and Zisman, W.A.:** "Surface Activity of Fluorinated Organic Compounds at Organic-Liquid/Air Interfaces: Part II. Surface Tension vs Concentration Curves, Adsorption Isotherms, and Force-Area Isotherms for Partially Fluorinated Carboxylic Esters," Naval Research Labs Report 5364, Surface Chemistry Division, Chemistry Division, October 8, 1959.
■ **Shafrin, E.G. and Zisman, W.A.:** "Constitutive Relations in the Wetting of Low-Energy Surfaces and the Theory of the Retraction Method of Preparing Monolayers," Naval Research Labs Report 5394, Surface Chemistry Branch, Chemistry Division, October 21, 1959.
■ **Jarvis, N.L. and Zisman, W.A.:** "Surface Activity of Fluorinated Organic Compounds at Organic-Liquid/Ari Interfaces: Part III. Equation of State of Adsorbed Monolayers and Work of Adsorption of a Fluorocarbon Group," Naval Research Labs Report 5401, Surface Chemistry Branch, Chemistry Division, November 17, 1959.
■ **Baker, H.R., Leach, P.B., Singleterry, C.R., and Zisman, W.A.:** "Surface Chemical Methods of Displacing Water and/or Oils and Salvaging Flooded Equipment: Part I. Practical Applications," Naval Research Labs Report 5606, Surface Chemistry Branch, Chemistry Division, Frebruary 23, 1961.

- Bascom, W.D. and Singleterry, C.R.: "The Adsorption of Oil-Soluble Sulfonates at the Metal/Oil Interface," Naval Research Labs Report 5623, Surface Chemicatry Branch, Chemistry Division, July 14, 1961.
- Bascom, W.D. and Singleterry, C.R.: "The Adsoprtion of Oil-Soluble Sulfonates at the Metal/Oil Interface," Naval Research Labs Report 5623, Surface Chemistry Branch, Chemistry Division, July 14, 1961.
- Zisman, W.A.: "Adhesion," Ind. Eng. Chem., 1963, 55(10), 19-37.
  Atwood, D.K.: "Restoration of Permeability to Water-Damaged Cores," SPE paper 905, 1964 SPE Annual Fall Meeting, Houston, October 11-14.
  Schmolka, I.R. and Raymond, A.J.: "Micelle Formation of Polyoxyethylene-Polyoxypropylene Surfactants," 1965 Annual Spring Meeting American Oil Chemists' Society, Houston, April 26.
- Kelly, W.R.: "Relationship Among 'Emulsion' Type, Detergency and Foam," J. Amer. Oil Chem. Soc., 1966, 43, 358-363.
- Wade, W.H.: "Spontaneous Imbibition of fluids in Porous Vycor," Soc. Pet. Eng. J., 1974, 139-144.
- Schechter, R.S., Wade, W.H., and Wingrave, J.A.: "Sorption Isotherm Hysteresis and Turbidity Phenomena in Mesoporous Media," J. Colloid Sci., Manuscript # 4880-2.
- Doe, P.H., El-Emary, M.M., Wade, W.H., Schechter, R.S.: "The Influence of Surfactant Structure on Low Interfacial Tensions," ACS Symp. Ser., 1979, 91, 17-34.
- Koukounis, C., Wade, W.H., and Schechter, R.S.: "Phase Partitioning of Anionic and Nonionic Surfactant Mixtures," SPE paper 8261, 1979 SPE Annual Fall Technical Conference and Exhibition of AIME, Las Vegas, September 23-26.
- Ziegler, V.M. and Handy, L.L.: "The Effect of Temperature on Surfactant Adsorption in Porous Media," SPE paper 8264, 1979 SPE Annual Fall Technical Conference and Exhibition of AIME, Las Vegas, September 23-26.
- Eakin, J.L., Johansen, R.T., Hopkins, A.D., and Taleaferro, R.W.: "How Chemical Treatment Gets Rid of Gas-Well Water Blocks," Oil Gas J.(November 26, 1962), 85-89.
- Carney, M.J. and Wieland, D. R.: "Stimulation of Low Permeability Gas Wells in the Rocky Mountain Area," SPE paper 4396, 1973 SPE Rocky Mountain Regional Meeting, Casper, May 15-16.
- Smith, C.F.: "Gas Well Fracturing Using Gelled Non-Aqueous Fluids," SPE paper 4678, 1973 SPE Annual Meeting, Las Vegas, Sept. 30-Oct. 3.
- Gras, E.H.: "Alcohol Comes on Strong as a Fracturing Fluid," Drilling-DCW(May, 1973).
- Tindell, W.A., Misak, M.D. and Gras, E.H.: "The Use of Alcohol-Water Mixtures in Fracture Stimulation of Gas Wells," Proc. Annu. Southwest. Pet. Short Course, 1974, 61-65.
- Tannich, J.D.: "Liquid Removal From Hydraulically Fractured Gas Wells," J. Pet. Technol.(November 1975) 1309-1317.
- Holditch, S.A.: "Factors Affecting Water Blocking and Gas Flow from Hydraulically Fractured Gas Wells," J. Pet. Technol.(December 1979) 1515-1524.
- Agarwal, R.G., Carter, R.D., and Pollock, C.B.: "Evaluation and Performance Prediction of Low-Permeability Gas Wells Stimulated by Massive Hydraulic Fracturing," J. Pet. Technol.(March 1979) 362-372.
- Bostic, J.N., Agarwal, R.G., and Carter, R.D.: "Combined Analysis of Postfracturing Performance and Pressure Buildup Data for Evaluating an MHF Gas Well," J. Pet. Technol.(October 1980) 1711-1719.
- Craighead, M.S and Hossaini, M.: "Foamed Anhydrous Methanol Stimulation," SPE paper 12315, 1983 Eastern Regional Meeting, Champion, PA, November 9-11.
- Penny, G.S.: "Method of Increasing Hydrocarbon Production from Subterranean Formations," US Patent 4,702,849(1987).
- Denekas, M.O., Mattax, C.C. and Davis, G.T.: "Effects of Crude Oil Components on Rock Wettability," paper No. T.P. 8095, Pet. Transactions, AIME, Vol. 216, 1959, 330-333.
- Graham, J.W., Monaghan, P.H., and Osoba, J.S.: "Influence of Propping Sand Wettability on Productivity of Hydraulically Fractured Oil Wells," T.P. 8094, Pet. Transactions, AIME, Vol. 216, 1959, 324-329.
- Wade, W.H., et al.: "Low Interfacial Tensions Involving Mixtures of Surfactants," SPE paper 6002, 1976 SPE Annual Technical Conference and Exhibition, New Orleans, October 3-6.
- Trogus, F., Sophany, T., Schechter, R.S., Wade, W.H.: "Static and Dynamic Adsorption of Anionic and Nontonic Surfactants," SPE paper 6004, 1976 SPE Annual Technical Conference and Exhibition, New Orleans, October 3-6.
- Wade, W.H., Schechter, R.S. and Jacobson, J.K.: "Low Interfacial Tensions Involving Mixtures of Surfactants," SPE paper 6002, 1976 SPE Annual Technical Conference and Exhibition, New Orleans, October 3-6.
- Newlove, J.C., McDougall, L.A., Walker, J.R., and Stockwell, J.R.: "Polymer Article of Manufacture," CA 1230219 A1(1987).
- Saliba, C.A.M. and Gonzalez, G: "Capillary Pressure, Wettability, and Corrosiveness of Stimulation Fluids Containing Alcohols," Bol. Tec. PETROBRAS, 1987, 30(1), 25-32, Chemical Abstracts 106(26):216585z(1987).

- Crema, S.C.: "Liquid Foaming Additives Used in the Stimulation of Oil and Gas Wells, US Patent 4,609,477(1986).
- Dunning, J.: "Surfactants Revisited," CHEMTECH, 1984, 14(1), 46-50.
- Schechter, R.S. et al.: "Modeling Crude Oils for Low Interfacial Tension," Soc. Pet. Eng. J., Dec. 1976, 351-357.
- Schechter, R.S. et al.: "Low Interfacial Tension Invovling Mixtures of Surfactants," Soc. Pet. Eng. J., April 1977, 122-128.
- Schechter, R.S. et al.: "Static and Dynamic Adsorption of Anionic and Nonionic Surfactants," Soc. Pet. Eng. J., Oct. 1977, 337-344.
- Schechter, R.S. et al.: "Interfacial Tension and Phase Behavior of Surfactant Systems," Soc. Pet. Eng. J., Aug. 1978, 242-252.
- Wade, W.H. et al.: "Interfacial Tension and Behavior of Nonionic Surfactants," Soc. Pet. Eng. J., Dec. 1979, 349-356.
- Wasan, D.T. et al.: "Observations on the Coalescence Behavior of Oil Droplets and Emulsion Stability in Enhanced Oil Recovery," Soc. Pet. Eng. J., Dec. 1978, 409-417.
- Saidi, A.M. and Hesselink, F.T.: "Method for Recovering Oil from an Underground Formation," UK Patent GB 2062063(1981), Chemical Abstracts 95(26):222643n(1981).
- Redford, D.A.: "Solvent Process for Developing Interwell Communication Path in a Viscous Petroleum Containing Formation Such as a Tar Sand Deposit," CA Patent 1015656(1977).
- Johnson, F.S., Jones, R.A., and Miller, J.S.: "Laboratory Investigation Using Solvent to Recover Heavy Oil from a Fractured Reservoir," Chemical Abstracts, 87(24):186798r(1977).
- Thompson, J.L.: "Composition and Method for Reducing the Surface Tension of Aqueous Fluids," US Patent 4,018,689(1977).
- Davis, J.A., Jr. and Rhudy, J.S.: "Hydraulic Fracturing of Subterranean Reservoirs Using Polyethylene Oxide-Based Fracturing Fluid," US Patent 3,747,681(1973).
- Redford, D.A.: "Developing Interwell Communication in a Tar Sand," US Patent 3,706,341(1972).
- Formation Damage at Prudhoe Bay, Alaska, by Inhibitor Squeeze Treatment," J. Pet. Technol., June 1985, 1019-1034.

RECEIVED February 21, 1989

INDEXES

Author Index

Affiliation Index

Subject Index

Production and indexing: Colleen P. Stamm
Acquisition: Cheryl Shanks

Elements typeset by Hot Type Ltd., Washington, DC
Printed and bound by Maple Press, York, PA

Other ACS Books

Chemical Structure Software for Personal Computers
Edited by Daniel E. Meyer, Wendy A. Warr, and Richard A. Love
ACS Professional Reference Book; 107 pp;
clothbound, ISBN 0–8412–1538–3; paperback, ISBN 0–8412–1539–1

Personal Computers for Scientists: A Byte at a Time
By Glenn I. Ouchi
276 pp; clothbound, ISBN 0–8412–1000–4; paperback, ISBN 0–8412–1001–2

Biotechnology and Materials Science: Chemistry for the Future
Edited by Mary L. Good
160 pp; clothbound, ISBN 0–8412–1472–7; paperback, ISBN 0–8412–1473–5

Polymeric Materials: Chemistry for the Future
By Joseph Alper and Gordon L. Nelson
110 pp; clothbound, ISBN 0–8412–1622–3; paperback, ISBN 0–8412–1613–4

The Language of Biotechnology: A Dictionary of Terms
By John M. Walker and Michael Cox
ACS Professional Reference Book; 256 pp;
clothbound, ISBN 0–8412–1489–1; paperback, ISBN 0–8412–1490–5

Cancer: The Outlaw Cell, Second Edition
Edited by Richard E. LaFond
274 pp; clothbound, ISBN 0–8412–1419–0; paperback, ISBN 0–8412–1420–4

Practical Statistics for the Physical Sciences
By Larry L. Havlicek
ACS Professional Reference Book; 198 pp; clothbound; ISBN 0–8412–1453–0

The Basics of Technical Communicating
By B. Edward Cain
ACS Professional Reference Book; 198 pp;
clothbound, ISBN 0–8412–1451–4; paperback, ISBN 0–8412–1452–2

The ACS Style Guide: A Manual for Authors and Editors
Edited by Janet S. Dodd
264 pp; clothbound, ISBN 0–8412–0917–0; paperback, ISBN 0–8412–0943–X

Chemistry and Crime: From Sherlock Holmes to Today's Courtroom
Edited by Samuel M. Gerber
135 pp; clothbound, ISBN 0–8412–0784–4; paperback, ISBN 0–8412–0785–2

For further information and a free catalog of ACS books, contact:
American Chemical Society
Distribution Office, Department 225
1155 16th Street, NW, Washington, DC 20036
Telephone 800–227–5558